Real Time Microcomputer Control of Industrial Processes

International Series on
MICROPROCESSOR-BASED SYSTEMS ENGINEERING

VOLUME 5

Editor

Professor S. G. Tzafestas, *National Technical University, Athens, Greece*

Real Time Microcomputer Control of Industrial Processes

edited by

SPYROS G. TZAFESTAS

Department of Electrical Engineering,
National Technical University of Athens,
Athens, Greece

and

J. K. PAL

Engineering Technology and Development Division
Engineers India Limited,
New Delhi, India

KLUWER ACADEMIC PUBLISHERS
DORDRECHT / BOSTON / LONDON

Library of Congress Cataloging in Publication Data

Real time microcomputer control of industrial processes / edited by
  S.G. Tzafestas, J.K. Pal.
       p.   cm. -- (Microprocessor-based systems engineering)
    Includes bibliographical references and indexes.
    ISBN-13:978-94-010-6761-4   e-ISBN-13:978-94-009-0609-9
    DOI: 10.1007/978-94-009-0609-9

    1. Process control--Data processing. 2. Microcomputers-
  -Industrial applications.   I. Tzafestas, S. G., 1939-   . II. Pal,
  J. K.   III. Series: International series on microprocessor-based
  systems engineering.
  TS156.8.R424 1990
  670.42'75433--dc20   90-4768

ISBN-13:978-94-010-6761-4

Published by Kluwer Academic Publishers,
P.O. Box 17, 3300 AA Dordrecht, The Netherlands.

Kluwer Academic Publishers incorporates
the publishing programmes of
D. Reidel, Martinus Nijhoff, Dr W. Junk and MTP Press.

Sold and distributed in the U.S.A. and Canada
by Kluwer Academic Publishers,
101 Philip Drive, Norwell, MA 02061, U.S.A.

In all other countries, sold and distributed
by Kluwer Academic Publishers Group,
P.O. Box 322, 3300 AH Dordrecht, The Netherlands.

Printed on acid-free paper

CONTRIBUTORS

AIO, T.	Chiyoda Corporation, Tsumuri-ku, Japan
DASGUPTA, S.	Gas Authority of India Ltd., New Delhi, India
DE KEYSER, R.M.C.	University of Ghent, Ghent, Belgium
GUPTA, A.	Sloan School of Management, MIT, USA
IHARA, H.	Hitachi Ltd., Yokohama, Japan.
KAWAI, S.	Fuji Facom Corp., Tokyo, Japan.
KOIKE, Y.	Fuji Electric Co. Ltd., Tokyo, Japan.
LAPPUS, G.	Technical University of Munich, Munich, F.R.G.
MAHALANABIS, A.K.*	Pennsylvania State University, U.S.A.
NAKAMURA, H.	Kyushu Denki Seizo (EM) Co., Fukuoka, Japan.
NOHMI, M.	Hitachi Ltd., Kawasaki, Japan.
PAL, J.K.	Engineers India Ltd., New Delhi, India.
PURKAYASTHA, P.	Desein (New Delhi) Pvt Ltd., New Delhi, India.
SAITO, K.	Hitachi Ltd., Ibaraki-ken, Japan.
SCHMIDT, G.	Technical University of Munich, Munich, F.R.G.
TANAKA, N.	Chiyoda Corporation, Tsumuri-ku, Japan.
TOONG, H.-M.D.	Sloan School of Management, M.I.T., U.S.A.
TZAFESTAS, S.G.	National Technical University, Athens, Greece.
UCHIDA, M.	Kyushu Electric Power Co., Inc., Fukuoka, Japan.
VAN OSTAEYEN, J.L.S.	I.M.D.C. n.v., Antwerp, Belgium.

* Passed away shortly after providing his contribution to the book

PREFACE

The introduction of the microprocessor in computer and system engineering has motivated the development of many new concepts and has simplified the design of many modern industrial systems. During the first decade of their life, microprocessors have shown a tremendous evolution in all possible directions (technology, power, functionality, I/O handling, etc). Of course putting the microprocessors and their environmental devices into properly operating systems is a complex and difficult task requiring high skills for melding and integrating hardware, software and systemic components.

This book was motivated by the editors' feeling that a cohesive reference is needed providing a good coverage of modern industrial applications of microprocessor-based real time control, together with latest advanced methodological issues. Unavoidably a single volume cannot be exhaustive, but the present book contains a sufficient number of important real-time applications.

The book is divided in two sections. Section I deals with general hardware, software and systemic topics, and involves six chapters. Chapter 1, by Gupta and Toong, presents an overview of the development of microprocessors during their first twelve years of existence. Chapter 2, by Dasgupta, deals with a number of system software concepts for real-time microprocessor-based systems (task scheduling, memory-management, input-output aspects, programming language requirements, etc). Chapter 3, by Tzafestas and Pal, provides a survey of major digital control algorithms for industrial microprocessor-based control (single-input / single-output and multi-input / multi-output). Chapter 4, by late Mahalanabis, gives the fundamental elements of minimum variance filtering and prediction algorithms referred to state-space and CARMA models. Chapter 5, by Purkayastha, discusses the evolution of distributed control systems from both the digital computer and the process control viewpoints. The various generations of distributed systems are considered, and the control system architecture is studied in relation to the plant structure. Section I closes with Chapter 6, by Pal and Tzafestas, where the state of art of industrial control system modelling and simulation is presented (mathematical and numerical methods, simulation software, simulation languages) along with an industrial case study offered by Aio and Tanaka.

Section II involves eight chapters dealing with real-time applications. Chapter 7, by De Keyser, provides a deep description of a microprocessor-based self-adaptive controller used for residence building control, greenhouse climate control, incubator control, and ventilation control in animal houses. In Chapter 8, by De Keyser and Van Ostaeyen, the latest developments of microprocessor-based hardware, software and control concepts are employed for the design and implementation of a self-learning control system for cutter suction dredging ships. The chapter includes detailed discussions of the various steps of this design together with the actual multi-microcomputer implementation on-board of the real dredgers. Chapter 9, by Uchida, deals with thermal power plant control using linear quadratic regulator and feedforward control, and provides all the hardware, software and systemic elements required. Chapter 10, by Nakamura, provides a status report of electric power system control (on-line control, hierarchical control, load frequency control, dispatching control, security assessment). Chapter 11, by Saito, presents an overview of the state-of-art of real-time control in the steel industry, involving the microcomputer as the main control element. This element has made possible an effective application of modern control and knowledge-based control in the steel industry. This point is explained in detail. Chapter 12, by Lappus and Schmidt, is devoted to the class of gas pipeline networks, and studies the application of microprocessors and microcomputers for process monitoring, fault detection / diagnosis and control. Chapter 13, by Kawai and Koike, describes in detail how the microcomputer-based control is applied in cement industry for kiln control, raw material blending control, and electric power demand control. New techniques (auto-regressive, predictive, auto-tuning PID control) are examined, together with the CRT panel and man-machine communication problems. The book closes with Chapter 14, by Ihara and Nohmi, where the current state of microcomputer based control applications in railway transportation systems is presented. Particular typical examples of Japanese transportation systems (Shinkansen, subways, monorails, private railways, etc) are described.

The book is appropriate for the senior undergraduate and the postgraduate student, but it is mainly intended for the researcher or professional who wants to see how microcomputer-based control was applied in specific modern-life processes. It reflects the experience and background of twenty experts working in academia and industry. No attempt was made to enforce a uniform style, but all chapters contain sufficient background material for independent reading. The reader can find advanced methodological concepts and techniques, in addition to the various practical implementation issues presented in the application chapters.

We are really indebted to all contributors for their enthusiastic support and their effort to offer excellent up-to-date manuscripts. Particularly, we are grateful to Mr. S. Dasgupta for his extra help in proof reading several retyped chapters, and in other editorial tasks. At this point we wish to express our grief for the death of our colleague, and contributor of the book, Professor A.K. Mahalanabis, an eminent and dedicated scientist.

We hope that this book will be an important addition to the current literature on microprocessor-based systems, and that the integrated issues and techniques included will help in the further progress and development of industrial systems of higher productivity and quality.

March 1990   S.G. Tzafestas
                                                              J.K. Pal

CONTENTS

SECTION I METHODOLOGY: HARDWARE, SOFTWARE AND SYSTEMIC TOPICS

Chapter 1
MICROPROCESSORS - THE FIRST TWELVE YEARS
A. Gupta and *Hoo-Min D. Toong*

Chapter 2
INTRODUCTION TO REAL-TIME SYSTEM SOFTWARE
S. Dasgupta

SECTION II   REAL TIME APPLICATIONS

Chapter 7
A LOW COST SELF-ADAPTIVE MICROPROCESSOR
CONTROLLER: APPLICATION TO HEATING
AND VENTILATION CONTROL
R.M.C. De Keyser

Chapter 8
SELF-ADAPTIVE MULTI-MICROPROCESSOR CONTROL
OF CUTTER SUCTION DREDGING SHIPS
R.M.C. De Keyser and J.L.S. Van Ostaeyen

Chapter 9
ADVANCED CONTROL IN THERMAL POWER PLANTS
M. Uchida

SECTION I

**METHODOLOGY: HARDWARE, SOFTWARE
AND SYSTEMIC TOPICS**

Chapter 1

MICROPROCESSORS - THE FIRST TWELVE YEARS

Amar Gupta and Hoo-Min D.Tcong,Member IEEE
Sloan School of Management,Massachusetts
Institute of Technology,Cambridge,MA 02139

ABSTRACT. During the first twelve years of their existence,microprocessors have evolved at a dramatic pace in terms of numbers,technology, power, functionality, and applications.This paper begins with a technical explanation of key microprocessor specifications.Four generations of microprocessors are considered in detail, alongwith case studies of popular chip families.Special-purpose processors,software issues, and performance evaluation techniques are other areas of discussion.Overall industry trends are presented.

1. INTRODUCTION

Looking at the wide array of sophisticated microprocessors available today to support virtually all conceievable applications, it is difficult to believe that microprocessors have been in existence for just over a decade. The term "microprocessor" was first used in 1972. However,the era of microprocessors commenced in 1971 with the introduction of the Intel 4004, a "microprogrammable computer on a chip" composed of an "integrated CPU complete with a four-bit parallel adder,16 four-bit registers, an accumulator and a push-down stack on a chip"[1]. The 4-bit 4004 CPU contained 2300 transistors and could execute 45 different instructions. Successive generations of microprocessors,using 8-bit, 16-bit and 32-bit chips, commenced in 1972,1974 and 1981,respectively. During their twelve years of existence, the number of devices per chip has increased by a factor of 200,the clock frequency by a factor of [50],and the overall throughput of the microprocessor has increased by two to three orders of magnitude.By all standards, the pace of the microelectronic revolution is unparalleled - it even outstrips the pace of advancement of the computer.The time elapsed between the original invention and the adoption of 32-bit word size is just one decade in the case of microprocessors, almost half the time lag as in the case of computers. "After only a single decade,in fact, microprocessor design has evolved from a situation in which it lagged far behind conventional computer design to a place where it is beginning to take the lead"[2].This is both in terms of computing power and architectural sophistication.

A microprocessor is the central arithmetic and logic unit of a computer scaled down so that it fits on a single silicon chip(some times several chips)holding tens of thousands of transistors,resistors and similar circuit elements[3]. Early microprocessors performed basic functions only.Additional chips were required to generate timing signals,to provide primary memory for program and data storage , and to

S. G. Tzafestas and J. K. Pal (eds.), Real Time Microcomputer Control of Industrial Processes, 3–47.

interface with peripheral units. During their first years of existence, microprocessors were used almost exclusively as microcontrollers on a dedicated basis in process control applications. They offered a limited instruction set and needed to be programmed directly in machine language. Over the years, the evolution of superior system architectures, the availability if improved higher level languages, and the increase in power and flexibility has greatly broadened the domain of microprocessor usage [73],[74]. Microprocessors have been the direct cause of the computer game explosion and the personal computer revolution[64].

Early microprocessors were fabricated using p-channel MOS technology. In 1974 RCA introduced the 1802, the first microprocessor fabricated in CMOS technology. Over the years microprocessors have been fabricated in almost all available semiconductor technologies. The Intel 8080 used n-channel enhancement-mode MOS, the Intel 8085 used n-channel depletion-mode MOS [18], the Fairchild 9440 used bipolar technology, and the Fairchild 16-bit processor uses I^2L technology [17]. In view of the large number of transistors, most manufacturers prefer using MOS technology in preference to bipolar transistor technology. Currently, the most popular MOS technology is n-channel MOS (NMOS) by virtue of its high packing density and fast switching speeds. To emphasize the high-performance aspect, Intel prefers to call it HMOS. The Intel 8086, introduced in 1978, contained 29000 transistors implanted in an area of 48,400 mil^2, the Intel 80286, introduced in 1982, has three times the circuit density with 130000 devices in an area of 70225 mil^2 [14]. Such design innovations will enable NMOS to remain popular at least until the mid 1980's [77],[78]. A leading competitor is complementary MOS (CMOS) technology [76],[80]which provides faster

speed and lower power consumption than circuits implemented with traditional PMOS and NMOS technology; the disadvantage of CMOS lies in its lower packing density. Classical CMOS logic designs have an equal number of n- and p-channel devices. Newer CMOS designs use a higher number of n-channel devices than p-channel devices in order to implement higher circuit"densitites; High-density CMOS microprocessor circuits have been implemented among others by Bell Laboratories [4], by Rockwell International [5], and by Hewlett-Packard [15]. In coming years, CMOS will become the most popular technology for fabricating microprocessors because of the advantages offered by CMOS in terms of faster speed, lower power consumption and higher noise immunity [6] [75], [76].

The most significant single criterion for evaluating and selecting microprocessors is the word size of the microprocessor which reflects the basic unit of work. A larger word size implies higher processing power and greater addressing capabilities. In the early years of microprocessor evolution, the size of register, the width of internal instruction paths and data paths, and the width of external instruction paths and data paths were almost always identical. This is rarely true now. Larger external paths require the chip package to have a large number of pins, which increases the total packaging and production costs. Thus chips nowadays tend to have larger internal paths than external paths. For example, the Motorola 68000 and the National NS

16032 have 32-bit internal paths and 16-bit external paths. A true
32-bit microprocessor has all paths and all internal units designed to
communicate or process 32 bits in parallel; some of these paths may,
in fact, be wider than 32 bits. For example, a 32-bit by 32-bit multiply
yields a 64-bit result, and some 32 bit microprocessors are designed to
store intermediate results with higher accuracy. If one assumes that a
generation ends with the introduction of the pioneer chip of the next
generation, then the 4, 8-, and 16-bit microprocessor eras lasted for
1,2 and 7 years, respectively.   However,although 32-bit chips are now
available, the 16-bit chips will continue to dominate the market until
the mid 1980's. Also, smaller word-width microprocessors are used for a
whole range of control applications, where there is no requirement for
either the higher power or the greater addressing capabilities of cost-
lier microprocessors with larger word width.

Microprogramming capabilities are now becoming a standard feature.
Unlike early microprocessors which used "hard-wired architectures", a
microprogrammed processor, though inherently slower offers greater
flexibility in terms of easier incorporation of changes or additions to
the instruction set. To minimize the size of the control store area, a
two-level control structure is frequently employed. First, the machine
instructions are converted into sequences of microinstructions in the
microcontrol store. These microinstructions serve as pointers to nano-
instructions in the nanostore. The nanocontrol store contains an arbitrarily ordered set of unduplicated machine-state control words, which
control the execution unit. In the case of the Motorola MC68000, the
two-level structure requires 22.5 kbits of control store, 50 percent
less than the control store requirements for a single-level implementation. However, the two-level structure increases total access time.
Even though microprogrammed architectures are now standard, very few
chips can be microprogrammed by the user. The NCR Corporation four-chip
3200 series 32-bit microprocessor [7] offers external microprogramming
capability; with this chip set, a Texas Instruments 9900 16-bit microprocessor operating system and instruction set can be emulated using
12 kbytes of external store and an optimized Cobol virtual-language
machine can be emulated with 25kbytes. With such microprogramming
capabilities, one can emulate popular minicomputer and mainframe
computer instruction sets at a small fraction of the cost involved
in buying an entire computer system.

In 1975, Moore [16] had predicted that by the end of the decade
the increase in complexity of VLSI chips would approximately double
every two years rather than every year as in the preceding decade. In
terms of chip complexity, microprocessors have been at the leading
edge of technology. An excellent example is the Hewlett-Packard 32-bit
microprocessor chip, the heart of an HP-9000 computer system, with
450 000 transistors on a single chip [8],[72]. One advantage of the
increased transistor density is the ability to implement instruction
repertoires that are much larger and much more powerful than the ones
they offered previously, enabling more compact programs [88].Frequently
the same chip uses instructions of varying sizes depending on instruction type, size of data, and addressing mode used; this variable size
increases architectural complexity but reduces code size. Also,indivi-

dual instructions are becoming increasingly powerful. On several systems a single instruction can control the transfer of an entire block of data from the memory, or manipulate several registers simultaneously. Such single instructions can replace several instructions of earlier systems. The contemporary instructions have close resemblance with instructions of higher level languages [19], facilitating compilation of such programs by performing in hardware an increasing number of functions traditionally done in software. As instruction size and complexity have increased, the complexity of the chip has increased and so has the design effort [110] from under a man-year to over 100 man-years of engineering time. Patterson [9] [89] has implemented reduced instruction set architectures to enable better utilization of the potential of evolving technology. In coming years commercial microprocessors are unlikely to offer larger instruction repertoires – instead the focus will be on increasing the power of individual instructions and on supporting instruction sets with an established base. The T-11 chip [11] emulates the PDP 11 instruction set and offers an execution speed comparable to that of the PDP 11/34. The IBM System 370 instruction set has also been implemented on a single chip [12] though additional chips are still needed for supporting general-purpose registers and control storage. Other vendors, too, are expected to adopt instruction sets that have an established user base.

Over the years, the addressing capability of microprocessors has increased very significantly. The number of auxiliary addressing modes (like indirect, indexed, autodecrementing) has also increased. Most vendors offer specialized memory management chips, like the Motorola 68451 [81] and the National Semiconductors NS16082, to facilitate access to larger memory sizes and to assure adequate memory protection. However, the Intel iAPX 286, unlike its predecessor, the 8086, incorporates the memory protection function on the microprocessor chip itself [13]. Almost all new chips support virtual memory, giving an overall address space of the order of gigabytes. This virtual memory enables execution of very large programs and assists in simultaneous support of multiple users; virtual storage techniques make use of disk storage to augment main memory (RAM's and ROM's) for program storage and execution. Cache memories, on the other hand, are very fast memory units of relatively small size; information needed by the processor is prestaged from the main memory (or possibly from the disk) to the cache memory to enable faster processor execution. The increase in throughput is determined by the "hit ratio" – the fraction of times the processor finds the desired information in the cache.

Microprocessors differ significantly in their ability to store and manipulate different types of data. Although data in the form of bytes and words are generally supported on contemporary microprocessors data in other forms, like bits, binary-coded-decimal (BCD) words, floating –point numbers, and character strings are not always directly supported. For example, data, in bit form, generally used in control applications, are not supported on the Intel 8086. BCD digits are not supported on the Intel iAPX 432. Floating-point numbers are not supported on the Bellmac-32A, and character strings, used in text processing applications, are not directly supported on the Motorola MC68000. The availability of support for greater variety of data types makes the microprocessor suitable for a larger range of applications. Unfortunately, supporting multiple data types involves increased design complexity and an increased number of devices. In some cases, additional data types can be supported by using auxiliary processor chips. For example, the Intel 8087, the Motorola 68881, and the NS16081 floating-point chips can be used with the 16-bit microprocessors of the respective manufacturers. The proposed IEEE standard P754 on floating point arithmetic is one data format that all chips attempt to support, either directly or through use of coprocessors.

The microprocessor revolution represents a trend towards implementing all the components of a computer on a small number of chips [63], [82]. Any computer system, irrespective of size, power, and capabilities, combines three classes of subsystems - CPU's (for arithmetic, logic, and control functions),"memories" (READ/WRITE (RAM) and/or READ-ONLY (ROM)), and input/output interfaces for peripheral control. Early microprocessors performed the basic CPU functions only. As better technology became available to enable integration of larger numbers of devices on the same chip, it became feasible to implement an increasing number of auxiliary functions on the microprocessor chip itself resulting in increased popularity of computers built with very few chips. A microcomputer [86] combines a microprocessor with memory and input/output capabilities on one or several chips. Single-chip microcomputers[83] constitute an important subset of microprocessors in which all functions including memory, are implemented on the same chip. In view of the chip area devoted to auxiliary functions, there is always a time lag between the introduction of a microprocessor chip of a given word size and the introduction of a microcomputer chip of an equivalent word size. For example, the first 8-bit single-chip microcomputer, the Intel 8048, was introduced in 1976, four years after the introduction of the first 8-bit microprocessor, the Intel 8008 [1].

There are several characteristics that distinguish single-chip microcomputers from conventional architectures. Because of the large volumes in which single-chip microcomputers are purchased real estate per chip must be minimized. For instance, a 25- percent savings on a $2 microcomputer can result in savings of $500 000 for a million units. This optimization requires careful consideration of architectural trade-offs, memory design factors, instruction sizes, memory addressing techniques, and other design constraints with respect to area, performance, and run-time parameters [65]. By the mid 1980's, a typical chip will contain about a million transistors offering a tremendous potential for implementing many sophisticated features in hardware. The processor section of the chip could include on-chip memory hierarchy, multiple homogenous caches for enhanced execution parallelism, support for complex data structures and high-level languages, a flexible instruction set, and communication hardware [63]. More than four fiths of the transistors on the chip are expected to be allocated for memory functions. Although, architecturally, this computer-on-a-chip will compare favourably with modern computes! in terms of throughput, implementation of processing power comparable to the CRAY-1 on a single chip is not likely to happen until the end of the 1980's.

At the other extreme of the spectrum, there are applications that
require higher computing power or better accuracy than that provided by
single-chip microprocessors. For such applications, bit-sliced organisation [66] enables linking several identical modular chips in parallel
to achieve higher accuracy. Also, since these chips are implemented in
either bipolar or emitter-coupled logic (ECL) technology bit-sliced
chips offer higher throughput than MOS chips. Thus by using multiple
4-bit chips of this kind , one can easily integrate systems offering
an effective word size of 8, 12, 16 bits, or even more. The 2900 series
of 4-bit chips manufactured by MOS Technology and Advanced Micro Devices
has been very popular. Note, however, that although the Intel iAPX 432
uses a three-chip set, it does not represent an example of bit-sliced
architectures, as the three chips are neither capable of operating
individually, nor are they identical to each other. With the advent of
wider word size, general-purpose microprocessors, the primary merit of
using bit-sliced microprocessors for higher accuracy has gradually
eroded over the years. Designers of these chips claim that they can
continue to boost speed by 30 percent every two years by using ECL
internal circuitry and advanced device processing techniques. It is
useful, however to note an example of a phenomenon that occurs in the
semiconductor industry - in 1972, just a year after the commencement
of the microprocessor era, American Micro-Systems Inc. announced the
AMI 7200, the first 8-bit processor slice, or "byte-slice"; such slices
became a commercial reality only with the introduction of the Fairchild
F100200 in 1979 [66]. The need for caution in depending on chips using
the leading edge of technology cannot be overemphasized.

2.   4-BIT MICROPROCESSORS

Although the Intel 4004 and the Rockwell PPS-4 were introduced
earlier, the TMS-1000 series introduced in 1974 has made Texas Instruments the leading manufacturer of 4-bit processors used by millions in
games, toys, calculators, and other low-end controller applications.The
low price of less than $1 per chip is made possible by two major factors
- first, the PMOS technology has become a mature technology and, second,
the widespread usage of such chips has reduced production costs through
economies of scale. Equivalent microprocessors implemented in CMOS(for
low power applications, for example, CMOS TMS 1000) or bipolar techno
logy (for higher performance) cost more. The 4-bit CMOS microprocessors
are used in conjunction with liquid-crystal displays (LCD's) for a wide
array of hand-held products. Low-cost microprocessor continue to have a
tremendous potential use in industrial (testing, process control, instrumentation, manufacturing), commercial, and consumer applications
(see [20]). Like the basic TTL NAND gate, it is unlikely that the basic
4-bit processor will ever disappear.

The wide spectrum of additional capabilities implemented on 4-bit
processor chips includes the following:-
  (a)   significant on-chip ROM and RAM;
  (b)   versatile instruction set;
  (c)   flexible I/O protocols;
  (d)   on-chip A/D converter as on TMS2100;
  (e)   multiple CPU's on same chip (e.g., the National Semiconductor

COP2440 contains two identical CPU's alongwith ROM, RAM, system timing, and internal logic);
(f)   ability to drive displays directly -NED (LCD) 79 , Hitachi (LCD) ITT (LCD), and TI (Vacuum Fluorescent).
The trend of supporting additional functions on the same chip sustains interest in 4-bit microprocessor chips.

3. 8-EIT MICROPROCESSORS

Although the second generation commenced with the introduction of the Intel 8008 in 1971, the domain of 8-bit microprocessors witnessed several significant improvements in hardware and system concepts with the introduction of the Intel 8080 and the Motorola 6800 in mid 1974. The Intel 8080 commenced the trend of using NMOS technology, of executing decimal and BCD arithmetic, and of the 16-bit address bus. Unlike other microprocessors that required multiple power supplies, the motorola 6800 was the first 5-V single-power-supply microprocessor. Also, instead of relying on TTL chips for interfacing, peripheral processors became available for supporting interface functions such as CRT and floppy disk control.

In sharp contrast to early 8-bit microprocessors with a structure supporting uniform 8-bit data paths, current 8-bit microprocessors frequently have 16-bit or even 32-bit internal data paths. The Intel 8088, the Texas Instruments 9980, and the Motorola 6809 all offer 8-bit external buses, but process data internally as 16-bit words. The NS 16008 uses an 8-bit external bus and a 32-bit internal architecture. Such a structure enables retaining compatibility at the bus level with earlier 8-bit microprocessors, while enabling users access to new software. The newer 8-bit microprocessor chips are versions of popular 16-bit or 32-bit architectures intended for low-cost applications.

Since the introduction of the first 8-bit, single chip microcomputer the Intel 8048 in 1976 [1], several families of 8-bit microcomputers have become available. In most cases (Motorola and Rockwell for example), these microcomputers are enhancements of the respective 8-bit microprocessor family. However, the Zilog Z8 is architecturally dissimilar to the Zilog Z80.

Depending on the intended application scenario, system designers demand microcomputers with one or more of the following characteristics:
(a) high speeds   (b) low power consumption
(c) large sized ROM   (d) large sized RAM
(e) microprogramming   (f) easy interfacing of I/O
(g) low cost
Since there is no global optimal solution, vendors now offer whole families of chips, each family with a particular architecture; different members of the same family offer different combinations of ROM, RAM, and auxiliary functional capabilities. Examples include Intel [21], Motorola [22], Rockwell [23], Zilog [24], and Texas Instruments [25].

4. 16-BIT MICROPROCESSORS

The era of 16-bit microprocessors began in 1974 with the introduction

of the PACE chip by National Semiconductor. The Texas Instruments TMS
9900 was introduced two years later. Subsequently, the Intel 8086
became commercially available in 1978, the Zilog Z8000 in 1979, and the
Motorola MC68000 in 1980. Several higher performance versions of the
original chips are now available [84]. It is difficult to analyze in
depth the characteristics of all 16 bit microprocessors that have been
developed or even the subset of chips [96]-[97]that is currently available.
able. We concentrate on the trends exemplified by the more popular
chips.

   The characteristics of leading 16-bit microprocessors are summarized in Table I [26]-[29]. The first five columns show the trend over
the years of increasing functional capabilities, the addressing range,
the diverse types of data supported, and clock speeds. The Intel 80286
is a higher capability version of the Intel 8086 [94] and displays the
problems inherent in using new technology while retaining compatibility
with an earlier chip. Upward compatibility often prohibits any major
changes in chip architecture; this, in turn restricts implementation
or support of newer functions and data types. The five-fold increase
in the number of transistors is used to increase performance six-fold,
and to perform memory management and protection fucntions on the chip
itself. Similarly, the Motorola MC68010, introduced in 1982, retains
the basic architecture of the original MC 68000, but provides additional
support for virtual memory [121]. This similarity of architectures enables many different 16-bit microprocessor chips to be studied through
evaluation and comparison of a few families of chips.

   The basic structure of the Texas Instruments 9900 is shown in Fig.
1. It is a very simple design with no instruction prefetch or pipelining. The processor contains three registers - the program counter, the
status register, and the workspace pointer; in addition 16 general
purpose "workspace" registers are set up in memory. By altering the
content of the workspace pointer, a new set of workspace registers is
obtained. This simple "context switching" mechanism enables handling
of interrupt and subroutine calls without need for saving or stacking

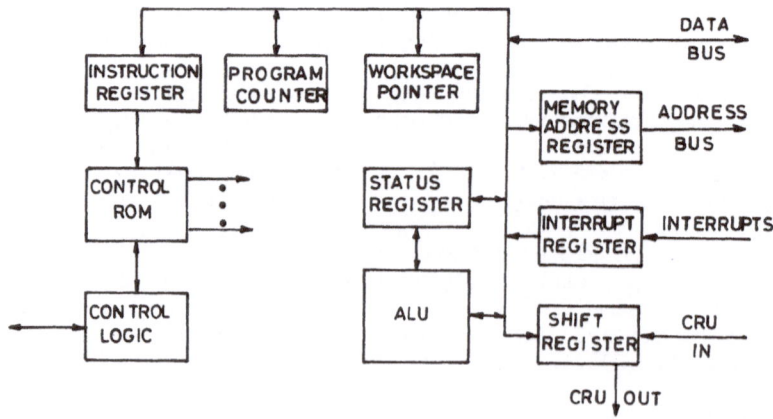

Fig.1 : Basic structure of the 9900

TABLE I: SPECIFICATIONS OF 16-BIT MICROPROCESSORS NOT INCLUDING FUNCTIONS PROVIDED BY COPROCESSORS OR AUXILIARY CHIPS

	TI 9900	Intel 8086	Zilog Z-8000	Motorola 68000	NS 16032	Intel 80286
Year of commercial Introduction	1976	1978	1979	1980	1982	1982
No.of basic instructions	69	95	110	61	82	121
No.of General-Purpose Registers	16	14	16	16	8	14
Pin Count	40	40	48/40	64	48	68
Direct Address Range (Bytes)	64K	1M	48M*	16M/64M	16M	16M
No.of Addressing Modes	8	24	6	14	9	
Basic Clock Frequency	3MHz	5MHz (4-8MHz)	2.5-3.9 MHz	5-8MHz	10MHz	8MHz
System Structures						
Uniform Addressability	X	X	X	✓	✓	X
Module Map and Modules	X	X	X	X	✓	X
Virtual	X	X	X	X	✓	✓
Primitive Data Types						
Bits	✓	X	✓	✓	✓	X
Integer Byte or Word	✓	✓	✓	✓	✓	✓
Integer Double-Word	X	X	✓	✓	✓	X
Logical Byte or Word	✓	✓	✓	✓	✓	✓
Logical Double-Word	X	X	X	✓	✓	X
Character Strings (Byte Word)	✓	✓	✓	X	✓	✓
Character Strings (Double-word)	X	X	X	X	✓	X
BCD Byte	X	✓	✓	✓	✓	✓
BCD Word	X	X	X	X	✓	X
BCD Double-word	X	X	X	X	✓	X
Floating-Point	X	X	X	X	X	X
Data Structures						
Stacks	✓	✓	✓	✓	✓	✓
Arrays	✓	X	X	X	✓	X
Packed Arrays	✓	X	X	X	✓	X
Records	✓	✓	✓	✓	✓	✓
Packed Records	X	X	X	X	✓	X
Strings	X	✓	✓	X	✓	✓
Primitive Control Operations						
Condition Code Primitives	✓	X	✓	✓	✓	X
Jump	✓	✓	✓	✓	✓	✓
Conditional Branch	✓	✓	✓	✓	✓	✓
Simple Iterative Loop Control	✓	✓	✓	✓	✓	✓
Subroutine Call	✓	✓	✓	✓	✓	✓
Multiway Branch	X	X	X	X	✓	X

contd...

Control Structure						
External Procedure Call	X	X	X	X	✓	✓
Semaphores	X	✓	✓	✓	✓	✓
Traps	✓	✓	✓	✓	✓	✓
Interrupts	✓	✓	✓	✓	✓	✓
Supervisor Call	✓	X	✓	✓	✓	✓
Compatibility with other Microprocessors	X	X	X	X	X	✓

*6 segments of 8M each

the contents of the old set of workspace registers. All memory is accessed directly, a 16-bit word at a time; thus the maximum memory size is 64 kbytes. However the 9900 has byte instructions to perform operations on either of the two bytes fetched from memory. The microprocessor, memory, and I/0 devices are interconnected by a local bus. The communications register unit (CRU) handles I/0 interface using special instructions to address external devices via the CRU lines and address bus. The CRU mechanism can address up to 4 kbits each of input and output. All CRU data transfers to and from I/0 devices are handled serially. There are eight addressing modes in all. In 1981, Texas Instruments introduced the 9995, with 256 bytes of internal RAM, higher I/0 workspaces, and an instruction trap (MID instruction interrupt) to enable simulation of new instructions and to trap on illegal opcodes. The new 99000 family [30],[85]offers single instruction prefetch, an internal oscillator and clock generator, long-word arithmetic, test-and-set primitives, and support for multiprocessor and DMA configurations.

The basic structure of the Intel 8086 is shown in Fig 2. The CPU consists of two separate processing units, the execution unit (EU) and the bus interface unit connected by a 16-bit ALU data bus and an 8-bit Q bus. The EU obtains instructions from the instruction prefetch queue, IQ maintained by the BIU and executes instructions using the 16-bit ALU. The EU accesses memory and peripherals through requests to the BIU, which is the second processing unit, performing all bus operations for the EU on a demand basis; the BIU is also responsible for prefetching instructions. The 6-byte instruction prefetch queue feeds instructions to the EU in 8-bit segments. A substantial number of the 95 basic instructions are only a byte long. In the few 16-bit instructions, only the first byte is used for operation codes; the second byte specifies data displacement. Instructions

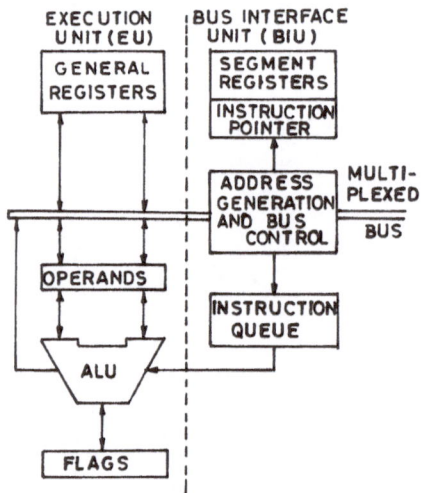

Fig 2: Basic structure of Intel 8086.

longer than 2 bytes use the remaining bytes for specifying data. The
8086 instruction set is an expanded version of the 8080 instruction set.
However, base segment registers have been added to enable programmers
to do process swaps with relative ease, and to increase memory addressing capabilities. The 1 Mbyte of real addressing space is treated as a
group of segments, each segment 64 kbytes in size. Four segments are
addressable at one time, providing up to 64 kbytes of code, 64 kbytes
for stack, and 128 kbytes for data. Multiple concurrent stacks are not
feasible. The Intel 8086 execution unit contains four 16-bit pointer
and index registers and four 16-bit data registers addressable on an
individual byte basis. These eight registers are used implicitly by the
instruction set, providing compact encoding at the cost of reduced
flexibility. The BIU contains additional registers for addressing
purposes. The 8086 utilizes two types of multimaster buses, the local
bus and the system bus. Microprocessors are always connected to a local
bus, and memory and I/O usually reside on a system bus. These two buses
are linked by interface components. The system bus is functionally and
electrically compatible with the Intel Multibus. The Intel 8086 has a
64-kbyte separate I/O space. A memory-mapped I/O capability that can
respond like a memory device is available for linking I/O devices. Any
memory reference instruction can be used to access an I/O device,
providing additional programming flexibility. High-speed I/O operations
can be carried out with traditional DMA controllers. Intel also offers
the 8089 IOP, an independent processor with two DMA channels and an
instruction set tailored for I/O operations. Overall, the 8086 instruction set provides automatic repetition of many instructions, decimal
operations, error traps, and a large I/O space through indirect addressing. Designed to be used as a coprocessor, the Intel 8087 Numeric Data
Processor [101] supports 32-, 64-, and 80-bit data structures and
floating-point arithmetic. The Intel 80186 introduced in 1982 offers
twice the performance of the standard Intel 8086, and offers twelve
additional instructions. The new instructions facilitate string operations, subroutine operations, context switching, and detection of out-
of-range values. The Intel 80286, also introduced in 1982, offers still
higher performance (up to six times that of the 8086), virtual memory,
on-chip memory management, a four-level memory protection hierarchy, and
is directed towards the multiuser environment [31],[32]; the 80286 uses
a superset of the 80186 instruction set with sixteen additional instructions.

The basic structure of the Zilog Z8000, architecturally different
from the Z80, is shown in Fig.3. The internal 16-bit data bus is used
for internal addressing and data communication. The instructions are
fetched through the Z-bus interface and executed by the instruction
execute control unit. Throughput is enhanced through "limited" pipelining, which allows prefetching of the next single-word instruction (or
the first word of the next multiword instruction) from the memory into
the instruction buffer. This occurs only during the execution of the
current instruction, provided the current instruction does not require
the bus to complete the execution cycle. Unlike the 8086, the Z8000
does not use implied registers. All sixteen 16-bit general-purpose
registers can be used as accumulators and all but one can be used as

14

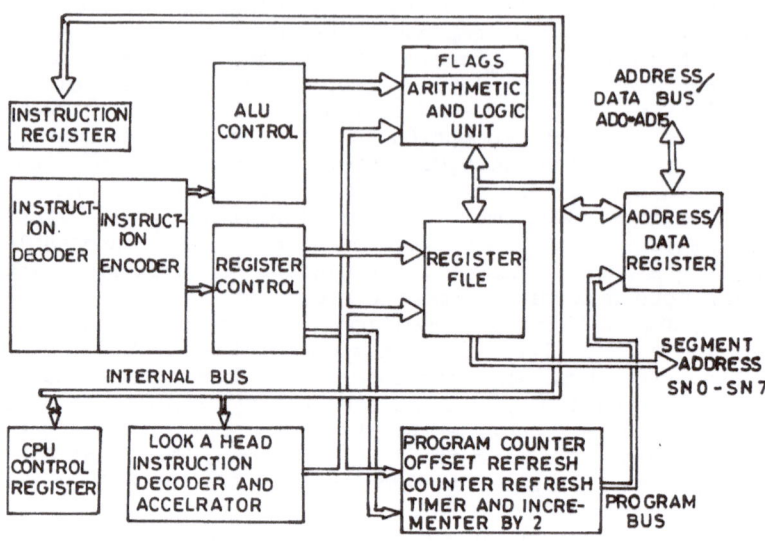

Fig.3: Basic structure of the Z8000

index pointers or memory points. The one exception is an escape mechanism for address changes. The general register architecture avoids
bottlenecks inherent in dedicated or implied registers. Memory addresses
are always expressed in bytes. The 8 Mbytes of directly addressably
memory is split up as 128 segments, each of 64 kbytes. To increase the
address range beyond the nominal limits, code, data, and stack spaces
in the CPU's system and normal modes can be physically separated using
the Z8010 memory management unit (6x8 Mbytes = 48 Mbytes). Unlike the
8086, the Z8000 permits multiple concurrent stacks; stacks can be
located anywhere in memory and are addressed via stack pointer registers. Any register, barring one, can serve as a stack pointer by means
of PUSH and POP. Call return, interrupts, and traps use implied stack.
The system stack can be accessed only in system mode, whereas the normal
stack can be accessed in both modes. The Z-bus interconnects Z8000 family components [33] in a master/slave fashion. The CPU obeys the Z-bus
protocol directly at the chip level, and no extra circuitry is needed
to generate bus signals. Multiplexing of addresses and data minimizes
pin count; multiplexing does not affect read times, but degrades write
times. Almost in all cases, reads outnumber writes three or four to one,
and as such, multiplexing does not result in any significant performance
loss. Demultiplexing is performed when necessary within the individual
modules. The daisy chain serial priority philosophy resolves interrupts/
traps, bus requests, and requests for shared resources. In the system
mode, two different I/O instructions trasfer data between the CPU and
peripherals, and special I/O instructions transfer data to and from
external support chips. Processor status information eanbles separation

of address spaces. The I/O addressing scheme is identical to the basic
memory addressing scheme. For DMA operations, two signals - bus request
and bus acknowledge - are available. Inhibited from controlling the bus
during DMA operations, the CPU must wait for the bus to be given up by
DMA controller. The instruction set facilitates multiprogramming through
a context switching facility. Other instruction highlights include
signed 32-bit multiply and divide, decimal operations, multiple load,
vector based instructions, and the test-and-set instruction for multiprocessor applications as well as support for floating-point operations
[87]. Addressing schemes include indexing, with or without displacement,
and multiple increment indexing.

The Motorola MC68000 (see Fig.4) is architecturally [34] quite
different from the TI9900, Intel 8086, and the Zilog Z8000 chips. The
microcode-based CPU is centered around a microprogram-controlled execution unit. Most internal data structures are 32 bits wide. To minimize
the control store area size, a two-level structure comprised of a micro-
store and a nanostore is used which, however, increases total access
time. An attempt has been made to overcome this by means of a pipelined
architecture, in which the instruction fetch, instruction decode, and
instruction execute cycles are fully overlapped across every macroins-
truction boundary. In branch conditions, the prefetch is associated with
the most likely branch condition. The MC68000 execution unit is a dual-
bus structure that performs both address and data processing. There are
eight 32-bit data registers, seven 32-bit address registers, and two
implied 32-bit stack pointers. The data registers can be addressed as
byte, word, or double-word registers. The address registers are used
for 32-bit base addressing, 32-bit software stack operations, and word
and long-word address operations. The implied stack pointers are used
for 32-bit base addressing and for word and long-word operations. The
MC 68000 views memory as a linear space of up to 16 Mbytes with no
internal segmentation. Although words are normally addressed , single
bytes can be read or written using upper and lower data byte strobes.
Instructions and multibyte data are always aligned on even byte boundaries. The memory management unit, MC68451, supports management and
protection of 32 variable-sized segments, ranging from 256 bytes to 16
Mbytes in increments of 256 bytes. Multiple concurrent user stacks and
queues can be created and maintained by employing the address register
indirectly with post-increment and
pre-decrement addressing modes. The
MC68000 possesses no separate I/O
space. All I/O is memory mapped
and all I/O protection must occur
at memory protection level. Three
signals - bus request, bus grant,
and bus grant acknowledge - allow
master devices to get control of
the bus for DMA operations. The
68000 family is supported by two
different master/slave-based multimaster buses for interconnection of

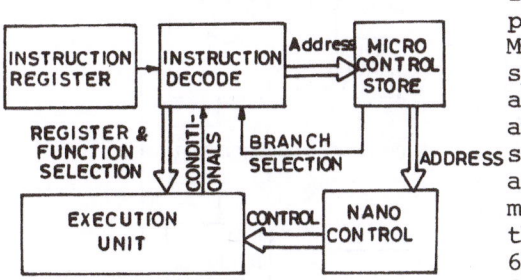

Fig.4:Basic structu-e of the MC68000. components - the local bus and the

16

global bus. The local bus connects microprocessor, memory, and I/O devices to form individual microcomputer modules, and the global bus interfaces to various local buses through bus arbitration modules. Versabus [95] is an extended version of the local bus. Overall, the MC68000 has a regular instruction set and provides multiuser support. It emphasizes space-efficient code through "quick" instructions and short jumps on loops. The MC68000 offers the advantage of excellent debug tools like single-step execution, traps on illegal instructions, and debug mode. Context switching facilitates multiprogramming, and the test-and-set instruction aids in multiprocessor and database applications. Other advantages include complex push and pop capabilities, the 32-bit internal structure, and instructions for multiple load and signed multiply and divide. Real-time control applications are aided by multilevel interrupt and seven auto-vector interrupt capabilities. The MC68010 [35], introduced in 1982, offers virtual memory facilities, and superior operating system support through relocatable exception vectors and additional system control registers. The MC68020 [35] offers full 32-bit capability. The MC6881 [35] floating point coprocessor can only be used in conjunction with the MC68020.

The National NS16032 (see Fig.5), sometimes classified as a 32-bit microprocessor [36],[90] ,contains several similarities to the MC-68000 architecture. It, too, uses 32-bit data paths and 32-bit address arithmetic to access memory without the overhead of segmentation registers; externally it uses a multiplexed address and data bus with 24 bits of external address of 16 bits of external data. The NS16032 also uses a two-level microcode that enables sharing of microcode routines by all instructions and avoids time-consuming subroutine calls and space-consuming repeated microcode flows. The instruction queue, in the case of the NS16032, consists of a double-ended 8-byte first-in first-out, with 16-bit input and output buses. As microinstruction n is executed, microinstruction n+1 is decoded, and microinstruction n+2 is selected by the microcode ROM address decoder. There are eight general-purpose registers, all 32 bits wide, and eight dedicated registers of smaller

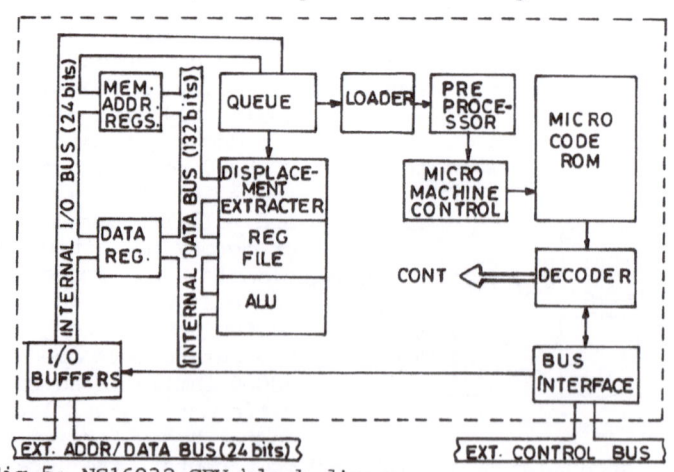

Fig.5: NS16032 CPU block diagram

size. The NS16032 provides a 16-byte uniform address space and nine types of address modes. The length of the address mode offset constants are encoded in the upper 2 bits of the offsets so that small offsets (-64 to 63) require only 1 byte in the instruction stream, and larger offsets take 2 or 4 bytes. Instructions are of variable length. Common zero-operand instructions, such as a branch instruction, have a 1-byte opcode. The one-operand and two-operand instructions use a 2-byte basic instruction. All instructions can use all applicable addressing modes - this is termed an orthogonal instruction set and reflects the general industry trend. Instructions are aligned on byte boundaries for code compactness. The NS16032 offers instructions to operate on packed decimal quantities, on arrays, and on blocks. The ability to operate on variable-length character strings makes the processor suited for text processing applications. The NS16081, which is the floating-point co-processor of the NS16032 provides 32-bit and 64-bit floating-point capabilities, and the memory management unit, the NS16082, provides demand paged virtual memory facilities. Virtual address to real address translation is accomplished through two levels of page tables and offset specifications. Each page, 512 bytes in size,is assigned a protection code, and this provides an access control mechanism. The memory management unit has eight registers and provides a flow-tracing facility for both sequential and nonsequential instructions. The NS16032 has a hardware backup mechanism that saves contents of the processor status register, the stack pointer, and the program counter, and automatically restores them to their original values in case of an instruction retry. In view of the architectural orientation for support of virtual memory, the NS16032 is quite similar to the Motorola MC68010. The 32-bit version the NS32032, uses the same architecture but provides 32-bit external data paths like the Motorola MC68020.

The 16-bit microprocessors considered so far are secondsourced by several companies outside the United States. A significant portion of the NS16032 design work was done by the National Semiconductors subsidiary in Israel. Some other design efforts deserve mention here.Philips Data Systems in Europe has implemented the SP 16C/10 microprocessor in NMOS; this chip also offers virtual memory and support for coprocessors and multiple processors[37]. Fujitsu of Japan has implemented in CMOS a 16-bit processor with on-chip error detection, virtual memory support, and four hierarchical levels of operation to efficiently handle multiprocessing [38]. Within the United States, single chips have become available that replace entire CPU's of popular minicomputers like the PDP 11[39] and the Data General Eclipse[40].

In this world of increasing variety of 16-bit microprocessor chips, the task of selecting a particular chip is a difficult one. Apart from hardware and architectural details which we have discussed so far, it becomes essential to evaluate system performance at multiple levels [41] as follows:

a) Instruction Level: At this level one compares timings on an instruction-by-instruction basis. This level is especially relevant when the chip is not physically available such as is the case of newly announced or newly released processors.

b) Routine Level: Sample user tasks coded in assembly language are often represented as competitive analyses between processors by manufacturers. The application programs frequently reflect advantages unique to the "winning" microprocessor.

c) Support Task Level: The sophistication of operating systems support is evaluated.An example of a support task would be a multilevel queue management used by the process scheduling algorithm of the operating system.

d) System Functions Level: Specific operating system and user tasks are encoded, e.g., memory management algorithms for segmentation.

e) User Job Level: This benchmark spawns many processes, thereby exercising the operating system and all lower level benchmarks.

16-bit microprocessors have been compared and evaluated by Heering [42], by Grappel and Hemenway [43],[44], by Rajulu and Rajaraman[117],by Toong and Gupta 26 , 41 , and by Prycher 120 . Table II, based on information from [26],[27],[36], compares performance at the instruction level. Comparison at higher levels of the evaluation hierarchy are summarized in Table III [41]based on the same clock frequency. As seen from this table, the performance is heavily dependent on the choice of algorithm. Comparison at higher levels is even more application dependent. Thus all general comparisons must be used with extreme caution.

TABLE II:EXECUTION SPEEDS(IN MICROSECONDS) OF 16-BIT MICROPROCESSORS[26] [27],[36](Note that different chips use different clock frequencies)

		Chips and Clock Frequency					
Operation	Data Type	TI9900 at 3MHz	Intel 8086 at 5MHz	Zilog Z8000 at 5MHz	Motorola MC68000 at 8MHz	National NS16032 at 10MHz	TI 99110 at 6MHz
Register-TO- Register Move	Byte/Word	4.60	0.40	0.75	0.50	0.30	0.50
	Double-Word	9.80	0.80	1.25	0.50	0.30	1.00
Memory-To- Register Move	Byte/Word	7.30	3.40	3.50	1.50	1.00	0.83/ 0.67
	Double-Word	14.60	6.80	4.25	2.00	1.50	1.33
Memory-To- Memory Move	Byte/Word	9.90	7.00	7.00	2.50	1.70	1.00/ 0.83
	Double-Word	19.80	14.0	8.50	3.75	2.50	1.67
Add Memory To Register	Byte/Word	7.32	3.60	3.75	1.50	1.20	0.83
	Double-Word	21.30	7.20	5.25	2.25	1.60	2.00
Compare Memo- ry-To-Memory	Byte/Word	9.90	7.00	7.25	3.00	1.70	1.00
	Double-word	19.80	14.00	9.50	4.00	2.50	2.00
Multiply Memory-To- Memory	Byte	21.90	13.00	20.25	N/A	3.50	4.17
	Word	21.90	23.00	16.00	8.75	5.10	4.17
	Double-word	180.64	115.20	85.75	43.00	8.30	26.38
Conditional Branch	Branch Taken	3.60	1.60	1.50	1.25	1.60	0.50
	Branch Not Taken	2.90	0.80	1.50	1.00	0.80	0.50

contd.....

| Modify Index Branch
Branch If Taken
Zero | 7.60 | 2.20 | 2.75 | 1.25 | 1.30 | 1.00 |
| Branch To
Subroutine | 7.90 | 3.80 | 3.75 | 2.25 | 2.00 | 1.00 |

Table III: COMPARISON OF PERFORMANCE OF 16-BIT MICROPROCESSORS (n, in the case of the hash algorithm, denotes the number of searches before an open slot is found).

| Level | Example Used | PERFORMANCE | | |
		Intel 8086	Z8000	MC68000
Routine Level	(a) Booth's multiplication algorithm	1.00	1.41	1.26
	(b) Polynomial evaluation	1.00	1.80	1.97
Support Task Level	(a) Stack exerciser	1.00	8.42	3.49
	(b) Hash algorithm	144+ 8*n	188+ 66*n	141+ 54*n

5. 32-BIT MICROPROCESSORS

Microprocessors with 32-bit internal paths and 16-bit external paths have been in existence since 1980. However, the era of true 32-bit microprocessors begins in 1981 with the commercial introduction of the Intel iAPX432. IBM's implementation of the IBM 370 on a single chip in 1980 [45] is generally overlooked because of the lack of commercial production of the device.

As mentioned in the previous section, the Motorola 68020 and the National 32032 are very similar to their respective 16-bit processors; as such, they are not analyzed here. We discuss general trends by considering three different 32-bit microprocessors chosen on the basis of their architectural innovativeness and the availability of good documentation [46]. These are the Bellmac-32A microprocessor from Bell Labs [47],[48], the "no-name" 32-bit CPU chip from Hewlett-Packard [49][50], and the Intel iAPX 432 [51]-[53]. Even though the former two are designed for internal use, they are harbingers of similar products in the public domain.

5.1 General Characteristics

The general characteristics of the three microprocessors are summarized in Table IV.

The Bellmac-32A single-chip CPU has been fabricated in twintub CMOS technology that dissipates less than 1 W of power and uses new "domino circuits" that operate at twice the speed of previous CMOS circuits and enable a single clock pulse to activate many circuits

TABLE IV: GENERAL CHARACTERISTICS OF 32-BIT MICROPROCESSORS   46

	BELLMAC-32A	HP 32-BIT CPU	INTEL iAPX 432
Yr.of Commercial Introduction.	1982*	1982*	1981
TECHNOLOGY	2.5um DOMINO CMOS	1.5/1.0umNMOS	HMOS
No.of Transistors	146,000	450,000	219,000 On three chips
Size of Chip	$160,000MIL^2$	$48,400MIL^2$	$100,000MIL^2$ each
Power Dissipation	0.7 Watt at 8MHz	4 Watts	2.5watts/chip
Pin count	63 Active 83 Total	83	64 per chip
Basic Clock Frequency	10MHz	18MHz	8MHz
Direct Address Range(Bytes)	2^{32}	2^{29} Real 2^{41} Virtual	2^{24} Real, 2^{40} Virtual
No.of General Purpose Registers	16 User-Visible	28(Not all Gen.Purpose)	No Registers Visible to User
No.of Basic Instructions	169	230	221
No.of Addressing Modes	18	10	5
* Currently for internal use only			

simultaneously. A single 40-MHz clock is used to generate two 10-MHz phase-shifted clock signals. The Bellmac-32A Chip has been designed to provide support for the programming language C.Single instructions can move blocks of data from memory to memory, or push and pop a group of registers with respect to the stack. Sophisticated hardware facilities include a barrel shift circuit that shifts 0 to 31 bits in a single cycle. The operating system can be included in the address space of every process, and includes a hardware interface for a process-oriented operating system and a set of exception-handling mechanisms. The exception structure provides four levels of execution privilege. It is intended for real-time control applications. Neither floating-point nor decimal arithmetic are supported. An auxiliary processor to perform such operations is at the investigation stage [48] . At present, "extension" instructions are provided to perform these functions. There is little compatibility with any existing microprocessor. The development of the Bellmac-32A chip was accomplished in a relatively short time using extensive computer-aided design techniques[91]. A mask generation program permits "technology updatability" enabling old mask sets to be updated to new rule(s) easily; the simulation files are generated automatically to benefit from the evolving technology that permits thinner linewidths. On the chip itself, access is available to most registers for test and debug purposes via special internal access features.Unlike the Hewlett-Packard chip (discussed next), there is no facility for automatic self-test during the power-up sequence.

The HP 32-bit CPU chip has the highest circuit density- 450 000 transistors implemented on a single chip using a 1-um pitch n-channel MOS double-layer-metal technology. This microprocessor uses two non-overlapping clocks each of 18-MHz frequency (generated from an external 36-MHz clock), and is microcoded using 9K (38-bit) words of ROM control store addressed via a set of 14-bit registers in the sequence stack.The microinstructions are decoded by PLA. Most microinstructions execute in a one clock cycle of 55 ns. Pipelining of memory operations permits initiation of 32-bit memory read every two states (110ns) even though the access time is longer. The self-test routine, executed by the CPU during the power-up sequence, automatically tests operations internal to the chip. This HP chip offers several sophisticated hardware and software features. The hardware implemented N-bit barrel shifter can shift a 32-bit quantity right or left 0 to 31 places in a single clock cycle. The load instruction includes automatic bounds checking and takes only 550 ns. The arithmetic and logical instructions can manipulate 32- and 64-bit IEEE standard floating-point operands. Text editing capabilities are inherent in the move and string instructions that manipulate byte arrays and string-type data. All communication to and from the chip involves use of the memory-processor bus. This 32-bit-wide multiplexed address/data bus permits pipelined data transfers at 36-Mbytes/s transfer rate. Facilities to communicate with the memory and with peripheral devices are provided by the memory controller chip and the I/O processor chip, respectively.

In many respects, the Intel iAPX 432 is very different from our previous examples. First, just as the Burroughs 5000 introduced a trend two decades ago of architectures especially designed to support high-level languages, the Intel iAPX 432 is designed to be programmed entirely in a high-level language, and the system architecture is consciously oriented toward supporting Ada, the programming language sponsored by the Department of Defense. Secondly, whereas the other 32-bit processors are single-chip CPU's, the Intel iAPX 432 is designed as a three-chip set [100]. These three chips are connected to each other through a processor-memory interconnection bus, which is different from the traditional Intel Multibus.TheGeneral Data Processor System, usually referred to as GDP, consists of two chips - the iAPX 43201 [53],[98], with 110 000 transistors, which is responsible for instructions decoding, and the iAPX 43202 [51],[93]with 49 000 transistors, which performs actual instruction execution. The iAPX 43203 I/O interface processor[52], 92 containing 60 000 transistors, maps I/O bus addresses into the main memory address space, and also provides attached I/O processors with a set of interprocess communication primitives. Two other chips, the 43204 Memory Control Unit (MCU) and the 43205 Bus Interface Unit (BIU), are used in larger configurations with multiple memories,processors, and buses. The Intel iAPX 432 is architecturually very different from its predecessors; namely, the 8080, 8085, and 8086; this difference, although providing technological enhancements, newer functions, and higher overall throughput, greatly limits the programming compatibility which exists among the earlier Intel products. On the Intel iAPX 432, the total physical address size is limited to 2^{24} bytes. The upper limit on the logical address space is 2^{40} bytes. However, at any instant the logical addressing environment of a program is restricted to 2^{32} bytes.

The instructions are of variable length ranging from 6 to 344 bits in size; each instruction's operator can have 0,1,2, or 3 operands. The general processor flexibility is accompanied by adequate built-in security mechanisms that restrict access to program and data on a "need-to-know" basis. Other highlights include hardware-implemented concurrent programming functions, self-dispatching processors with hardware-implemented process scheduling, Ada high-level language support, and the capability to perform specialized functions, for example, floating-point operations and string operations, without the need to attach auxiliary, specialized chips.

5.2   Technology

All the chips reflect a conscious attempt toward the integration of an enormously large number of transistors. Even with a significantly smaller chip size, Hewlett-Packard has packaged the largest number (450000) on a single chip, almost twice what Intel has packaged on three chips put together. In order to achieve this intense packaging density, Hewlett-Packard uses an electron beam to generate masks that provide 1.5-μm wide lines and 1.0-μm spaces with +0.25μm tolerances. The large number of devices and the low width of the lines contribute to the higher power dissipation (4W). The packaging technology used by Hewlett-Packard consists of a copper core, on which the CPU and auxiliary chips are directly mounted and four layers of interconnect, separated by low-capacitance teflon dielectrics. The Intel iAPX 432 is also implemented in NMOS (Intel prefers to call it n-channel silicon-gate HMOS technology). However, as compared to Hewlett-Packard, the transistor density is much lower, permitting implementation of wider 3.5-μm structures.
     Because of the CMOS implementation, the Bellmac-32A chip is the least power-consuming microprocessor. The twin-tub CMOS process (Fig.6) gives high switching speed in both n- and p-channel devices, because each tub is separately implanted for optimum doping. The process takes advantage of reduced geomtries for denser design and the n^+ substrate layer avoids the thyristor-like latchup characteristic usually found in CMOS circuits. The optimization of percentages of n- and p-channel components has enabled implementation of high circuit densities while retaining the CMOS advantages of fast speeds and low power consumption.

5.3   Basic Principles of Operation

The structure of the Bellmac-32A microprocessor is shown in Fig.7. It consists of two distinct functional units - a fetch unit which controls interactions with the external memory and an execution unit which controls the manipulation and processing of data. Both units, as well as the bus, have full 32-bit capability. The instruction stream is byte oriented. The first byte specifies the addressing mode and the register and the subsequent bytes specify additional data. All byte and halfword operands are sign or zeros extended to 32 bits when they are fetched. Instructions are monadic, dyadic, or triadic depending on the number of operators being 1,2, or 3, respectively. Instructions fetched from the memory are stored in the instruction queue, and translated into a series of microinstructions using a PLA. The arithmetic address unit

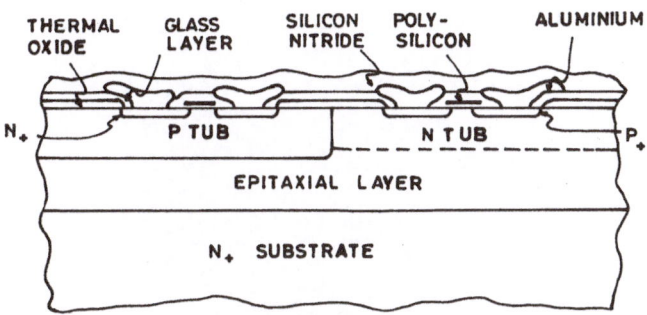

Fig.6: The Bellmac-32A implemented in domino CMOS   54

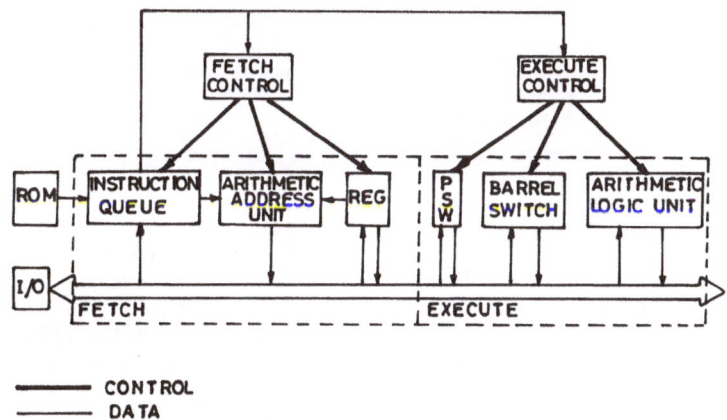

Fig.7: Bellmac-32A CPU architecture

performs all address calculations. The arithmetic logic unit in the
execute block performs the actual execution of microinstructions. The
emphasis on support for a process-oriented operating system generates a
need to store all instructions, data , and register values associated
with a process at times of switching from one process to another; the
Bellmac-32A microprocessor performs these functions in hardware.

The Hewlett-Packard 32-bit chip, shown in Fig.8, uses a microcode
control store ROM of 9216 words, each of 38 bits. Microinstructions
accessed from the ROM are decoded by the PLA and drive control lines
which determine the operations of the 32-bit register stack and the
ALU. The flow of instructions to the PLA is controlled by the sequencing
machine which contains a microprogram counter, a set of incrementers,
three registers for microcode subroutine return addresses, and a machine
instruction opcode decoder. The opcode decoder generates the starting
address in control store for the microcode routine that implements each
machine instruction. The test condition multiplex facilitates conditional jumps and skips in the microcode. The ALU contains an n-bit shifter, a
32-bit logical selector, and a 32-bit full look-ahead adder, which also

performs integer multiplication and division using special hardware.

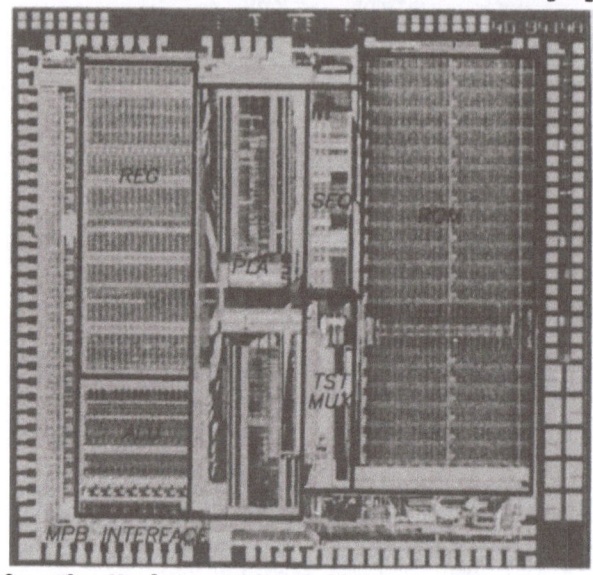

Fig.8: The Hewlett Packard 32 bit microprocessor

The operation of the iAPX 432 involves communication between the iAPX 43201 instruction fetch and decode unit and the iAPX 43202 instruction execution unit via a microinstruction bus. The iAPX 43203 I/O interface unit is connected through the processor-memory interconnect bus as shown in Fig. 9. The characteristics of these chips are summarized in Table V. Data can be manipulated in the form of 8-bit characters, 16/32-bit ordinals, 16/32-bit integers, 32/64/80-bit floating-point variables, bit strings, arrays, records, and as "objects" which are data structures

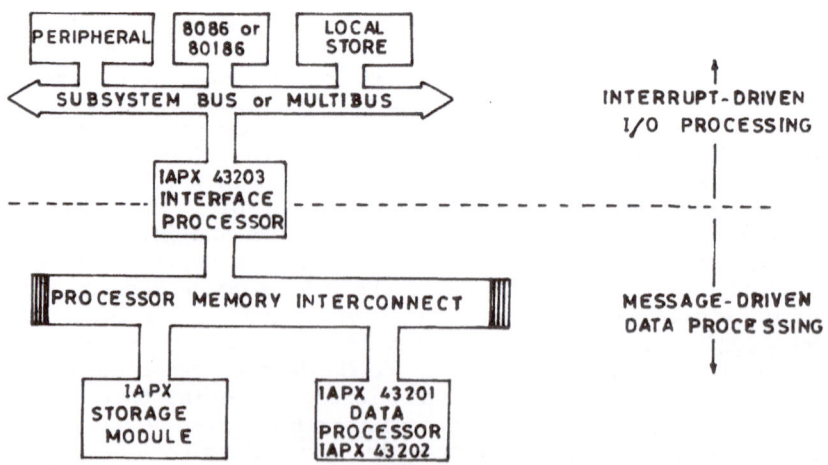

Fig: 9- Intel's iAPX 432 three-chip 32-bit microprocessor

TABLE V: CHARACTERISTICS OF THE GENERAL DATA PROCESSOR

CHARACTERISTICS	iAPX 43201	iAPX 43202	iAPX 43203
DIE SIZE (IN μm)	318x323	366x313	358x326
TOTAL DEVICE PLACEMENTS	110,000	49,000	60,000
FUNCTION OF UNIT	INSTRUCTION FETCH AND DECODE	INSTRUCTION EXECUTION	I/O INTERFACE
FUNCTIONAL SUB-UNITS	INSTRUCTION DECODER; MICROINSTRUCTION SEQUENCER (MIS)	DATA MANIPULATION UNIT (DMU); REFERENCE GENERATION UNIT (RGU)	DATA ACQUISITION UNIT (DAU); MICROEXECUTION UNIT (MEU)
FUNCTION OF SUB-UNIT	DECODES VARIABLE LENGTH BIT ALIGNED INSTRUCTIONS; SEQUENCES VERTICALLY ENCODED MICROINSTRUCTIONS AND INPUTS THEM TO THE MEU	CONTAINS OPERAND AND UNITS TO IMPLEMENT THE MACROINSTRUCTION SET EFFICIENTLY; CONTAINS REGISTERS AND FUNCTIONAL UNITS FOR LOGICAL-TO-PHYSICAL ADDRESS TRANSLATION AND ACCESS RIGHT VERIFICATION	PERFORMS PREFETCH AND POST-WRITE BUFFERING OF DATA AND GENERATES MAIN SYSTEM MEMORY ACCESS; PERFORMS SYSTEM ACCESS ENVIRONMENT MANIPULATION, INTER-PROCESSOR COMMUNICATION, AND ADDRESS MAP SETUP

containing information in an organized manner. These objects can be referenced as a single entity, and their internal organization is hidden and protected from all other procedures by hardware mechanisms. Each object has defined for it a set of operations (procedures or instructions) that are permitted to manipulate it directly. Examples of hardware-defined are as follows:

a) processor objects :   represent the physical processors;
b) process objects: represent the individual computing tasks;
c) context objects: represent the activation of a program unit;
d) dispatching port/   provide a stream of work for a set of
    objects: processors;
e) communications port   supports interprocess communication and
    objects: synchronization.

The iAPX 432 instruction set supports objects through messages (SEND,WAIT), and context (CALL,RETURN), storage pools (ALLOCATE,TYPE), processes (SCHEDULE,DISPATCH). Objects permit each program module (e.g., an Ada package) to operate as an individualized "virtual machine" for more reliable operation, since errors in the module are confined to its virtual machine, the provide simpler compilation, since each virtual machine has its own address space.

5.4   Register Organization

The Bellmac-32A microprocessor contains a special program counter register and fifteen other registers (32 bits wide) that can be referenced in any addressing mode. Of these fiteen registers, three are used to support operating system functions (as interrupt stack pointer, process control block pointer, processor status word) and can be written when the processor is in kernel execution level. Another three registers are used as a stack pointer, a frame pointer, and an argument pointer by certain instructions.

At the heart of the Hewlett-Packard chip, there is a register stack containing 28 identical general-purpose registers (not all accessible by software), each 32 bits wide and an ALU with four operand/result registers. The register stack uses two databases and contains auxiliary logic such as top of stack registers and instruction registers. In any register, each of the 32-bit cells can receive data from or dump data to either of the two data buses, as determined by the PLA outputs.

On account of its multichip organization, the register structure of the iAPX 432 cannot be compared with the others on an equitable basis. The microinstruction execution unit performs several functions traditionally associated with registers. Its functional subunit, the Reference Generation Unit (RGU) contains a 43-bit by 20-entry register array to support logical-to-physical address translation and access rights verification. The other functional unit, the Data Manipulation Unit(DMU) contains its own set of operand registers, implemented as double-ended queues, to optimize arithmetic calculations on variable length operands with a fixed length (16-bit) bus. Intel claims that the compiler complexity is reduced by keeping registers "behind the scenes" rather than as visible features of the architecture [99].

5.5 Instruction set

The 32-bit processors offer powerful instruction sets and support a wide spectrum of distinct data structures. These capabilities are summarized in Table VI.

The Bellmac-32A offers 169 instructions, more than twice the number on the NS16032. It supports bytes, halfwords, words, and bit fields. Strings are supported by special block instructions, and the string format conforms to the C language. This design objective is manifest in the implementation of the instruction repertoire. The result of unary operations Negate and Complement (implemented as a move instruction) can either replace the existing datum or be placed in a new location. The dyadic form stores the result in the second operand and the triadic form places the result in the third operand, with the first two unaltered: these dyadic and triadic forms of instructions are available for all operators. High-level procedure linkage operations assist in manipulating the stack frame, saving registers, and in transferring control between procedures. Also, explicit instructions (Call Process and Return to Process) are provided for switching processes by the operating system. On the negative side, the Bellmac-32A does not support floating-point or decimal arithmetic.

The Hewlett-Packard 32-bit chip offers a still larger repertoire of 230 instructions and 32- and 64-bit floating-point arithmetic capability. The Load and Store instructions can transfer double words in addition to the bits, bytes, halfwords, and words. Move and String instructions manipulate both unstructured byte arrays and structured string data. Hardware support includes four top of stack registers ·to handle push and pop operations and to provide "data valid" indication. The large number of transistors enables hardware implementation of features that have traditionally been done by software. For example, there is support for run-time bounds checking of addresses performed on all memory accesses.

The iAPX 432 processor supports integer data in the form of features that have tradtionally been done by software. For example, there is support for run-time bounds checking of addresses performed on all memory accesses. The iAPX 432 processor supports integer data in the form of halfwords and words, and floating-point data in the form of words, double words, and as 80-bit quantities. The usual 80-bit-wide quantities, called temporary reals, are used to store intermediate results to improve accuracy of final results. Instructions are bit-variable in length and are not constrained to coincide with byte or word boundaries. The total number of instructions is 230 and the longest is 344 bits long. By allowing both stack arithmetic and memory-to-memory arithmetic, the iAPX 432 has the potential of providing denser code and faster speeds at evaluating expressions. An instruction contains four fields. The first two fields, the class field and the format field, specify how many operands are in the instruction and how they are to be accessed. The third field, the reference field, contains the logical addresses of up to three operands. The last field specifies the operator itself. The processor reads an instruction segment in units of 32 bits. The instructions are decoded by the 43201 decoding chip, and

TABLE VI: CHARACTERISTICS OF DIFFERENT MICROPROCESSORS (WITHOUT THE USE OF COPROCESSORS OR AUXILIARY CHIPS)

	BELIMAC 32-A	HP 32	INTEL iAPX 432
SYSTEM STRUCTURES			
UNIFORM ADDRESSABILITY	✓	✓	✓
MODULE MAP AND MODULES	x	x	✓
VIRTUAL	✓	✓	✓
PRIMITIVE DATA TYPES			
BITS	✓	✓	✓
INTEGER BYTE OR HALFWORD	✓	✓	✓
INTEGER WORD	✓	✓	✓
LOGICAL BYTE OR HALFWORD	✓	✓	✓
LOGICAL WORD	✓	✓	✓
CHARACTER STRINGS (VARIABLES)	✓	✓	✓
BCD BYTE OR HALFWORD	x	x	x
BCD WORD	x	✓	x
32 BIT FLOATING POINT	x	✓	✓
64 BIT FLOATING POINT	x	✓	✓
80 BIT FLOATING POINT	x	x	✓
DATA STRUCTURES			
STACKS	✓	✓	✓
ARRAYS	✓	✓	✓
PACKED ARRAYS	x	x	x
RECORDS	x	x	✓
PACKED RECORDS	x	x	x
STRINGS	✓	✓	✓
PRIMITIVE CONTROL OPERATIONS			
CONDITION CODE PRIMITIVES	✓	✓	✓
JUMP	✓	✓	✓
CONDITIONAL BRANCH	✓	✓	✓
ITERATIVE LOOP CONTROL	x	✓	✓
SUBROUTINE CALL	✓	✓	✓
MULTIWAY BRANCH	✓	✓	✓
ORTHOGONAL INSTRUCTION SET	✓	✓	✓
CONTROL STRUCTURE			
EXTERNAL PROCEDURE CALL	✓	✓	✓
SEMAPHORES	✓	✓	✓
TRAPS	✓	✓	✓
INTERRUPTS	✓	✓	✓
SUPERVISOR CALL	✓	✓	✓
OBJECTS	x	x	✓
HIERARCHICAL OPERATING SYSTEM	✓	x	✓
OTHERS			
USER MICROCODE	x	x	x
DEBUG MODE	✓	✓	x
COMPATIBILITY WITH OTHER MICROPROCESSORS	x	x	x
SELF-TEST DURING POWER-UP	x	✓	x

the resulting stream of microinstructions are executed by the 43202 execution chip.

5.6 Memory Organization

The Bellmac-32A chip offers several addressing modes: literal, byte/ halfword/word immediate, register,register deferred, short offset (for frame and argument pointers), absolute, absolute deferred, byte/half- word/word displacement deferred, and expanded operand type. The Bellmac 32A chip includes hardware interface for a process-oriented operating system and a set of exception-handling mechanisms; this operating sys- tem can be included in the address space of every process, enabling each process to execute independently.

The Hewlett-Packard 32-bit processor views the memory space for each program as an active code segment (one of 4096 code segments), a stack segment, a global data segment, and a set of 4096 external data segments. Segment pointers, maintained in 32-bit on-chip registers, include base and limit register for the code, stack, and global data segments, the current instruction address in the code segment, the address of the most recent stack marker in the stack segment, and the address of the top of stack in memory. External data segments are acce- ssed via a set of memory resident tables. A memory controller chip is used to control 256 kbytes of RAM (20 chips each of 128K) or 512kbytes of ROM (8 chips each of 640K), as shown in Fig.10. The memory processor bus has a transfer rate of 36 Mbytes/s, and the overlapped access method permits high throughout. The memory controller maps logical to physical addresses in 16-kbyte blocks, and permits byte,halfword, word, and sema- phore operations.

The iAPX 432 uses a segmented memory scheme with up to 2^{24} segments, each segment 2^{16} bytes long yielding a total virtual space of 2^{40} bytes. A two-step mapping process separates the relocation mechanism from the

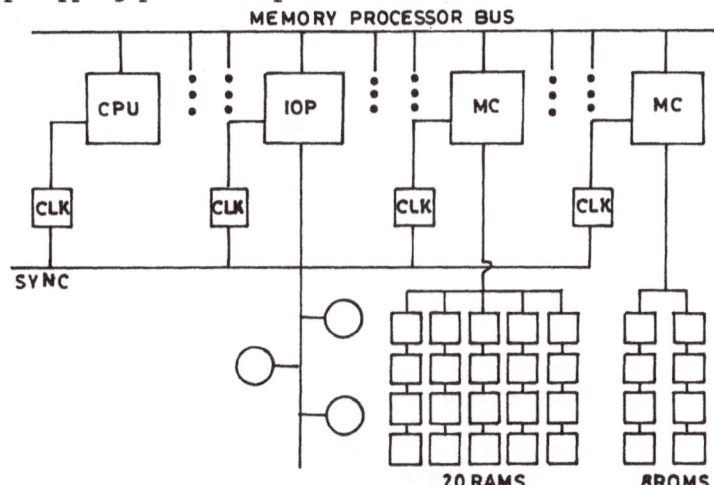

Fig.10: Hewlett-Packard 32-bit system configuration

access control mechanism. Segments are of two types: access and data.The
latest release of 432 microcode allows segments to contain both access
and data. Access quantities are protected by means of an addressing
fence in each segment. Values below the fence are access, while those
above the fence are data. The location of the fence is determined by the
object represented. The iAPX 432 provides four addressing modes: the
base and index direct is used to access scalers; the base indirect,index
direct is used to access records; the base direct,index indirect is used
to access static arrays; the base and index indirect is used to access
dynamic arrays. The emphasis is on addressing objects through use of
access descriptors in the form of directory index and segment index.
This object-oriented architecture facilitates implementation of high-
level languages like Ada [102],[103], Pascal, and PL/I.

5.7 Performance Estimates

The timing estimates for several elementary operations summarized in
Table VII must be interpreted with caution. Actual throughput is a
function of the exact instruction sequence, displacements, data lengths,
clock frequency, and other factors. Also, since the figures have been
provided by the respective manufacturers, it is appropriate to assume
the figures to reflect optimal performance estimates. The NS16032,though
not a true 32-bit microprocessor, has been included for comparison purposes. Notice that the comparison is being made at dissimilar clock
frequencies, and that in the Intel case several chips are involved,and
the timing includes some operating overhead.
    Hansen et al [55] have used four programs (string search, sieve,
puzzle, and Ackermann's function) to evaluate iAPX 432 performance in
comparison to 16-bit microprocessors and VAX 11/780. Using a VAX 11/780
operating in a VMS Pascal environment as the base, the relative code
size is summarized in Table VIII. These figures indicate that in spite
of bit-variable length instructions, the code size was larger on the
iAPX 432 as compared to the 68000, probably because of the inability of
the former to refer to a local variable or constant using fewer than
16 bits of address. The execution timings are summarized in TAbles IX
and X [56]. In all these performance evaluation exercises, the iAPX 432
is tested as a high-level language uniprocessor for integer and character programs; thus potential benefits of transparent multiprocessing,
data security, and increased programmer productivity are not reflected
[55]. Also, the timings for the Hewlett-Packard and National chips must
be taken with caution as they have not been verified by any independent
organization.
    Overall,the performance of the newer microprocessor approaches the
performance of mainframes. The Intel 432 at 8-MHz clock frequency takes
6.375μs for a 32-bit integer multiply and 27.875 μs for an 80-bit floating point multiply. The equivalent figures for the IBM 370/148 are 16.0
and 38.5 μs, respectively. In terms of basic computational power, the
iAPX 432 is superior to an IBM 370/148, and the Bellmac-32A and the
Hewlett-Packard processors are expected to be superior compared to an
IBM 370/158. However, whereas the IBM 370 family is supported by

TABLE VII: TIMING ESTIMATES IN MICROSECONDS UNLESS OTHERWISE
          SPECIFIED   46

Operation	NS 16032	Bellmac- 32A	HP	Intel 432
GENERAL: Clock Speed:	10MHz	10MHz	18MHz	8MHz
MOVE: Memory-to-Register	1.5	0.95	0.56	0.75
ADD:32-Bit Integer	1.6	0.4	0.055(hardware time) 0.275(instruction time)	0.5
ADD: Floating Point	9.3 with NS16081	Not supported in hardware	6.0(for 64-Bit) 4.7(for 32-Bit)	19.125(for 80-bit)
MULTIPLY: 32-bit Integer	8.3	1.8-9.5 depending on operands	1.8(hardware time) 2.9(instruction time)	6.375
MULTIPLY: Floating Point	7.1(for 64-bit) with NS 16081	Not supported in hardware	10.4(for 64-bit) 5.1(for 32-bit)	27.875(for 80-bit)
DIVIDE:32-bit Integer	9.2	Dependent on Operands	9.4(64-bit/ 32-bit) 5.2(32-bit/ 32-bit)	10.625
DIVIDE:Floating Point	10.8 with NS 16081	Not supported in hardware	16.0(for 64-bit) 6.5(for 32-bit)	48.25(for 80bit)

TABLE VIII: RELATIVE CODE SIZE (Numbers smaller than 1.0 indicate more
            compact code than on VAX)

MACHINE	LANGUAGE	WORD SIZE	RATIO TO VMS PASCAL(<1=> SMALLER)				
			SEARCH	SIEVE	PUZZLE	ACKER	AVG±SD
VAX-11/ 780	C	32	0.60	0.38	0.77	0.5	0.5 ±0.2
	PASCAL (UNIX)	32	0.95	1.24	1.49	0.72	1.1±0.3
68000	C	32	0.79	0.55	1.01	0.50	0.7±0.2
	PASCAL	16	0.72	0.29	0.60	0.36	0.5±0.2
	PASCAL	32	0.74	0.31	0.64	0.38	0.5±0.2
8086	PASCAL	16	0.94	0.85	0.79	0.91	0.9+0.1
432(REL.3)	ADA	16	0.76	0.44	0.84	0.42	0.6±0.2

TABLE IX: EXECUTION   TIMES

MACHINE	LANGUAGE	WORD SIZE	TIME (MILLISECONDS)			
			SEARCH	SIEVE	PUZZLE	ACKER
	C	32	1.4	250	9400	4600
VAX-11/780	PASCAL (UNIX)	32	1.6	220	11,900	7800
	PASCAL (VMS)	32	1.4	259	11,530	9850
	C	32	4.7	740	37,100	7800
68000 (8MHz)	PASCAL	16	5.3	810	32,470	11,480
	PASCAL	32	5.8	960	32,520	12,320
68000 (16MHz)	PASCAL	16	1.3	196	9180	2750
	PASCAL	32	1.5	246	9200	3080
8086 (5MHz)	PASCAL	16	7.3	764	44,000	11,100
432/REL.3 (8MHz)	ADA	16	4.4	978	45,700	47,800
80286 (8MHz)	PASCAL	16	1.4	168	9138	2218
80286 (10MHz)	PASCAL	16	1.1	135	7311	1774
HP 32-BIT CPU* (18MHz)	PASCAL	32	NA	NA	7450	2590
NS 16032* (7MHz)	PASCAL	32	NA	NA	24,000	9900

*Indicates experimental prototype
**Indicates vendor-provided information.

TABLE X: PERFORMANCE   AT   8 MHz

WAIT STATES	MACHINE	LANGUAGE	TIME (MILLISECONDS)			
			SEARCH	SIEVE	PUZZLE	ACKER
4	68000	PASCAL	5.3	810	32,470	11,480
	432 (REL.3)	ADA	4.4	978	45,700	47,800
	8086	PASCAL	4.6	448	27,500	6938
0	68000	PASCAL	2.6	392	18,360	5500
	80286	PASCAL	1.4	168	9138	2218

extensive software in terms of compilers and application programs, it
would take some time for similar facilities to be available on 32-bit
microprocessors. A typical end user has to decide among several options:
wait for the desired application software to become commercially available; develop the software in house; or use an earlier generation microprocessor that provides the software needed.

6.   SYSTEM ISSUES

Concurrent with the development of basic processors, manufacturers have
developed sophisticated chips for auxiliary functions - memory mangement,

control of DMA operations, control of peripherals, bus management and arbitration, control of communications, and control of input and output functions. However,a particular chip of this kind is designed to support only a particular   family of microprocessor chips of a particular vendor. Exhaustive lists of chips are published annually in several trade magazines. Usually there is a time lag between the introduction of the processor chip itself and the introduction of support chips.This results in extra effort in implementing systems based on recently intro-duced microprocessors.

In order to take over applications previously handled by mini-computers and mainframes, users demand that microprocessor based systems offer high throughput, ease of use, and friendliness of the system[104]. To optimize utilization of resources and to minimize user effort, an operating system is used. It is used for some or all of the following functions:

   a) processor management
   b) memory management
   c) peripheral management
   d) file management
   e) task scheduling and process management
   f) user-oriented facilities like command-line interpreter
   g) miscellaneous features to support networking,utilities, and
      high-level languages

In order to fully comprehend the emerging trends in microcomputer software, it is relevant to know a little about the history of the pop-ular operating systems.

The earliest uses of microprocessors were in embedded control app-lications. Programs were written on mainframe computers, cross-assembled for the micro, and loaded as object code into the micro's memory for execution. To aid the writing of such control programs, Intel, in 1972, hired MAA (Microcomputer Applications Associates, later to become Digital Research) to design and implement a systems programming langu-age. This language, called PL/M (Programming Language for Microcompu-ters) used ideas from PL/I, Algol, and XPL, the command-writing langu-age. PL/M became quite popular and is still used.

Along with PL/M, MAA proposed a small operating system, called CP/M (Control Program for Microcomputers), to enable applications to be written and compiled on the Intel 8080-based microcomputer. As Intel was reluctant, MAA developed the product independently in 1974. CP/M subsequently became the most popular operating system for microcomputers; now there are 200,000 installations using a wide spectrum of hardware configurations. It has become a de facto standard with most vendors in both the United States and abroad supporting it on their 8-bit and 16-bit microprocessors [105]-[107]. CP/M dominates the single-user envi-ronment.

UNIX, developed during the 1970's at Bell Laboratories is the premier example of an operating system optimized for program development by professional programmers in a multiuser interactive environment.The popularity of UNIX can be judged by the vast number of look-alike operating systems, such as: Coherent and Xenix on the 8086; Zeus, Onix, and Xenix on the Z-8000; Uniflex, Idris, Coherent, and Xenix on the

68000; Cromix, UNIX, and Idris on the Z-80; and Idris, Xenix, and Coherent on LSI-11 and PDP-11 systems.

According to Kenneth Thompson, the principal architect of the UNIX operating system, "the UNIX kernel consists of about 10 000 lines of C code and about 1000 lines of assembly code. The assembly code can be further broken down into 200 lines included for the sake of efficiency (they could have been written in C) and 800 lines to perform hardware functions not possible in C. The assembly code represents 5 to 10 percent of what has been lumped into the broad expression the UNIX operating system. The kernel is the only UNIX code that cannot be substituted by the user to his own liking" [57]. Some companies like Microsoft have adapted this hardware-dependent code for several microprocessors. Others like Mark Williams Company have chosen to rewrite the entire code based on the UNIX design.

Inspired by UNIX, Digital Research, the originator of CP/M, has developed the MP/M operating system. Similar to UNIX, MP/M is a multiuser, multitasking operating system. But unlike UNIX, MP/M has a realtime kernel that can be either interrupt-driven or dependent upon device polling. Also, MP/M uses multilevel directories rather than the tree structure provided by UNIX. In the UNIX environment, a fast disk file is used as a buffer to communicate between two processes. Such "pipes" are opened and closed as standard files, but are limited to character I/O only. MP/M permits variably sized messages to be written to and to be read from an unlimited number of processes, through buffers maintained in the memory. Each such "queue" has a name and is treated like other disk files. Further, queues can be optimized for message sizes. UNIX, on the other hand, offers a large array of excellent system development tools, and the ability to link utilities through a single command. The advent of personal computers has resulted in growing popularity of MS-DOS operating system developed by Microsoft. This operating system is supported on many leading computers of U.S. and Japanese make.

Unfortunately in all these cases, even though the same operating system is supported on several computer systems, the differences in hardware make it essential to modify application programs to execute under the same operating system on different systems. New techniques are evolving that enable such differences to be transparent to end users. For example, the UCSD p-system [116], originally developed by the University of California at San Diego, permits maximum level of software portability through use of intermediate code, called p-code, into which all the high-level languages are compiled. When a new processor is introduced, the p-system is implemented simply by writing an interpreter that translates p-code into the new processor's native code. The penalty is in terms of reduced execution speed of interpreted code. The advantage is in terms of ability to execute the same program under different environments like 8086, Z8000, 68000, TI 9900, and others. As microprocessors offer increased speeds, and as programming costs continue to escalate, it is likely that more users will accept the penalty of reduced execution speeds, and opt for using such techniques of intermediate code to aid conversion of programs to achieve software compatibility.

In the domain of large computer systems, Fortran and Cobol became industry standards by virture of their availability on a large number of systems. This does not, however, imply that all vendors offered identical languages. The traditional problem of incompatible higher level languages has been carried over to the microprocessor area. Although Basic is supported on almost all systems, the different vendors offer significantly different versions, depending on word size, memory addressing, and other factors. Frequently , multiple versions of Basic are available for the same machine (for example, on the IBM Personal Computer alone there are three versions of Basic available). The current trend in new operating systems is to support Pascal and Ada. These two languages are likely to be major forces in coming years. The increasing use of higher level languages will mitigate the problems of soaring software development costs [69]. The operating systems themselves are also written entirely in higher level languages, e.g., the iMAX for iAPX 432 is written in Ada [118].

Unlike the early years of microprocessors when stand-alone applications dominated the scene, the need to transmit and receive data and programs is an important characteristic of current generation systems [108],[109]. Communication between microprocessor-based systems can be analyzed at different levels. The International Organization for Standardization has developed a Reference Model of Open Systems Interconnection (ISO-OSI) comprised of seven layers as follows [58],[59]:

   i) the physical layer,
  ii) the data link layer,
 iii) the network layer,
  iv) the transport layer,
   v) the session layer,
  vi) the presentation layer,
 vii) the application layer.

The physical layer deals with transmission of raw bit stream, and the electrical protocols. For example, RS 232 is a physical link protocol that specifies the required voltages, number of wires, and transmission speeds over a serial communication link. The data link layer deals with issues of converting unreliable transmission links into reliable ones by using techniques like checksums to validate information received over the line. Although virtually every leading vendor is proposing a somewhat different protocol, two protocols (Carrier Sense Multiple Access (CMSA/CD) and Tokenpassing) currently dominate the consideration for standards embodied in IEEE 802 specification.The network layer deals with conventions that govern the transmission of data messages over the communication highway, for example, X.25. The transport layer is used to shield the customer's portion of the network from the carrier's portion; thus a change in carrier should be transparent to the computers at the two ends of the link. The session layer deals with setting up, managing, and splitting down process-to-process connection. The presentation layer deals with transformations (like data compression) on the data to be transmitted. The application layer is at the discretion of the users and refers to the ability of application programs involved in communication to freely exchange data and programs. Although the goal is to enable efficient communication at the

application level, this goal involves efficient communication at the
other levels as well. Efficient communication, especially between
dissimilar systems, is critically dependent on the protocols and standards for interconnection. Interconnection standards [113], like other
standards, are usually based on the design adopted by a large-vendor.
For example, the IEEE 802 [114] is based on the Hewlett-Packard design,
and the IEEE P796 is based on the Intel Multibus [115]. With multiple
standards, conversion of information from one standard to another involves an unproductive overhead.

Note that multiple communication standards are inevitable especially because interconnection of computing resources can be at several
distinct levels. At the basic hardware level, backplane busses such as
Multibus, S-100 and VME are used to communicate (over very small distances, usually of less than 1 m) data and information between processors, processor and memory, and between processors and fast peripheral
units. At the next level, I/O busses such as IEEE 488 bus standard are
used to communicate between the processor and I/O Units. Then, Local
Area Networks (LAN's) are used to provide data communication capabilities in buildings and over a few square kilometers of area. In this
paper, we only consider the role of backplane busses used in multiprocessor configurations.

7.   MULTIPROCESSOR CAPABILITIES

To increase computational bandwidth and/or system resilience,
integration of several microprocessors in a single system frequently
becomes necessary. The overall throughput and efficiency of such systems is directly dependent on the hardware and software interconnection
mechanisms supported by the basic microprocessor chips. Many different
interconnection systems have evolved over the years. The single time-
shared bus offers distinct advantages as an interconnection mechanism
for multi-microprocessor systems. Under such a scheme, different modules
can share the bus resource equally on a time-multiplexed or demand-
multiplexed basis. We concentrate on the Intel multiprocessing strategy
to illustrate these interconnection characteristics.

A typical iAPX 432-based multiprocessor configuration is shown in
Fig.11. Different iAPX 432 processor pairs (43201, 43202) are connected
to a single processor-memory interconnect bus which Intel refers to as
a Packetbus. It operates on a split-transaction basis. For example, if
a processor needed to access some data from memory, it would send a
message to the appropriate storage module. The actual transfer of the
message on the bus will occur when the arbitration mechanism grants
the bus to the particular CPU. When the request for the data is
received by the storage module, it accesses the data, but during this
period of data access, the bus is freed up for use by others. And,
finally, when the storage module is ready with the data, it requests
the bus, gets them, and sends the "reply" to the CPU. By freeing up the
bus during the period of memory access, which is significantly longer
than the bus service time, more processors can communicate on the same
bus in a given span. On the iAPX 432, variable length (1 to 10 bytes)

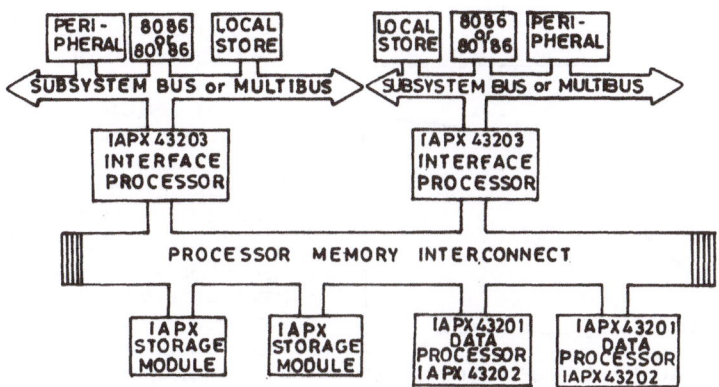

Fig.11: A typical multiple iAPX 432 based system.

of data messages are used for request/reply packet, and a 32-bit word
can be transferred in 250 ns. The split-transaction protocol used by
Intel offers maximum benefits in the "homogeneous" multiprocessing case
with identical processors being used to permit any processor to handle
a given task. "Heterogeneous" multiprocessing, for I/O or special-purpose applications, involves use of different types of microprocessors.
Unfortunately, no 16-bit microprocessor can be interconnected directly
to the Packetbus. One option to interconnect an Intel 8086 is through a
Multibus and an iAPX 43203 interface processor as shown in Fig. 11.
Although the iAPX 43203 can undertake additional responsibilities,
besides the interconnect function, the overhead of code conversion and
differences in protocol between the Multibus and the Packet makes such
heterogeneous multiprocessor configurations less efficient than homogeneous multiprocessor configurations. The iAPX 43204 and iAPX 43205
support multiple busses to reduce contention problems. However, multiple
busses introduce time delays and impose higher overhead than a single
bus configuration.

Hewlett-Packard also uses a demand multiplexed bus [60] with a
single-CPU using 30 percent of the total bus bandwidth for typical
instruction mixes. In general, as the number of processors increases,
there are more bus users leading to increased bus contention. Also,
there is additional overhead in terms of control and coordination of
multiple resources. The latter overhead is difficult to estimate, and
most studies simply neglect to take it into account. Fig. 12 reflects
Intel's estimates of effective number of processors versus actual number
of processors using a single bus. Since a dual-processor configuration
is indicated as having twice the processing power of a single one, the
exclusion of software overhead is evident. Notice that the curve

38

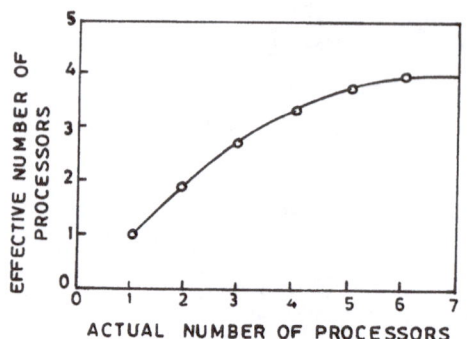

EFFECTIVE NUMBER OF PROCESSORS

ACTUAL NUMBER OF PROCESSORS

Fig.12: Intel estimated iAPX 432 bus
efficiency for a single memory bus multiprocessor system.

flattens off quickly, and
irrespective of the number of
physical processors used, it
is impossible to get an aggregate performance exceeding
four times the power of a
single one using a single
memory bus. Hewlett-Packard
60 claims that with four
processors the overall performance with multitasking
ranges from 2.9 to 3.7 times
the uniprocessor performance,
depending on the instruction
mix. According to an anonymous reviewer, systems using
eight iAPX 432 processors
have been designed. With
proper interface circuitry
and bus protocols, one can integrate up to 25 processors on a single
bus with with positive incremental increases in overall system throughput [61], and facilitate enhanced concurrent operation of all system
resources [62]. Also, the minimal cost of processing elements enables
duplication of system resources, and peer or self-testing of these
resources. Thus using multimicroprocessors and efficient bus operation
techniques, several recent computer systems offer both high throughput
and significant fault tolerance at much lower prices than before. In
view of these advantages, these systems are becoming increasingly
popular for transaction processing applications.

8. SPECIAL-PURPOSE PROCESSORS

During their first decade of existence, the focus was on general-
purpose microprocessors and microcomputers. In 1981, Intel announced
the 2920 analog processor with an on-chip D/A converter, which could
also be used for A/D conversion using the successive-approximation
algorithm. The precision of conversion is 9 bits, but internal arithmetic is done with 25 bits. An on-chip PROM can hold a 192-instruction
program, which is executed in sequence with no program jumps. Since an
instruction takes 400 ns, the program execution time is 76.8 μm or less,
corresponding to a sampling rate of 13kHz. This enables the processor
to sample and process analog waveforms with a bandwidth of 6.5kHz which
is quite adequate for high-quality speech applications.

In 1982, Texas Instruments announced the TMS320 single-chip microcomputer for digital signal-processing applicatons [68]. Unlike conventional von Neumann architecture, the Harvard architecture of the TMS320
uses a dual-bus design for parallel fetching of code and data to support
high-speed complex arithmetic. This chip has been optimized for speed,
rather than for size or power consumption. As such, significant parallelism has been implemented using a large number of devices. A fully

parallel 2's complement 16-by-16 multiplier is incorporated
enabling multiplication in just one cycle of 200 ns, plus
another 200 ns to latch the result. It does not, however,
have an on-chip A/D or D/A converter. Intended applications
include digital filtering, signal handling, data compression
and fast Fourier transforms. Microprocessors for signal-
processing applications have also been developed by AMI
(S28211 family) and by NEC (μ PD7720). The AMI 29500 signal
processor uses the 2900 type bit-slice architecture to
implement a sum-of-products number crunching capability for
array processing applications. It has been designed to perform a 1024-point complex "fast fourier" transform in 2 ms
(400 ns per butterfly).
    The trend in the special-purpose processor domain is
towards systolic data architectures to increase computational throughput without increased memory bandwidth. Unlike
traditional architectures, where data are moved one at a
time through a processing element, a systolic architecture
relies on an array of processing cells. Data are pumped
through the array, undergoing changes at each cell, before
returning to memory, similar to the manner in which blood
circulates from and to the heart. Since speech- and video-
processing techniques rely heavily on fast Fourier transforms, which can be implemented as an array of definite
processing operations, systolic data architectures offer
great potential for use in newer processor chips specifically
geared towards speech and video applications. An excellent
summary of research on alternative design methodologies for
special-purpose processors can be found in[70].

9. CONCLUSION

The ability to implement increasing numbers of devices on
the same chip has enabled microprocessors to offer increased
functional capabilities at diminishing costs over the years.
The enhanced capabilities are in the areas of processing
power, peripheral support, and in terms of software. These
trends are likely to continue. The domain of microprocessors constitutes a dynamic world, and we can always expect
newer and more powerful chips.

10. GLOSSARY

Accumulator: A register that is used to store results of
arithmetic and logical operations.
Arbitration: The process of selecting who will use a resource next in the light of requests from several entities
(typically processors).

Arrays: Refers to a set of data that is organized in a matrix format.

Barrel shift circuit: A circuit that enables all data stored in a particular register to be shifted left or right by an arbitrary number of bits - such circuits are used in implementing fast multiplication techniques.

BCD: Binary-coded decimal, a technique to store each decimal digit as a combination of four binary bits.

Bit-sliced organization: Refers to use of multiple, but identical, microprocessors, each with "small" word sizes to genrate a processor with larger effective word size.

Cache memory: Very fast, but limited capacity, memory used for prestaging operands and data needed by the processor. By reducing effective memory response time, the overall computer system throughput is enhanced.

CMOS: Complementary metal-oxide-semiconductor technology uses both n- and p-channel devices. This combination yields faster devices and lower power consumption than either N- or PMOS alone.

Concurrent user stacks: In earlier systems, none or only one user-defined stack was supported. New chips allow multiple stacks to be used concurrently. This development allows users greater flexibility.

Context switching: Refers to the processor ceasing execution of one program, and initiating execution of another program in a multiprogrammed environment. A context switch incurs the overhead of saving the status of the discontinued program and of loading the status, or context of the new program to be run.

DMA: Direct memory access enables direct transfer of information between memory and peripherals without information passing through the processor.

ECL: Emitter-coupled logic used for implementing systems with very fast speeds. This technology suffers higher power consumption and lower device densities than MOS techniques.

Hardware-implemented process scheduling: Scheduler algorithm hardwired or in firmware.

Interface processor: A processor used to convert codes and protocols to allow interfacing of two entities, such as communications lines, computers, termianls, etc.

Microprocessor: The semiconductor central processing unit manufactured usually as a single physical device.

MOS: Metal-oxide-semiconductor, refers to a device that is the result of processing a semiconductor material in various stages with deposited metals such as aluminium and oxide layers to form basic switching and electrical circuit elements.

Multilevel memory protection hierarchy. Refers to the ability to distinguish between several levels of requests for memory operations and to enable enhanced degrees of memory protection.

Multiple homogeneous caches: Refers to the ability to support several cache memories of identical characteristics.

Multiplexing of address and data: Refers to the use of the same physical pins or bus wires to carry both address information and actual data on a time-multiplexed basis.

NMOS: n-channel MOS: in such devices electric current is constituted by a flow of negative charges.

On-chip memory hierarchy: Storage of different types of memory organi-
zations on the same (processor) chip.
Orthogonal instruction set: An instruction set that allows all (relevant) addressing methods to be used with all instructions.
Packed arrays: Arrays that are compressed to contain all information
in the original array in the minimum storage space.
Packed records: Records that are comparessed to contain all information in the minimum storage space.
Pipelining: The mechanism of fetching next sequential program operand
or data while the present instruction is being executed.
PLA: Programmed logic array, a structured matrix of chip interconnections accomplished by mask programming. Often used to implement
instruction decode and control logic.
PMOS: p-channel MOS; in such devices electric current is constituted
by a flow of positive charges.
Self-dispatching processors: Processors, in a multiprocessor environment, that are seeking tasks on a continuous basis; as soon as a new
task enters the system, an idle self-dispatching processor will
immediately direct itself to start executing the task.
Stack: A last-in-first-out (LIFO) buffer used for storing data and
interrupts. Also referred to as a FILO (first-in-last-out).
Trap: A special condition during program execution which, if encountered, will cause special action or exception processing to occur.
Vector-based instructions: Powerful instructions that enable many
sequential operations to be carried out atomically. A vector instruction may take an interrupt, store the machine state, branch to the
appropriate device or handle stacks and allocate a stack frame in a few
microseconds. This process would take several sequential subroutines if
it were executed using traditional operating systems.
Virtual machines: Multiple copies of the system's complete hardware-
software interface are efficiently replicated through a combination of
hardware and software.

ACKNOWLEDGEMENT

The authors thank the two anonymous reviewers of this paper. Their
comments have assisted in enhancing the quality of this paper.

REFERENCES

1   R.N.Noyce and M.E.Hoff, Jr.,"A history of microprocessor develop-
    ment at Intel", IEEE Micro, vol.1, No.1, pp.8-21,Feb.1981.
2   D.Moralee,"Microprocessor architectures: Ten years of develop-
    ment", Electronics and Power, pp.216-221, March 1981.
3   H.D.Toong,"Microprocessors", Sci.Amer., pp.146-161, Sept.1977.
4   R.H.Krambeck, C.M.Lee, and H.F.S.Law,"High-speed compact circuit
    with CMOS", IEEE J.Solid-State Circuits,vol.SC-16, pp.614-618,
    June 1982.
5   D.W.Best et al.,"An advanced-architecture CMOS/SOS microprocessor",
    IEEE Micro,vol.2,no.3,pp.10-26,Aug. 1982.

6   D.L.Wollesen, "CMOS LSI – The computer component process of the 80's", Computer, pp. 59–67,FEb. 1980.

7   W.R. Iversen, "32–bit Chip set will offer huge microprogram store" Electronics, pp. 47–48, Sept.8, 1982.

8   J.W. Beyers et al., "A 32–Bit VLSI CPU chip", IEEE J.Solid State Circuits, Vol. SC–16, no.5, pp.537–542, Oct. 1981.

9   D.A.Patterson and C.H.Sequin, "A VLSI RISC", Computer, vol.15, no. 9, pp. 8–21, Sept. 1982

10   G.Radin, "The 801 minicomputer", in Proc.Symp.Architectural Support for Programming Languages and Operating Systems (Mar.1–3,1982), pp. 39–47.

11   R. Ochester, "Low–cost 16–bit microprocessor has performance of midrange minicomputer", Electronics, pp.129–133, Nov.3, 1981.

12   C.Davis et al., 'Gate array embodies System/370 processor," Electronics, pp.140–143, Oct. 9,1980.

13   P.Feller et al.,'Memory protection moves onto 16–bit microprocessor chip," Electronics, pp.133–137, Feb. 24,1982.

14   S.M.S.Liu et al., "HMOS III Technology", IEEE J.Solid–State Circuits, vol.SC–17, no.5, pp.810–815, Oct. 1982.

15   F.A.Ware et al., "64–bit monolithic floating point processors", IEEE J.Solid State Circuits,vol.SC–17,no.5, pp.898–907,Oct.1982.

16   G.E.Moore,"Progress in digital integrated electronics", in Proc. Int. Electron Devices Meet., pp.11–13, Dec.1975.

17   H.Hingarh et al., "A 16–bit microprocessor for realtime applications", in Proc.Int.Solid–State Circuits Conf.Feb.23–25,1983.

18   S.P.Morse, et al.,"Intel microprocessors – 8008 to 8086",Computer pp.42–60,Oct. 1980.

19   J.Rattner and W.W.Lattin, "Ada determines architecture of 32–bit microprocessor", Electronics,pp.119–126, Feb.24,1981.

20   P.M.Russo,'VLSI impact on microprocessor evolution, usage and system design", IEEE Trans. Electron Devices, vol.ED–27,no.8, pp.1332–1341, Aug.1980.

21   _____, "Intel 8048 and 8051 family of chips", EDN,pp.126–133, Oct. 27,1982.

22   E.Peatrowsky,"EPROM MCU's reduce hardware liability, " in Proc. MIDCON 1981, session 21, paper 2, pp.1–3 Nov.1981.

23   D.E.Smith,"New high performance one chip microcomputer", Electron Eng., pp 56–65, Feb.1982.

24   P.R.Brown,"Advanced hardware features of the Z8 microprocomputer family in Proc.MIDCON 1981, session 21, paper 5,pp 1–10,Nov.1981.

25   M.Patrick and J.Millar,"An innovative microcomputer for the 1980's", in Proc.MIDCON 1981, session 21, paper 4, pp.1–13 Nov. 1981.

26   H.M.D. Toong and A.Gupta, "An architectural comparison of contemporary 16–bit microprocessors", IEEE Micro, vol.1, no.2, pp.26–37, May 1981.

27   R.V.Orlando and T.L.Anderson,"An overview of the 9900 microprocessor family", IEEE Micro,vol.1,no.3,pp.38–44,Aug.1981.

28   ____,Introduction to the iAPX 286. Intel Corp.,Feb.1982.

29   S.Bal et al., "The NS16000 family – Advances in architecture and hardware", Computer, pp.58–67, June 1982.

30   D.Laffitte,"New-generation 16-bit microprocessors - Fast and
      function-oriented",Electron.Des.,vol.29, no.4,pp.111-117, Feb.
      19,1981.
31   R.J.Markowitz,"iAPX 286: Virtual memory and distributed computing"
      in Proc.WESCON 1981,session 9,   paper 2, pp.1-7.
32   G.Louie et al., "A 16-bit microprocessor with on-chip memory pro-
      tection," in Proc.ISSCC 83,session II, Feb.1983.
33   R.Mateosian,"Benefits of Z8000 family planning", in Proc. of
      WESCON   81, session 1, paper 5, pp.1-7.
34   J.W.Browne,Jr.'MC68000-Break away from the past", in Proc.WESCON
      1981, session 1, paper 4, pp 1-8.
35   ____,68010,68020 Product Preview. Motorola Semiconductors,1982.
36   A.Kaminker et al.,"A 32-bit microprocessor with virtual memory
      support", IEEE J.Solid-State Circuits, vol.SC-16, no.5, pp.548-
      557,Oct. 1981.
37   J.Beekmans et al.,"Chip set bestows virtual memory on 16-bit
      minis", Electronics, pp.134-138, June 2, 1981.
38   N.Inui et al., "16 bit C-MOS processor packs in hardware for
      business computers, Electronics, pp 182-186, June 16,1981.
39   R.Ochester,"Low-cost 16-bit microprocessor has performance of
      midrange computer", Electronics, pp.129-133, No.3,1981.
40   R.Rubinstein et al.,"Compatibility about μeclipse and speed;goals
      of a small machine", Comput.Des.,pp.69-76, Aug.1982.
41   H.D.Toong and A.Gupta,"Evaluation kernels for microprocessor
      performance analyses, " Perform.Eval.,vol.2,no.1,pp.1-8,May 1982.
42   J.Heering,"The INtel 8086, the Zilog Z8000 and the Motorola MC
      68000 microprocessors",EUROMICRO J.,vol.6, pp.135-143,1980.
43   R.Grappel and J.Hemenway,"Evaluating the 16-bit chips", Mini-
      Micro Syst.,pp.152-162,Dec.1980.
44   ____,"A tale of four μPs: Benchmarks quantify performance",
      EDN, pp.179-185, Apr.1,1981.
45   C.Davis et al.,"Gate array embodies system 1370 processor",Elec-
      tronics,Oct. 9,1980.
46   A.Gupta and H.D.Toong.,"An architectural comparison of 32-bit
      microprocessors", IEEE Micro,vol.3, no.1, pp.9-22,Feb.1983.
47   B.T.Murphy et al., "A CMOS 32b single chip microprocessor",
      in Proc.IEEE Solid-State Circuits Conf.,pp.230-231,Feb.,1981.
48   A.D.Berenbaum et al.,"The operating system and language support
      features of the BELLMAC-32 microprocessor", in Proc.Symp.Archi-
      tectural Support for Programming Languages and Operating Systems,
      pp. 30-38, Mar. 1982.
49   J.W.Beyers et al.,"A 32-bit VLSI CPU chip", IEEE J.Solid-State
      Circuits, vol.SC-16, pp.537-542, Oct. 1981.
50   J.M.Mikkelson et al.,"An NMOS VLSI process for fabrication of a
      32-bit CPU chip", IEEE J.Solid-State Circuits, vol.SC-16,pp.
      542-547, Oct.1981.
51.   D.L.Budde et al.,"The execution unit for the VLSI 432 general
      data processor", IEEE J.Solid-State Circuits,vol.SC-16,pp.514-
      521,Oct.,1981.
52.   J.A.Bayliss et al.,"The interface processor for the Intel 432
      32-bit computer", IEEE J.Solid-State Circuits,vol.SC-16,pp.522-

530,Oct. 1981.

53   J.A. Bayliss et al.,"The instruction decoding unit for the VLSI
     432 general data processor",IEEE J.Solid-State Circuits,vol.
     SC-16, pp.531-537, Oct. 1981.
54   A.F.Shackil,'Microprocessors", IEEE Spectrum, pp.32-33,Jan.,1982
55. P.M.Hansen et al., "A performance evaluation of the Intel iAPX
     432", Comput.Architecture News,vol.10, no.4,pp.17-26,June 1982.
56   D.A.Patterson,"A performance evaluation of the Intel 80286",
     Comput.Architecture News, vol.10, no.5, pp.16-18,Sept. 1982.
57   K.Thompson, Bell Syst.,Tech J.,vol.57, no.6, pt.2,p.1931,July-
     Aug. 1978.
58   H.Zimmermann,"OSI reference model - The ISO model of architecture
     for open systems interconnection", IEEE Trans.Commun.,vol.COM-28,
     pp.425-432, Apr. 1980.
59   A.S.Tanenbaum, "Network protocols", ACM Computing Surveys,vol.
     13,no.4,pp. 454-489,Dec.,1981.
60   D.Seccombe, Hewlett-Packard, Nov.8,1982, personal communication.
61   H.M.D.Toong,S.O.Strommen and E.R.Goodrich II, " A general multi-
     microprocessor interconnection mechanism for non-numeric process-
     ing", in Proc. the 5th Workshop on Computer Architecture for
     Non-numeric Processing, pp.115-123, 1980.
62   A.Gupt and H.M.D. Toong, "Increased concurrency in m-n multi-
     processor systems", in Proc. 3rd Int.Conf. on Distributed Comput-
     ing Systems (Miami/Ft.Lauderdale,FL,Oct.18-22,1982),pp.146-151.
63   D.A.Patterson and C.Sequin, 'Design considerations for single-
     chip computers of the future",IEEE J. Solid-State Circuits,vol.
     SC-15, pp.44-52,1980.
64   H.M.D.Toong and A.Gupta, "Personal Computers", Sci.Amer.,pp.
     88-99, Dec. 1982.
65   H.Cragon, 'The elements of single-chip microcomputer architecture",
     Computer,pp.27-41,Oct. 1980.
66   J.P.Hayes, 'A Survey of bit-sliced computer design", J.Digital
     Syst., vol.V, no.3,pp.203-250,1981.
67   K.Thompson, "_The UNIX operating system," Bell Syst.,Tech.J.,
     vol.57,no.6,pt.2,p.1931, July-Aug. 1978.
68   K.McDonough et al., 'Microcomputer with 32-bit arithmetic does
     high precision number crunching", Electronics,pp.105-110,Feb.
     24, 1982.
69   R.Bernhard, 'More hardware means less software",IEEE Spectrum,
     pp.30-37, Dec.1981.
70   P.C.Treleaven, 'VLSI processor architectures",computer,vol.15,
     no. 6,pp. 33-45, June 1982.
71   A.Gupta and H.M.D. Toong, Advanced Microprocessors.New York:
     IEEE Press, 1983, IEEE reprint book.
72   J.W.Beyers, et al., "A 32-bit VLSI CPU chip", in Dig.Tech papers
     1981,IEEE Int. Solid-State Circuits Conf pp. 104-105.
73   ____, 'Microprocessors", Electron.Eng.,pp.60-85, Sept.1981.
74   H.M.J.M.Dortmans ,"Application of microprocessors",J.Phys.E.
     Sci.Instrum.,vol.14,pp.777-782, 1981.
75   D.G.Fairborn,   'VLSI technology", Computer,pp.87-96,Jan.1982.
76   C.M.Lee and C.G.Lin-Hendel,"Current status and future projection

of CMOS technology", in Proc. COMPCON, Fall (Sept. 20-23,1982), pp. 716-719.

77   D.J.McGreivy and K.A.Pickar, VLSI Technologies Through the 80s and Beyond. New York: IEEE Computer Soc.Reprint Book, 1982.

78   R.Rice,VLSI Support Technologies - A Tutorial. New York: IEEE Computer Soc. Reprint Book, 1982.

79   T.Knowlton, "μPD7500 family of 4-bit microcomputers - big jobs with small programs",in Proc.MIDCON 1981, session 24, paper 5, pp.1-5.

80   R.M.Cushman,"CMOS microprocessor and microcomputer ICs",EDN,pp. 88-100, Sept. 29,1982.

81   J.F.Stockton,"A virtual breakthrough for micros", Comput.Des,pp. 153-162, Aug.1982.

82   H.W.Lawson, Jr.,"New directions for micro- and system architectures in the 1980s", Proc.Nat.Computer Conf.,1981,pp.57-62.

83   ___,"Single-chip microcomputers: Performance and features",Electron.Des.,pp 130-139,Oct.14,1982.

84   D.Bursky, "16-bit families swell with greater integration", Electron. Des. pp 103-112, Oct. 14,1982.

85   W.D.Hopkins,"Speed/cost tradeoffs of using the TMS99110 micro - processor", in Proc.ELECTRO 1982,session 18,   , paper 3, pp. 1-7.

86   W.D.Huston,"Microcomputers-heritages and the future", IEEE Trans.Consum.Electron., vol.CE-26,pp.129-141,Feb. 1980.

87   D.Stevenson,"Floating-point processing with Zilog's Z-8000 CPU", in Proc.ELECTRO 1982,session 18, paper 4,pp.1-2.

88   J.T.Twardy,"Fourth generation architecture allows performance of larger jobs with smaller programs", in Proc.ELECTRO 1982,session 25,paper 4, pp.1-8.

89   D.A.Patterson and R.S.Piepho, "Assessing RISCs in high-level language support", IEEE Micro, pp.9-19, Nov. 1982.

90   L.Kohn, "A 32b microprocessor with virtual memory support", in Proc. IEEE Solid-State Circuits conf.,pp.232-233,Feb. 1981.

91   H.F.S.Law,"Layout technology for high performance VLSI",in Proc. COMPCON,Fall (Sept. 20-23,1982),pp 40-43.

92   J.A.Bayliss et al.,"The interface processor for the 32-bit computer", in Dig. Tech papers, 1981, IEEE Int.Solid-State Cricuits Conf., pp.116-117,263.

93   D.L.Budde et al., "The 32-bit computer execution unit," in Dig. Tech papers, 1981 IEEE Int. Solid-State Circuits Conf., pp.112-113, 261.

94   J.Klovstad, G.M.Catlin, and T.Zingale,"16-bit μP crams peripheral support on chip," Electron.Des.,pp.191-196, June 10,1982.

95   J.Black and J.Kister,"VERSA bus: A powerful structure for multiprocessing applications", in Proc.MIDCON 1981, session 25,paper 6, pp.1-6.

96   W.Twaddell,"EDN's ninth annual μP/μC chip directory, "EDN,pp. 98-204,Oct. 27,1982.

97   ___,"General-purpose microprocessors: Performance and features, " Electron.Des.,pp.118-139,Oct.14,1982.

98   W.S.Richardson et al.,"The 32 bit computer instruction decoding

unit," in Dig. Tech Papers, 1981 IEEE Int.Solid-State Circuits Conf., pp.114-115, 262.

99 ____," Intel432 System Summary: Manager's Perspective, Intel Corp. Manual Number 17186-001, 1981, p.29.

100 W.W.Lattin et al., "A-32bit VLSI micromainframe computer system", in Dig.Tech Papers, 1981, IEEE Int.Solid-State Circuits Conf., pp.110-111.

101 T.Zingale, "Broadening the scope of microcomputer numeric applications with the 8087 numeric processor extension", in Proc. ELECTRO 1982,session 18, paper 1,pp.1-7.

102 S.Ziegler et al., "Ada for the Intel 432 microcomputer",Computer pp. 47-56, June 1981.

103 F.J.Pollack, et al., "Supporting Ada memory management in the iAPX-432", in Proc.Symp. on Architectural Support for Programming Langua ges and Operating Systems, Mar. 1982.

104 M.Schindler,"Operating systems help micro act like minis", Electron. Des.,pp.SS41-SS45, Mar. 18,1982.

105 __,'The latest in microcomputer operating systems", Electron.Des. pp.SS46-SS64, Mar. 18,1982.

106 G.Kotelly, "EDN's third annual μC operating systems directory", EDN, pp.80-157, Sept. 15, 1982.

107 R.C.Johnson,"Operating systems hold a full house of features for 16-bit microprocessors", Electronics, pp.113-120, Mar. 24,1982.

108 E.A.Freeman and K.J.Thurber Eds.,Microcomputer Networks - A Tutorial, New York: IEEE Computer Soc., 1981.

109 P.L.Borrill,'Microprocessor bus structures and standards",IEEE Micro, vol.1, no.1, pp.84-95, Feb. 1981.

110 E.H.Frank and R.F.Sproull,"Testing and debugging custom integrated circuits". ACM Comput. Surv., vol.13, no.4, pp.425-451, Dec. 1981.

111 S.P.Joshi,"Ethernet controller chip interfaces with variety of 16-bit processors",Electron. Des., pp.193-200, Oct. 14,1982.

112 R.Gilbert,"The general-purpose interface bus",IEEE Micro,vol.2, no.1, pp.41-51, Feb.1982.

113 M.Graube,"Local area nets: A pair of standards", IEEE Spectrum, pp.60-64,June 1982.

114 T.J.Harrison,"IEEE project 802: LOcal area network standard", in Proc. ELECTRO 1982,session 17, paper 1, pp.1-11.

115 R.Dilbeck and J.Barthmailer,"The multibus/IEEEP796 bus standard and microcomputer system architecture for the 80's",in Proc.MIDCON 1981,session 25,paper 2, pp.1-8.

116. C.A.Irvine,"UCSD system makes programs portable", Electron.Des., pp.113-118,Aug.1982.

117 R.G.Rajulu and V.Rajaraman,"Execution-time analysis of process control algorithms on microprocessors", IEEE Trans.Ind.Electron, vol.IE-29,no.4, pp.312-319, Nov. 1982.

118 K.C.Kahn et al.,"iMAX: A multiprocessor operating system for an object-based computer", in Proc.8th Symp.on Operating Systems Principles (ACM, Dec.,1981)pp.127-136.

119 C.B.Peterson et al.,"Two chips endow 32-bit processor with fault-tolerant architecture",Electronics,pp.159-164,Apr.7,1983.

120   M.De Prycher,"A performance comparison of three contemporary
       16-bit microprocessors", IEEE Micro, vol.3, no.2, pp.26-37,Apr.
       1983.
121   D.MacGregor and D.S.Mothersole, 'Virtual memory and the MC68010",
       IEEE Micro, vol.3, no.3, pp.24-39, June 1983.

Chapter 2

INTRODUCTION TO REAL-TIME SYSTEM SOFTWARE

Sanjay Dasgupta
Gas Authority of India Ltd.
Samrat Hotel, Chanakyapuri
New Delhi, 110021
INDIA

ABSTRACT. This chapter introduces system software concepts for real-time systems. The characteristics and evolution of the Real-Time Multi-tasking Operating System (RMOS) is covered, followed by a description of different types of RMOS. Concepts of task scheduling, memory management and input/output in the context of real-time operation are discussed. The concepts covered are illustrated with examples of code, data structures and mechanisms from practical real-time systems. Programming language requirements for real-time systems are covered together with a discussion of major languages.

1. INTRODUCTION.

Any computer-based system has two parts : hardware and software. The hardware consists of the physical parts such as the CPU, memory and storage devices, while the software consists of computer programs, or sequences of machine instructions that can be loaded into memory to be executed by the CPU.

The software is also made up of two parts. The first part -- applications code -- customises the hardware for a specific application. It determines whether a computer system will perform as a message switch, a mathematical simulator, or a process controller in much the same way that the content of a human memory (loaded during years spent in class rooms and labs) determines whether its owner will function as a dentist, an architect, or a lawyer. The second part of the software -- the system software -- functions in close association with the hardware and creates (for the applications software) a run-time environment with more effective and secure high level facilities than can be obtained from the raw hardware. It also provides a set of utility procedures that can be used to develop applications programs, and to effectively manage the computer systems operation.

1.1. The Run-time Environment.

The run-time environment (often loosely referred to as the OS or the

49

S. G. Tzafestas and J. K. Pal (eds.), Real Time Microcomputer Control of Industrial Processes, 49–80.
© 1990 *Kluwer Academic Publishers.*

executive) consists of a set of routines that control the operation of hardware devices and manage the allocation and utilisation of system resources such as memory and CPU time. The executive views the applications software as a set of independant, uniquely identified programs or tasks, and it controls the allocation of system resources to them in accordance with defined operating policies, task priorities and priviledges. It also provides a set of high level instructions or ´services´ (such as file manipulation services) which the tasks may use.

The set of run-time services determine the type of applications that may be run, and is the primary software consideration that determines whether real-time operation can be supported. The utilities play a supporting role, and are not strictly necessary for running the applications software.

1.2. The Utility Procedures.

The utility procedures are usually implemented as tasks that are run in response to specific commands to the executive. The utilities can be functionally divided into several groups:

The system management utilities enable the computer system to be properly started up and shutdown, system performance indices to be determined, and parameters that affect performance to be modified.

The software development utilities enable the development of new computer programs. The utilities in this category generally include text editors for creating the programs source code modules, compilers and assemblers for converting the source code modules into object code, and linkage editors that enable different object modules to be linked together into bodies of executable machine code.

A set of data and file management utilities are also provided to enable users to keep track of their data and program files and to enable the data to be transported to and from other systems.

2. THE REAL-TIME PROGRAM ENVIRONMENT.

The concept of parallelism (more than one thing being done at one time) is fundamental to real time systems. But in the early days there were no computer programming languages which could be used to express concurrent activities. On the other hand multi-programming had developed from a concept into a widely used feature. The need to express parallelism therefore, naturally found expression in multiple programs running concurrently under a multi-programming operating system.

The multi-programming operating system (MPOS) was, however, not an unmixed blessing. A major hinderance was the ceveat that each task proceeds without any knowledge of the state of, and unaffected by the actions of, any of the other tasks. This meant that a group of tasks would not be able to act in unision, because this requires a continuous exchange of knowledge and synchronisation of actions to produce the desired effect on the controlled system.

Clearly, the MPOS was a design based on extremely sound principles, and to undo some of them so as to let tasks in a real-time system meddle with each other would be a step backwards. A secure method by which the required interactions could be performed within the framework of an MPOS, had to be developed.

2.1. Executive Services for Real-Time Support.

One of the fundamental facilities in a MPOS was the Executive Service Request (variously known as supervisor call, executive directive, system service etc). The Executive Service Request (ESR) provided access to functions of the underlying hardware (such as I/O device operation) that all tasks needed, but were only allowed to use through a gateway mechanism supervised and controlled by the executive. All that was needed then, was to generalise and extend this mechanism to create new facilities, which could be used to effect the necessary interactions with the executive, and between tasks. The Real-Time Multi-tasking Operating System (RMOS) was thus born.

The implementation of parallel actions as individual programs running under a RMOS is an extremely elegant solution. It means that real-time programs can, for the most part, be written in any language, and factors such as portability and availability of efficient compliers can be taken into consideration.

2.2 Types of Real-Time Operating Systems.

RMOSs come in a variety of sizes, and with widely different characteristics and capabilities. At the low end of the size spectrum are operating systems for the 8-bit microprocessors used in small data acquisition units, programmable controllers, RTUs etc. At the other end are large operating systems for 32-bit computers that provide general purpose capabilities for a development or target environment. The ´character´ of an OS is determined by two aspects : the command language provided to users, and the services (ESRs) that running tasks use to interact with it.

The repertoire of OS commands available to a user at a terminal is determined by what devices are supported, and what kind of applications the OS is designed to handle. In the context of real-time systems, the command language is only of interest during the software development effort, and for system startup and software maintenance purposes. But when a RMOS hosts a running real-time system, there is little or no use for the command language. Most of the small RMOSs, therefore, have no command language, or command device, and any explicit control is exercised from a higher level entity such as an operator control station via the process communication links.

The real-time tasks in the target environment interact with the RMOS through the ESRs. A repertoire of well designed and flexible ESRs is therefore the hallmark of a RMOS.

2.2.1. Small (8-bit) systems.

Microprocessor based systems are often used to build the smaller components (programmable controllers, RTUs etc) of real-time control systems. These devices are placed close to the process being controlled, and interact directly with it. A communications link or data highway is provided through which these units may communicate with a higher level entity such as a central control station. The control station uses this communication link to send commands and information to, and recieve reports from, the remote units.

Such units perform a limited set of functions through the actions of a few tasks. The total size of code and data is relatively small, so it is possible to load and fix them in memory once for all. The hardware configuration of such a system therefore contains a considerable quantity of Read Only Memory (PROM) into which the RMOS and tasks are loaded. Sufficient Read/Write memory (RAM) is also provided for any storage needs of the tasks and the RMOS. The data storage requirements are also small enough to be accomodated in memory, so no file system and bulk storage devices are needed.

No manual intervention or attention is necessary to operate these systems. Once the power is turned on, a start up sequence is initiated that requests the higher level entity to ´down-load´ any required information (such as the current time, and instructions on what functions are to be performed). After this information is received, the unit can carry on its assigned functions unaided. Sometimes a facility is provided to down load tasks, which are then installed properly in memory and run. This is very useful, because memory space does not have to be individually reserved for seldom used functions such as debuggers and diagnostic programs.

The abscence of a file system, utility software, and a command language, makes these systems unsuitable for development work. Applications development for such systems is therefore performed on larger machines equipped with cross assemblers or cross compilers (which generate object code suitable for a processor different from that of the host). The developed tasks are then written into the PROM of the target system by using a special loader.

2.2.2. Systems for 16-bit and larger machines. Formerly, the only 16-bit systems to be used in the real-time applications area were the mini-computer class of machines. These machines feature extended memory support (between 256 kilobytes and a few megabytes), a wide variety of I/O devices including real-time process interfaces, mass storage devices, and a comprehensive command language.

The provision of extended memory support makes it possible for these machines to host large software systems, and thus be placed at the core of large control systems. Because memory is a relatively free resource, high level languages (which are less economic than assembly language in their use of memory) have come into increasing use. The operating systems are also larger and offer more than bare minimum facilities.

RMOSs for mini-computers are fairly complex pieces of software, and are always supplied by the manufacturer. The user only configures it to meet his needs through a process commonly referred to as [1] SYSGEN (system generation). During SYSGEN the developer selects software components that are actually needed in the target system. The options include drivers for a variety of I/O devices, a large set of ESRs, and

other executive features. Generally, not all the ESRs and executive
features offered are necessary to support a given application. Making a
judicious choice helps to keep the executive small, thus maximising the
space available for other purposes. SYSGEN also provides an opportunity
for the addition of features (such as a driver for a non-standard
device, or a new ESR), or to modify the standard software to make it
more suitable for the target application.

The presence of mass storage devices, and a comprehensive command
language means that these systems can also be used for development work.
Compilers for a range of languages, and other program development tools
are available for these machines, and it is usual for software
development to be performed on the target machine itself.

All these features, and much more, are available on the 32-bit
systems. The major advantage of these systems is the use of virtual
memory. Virtual memory creates an illusion of an unlimited memory
space, thus freeing the system designer from the chores of planning out
memory allocation maps; and sparing programmers the bother of having to
overlay tasks. Generalised computer nerworking capability, which made
its debut with the larger of the 16-bit systems, is offered as a
standard feature built into the 32-bit systems.

2.2.3. <u>Silicon operating systems.</u> A relatively new entrant to the
16-bit field is the 16-bit microprocessor. These devices are comparable
in power with the smaller mini-computers, and are finding increasing use
in the construction of standalone devices like programmable controllers
and RTUs. Like in the 8-bit applications described above, no bulk
storage is needed, so a solid-state operating system increases both
speed and reliability. But RMOSs for 16-bit processors are fairly
large, and tend to occupy much memory in ROM-based implementations.

This disadvantage is removed when OS primitives are implemented as
an extension to the processors instruction set, either within the
processor itself or in a separate co-processor. An example of such an
implementation is the 80130 OS firmware [2] which executes many of the
iRMX 86 primitives coded as an extension to the Intel 8086
micro-processors instruction set.

The implementation of OS primitives in this way is not limited to
micorprocessors alone. In many of the larger 32-bit mini-computers, OS
functions such as task despatching, input/output control and ESR service
routines are implemented in firmware.

3. TASK SCHEDULING

A task in any uni-processor multi-progamming system is always in one of
two states -- running (using the CPU to execute code) or waiting for its
turn to be able to use the CPU again. This is a fundamental
characterstic because there is only one CPU available to be used by many
tasks, and it must be alloted to each one by turns.

In a general purpose time sharing OS, this allocation is done on a
sequential, time-sliced basis. Each task is allowed to use the CPU for
a small interval of time called the time-quantum or the round-robin

54

scheduling interval. After its expiry, the next task in turn is alloted
the CPU and so on. The time slice for all tasks is either equal (for
each task to receive an equal share of sytem resources), or in
proportion to each tasks priority (for each task to receive a specified
share of system resources). Such a scheduling strategy is sufficient
for time sharing general purpose systems because it provides for
complete control of the only factor that is variable : task urgency.
Round robin scheduling is illustrated in Fig-1. All tasks are placed in
a circular queue which defines the order in which they become eligible
to use the CPU. While the time slice for each task may differ, the
order of execution is fixed.

3.1. Real-Time Task Scheduling.

A real-time system has quite different requirements. The tasks must
react quickly to events occuring in the external controlled system, and
must be randomly schedulable by an event-triggered and/or a time
synchronised mechanism. The state of the controlled process, therefore,
determines which of the tasks is required to be active at any time. The
following piece of code from a hypothetical data acquisition system
illustrates these concepts.

```
      .  .  .
      REPEAT
         Wait_till_next_interrogate_cycle;
         {
            Task is blocked here till a certain time.
         }
         FOR N := 1 TO Number_of_remotes DO
         BEGIN
            Interrogate_remote (N);
            Wait_for_reply (N);
            {
               Task is blocked here till reply arrives,
               or a time-out is detected.
            }
            Update_data_base(N)
         END
      UNTIL System_shutdown;
      .  .  .
```

The unusual statements are not special real-time commands, but
calls to procedures that perform these functions. (The idea is to hide
implementation details, and highlight the concepts involved.) The code
in this example interrogates a set of remote data acquisition units at a
defined rate, and updates a data base using the information received in
the replies. There are two sites in this code (marked with comments) at
which the task can not proceed further, till it receives a synchronising
stimulus from an external source. A task in such a state is termed
'blocked'. When an executing task becomes blocked, it relinquishes
control of the CPU so that any other task that is executable (not

blocked) may use it. When the external stimulus arrives, the blocked task becomes executable again.

In general a real-time system will, at any time, contain a group of tasks which are blocked (do not need the CPU), and another group of tasks that are executable (are actively competing for use of the CPU). The real-time scheduling problem, therefore, reduces to deciding which of the executable tasks is to be allowed to use the CPU, and for how long. A strategy that has been found to be effective for a wide range of applications is as follows:

Each task is alloted a relative priority number, and task scheduling is reviewed at each scheduling event. A scheduling event is declared whenever an executing task becomes blocked, or a blocked task becomes executable, and at every such event the highest priority executable task becomes the next running task. A running task will therefore continue running till it blocks itself (as in the example code above), or a higher priority blocked task becomes executable. In the former case the task returns to the blocked state. In the latter, it is ´pre-empted´ by the higher priority task, and returns to the executable state. This strategy is often termed pre-emptive, priority-based and event driven scheduling, and is illustrated in Fig-2. Although this scheme may appear to be unbalanced in its distribution of system resources, it is extremely effective for real-time sytems.

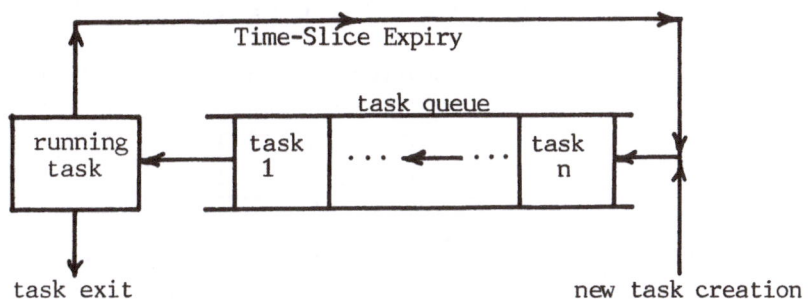

Figure 1. Round robin scheduling.

Fig-3 shows the typical task configuration in a real-time system. The tasks are organised in a pyramidal structure with the most time-critical tasks at the top and the least critical ones at the bottom. At the top of the pyramid is one (or a few) task(s) that are activated on an event-triggered or a time-scheduled basis. These tasks acquire the data that is needed for the rest of the tasks, and are central to the entire system in the sense that none of the other functions can proceed until data acquisition is complete. At each level down the pyramid are found increasing numbers of tasks of diminishing criticality, that use data produced by the level above, and create data needed by the level below. The tasks at the top of the pyramid are short and cyclic, and execute to completion quickly. But many of the tasks at the bottom perform extremely lengthy computations. These tasks might need to be stopped several times in the course of each activation

to permit tasks higher up in the hierarchy to run. The scheduling
requirements of such a system are easily met by the strategy described
above by assigning the tasks at the top the highest priority, and those
at the bottom the lowest.

When there are many tasks at a single priority level, special care
must be taken to even out the distribution of CPU time among the tasks.
This is usually done by simulating a circular queue of tasks at each
priority. Whenever a running task becomes blocked or is pre-empted, it
is returned to the tail of the queue for its priority class. Many RMOSs

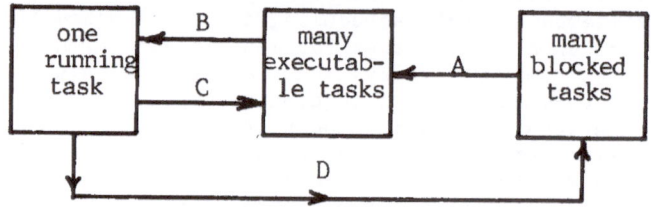

A,B & C - a blocked task becomes executable and pre-empts a
            lower priority running task

D & B   - the running task becomes blocked, thus relinqui-
            shing the CPU to the highest priority executable
            task

Figure 2. State transitions in real - time scheduling

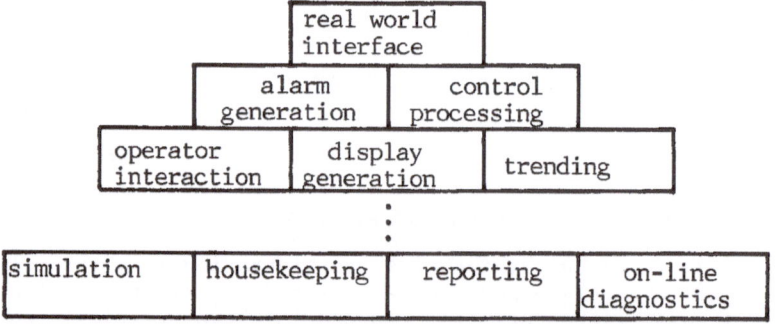

Figure 3. Structure of tasks in real-time system.

also provide a completely general round-robin scheduling regime at a
range of priorities at the lower most levels. This is useful in large
systems where interactive or offline facilities are needed for program
development, detailed report generation and other system housekeeping
jobs.

3.2. Implementation of Task States.

To manage the scheduling of tasks in a uni-processor system, a RMOS must

record and keep track of the status of each task. Typically, a block of
memory called a Task Control Block (TCB) is used to store scheduling and
other related information about each task. Information in the TCBs is
managed by the various synchronising mechanisms available in the RMOS,
and is used by a system process called the task despatcher to schedule
tasks for execution at each scheduling event.

When the task despatcher is activated, it scans all TCBs, tests
their status indicators and examines their priorities to determine which
task is to be despatched for execution next. The efficiency of the
despatching process can be improved by segregating TCBs into two groups
: executable and blocked, and designing the despatcher to scan only the
group containing TCBs of executable tasks. Searching for the highest
priority task can also be speeded up by organising the list of
executable task TCBs ordered by priority, or by keeping separate
sub-lists for each priority class.

When task TCBs are segregated in this way, a task's state may be
changed from blocked to executable or vice-versa merely by moving its
TCB from one group to the other. The TCB groups are stored as linked
lists to facilitate such transitions, and to move a TCB from one group
to another, only a few pointers need be changed.

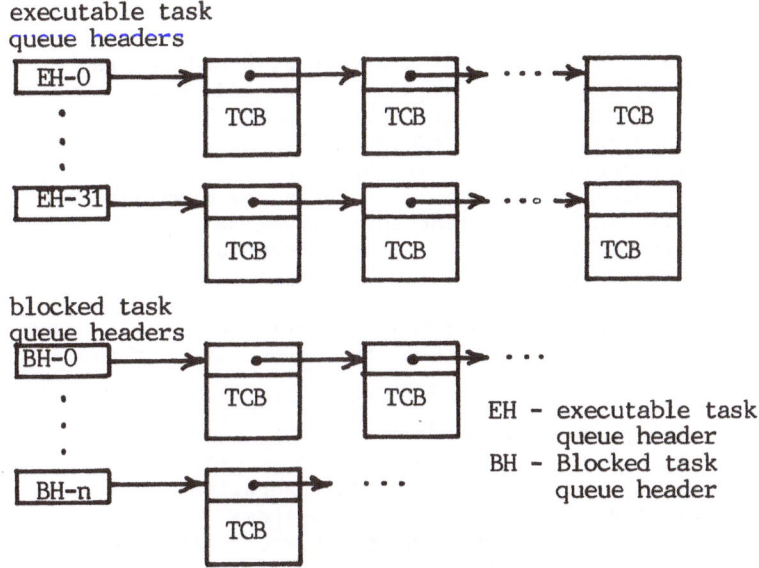

Figure 4. TCB grouping scheme of VAX/VMS.

Fig-4 illustrates the TCB grouping scheme used in the VMS operating
system [3] which runs on the 32-bit VAX-11 series of computers
manufactured by M/s Digital Equipment Corporation. There are 32 task
priority levels (0 to 31), and there is one executable-task queue for
each priority. The blocked tasks are also segregated into different

queues depending on the cause for blocking.

3.3. Task and Event Synchronisation.

Inter-task synchronisation is commonly implemented with ´wait´ and ´signal´ operations on semaphores. A semaphore is a software object consisting of a non-negative value and a queue of waiting tasks.

When a task executes a wait on a semaphore, the RMOS tests the value of the semaphore to check if it can be decremented. The value of a semaphore can not be made negative, so if its value is 0, the task is blocked, and a pointer to its TCB is added to the semaphores queue of waiting tasks. But if the value of the semaphore is greater than 0, its value is decremented and task execution continues unaffected. A task that becomes blocked by executing a wait on a semaphore is referred to as ´waiting for an event´. This event can be signalled by another task as described below.

When a task executes a signal on a semaphore, its wait queue is checked to see if there are any tasks waiting. If any TCB pointers are found, the first one is de-queued and the task pointed to is restored to the executable state. But if there are no tasks waiting, the value of the semaphore is incremented. The state of the task performing the signal operation is never affected.

Semaphores are also used in another form called event flags. Event flags can only have one of two values, 0 and 1. The signal operation is replaced by two operations -- clear and set -- which may be used to change its value. The wait operation in this case blocks a task if the specified event flag is clear, till such time as it becomes set, but does not affect the value of the flag.

The use of semaphores in this way provides means for tasks to synchronise actions amongst themselves. But real-time tasks also need to synchronise execution with external events. Such synchronisation can be achieved by using special I/O functions of real-time I/O devices (see section 5.1.3).

Since the semaphore and event-flag operations manipulate TCBs and other data structures within the executive, they are implemented as ESRs.

3.4. Time Synchronisation.

The ability to synchronise the execution of a task with time is extremely important in real-time systems, and an RMOS must provide ways in which tasks may determine the current time and generate programmable delays in execution paths.

Time keeping in a RMOS is performed by a real-time clock which causes interrupts at a regular rate. The associated interrupt service routine increments a time field within the RMOS at each interrupt. Since the time field is initialised at startup with the current date and time, it always contains the absolute current time. The time is represented as the number of clock ticks since a certain base date and time used as a convention for time measurements.

When a task makes a time-based delay request, it is blocked, and the magnitude of the requested delay is used to create a timer request packet, which is then threaded into a clock-queue. Each timer request packet contains the absolute time when it is to mature, and a pointer to the TCB of the requesting task. The interrupt service routine for the real-time clock inspects the clock queue at each clock tick after updating the current time field. If it finds any request packets with

TABLE - I. Example of Task Scheduling ESRs in a RMOS.

ESR	Function	ESR	Function
Wait	Blocks task till occurence of an event.	Delay	Blocks task for a certain period of time.
Post	Marks the occurence of an event.	Susp	Suspends execution of a task.
Mwait	Blocks task till any event in a group occurs	Rsum	Resumes execution of a suspended task.
Mwtand	Blocks task till all events in a group occur	Asusp	Suspends all other tasks.
		Arsum	Releases ASUSP status.
Timer	Puts task on scheduled status.	Ctime	Removes task from scheduled status

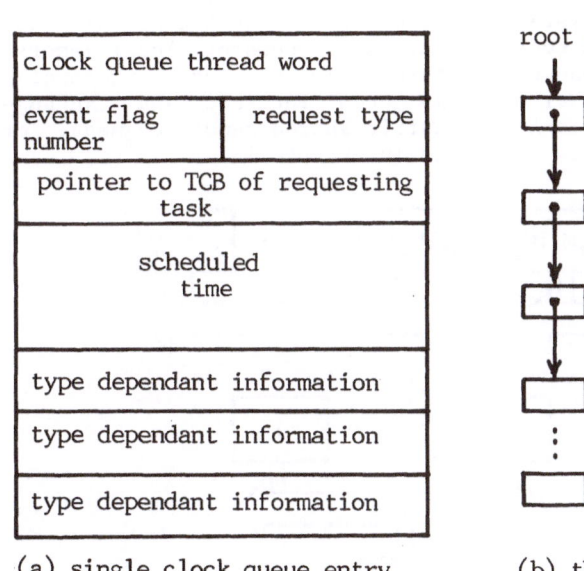

(a) single clock queue entry

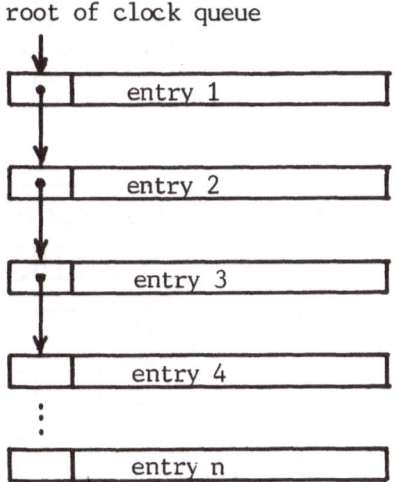

(b) the clock queue

Figure 5. The clock queue in RSX-11M.

the scheduled time equal to the current time, it uses the request packet's TCB pointer field to identify the task that requested the delay, and makes it executable again. The request packet itself is delinked from the clock queue and disposed. Fig-5 illustrates the clock queue [4] as used in RSX-11M (which runs on the 16-bit PDP-11 series computers manufactured by Digital Equipment Corporation.

Clock based services are also available in other forms. Tasks can request programmed timeouts (while waiting for other events) or request to be interrupted through an asynchronous trap mechanism at a defined time. Time synchronisation and other related facilities are implemented as ESRs.

3.5. An Example of ESRs for Task Scheduling.

The ESRs available for scheduling in various RMOSs differ slightly in form and method of use. Table-I is an example of the facilities available under the Hitachi HIDIC V90 series [5] process control computers.

3.6. Scheduling in Multi-processor Systems.

The implementation of multiple tasks by multi-programming a single processor is adequate for many systems. But for applications with severe response time requirements, it requires the use of extremely fast (and inordinately expensive) hardware, and it becomes more economical to use multiple processing. A multi-procesor system consists of several independant uni-processor systems tied together through a global bus. The individual uni-processors contain their own processor, local memory, and I/O device interfaces; and can perform their own local processing independantly. The individual systems can interact and exchange information [6] through a commonly accessible shared memory segment, or through directly connected data ports and interrupt lines.

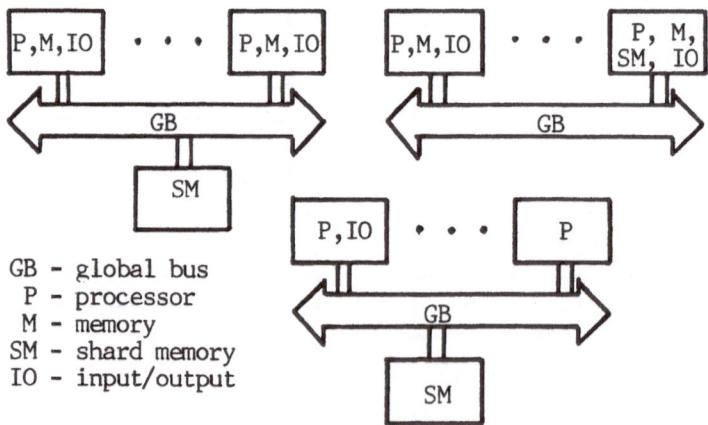

GB - global bus
P - processor
M - memory
SM - shard memory
IO - input/output

Figure 6. Multiprocessor configurations.

Multi-processor systems can be configured in several different ways, as depicted in Fig-6. The individual uni-processors can either all be identical, and perform under a load sharing regime; or be designed to perform predefined dedicated functions. In the former case, the number of tasks may be greater than the number of processors, and a scheduling mechanism not unlike that described above must be used. The executive data structures are placed in shared memory, and a convention that defines which processor is to perform schedule management must be implemented into the software. In the latter case each processor is assigned a specific function, and executes a fixed task. There is therefore no ´executable´ task state, and a task never has to wait for a CPU to become available. Inter task synchronisation and communication is performed through shared memory, access to which is regulated by the use of semaphores.

4. MEMORY MANAGEMENT

Memory management has come to mean different things to different people. The original motivation for introducing memory management in small computer systems was to extend the addressing range of CPUs which had small address words. Even today, providing better ways of running large programs on machines with limited memory forms a major component of the memory management system. Apart from this, memory management provides services that can be used for inter-task communication, optimising the use of memory, and creating other infrastructural tools needed for real-time systems development.

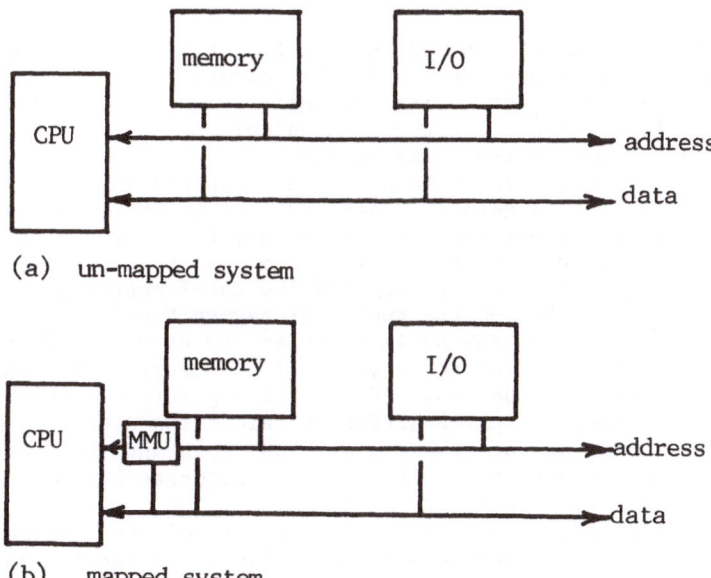

(a) un-mapped system

(b) mapped system

Figure 7. Difference between mapped and un-mapped system.

Fig-7 illustrates the essential conceptual difference between a mapped system architecture (one that uses memory management) and an unmapped one. For the sake of simplicity, we assume here that there is only one system bus that is used to connect the CPU to the memory and I/O systems.

Unmapped systems are characterised by the direct connection of the CPU address lines to the memory and I/O systems. They are simple, cheap, and fast in operation, but have the disadvantage that the physical memory locations accessed correspond with the logical addresses generated by the executing task. Since every executing task is able to access all of memory, it is not possible to implement any form of memory protection. Unmapped systems are therefore only used for small applications such as simple controllers, or data acquisition units at the very low end of real-time systems.

In a mapped system the logical address generated by the executing task is input to the Memory Management Unit (MMU). The MMU creates a physical address using the logical address and other information from the context of the executing task, which is then output to the system bus. The set of physical memory locations accessed by the task therefore depends on the logical to physical memory translation performed by the MMU. Mapped systems are relatively complicated, and slightly slower in operation. But the flexibility of operation that results from the use of memory management, more than offsets these disadvantages.

4.1 Memory Extension

Memory extension capability is needed to utililize a memory system containing more locations than can be directly addressed by the CPU address bus. The need for address extension was specially important in the early 16-bit systems, because a physical memory space of 64K bytes is too small for modern real-time systems.

A simple technique is to have a large physical memory divided into a number of banks or segments, each one of which is allocated to one task. The memory segment that is used at any time depends on which task is active. The memory addressing limitation is then transformed into a limit on the size of each segment, and the total amount of memory that can be used depends only on the number of segments used.

Memory segmentation can be implemented by using a MMU that creates physical addresses by adding a relocation constant to the logical addresses generated by the CPU. Different segments can then be created for different tasks by using different relocation constants for each task. Since the value of the relocation constant becomes the base address of the memory segment, the MMU register that holds its value is also known as the base register.

Since all segments are not necessarily of the maximum size, a method of defining the segment size must also be avilable. This is done by providing the MMU with a segment length register to hold this value. The MMU can then check all addresses generated by the CPU to ensure that tasks only refer to locations within their own segments and do not corrupt data or code belonging to other tasks. Fig-8 illustrates such a memory management scheme.

When such a scheme is used, each task is compiled and linked as if it were to use memory locations starting at location 0. Address references generated by the CPU at run time are interpreted by the MMU as an offset into the appropreate memory segment. This is known as run-time or dynamic relocation. A task in a mapped system can therefore be loaded any where in memory, and is not bound to a specific address.

4.2. Memory Protection

The successful operation of any multi-tasking system depends on the enforcement of a system of memory protection that prevents tasks from interfering with each other, and with the executive. On a mapped system, this is made possible by isolating tasks into mutually exclusive segments of physical memory through techniques such as those discussed above.

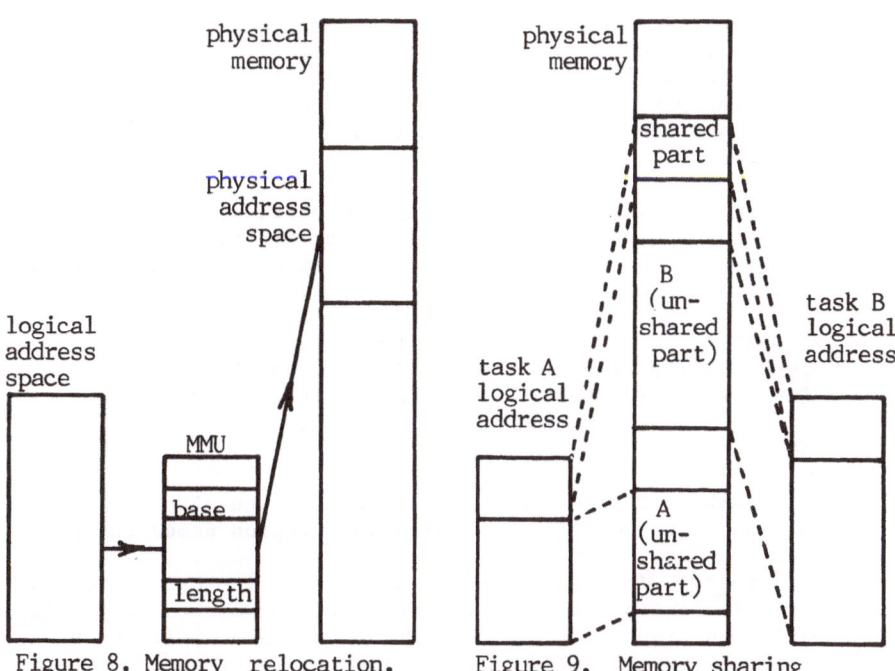

Figure 8. Memory relocation.   Figure 9. Memory sharing

The assignment of separate memory segments to each task can still not guarantee proper protection if the tasks are able to manipulate the MMU registers. The MMU registers are therefore designed to appear as memory locations within a segment which is only 'visible' to the OS, or are accessible through priviledged instructions only available to the executive.

4.3. Task Address Space Mapping

The relocation of logical address spaces to extend system memory has been discussed in earlier sections. Such mapping is entirely transparent, and the task is neither aware of, nor affected by the relationship between its logical address space and the assigned physical address space. This section deals with address relocation used to map parts of a task's logical address space to OS-defined objects in specific locations of system memory.

4.3.1 Shared data.

It is common in real-time systems for a group of tasks to access and manipulate a common data pool. Typically, one task in the group acquires the information from an external source such as a programmable controller, and places it in the common pool. The other programs can then access it for purposes of display, control, archiving, etc.

Fig-9 illustrates the allocation of memory to two tasks with a shared data region. The common data pool is allocated space at some convenient location in physical memory as an entity in its own right. The tasks using the pool are allocated only enough physical memory to hold their code and non-shared data. A block of their logical address spaces representing shared data is relocated so as to access the shared memory segment.

Some systems provide high level language based mechanisms for the implementation of data sharing. One such facility is the mapping of a named COMMON block in a FORTRAN program to an external shared region, as in the following example :

```
PROGRAM A                       PROGRAM B
DIMENSION IBUF(1000)            DIMENSION JBUF(1000)
COMMON /SHRDAT/ IBUF            COMMON /SHRDAT/ JBUF
    . . .                           . . .
END                             END
```

It is possible to make the two arrays IBUF and JBUF in the two separate programs refer to a common shared data region named SHRDAT.

4.3.2. Shared code.

An examination of any group of programs will show that they contain sections of common code. This common code, designed to handle basic functions such as conversion between ASCII and binary, file I/O, etc, is incorporated into the task image by the linker from the object module library. A great deal of memory space can therefore be saved if all tasks could use a single copy of such utility routines, installed in a memory area designated as a resident sharable library. As such, code sharing is not specifically a RMOS facility, but a real-time system needs to have as much free memory as possible to reduce program swapping and thus ensure fast response to events. Hence it would be a pity not to be able to take advantage of code sharing. Moreover, the tasks in a real time system are likely to obtain greater benifit from code sharing as they operate on the same set of data and objects, and therefore have many more sequences of common code.

Code sharing is physically similar to data sharing. One common code segment is loaded into memory as an independant entity, and the tasks that use the shared code are only allocated enough memory to hold their non-shared code and data. The part of each task's logical address space representing the shared code is relocated so as to be mapped to the single copy of the shared library. Since the shared segment contains code it is marked as 'read-only' to prevent its accidental corruption by any of the user tasks.

Because a resident library may be logically a part of more than one task, the routines in it must be re-entrant (capable of being interrupted in the context of one task, executed in the context of another, and then continued in the original context). Subroutines written in languages such as Pascal and C are naturally re-entrant. But when other programming languages are used, resident library routines often have to be written in assembly language to ensure re-entrancy.

4.3.3. Dynamic regions. A dynamic region is a segment of physical memory allocated from the system by a task for its exclusive use as an extension to its normal static data. (Static data is the part which is defined in the programs source text, and for which space is allocated at compile time.). Dynamic regions derive their name from the fact that they are created at run time and can expand or contract as needed.

A task creates a dynamic region by calling an ESR to allocate the specified quantity of memory. It must set aside a block of its logical address space to be used as a window for mapping a corresponding length of the dynamic region. The length of window can be smaller than the length of the region, so at any time a task may only access a part of it. This aspect is an advantage because the size of the region is not constrained by the free logical address space available in a task. A dynamic region can therefore be made as large as necessary to store the required data.

The use of a memory region not all of which is immediately accessible, may seem questionable for conventional purposes. But it is an extremely effective way of manipulating linked lists containing dynamic data structures often used in real-time systems. In such an application only small parts of the entire data structure are accessed at a time (to add or delete an alarm, change the value or status of one point, etc). The 'live data-bases' of real time systems are usually implemented as dynamic regions.

4.3.4. Further MMU features. It is evident that a MMU which relocates a tasks logical address space onto a contiguous range of physical addresses can not be used to implement the features described above. To make arbitrary memory sharing possible, the MMU must be able to divide a tasks logical address space into separate blocks which can be relocated independantly. A range of address blocks can then either be mapped to a segment containing the tasks own code and data, or can be used as a window to map an external segment anywhere in physical memory. The following example based on the scheme [7] used in the Toshiba TOSBAC 7/40 E series illustrates this concept :

A logical address is 16 bits wide so each task can access a

maximum of 64 Kbytes. The 64 Kb logical address space is conceptually
divided into eight 'logical blocks' each of 8 Kbytes by splitting the
logical address into two fields as in Fig-10. The most significant 3
bits represent the logical block number (0 to 7), while the least
significant 13 bits represent the offset (0 to 8 Kb) within a logical
block.

The physical address space is also divided into 'physical blocks'
of 8 Kbytes each, so that a physical block may accomodate one logical
block of any task. The relocation information for each task is
contained in a Convertion Table which contains the base addresses of the
physical blocks allocated to each of the eight logical blocks. Logical
to physical address translation is performed by using the logical block
number to index into the Conversion Table and obtain the base address of
the physical block. The base address is added to the offset to obtain
the physical address. This is illustrated in Fig-10.

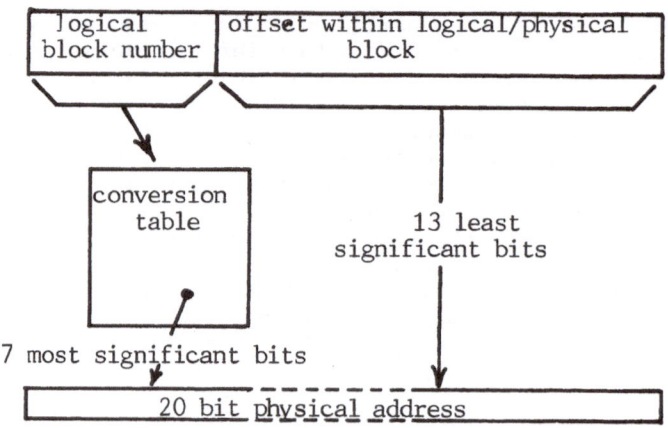

Figure 10. Memory relocation in TOSBAC sytem.

Observe that a 20 bit wide physical address is built up, so 1024
Kbytes of memory may be supported. Since a physical block base address
is a multiple of 8K, its least significant 13 bits are always zeroes.
It is therefore only necessary to store the most significant 7 bits in
the conversion table.

Dividing a tasks logical address space into separate blocks also
has other advantages. Each block can be characterised by an access-code
that the MMU can use to regulate memory access. Blocks containing code
can be marked as read-only to prevent accidental modification, while
data blocks can be marked as read-write.

4.4 Software Support for MMU Functions.

The discussion of memory management so far has been hardware oriented
because it was necessary to understand the MMU's role in providing these
facilities. But, as was mentioned earlier, a task itself has no direct
control on the MMU's operation. It must therefore, use the RMOS's

services, provided in the form of ESRs, to obtain these facilities.

4.4.1. <u>Object oriented functions.</u>   ESRs can provide comprehensive and
meaningful services only if they are defined in terms of concepts which
are a level above the primitive MMU functions of address blocking, block
relocation and access checking.   Each allocated block of memory must be
treated as an object identified by a unique name, and characterised by a
type code. The type (task segment, sharable data, resident library,
dynamic region etc), defines the kind of access that may be requested.
An RMOS must have a directory of memory objects (see section 4.5.1)
containing object name, type, status and memory location; so an object
may be referred to by name without any knowedge of its location in
memory.
     The ESRs can then be designed to provide object level services and
may be used to create or delete objects, and to establish sharing
relationships   between them.   Section 4.6 illustrates the facilities
generally available.

4.4.2. <u>Linkers and loaders for a shared environment.</u> The use of shared
regions by tasks in a real time system is extremely common.   While ESR
calls from within tasks can be used to map external shared objects, it
is far more convenient to let the loader and linker handle these
functions.
     When a task that needs to use a shared region is linked, the linker
creates separate task sections for the non-shared and shared parts.   The
non-shared part contains the tasks own code and data.   But the shared
part is only represented by a tag that identifies the name of the
region, the logical address range to use to map it, and the permitted
type of access.   When the task is activated, the loader allocates enough
memory only for the non-shared region, and sets up this part of the MMU
tables.   Information from the   shared   section tag and the RMOS´s object
directory is then used to create and add other entries to the MMU tables
to enable the task to map the shared object.

4.5. Physical Memory Allocation

Memory allocation is concerned with how (and how much) memory is
distributed among the tasks in a running system. A large and complex
real-time system has many tasks of varying priorities cycling through
different states of executability.   The total size of all the tasks is
often greater than the available physical memory, so a dynamic memory
allocation scheme must be used.   The allocation algorithm must itself be
linked with the task state transition mechanisms to ensure that
executable tasks are not kept waiting for want of memory.

4.5.1. <u>Segmented memory systems.</u>   Memory allocation by contiguous
segments can be realised with simple data structures and straight
forward supporting code, and is common on the smaller (16-bit) systems.
In a segmented system, the quantum of memory allocation is a partition :
a contiguous block of memory identified by a unique name and allocated

for a specific purpose. A RMOS needs two kinds of partitions, dedicated and transient. A dedicated partition reserves memory for a specific, long-term purpose. A transient partition is created to meet adhoc requests (activation of a new task, creation of a dynamic region etc), by allocating blocks from a pool of available memory. The following paragraphs describe the use of some of these principles under the RSX-11M operating system.

Fig-11 illustates a hypothetical memory configuration. The executive occupies 20 K words of the lowest memory locations. This area contains the executive code and data as well as the dynamic storage region where the executive builds data structures representing its current state (TCB lists, clock queues, PCB chain, etc.).

The rest of physical memory is available to be divided into partitions and allocated for specific purposes. The first few partitions are generally dedicated to OS-related functions and are created automatically when the RMOS itself is generated and configured. Dedicated partitions are allocated contiguously at the low end of memory to avoid fragmentation and wastage of memory space.

All the free memory beyond the end of the last dedicated partition (COMMON in Fig-11) is automatically placed in the GEN partition. GEN is infact the pool of free memory from which segments for transient purposes (sub-partitions in RSX-11M terminology) are allocated.

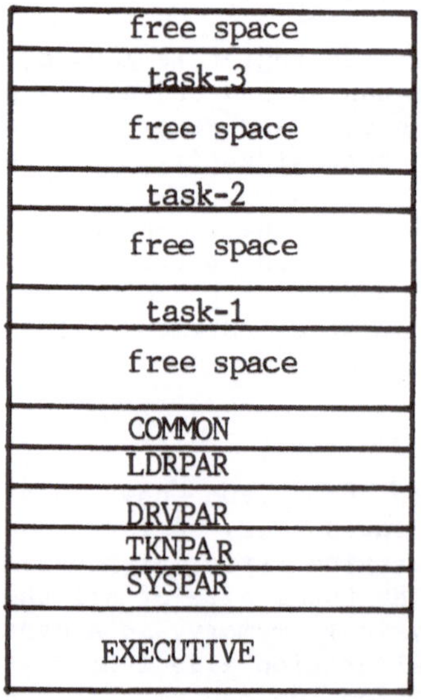

free space
task-3
free space
task-2
free space
task-1
free space
COMMON
LDRPAR
DRVPAR
TKNPAR
SYSPAR
EXECUTIVE

Figure 11. Typical memory configuration in RSX-11M

The RSX-11M executive keeps track of the current partition configuration in a dynamic data structure (the objects directory) like that illustrated in Fig-12(a). Each node in the structure is a Partition Control Block (PCB). Each PCB represents one partition or sub-partition, and contains all the information that the executive needs to know about it. The memory allocation algorithm works as follows:

The PCBs of all sub-partitions of GEN are scanned to find a large enough unallocted block of memory. The search is simplified by the fact that the PCBs are always linked together in ascending order of partitiion base address. If such a block is found, memory is allocated for the new sub-partition, and a new PCB is created and threaded into the PCB tree at the appropriate place to mark the allocation of the block of memory.

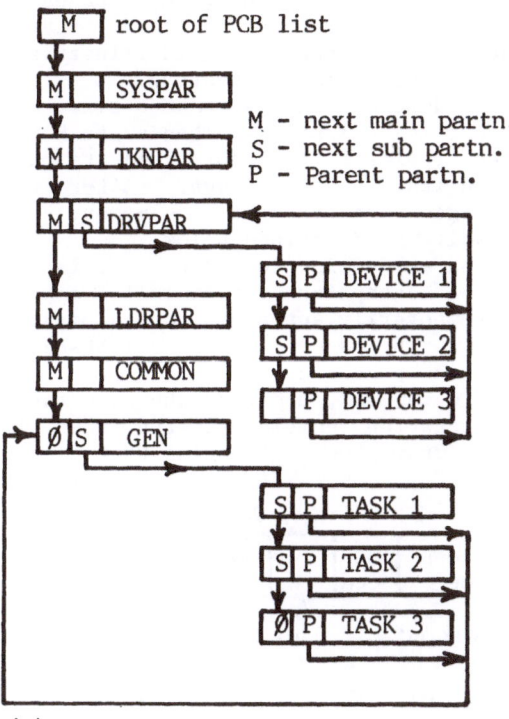

(a) PCB list

(b) Partition control block

Figure 12. Partition control block(PCB) and PCB list.

If a sufficiently large unused block of memory is not found, then the total size of all unallocated blocks (if any) is determined. If the aggregate size of all such blocks is sufficient then a special system program called the shuffler is activated. The shuffler physically moves all partitions towards one end of memory thus compacting all unallocated fragments into one large contiguous block.

If the total size of all unallocated fragments is also insufficient, more free memory can be created by swapping. This procedure consists of finding a large enough sub-partition that is occupied by a blocked task, copying its contents to a swapping file on disk, and releasing the memory that was allocated to it.

After sufficient free memory has been created by swapping or shuffling, a sub-partition is allocated as described above.

4.5.2 <u>Virtual Memory Systems</u>. In a virtual memory system, a task's logical address space is divided into relatively small blocks of a fixed size called pages. The physical memory and backing store are also segmented into pages or blocks of the same size. Mapping of virtual pages to physical pages is performed dynamically at run-time on a per-page basis. The mapping information for each task's pages is maintained in page tables which contain relocation and status information for each page of virtual address space. Fig-13 illustrates the format of a page-table entry in the VMS operating system.

The size of a task's page table depends on the task's size, and contains as many slots as the task has logical pages. When a task is first activated there are no pages resident in memory, and its page table contains pointers to blocks in its disk file image. After the task starts running, the required pages are brought into memory by a process known as demand paging, or faulting, described below.

Whenever a task attempts to access a location in a page that is not resident in memory, the MMU raises an exception (often called a page fault). The faulting task is then temporarily suspended, and a system process called the pager is activated. The pager examines the state of the suspended task and determines which page needs to be brought into memory. It then issues the required I/O command, modifies the page table entry, and exits. When the I/O operation completes, the task is re-started and permitted to continue from the instruction that caused the page fault.

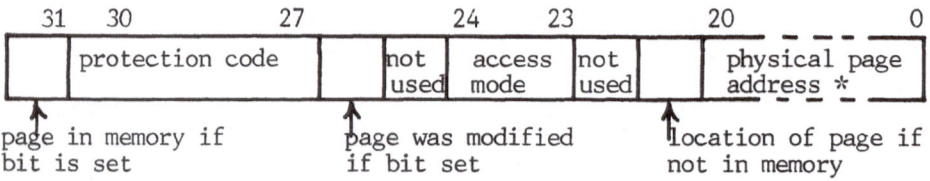

* the least significant 9 bits are always 0, and not stored.

Figure 13. Format of a page table entry in VAX/VMS.

This process continues, with increasing numbers of task pages being brought into memory, until the number of in-memory pages reaches a system specified limit (the working set limit). After this point, a task must surrender one page of memory back to the system for every page that is faulted into memory. But a task may fault in and surrender the same page several times over, so to maintain continuity any changes made must be preserved. Whenever the page table entry of a page indicates that it was modified, its contents are copied to a paging file before it is surrendered. The page-table entry is then modified so that if a fault for that page occurs again it is fetched from the paging file instead of from the tasks image file.

Physical memory allocation in a virtual memory environment is performed under control of several system parameters. Some of these are determined by the amount of physical memory, the number of active tasks, the rate at which each task generates page faults, and the quantity of unallocated free memory. Some of these parameters are adjusted by the virtual memory system at run time so as to optimise system throughput.

TABLE - II. ESRs for Memory Management.

$CRETVA	Creates a range of virtual addresses for a task. In effect the service creates page table entries that enable a range of virtual addresses to be used. The greatest use of this service is by the VMS process that activates tasks. However, any task may explicitly call this service to create additional addressing windows.
$DELTVA	Deletes a range of virtual addresses from a tasks page tables.
$CRMPSC	Creates a memory region known to VMS as a private section (such as for a dynamic region) or a global section (such as a sharable segment). The service optionally maps the created region to a specified range of virtual addresses.
$MGBLSC	Maps a named memory region to a specified range of task addresses. The region must have been created earlier by another task by using $CRMPSC.
$DGBLSC	Deletes a named global section.
$LKWSET	Locks the specified pages in the working set. This ensures that the pages are never replaced in the working set by other pages, and are therefore always in memory (except of course when the whole task is swapped out).
$ULWSET	Unlocks range of pages from the working set.
$LCKPAG	Locks a range of pages in physical memory. Pages that are locked in physical memory are neither paged nor swapped out.
$UNLPAG	Unlocks range of pages from memory.
$SETPRT	Allows a task to set or change the protection parameters of a range of addresses in its virtual address space.
$SETSWM	Allows a task to control whether it can be swapped out of memory.

While such a strategy is adequate for most purposes, a virtual memory system for a RMOS should permit some of these parameters to be pre-determined or tailored on a per-task basis to ensure that time critical tasks are never delayed by the run time overheads of memory management techniques.

4.6. An example of ESRs for Memory Management.

ESRs that affect the utiltsation and allocation of memory vary considerably from one RMOS to another because of differences in conventions used. The example in Table-II is taken [8] from VMS which uses a virtual memory management scheme.

5. THE INPUT OUTPUT SUB-SYSTEM

All Computers, whether real-time systems or otherwise, use various external input and output devices such as terminals, printers, magnetic tape and disk units, and communications interfaces. All these devices are designed for different functions, and so, have different characterstics. It would, however, be very distracting for any applications programmer, if the handling of different devices had to be done differently within application programs. Therefore, early in the development of operating systems, the I/O sub-system concept was brought in to provide device independance. The I/O sub-system provides an abstraction that makes any I/O device look like a collection of records organised into files, so that application programmes only appear to exchange blocks of data with logical files. This abstraction of devices as files is widely used in operating systems today and fits in extremely well with I/O statement concepts in contemporary high level languages.

5.1. Special Requirements Of Real-Time Systems

An I/O sub-system for a RMOS has certain special requirements that set it apart from one for a general purpose OS. The treatment of I/O as files leads to concise and clear code in traditional programs, but has many disadvantages for real-time programming.

High level language I/O constructs are designed more for the convenience of the programmer, with efficiency (a prime requirement for real-time work) being relegated to a secondary position. Consider the following lines in FORTRAN as an example:

```
        READ (1, 10) BUFFER
   10   FORMAT( ... )
```

The data stream that is read into BUFFER can come from a character oriented device such as a terminal, a block-structured device like a disk unit, or even from a program on another computer connected via a network link! Since the programmer has no way of specifying the characteristics of the data source in sufficient detail, the run time system associated with such a task must contain file control blocks and routines capable of handling the full range of possibilities. The resulting overhead in terms of memory utilisation and processing time can be too high for real-time operation. A faster and more direct way of accessing I/O devices must be available.

Moreover a device disguised as a file may only have a limited set of operations performed on it : Open, Read, Write and Close. These operations are adequate for devices which only store data, but real-time systems have to interact with special devices which have to be set up and controlled in the course of data transfer. Examples of such operations are setting communication line characteristics, controlling a modem on a serial line, setting the sampling frequency or gain of an analog to digital converter etc. The file oriented I/O operations are inadequate for performing these functions.

TABLE - III. Example of real-time BASIC I/O statements.

Statement	Description
X=DIN(4,2)	Sample digital status value on channel 2 of card 4 and place it in variable X.
DOT(2,1)=1	Output state ´1´ to channel 1 of card 2.
T=FIN(2,3,1)	Obtain frequency value on channel 3 of card 2 using timebase 1, and place its value in variable T
V=AIN(2,1)	Sample analogue input on channel 1 of card 2 and place its value in variable V.
AOT(4,1)=X	Output value in variable X to channel 1 of card 4.

There are some versions of extended Real-Time BASIC that have statements for performing device specific functions. Table-III lists some examples of special I/O statements from the ´MACSYM´ data

TABLE - IV. Example of device specific I/O function codes.

Device type	Function mnemonic	Description
Disk unit	IO.CRE	Create a new file.
	IO.DEL	Delete a file.
	IO.ACE	Access (open) a file.
	IO.DAC	Close a file.
	IO.RVB	Read specified blocks of file.
	IO.WVB	Write specified blocks of file.
Communic-ations line unit	IO.FDX	Set to full duplex mode.
	IO.HDX	Set to half duplex mode and primary or secondary status.
	IO.INL	Initialise line and set device characteristics.
	IO.TRM	Terminate commuications and disconnect channel.
	IO.RLB	Input a message from the line.
	IO.WLB	Output a message to the line.
Analog to digital converter	IO.RBC	Initiate multiple conversions of specified input channels with specified gains.
	IO.KIL	cancel all pending conversion requests.

acquisition system [9] manufactured by M/s Analog Devices. Such extensions are not standard, and may only be used to program products from specific vendors. Again, since BASIC can only be used for programming relatively small and simple applications, such facilities can not be considerded as general purpose tools for programming real-time systems.

5.1.1 ESR´s for performing Input/Output. The only facility that can be used to perform such special functions in a moderately language and OS-independant way is an Executive Service request that specifies an I/O function. Such an executive request appears as a subroutine call with the following parameters :
   Function-code and function-dependant parameters.
   Identification of device.
   Variable to receive status and completion codes.
   Address and length of data buffer.
The function code is device specific, and may be used to indicate any function that the device can perform. In addition to the data transfer operations of traditional storage devices, special operations of the type described above may also be invoked. Table-IV lists some I/O functions for a range of devices [10] available on the RSX-11M operating system.
   The following example illustrates the use of an ESR in a FORTRAN program under RSX-11M for setting the speed of a terminal line to 9600 baud:

```
DIMENSION IPARAM(6)
BYTE SETSPD(4)
DATA SETSPD / 3, 18, 4, 18 /
      .  .  .
CALL GETADR(IPARAM, SETSPD)
IPARAM(2) = 4
CALL QIO(1312, 1, , , ,IPARAM, ICODE)
      .  .  .
```

The subroutine QIO is an ESR used [11] to invoke device specific functions. The first parameter is the function code and in this case indicates ´set multiple characteristics´. The second parameter is a logical unit number, and is used to indicate the device on which the operation is to be performed. IPARAM is an integer array whose first two elements must contain the address and length respectively of a byte array (SETSPD) detailing the characteristics to be set. The bytes in the SETSPD array are interpreted in pairs. The first byte of each pair selects one of a set of device dependant characteristics, and the second supplies a value for it. In this example, 3 and 4 select receive-speed and transmit-speed respectively, while 18 is an index into an implied table of possible speeds whose 18th item is 9600. The consecutive commas are place holders for parameters not needed for this function.
   Obviously, this kind of coding is not portable across operating systems, but it illustrates the type of facility that must be available to enable high level programs to perform special I/O functions. Such

facilities are essential because high level programs generally do not have direct access to the device registers. This method also has other advantages. The consequence of a programming error while using an ESR call (such as specifying an illegal function code or illogical combination of parameters) is only limited to the erring task because the executive checks the validity of all such requests before attempting to perform the operation. But a programming error in a task that directly manipulates device registers could well be catastrophic for the entire system.

5.1.2. <u>User-written drivers for special devices.</u> A real-time system often uses special devices for process I/O and communications. In these cases, it is necessary to write a new device driver as part of the overall real-time system development effort. The I/O sub-system must therefore be a uniquely indentifiable and accessible part of the RMOS with a well documented functional modularity. The interfaces between the modules of the driver, and between the driver and the executive must also be well documented and defined. Procedures and techniques for testing and debugging a new device driver, and procedures for removing a faulty driver and replacing it with a corrected one without affecting the rest of the OS must also be available.

5.1.3. <u>Synchronous and Asynchronous Input/Output.</u> Asynchronous I/O (the ability of a task to issue an I/O request and then proceed without having to wait for the request to complete) is often used in real-time systems, and must be supported. When a task issues a synchronous I/O request, the executive makes it ineligible to receive further CPU time by blocking it. I/O request processing is however continued by the driver, and after the I/O operation is complete, the executive again restores the task to executable status, enabling it to resume execution. But in the case of an asynchronous request, the task continues executing, in parallel with processing by the driver. The task can subsequently re-synchronise with I/O completion or obtain status information through event flags or status blocks declared in the ESR call.

Contemporary high level languages do not support asynchronous I/O, and ESR's must be used to invoke the appropreate I/O functions and synchronising mechanisms.

6. PROGRAMMING LANGUAGES.

There are three major trends in languages used for programming real-time systems:
> Assembly language.
> General purpose higl-level lagnuages.
> Real-time languages.

The advantage of coding in assembly language is that it is efficient in execution and economic in the use of memory. The entire instruction set (except priviledged instructions of course) can be used, so full advantage can be taken of trapping and interrupt mechanisms; and limited forms of concurrency can be implemented through the use of asynchronous trap service routines. The major disadvantages are that assembly code is not portable, is difficult to structure (both code and data), and programming productivity is extremely low. But in the early systems, execution speed and economy in the utilisation of memory were overriding considerations, and there was no other alternative.

Assembly language is still used in the programming of small microprocessor based systems, and for creating simple RMOSs for these devices. But in the realm of large scale systems implementation, assembly language is now limited to the writing of resident libraries, and to take care of those odd jobs which can only be tackled by high level languages very awkwardly and inefficiently. It is estimated that upto 10 or 15 percent of the total code in large real-time systems can consist of assembly language inserts and subroutines.

6.1. General Purpose High-level Languages.

Because of the disadvantages of assembly language mentioned above, high level languages came into use as soon as sufficiently capable hardware became available. While these languages did not have all the desirable characteristics, it was advantageous to write as much of the code as possible in the high-level language, and leave the intractable parts to be dealt with by assembly language inserts or subroutines.

FORTRAN has been widely used because it is readily available and results in efficient object code. The main difficulty with FORTRAN is its lack of code and data structuring capabilities. In particular, the inability to handle data objects smaller than the computer word has to be overcome by the use of assembly language subroutines.

The need to use the OS´s ESRs directly in application programs introduced obscure and error prone logic into the programs. In an effort to remove this disadvantage software vendors formulated ´real-time extensions´ to the language. This movement led to a plethora of different dialects, and the Instrument Society of America (ISA) made an attempt to bring some order by proposing a set of standards [12] that specified the source forms that certain real-time functions should take. But FORTRAN was already on its way out as a real-time language, and the ISA standard procedures were still too system-dependant to become a guiding influence on the industry.

The advent of Pascal [13] had a marked effect on the real-time systems industry. The data structuring capability of Pascal, along with the ability to create user defined data types, made it possible to implement data-base routines completely in it. This is one of the areas in which FORTRAN needs assembly code assistance, so coding directly in Pascal makes programs more secure and reliable. Pascal supports dynamic data (see section 4.3.3) as a language feature, so manipulation of in-memory data-bases is also directly possible. The problems in using Pascal are mainly related with performing special purpose input/output functions (because the language only supports sequential files) and in separate compilation of source modules.

Like in the case of FORTRAN, a number of real-time extensions to Pascal have also been created. Of particular interest are the small data acquisition and control packages available from certain vendors. Some of these systems have language extensions that enable multiple concurrent ´processes´ to be defined [14] within a same program.

The C language [15] was developed along with the Unix operating system as a systems implementation language. C is an intermediate level language, but has a wealth of data and program structuring tools that can be used to express complex program structures concisely. Studies have also shown that object code generatd by many C compilers is more efficient and uses less space than that generated by other language compilers. C is a very compact language, so is easy to learn, and never ambiguous. Although originally intended to go with the Unix operating systems, it is rapidly becoming available on other operating systems, for processors of all sizes, and it is fast becoming a competitor to Pascal as a real-time systems implementation language.

6.2. Real-time Languages.

There are three features desirable in a language suitable for real-time work -- concurrency, efficiency and security. Languages with the ability to express concurrent operation have only appeared relatively recently. In the early real-time languages the main emphasis was on facilities for secure data manipulation (at a time when data structure and user-defined types were still not popular), and on creating compact forms of expression which could to compiled into efficient machine code. The other main motive for creating a class of real-time languages was to avail of the benefits that come from standardisation.

However, because of the relatively limited field of application it is difficult to find compilers and programmers for these languages, so the pressure to use general purpose languages with real-time extensions has always been strong.

CORAL 66 was developed [16] at the Royal Radar Establishment, and has received strong support in the UK, specially for writing software for military applications. It supports bit twiddling operations that can be used to manipulate data items stored and packed into parts of a computer word. But the facility for defining and allocating space for such variables is not fully general. It also supports the abitlity to address arbitrary memory locations, which combined with the bit twiddling operations can be used to perform low-level I/O operations.

RTL/2, which was developed within ICI [17] but later made publicly available, is based on ALGOL 68, and so has good program structuring tools. But like other languages of its time it has poor data structuring ability, and no language defined facilities for multi-tasking and synchronising. However, a standard set of real-time facilites was later defined which resulted in the development of a specification for RMOSs to support RTL/2. RTL/2 has, nevertheless, been widely used in the implementation of both, large scale systems and (more recently) microprocessor-based systems.

Modula [17] was the first language in the real-time field to implement multi-tasking as a language feature. Modula is largely based on Pascal and a subsequent derivative -- Concurrent Pascal. In addition

to supporting concurrent 'processes' within a program, it also
introduced the concept of the module. The module encapsulates data and
procedures in such a way that only the intended data and procedure call
interfaces are visible to the rest of the program. The encapsulation of
facilities into modules makes it possible to provide a high level of
abstraction, and in effect makes programs more secure. Another
important aspect of Modula is that it is used to write the application
as well as any underlying OS components (such as device drivers). There
are facilities to define the structure and location of device registers,
and to define entry points for device generated interrupts. Modula has
not been widely used, but many of its concepts have found their way into
Ada.

Ada [17] is a language developed as a result of the efforts of the
US Department of Defense. While basically intended for development of
software for embedded microprocessor applications, it is fast becoming
available on machines of all sizes, and appears set to become a standard
for professional software of all kinds. Ada is a large and complex
language into which have been distilled desirable features from many of
its predecessors. Of particular interest to real-time systems is the
fact that features such as multi-tasking, mutual exclusion, inter-task
synchronisation and communication, and time synchronisation are all
available as built in language features.

7. SYSTEM PERFORMANCE MONITORING.

A prime design objective of real-time systems is short response time,
and efforts must be made to keep it so even when the system is expanded.
Since these systems work in close association with process plants the
economic consequences of a system failure or performace degradation can
be grave, and tools must be available to monitor performance indices,
and locate problems before a total service breakdown. Most ROMSs provide
facilities that may be used to monitor the system software and obtain
reports on its performance.

7.1. Software Performance Monitors.

A software performance monitor is a set of routines within the executive
which collect run-time performace information. These routines have
access to all of the executive's internal data structures, and can
collect system-wide data such as the gross utilisation of memory, CPU
time and disk space; or can be used to focus on the impact of a single
task or a group of tasks. The reports generated can be used to determine
performance bottlenecks, and improve response by reassigning priorities
and altering memory allocation strategies.

The importance of these activities is proportional to the size and
complexity of the RMOS. Large virtual memory operating systems have
many modifiable system parameters that affect system performance, and
the reports can be used to 'tune' them to obtain the most from the
hardware.

7.2. Online Error Logging.

Online error logging is the continuous monitoring and logging of errors in the operation of memory and I/O devices. Some such faults (typically in the case of magnetic media storage devices) are transient in nature, and the failed operation often succeeds when retried. Such faults are called ´soft´ errors, and do not necessarily indicate a device failure. But an increasing frequency of such errors is often a precursor to actual failure, and error logging can provide a warning well before failure actually occurs. Error logging reports may be generated periodically, and be used to plan and schedule maintenance activities.

7.3. Crash Dump Analysis.

Crash dump analysis is a post-mortem technique used to investigate catastrophic failures of the executive. A crash dump is a copy (on magnetic media) of the contents of system memory that is generated automatically at the time of a crash (a fatal software error such as stack overflow occuring within the executive). The crash dump analyser converts the binary information in the dump into a formatted and annotated report containing full details of the status of the system at the time of the crash. The report identifies and decodes all executive data structures, lists values of hardware registers and identifies all objects resident in memory. It also identifies the cause of the crash and the task or RMOS component which was running at that time.

References.

(1) RSX-11M System Generation Manual, Digital Equipment Corporation, 1985.
(2) Using Operating System Processors to Simplify Microcomputer Designs (Order No 230786-001), Intel Corporation.
(3) VAX Software Handbook, Digital Equipment Corporation, 1980.
(4) RSX-11M Crash Dump Analyser Reference Manual, Digital Equipment Corporation, 1983.
(5) HIDIC V90 Series System Manual, Hitachi Ltd., 1985.
(6) Paker Y., Multi-microprocessor Systems, Academic Press Inc., London, 1983.
(7) TOSBAC 7/40 E Series Reference Manual, Toshiba Ltd.
(8) VAX/VMS Utility Routines Reference Manual, Digital Equipment Corporation, 1982.
(9) MACSYM 2 System Digest, Analog Devices Inc.
(10) RSX-11M I/O Drivers Reference Manual, Digital Equipment Corporation, 1983.
(11) RSX-11M Executive Reference Manual, Digital Equipment Corporation, 1983.
(12) Standards 61.1(1975) and 61.2(1976), Instrument Society of America.
(13) Jensen K., Wirth N., Pascal User Manual and Report, Springer Verlag, 1979.
(14) Micropower Pascal Language Reference Manual, Digital Equipment

(14) <u>Micropower Pascal Language Reference Manual</u>, Digital Equipment Corporation, 1983.
(15) Kernighan B.W., Ritchie D., <u>The C Programming Language</u>, Prentice Hall, 1976.
(16) Webb J.T., <u>CORAL 66 Programming</u>, NCC Publications, 1978.
(17) Young S.J., <u>Real Time Languages</u>, Ellis Horwood Ltd., 1982.

Chapter 3

DIGITAL CONTROL ALGORITHMS

S. G. TZAFESTAS
Control and Robotics Group
Computer Engineering Division
National Technical University of Athens
Zografou, Athens 15773, Greece

and

J.K PAL
Engineering Technology and Development Division
Engineers India Limited
1, Bhikaiji Cama Place
New Delhi 110 066, India

ABSTRACT. The last three decades have seen rapid development in modern control theory and microprocessors. Application of newly developed control algorithms in Industrial process control has gained popularity due to microprocessor based control systems. In this chapter all categories of control algorithms which have gained popularity amongst industrial users or which are industrially implementable are covered. We have first covered control algorithms for single-input single-output (SISO) systems and then multi-input multi-output (MIMO) systems. Three term controllers, time delay compensation techniques and fuzzy logic controllers, etc., have been discussed for SISO plants. Predictive, adaptive, optimal and state feedback controllers have been covered for both SISO and MIMO plants. We have discussed auto tuning techniques for SISO PID controllers and also briefly expert control systems. Algorithms have been described in their digital implementation form. Implementation of control algorithms in industrially available distributed digital control system has been discussed. A brief report has been given for actual implementation of some algorithms in industrial plants and benefits due to this implementation have been highlighted. In particular applications of optimal control and optimal filtering in power industry in UK and power, cement and steel industry in Japan and predictive control techniques in petroleum industries in North America have been mentioned while describing the particular algorithm. Recent research results to pool resources of linear quadratic gaussian design popular in power, cement and steel industry and predictive control (model algorithmic control and internal model control) popular in petroleum industries to evolve a better approach to controller design have been highlighted.

81

S. G. Tzafestas and J. K. Pal (eds.), Real Time Microcomputer Control of Industrial Processes, 81–138.
© 1990 *Kluwer Academic Publishers*.

I. INTRODUCTION

In this chapter we will discuss various control algorithms that can be implemented in industrial digital control systems. Emphasis will be placed only on those algorithms which have been implemented in some industrial plant. We will cover both single-input and single-output and multi-input multi-output forms. Seborg [1] has classified process control strategies under three categories given below:

Category I	**Conventional Control Strategies**
	• PID Control
	• Ratio Control
	• Cascade Control
	• Feed Forward Control
Category II	**Widely used Techniques**
	• Gain Scheduling
	• Time Delay Compensation
	• Decoupling Control
Category III	**Advanced Control Methods**
	• Adaptive Control
	• Predictive Control

We will discuss briefly all these techniques and also the popular linear quadratic control problem widely used in power industry [2]. We will also cover fuzzy logic controllers proposed by a class of authors [3].

First we will discuss various process models used to derive process control laws and include mathematical formulation of various control strategies. We will also cover implementation aspects of these strategies in industrially available digital control systems. Some examples of industrial implementations will be discussed.

2. PROCESS MODEL

We will introduce here five popularly used linearised models of processes [4,5].

2.1 Transfer Function

Time and frequency domain expressions for transfer function models both for single-input single-output (SISO) and multiple-input multiple-output (MIMO) systems are given by [5].

Time domain

SISO   MIMO
$y(k) = H(q) u(k)$ $\underline{y}(k) = k(q) \underline{u}(k)$

Frequency domain

SISO   MIMO
$y(z) = H(z) u(z)$ $\underline{y}(z) = k(z) \underline{u}(z)$

Here H(z) is given by

$$H(z) = \frac{b_0 z^n + b_1 z^{n-1} + \cdots + b_n}{z^n + a_0 z^{n-1} + \cdots + a_n}$$

And k(z) is given by

$$k(z) = [k_{ij}(z)]$$

(matrix qxp)

2.2 Impulse Response

Time and frequency domain expressions for impulse response models both for SISO and MIMO systems are given by [5].

Time domain

SISO

$$y(k) = \sum_{j=0}^{\alpha} h(j)\, u(k-j)$$

MIMO

$$\underline{y}(k) = \sum_{k=0}^{j} M(j)\, \underline{u}(k-j)$$

Frequency domain

SISO

$$H(z) = \sum_{j=0}^{\alpha} h(j) z^{-j}$$

MIMO

$$k(z) = \sum_{j=0}^{\alpha} M(j) z^{-j}$$

M(j) - Markov parameter
qxp   Sequences of impulse responses

2.3 Difference Equation Model

Time and frequency domain forms both for SISO and MIMO Systems are given by

Time domain

SISO
$$A(q)\, y(k) = B(q)\, u(k)$$

MIMO
$$A(q)\, \underline{y}(k) = B(q)\, \underline{u}(k)$$

Frequency domain

SISO
$$A(z)\, y(z) = B(z)\, u(z)$$

MIMO
$$A(z)\, \underline{y}(z) = B(z)\, \underline{u}(z)$$

A and B are polynomials
(ARMA description)

A , B are polynomial matrices
$A(z)$ – q x q matrix
$B(z)$ – q x p matrix
$k(z) = A^{-1}(z)\, B(z)$

$$A(z) = \sum_{i=0}^{q} A_i z^{g-1}$$

g is order of the auto regressive part

$$B(z) = \sum_{i=0}^{1} B_i \, z^{1-i}$$

1 is order of the moving average part.

2.4 Frequency Response

In order to use frequency response model a bilinear transformation has to be used as given below:

$$z = (1 + \omega)/(1-\omega) \text{ or}$$
$$z = (1 + \omega'' \, Ts/2) \, / \, (1-\omega'' \, Ts/2) \tag{4}$$

In sampled data system the frequency range which is important lies in the range $0 < \omega < \omega_s/2$ and the choice of sampling time Ts is important.

2.5 State Space Model

In the state space form the process model is described by

$$x\,(k+1) = A\,x\,(k) + Bu\,(k) \tag{5}$$

$$y\,(k) = C\,x\,(k) + Du\,(k) \tag{6}$$

Where x is an n-dimensional state, u a m-dimensional input and y a r-dimensional output. A,B, C and D are nxn, nxm, rxn and rxm dimensional matrices, and k denotes the sampling instant.

Equations (5) and (6) takes the following form when the system has disturbances.

$$x\,(k+1) = A\,x(k) + B\,u(k) + E\,\omega(k) \tag{7}$$

$$y\,(k) = C\,x(k) + D\,u(k) + F\,\omega(k) \tag{8}$$

Where $\omega(k)$ is a q-dimensional disturbance vector.

3.0 PROCESS CONTROL STRATEGIES

Having known the desired system and its model, the design and implementation of a controller can be taken up. The structure of the controller is described generally by terms like feedback and feedforward control, predictive control, multiloop control with or without decoupling, etc. There are excellent text books [6,7,8,9,10] which discuss the fundamental issues in digital control system design. The performance measures of the control system both in time and frequency domain have been discussed in these references. In this chapter popular single input controllers will be covered first. We will concentrate on their mathematical form in this section and subsequently we will briefly discuss their implementation in stand alone microprocessors or in controller subsystem in industrially available distributed digital control systems. After the single input controllers are covered, we will discuss the multi input versions. Our discussion will be concentrated on the class of controllers which have been implemented in industrial plants.

3.1 Controllers for Single-Input Single-Output System

We start with popularly used simple three term controllers (P, I, PI, PD, PID). The block diagram of a DDC loop is shown in fig. 1.

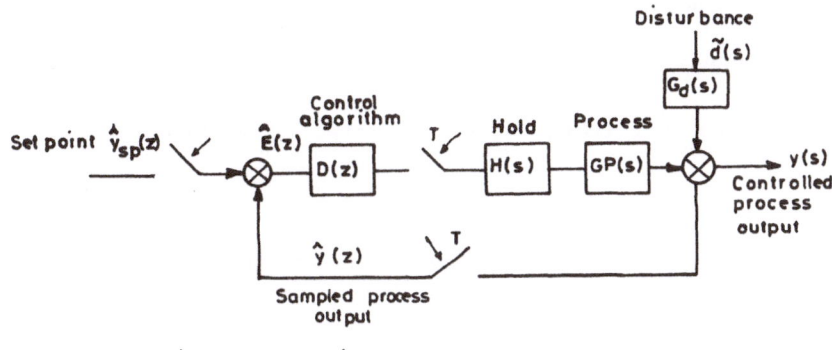

Fig.1   Block diagram of DDC loop

3.1.1 *Three term controller.* Proportional integral and derivative (PID) controllers, also known as three term controller have been used widely in the process industries for over four decades. The merits and demerits of PID controller have been discussed in [6-11].
The continuous time form of PID controller is written in the form :

$$u(t) = K_c\,[e(t) + \frac{1}{T_i}\int e(t)\,dt + T_d\,\frac{de}{dt}] + u_s \qquad (9)$$

Where K_c is the controller gain, T_i is the integral or reset time and T_d is the derivative or rate time. The controller output is u(t) and the error signal e (t) is defined by

$$e(t) = y_r\,(t) - y\,(t) \qquad (10)$$

Where y_r (t) and y (t) are process reference signal and output respectively. The controller bias u_s indicates the controller output when the error is zero.

A Laplace form of (9) with u_s =o is given by

$$U(s) = k_c\,[1 + \frac{1}{T_i s} + T_d\,s]\,E(s) \qquad (11)$$

Derivation of discrete PID approximation is available in [7, 8, 11].
The discrete forms of (9) and (11) can be obtained as

$$u(t) = K_c\,[\,e(t) + \frac{T_s}{T_i}\sum_{t=0}^{t} e(t) + \frac{T_d}{T_s}\,[e(t) - e(t-1)]\,] \qquad (12)$$

$$u(z) = K_c\,[1 + \frac{T_s z}{T_i(z-1)} + \frac{T_d(z-1)}{T_s z}\,]\,E(z) \qquad (13)$$

where z is the complex z - plane operator.

The above two forms of PID controllers are called proportional forms as the total output of the controller is calculated.

A velocity or incremental form can be derived if a change in control signal is defined as

$$\Delta u(t) = u(t) - u(t-1) \tag{14}$$

A velocity form of eqn. (12) then becomes

$$\Delta u(t) = K_c \left[(1 + \frac{T_s}{T_i} + \frac{T_d}{T_s}) - (1 + 2\frac{T_d}{T_s}) z^{-1} + \frac{T_d}{T_s} z^{-2} \right] e(t) \tag{15}$$

or

$$\Delta u(t) = K_c \left[(e(t) - e(t-1)) + \frac{T_s}{T_i} e(t) + \frac{T_d}{T_s} (e(t) - 2e(t-1) + e(t-2)) \right] \tag{16}$$

When z^{-1} is the backward shift operator.

This is a recursive algorithm more suitable for computer implementation.

The basic form of PID controller discussed so far is called the "setpoint on P&I&D" structures as the error signal drives the proportional derivative and integral term. Two more forms have been derived. One is by removing the setpoint from the derivative term referred to as "setpoint on - P&I" and is of the form

$$\Delta u(t) = K_c \left[e(t) - e(t-1) + \frac{T_s}{T_i} e(t) + \frac{T_d}{T_s} (-y(t) + 2y(t-1) - y(t-2)) \right] \tag{17}$$

A less commonly seen form is "setpoint on I only" which removes the setpoint from proportional and derivative term. This is given by

$$\Delta u(t) = K_c \left[-y(t) + y(t-1) + \frac{T_s}{T_i} e(t) + \frac{T_d}{T_s} (-y(t) + 2y(t-1) - y(t-2)) \right] \tag{18}$$

Other forms of PID controllers are also possible [6-11]. It should be noted that the structure of the PID control algorithm affects its closed-loop performance.

Note : a) Two Term Controller:
The velocity form of a two term controller i.e., proportional plus integral controller is given by:

$$\Delta u(t) = K_c \left[(e(t) - e(t-1)) + \frac{T_s}{T_i} e(t) \right] \tag{19}$$

b) 3 P Controllers:
3 P algorithms have been found to yield a control performance very similar to PI and PID algorithms for load disturbances. This algorithm has the time domain form:

$$u(t) = P_0 e(t) - P_1 e(t-1) + (1 - P_2) u(t-1) + P_2 u(t-2) \tag{20}$$

Where P_0, P_1 and P_2 are tuning parameters of 3P algorithm.

3.1.2 _Dahlin algorithm._ Method suggested by Dahlin [12-13] requires that the closed-loop response of a DDC loop behave like the response of a first order system with dead time to a unit step change in the set point

i.e.,
$$y(s) = \frac{e^{-\theta s}}{\mu s + 1} \frac{1}{s} \tag{21}$$

Where μ is the time constant of desired response and θ is the dead time of the response assumed to be $\theta = KT$ (K being an integer and T is the sampling time).

The transfer function and time domain forms of this controller are given by

$$D(z) = \frac{1}{HG_p(z)} \frac{(1 - e^{-t/\mu}) z^{-k-1}}{1 - e^{-T/\mu} z^{-1} - (1 - e^{-T/\mu}) z^{-k-1}} \tag{22}$$

$$u(k) = \frac{1}{\gamma_4} [\frac{1}{k_5} \{\gamma_2 e(k) + k_1 e(k-1) - k_2 e(k-2)\} - k_3 u(k-1)$$
$$+ k_4 u(k-2) + k_5 u(k-N-1) + k_6 u(k-N-2) + k_7 u(k-N-3)] \tag{23}$$

Here N is the closest integer number of sampling times in the loop's dead time, and γ_4, $k_1 - k_7$ are constants and have been defined in [13].

The above expression for u (k) has been derived for process transfer function being first order lag plus dead time.

3.1.3 _Kalman algorithm._ The Kalman algorithm [12] is a minimum settling time approach to the design of digital controllers. In using this technique restrictions are placed on the manipulated and controlled variables. From these restrictions expressions for the controlled and manipulated variables can be found as

$$c(z) = \sum_{k=0}^{\alpha} C_n z^{-n} \tag{24}$$

$$M(z) = \sum_{k=0}^{\alpha} m_n z^{-n} \tag{25}$$

$$R(z) = \frac{1}{1 - z^{-T}} \tag{26}$$

$$P(z) = \frac{C(z)}{R(z)} = p_1 z^{-1} + p_2 z^{-2} \tag{27}$$

$$Q(z) = \frac{M(z)}{R(z)} = q_0 + q_1 z^{-1} + q_2 z^{-2} \tag{28}$$

Controller transfer function is given by

$$D(z) = \frac{Q(z)}{P(z)} \frac{P(z)}{1-P(z)} = \frac{Q(z)}{1-P(z)} \tag{29}$$

3.1.4 _Dead beat controller._ We require that the response of the process to a unit step change in the setpoint exhibit no error at all sampling instants after the first. The dead beat algorithm is physically realizable if the time delay in the HG_p, (z) is not larger than one sampling period and the transfer function form is given by

$$D(z) = \frac{1}{HG_p(z)} \frac{z^{-1}}{1-z^{-1}} \tag{30}$$

3.1.5 _Pole positioning controller._ The aim of this design is to place the closed-loop poles in prespecified locations. Any negative pole of the controllers discrete transfer function will cause ringing. Dahlin has suggested elimination of ringing poles. We should design a digital controller so that all its poles are positive preferably with absolute values (0.4 to 0.6).

We will discuss pole positioning controller in state space form in next section.

3.1.6 _Nonlinear algorithm._ Here the control law is nonlinear function of control error e and can be described as

$$u = \begin{array}{ll} w \, sign \, (e) & \text{if } |e| > T_B \\ \dfrac{w \, (|e| - T_A) \, sign \, (e)}{T_B - T_A} & \text{if } T_A < |e| < T_B \\ o & \text{if } |e| < T_A \end{array} \tag{31}$$

Where w is controller gain, T_A is dead zone value and T_B is saturation value $(T_A \leq T_B)$

Such algorithm realises all possible functions e.g., Two state relay if $T_A = T_B = o$, Three state relay if $T_A = T_B \neq o$ and saturation if $T_A = o$, $T_B \neq o$

In order to reach a distant desired position in a gentle and smooth way, a square root control law may be applied

$$u_n = \begin{cases} W \, sign \, (e_n) \, sqrt \, (|e_n|) \text{ if } |e_n| > T_B \\ k_1 \, u_{n-1} + k_2 \, e_n + k_3 \, e_{n-1} \text{ otherwise} \end{cases} \tag{32}$$

where $\quad k_1 = 1/(1+D.T_B), k_2 = W \, (1+D.T_A) \, / \, (1+D.T_B)$
$\qquad k_3 = W \, / \, (1+D.T_B)$

3.1.7 _Feed forward control._ For measurable external disturbance v on the process, the control performance with respect to disturbance can be improved by feedforward control.

If we assume mathematical model for the process behaviour

$$G_p(z) = \frac{y(z)}{u(z)} = \frac{B(z^{-1})}{A(z^{-1})} \times z^{-d} = \frac{b_1 z^{-1} + \cdots + b_m z^{-m}}{1 + a_1 z^{-1} + \cdots + a_m z^{-m}} \tag{33}$$

and disturbance behaviour

$$G_d(z) = \frac{n(z)}{v(z)} = \frac{D(z^{-1})}{C(z^{-1})} = \frac{d_0 + d_1 z^{-1} + \cdots + d_q z^{-q}}{1 + c_1 z^{-1} + \cdots + c_q z^{-q}} \tag{34}$$

are known, for ideal feedforward control we have

$$G_s(z)\, G_p(z) = G_d(z) \tag{35}$$

Here $G_s(z)$ is the transfer function of the feedforward controller and is given by

$$G_s(z) = \frac{G_d(z)}{G_p(z)} = \frac{A(z^{-1})\, D(z^{-1})}{B(z^{-1})\, z^{-d}\, C(z^{-1})} \tag{36}$$

Ideal feedforward control is impossible for processes with dead time. One can use parameter optimised feedforward controller design method of Isermann [7]. We will discuss approach of Davison [18] using state space technique in section 3.1.10.

3.1.8 *Cascade controller.* Fig. 2 shows the block diagram of cascade control system.

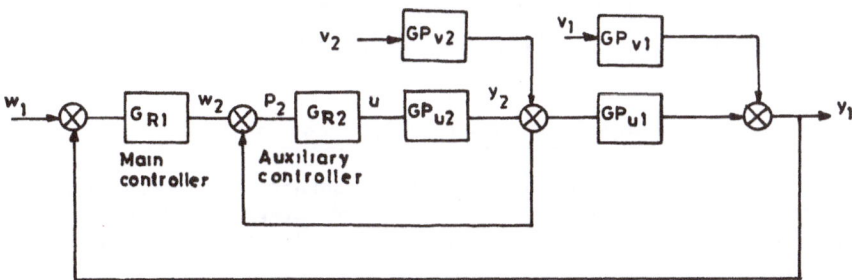

Fig. 2 Cascade controller

This system has the following benefits:

a)   Disturbances which act on the process plant G_{p2} are already controlled by auxiliary control loop before they influence the controlled variable y_1;

b)   By presence of auxiliary feedback system parameter variations in G_{p2} are attenuated;

c)   The behaviour of controlled variable y, is quicker if auxiliary control leads to faster mode than the process plant G_{p2};

For a PI controller as auxiliary controller and a PID controller as main controller the control algorithm takes the form [7]

$$e_1(k) = w_1(k) - y(k) \tag{37}$$

$$w_2(k) = w_2(k-1) + q_1 e_1(k) + q_2 e_1(k-1) + q_3 e_1(k-2) \tag{38}$$

$$e_2(k) = w_2(k) - y_2(k) \tag{39}$$

$$u(k) = u(k-1) + q_4 e_2(k) + q_5 e_2(k-1) \tag{40}$$

Any variety of controller for single input single output system is suitable as auxiliary controller and main controller. Any combination is possible.

3.1.9 *Time delay compensation (Smith-predictor).* Time delays are common occurence in the process industries due to recycle loops, distance-velocity lags in fluid flow and the "dead time" inherent in many composition analyses. To compensate for time delays special control techniques have received attention in literature. Of various techniques, Smith-Predictor has received most attention. A block diagram of the discrete form of the Smith Predictor including a zero order hold is shown in Fig.3.

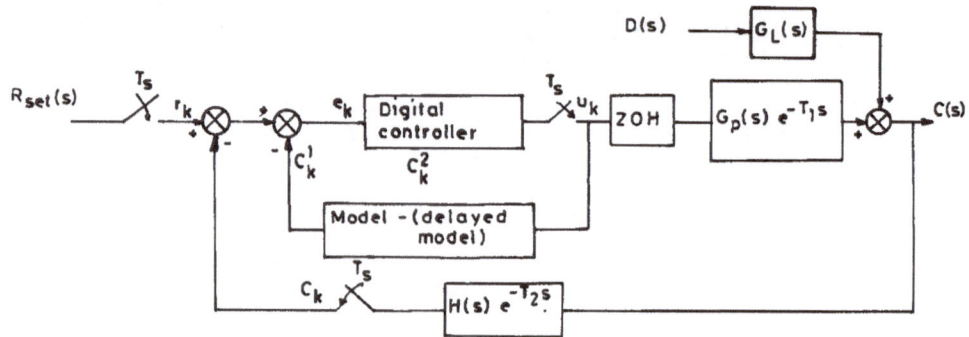

Fig.3 Smith predictor (SP) control system

The feedback loop around the digital controller contains a block whose output represents the difference between two model outputs: the response of a system without time delays, C_k^1 minus the response of the system which contains time delays C_k^2. If the process model is considered to be a first order system plus time delay,

$$\tau_p \frac{dc}{dt} + c(t) = k_p u(t-T) \tag{41}$$

where T is the total time delay, i.e., $T = T_1 + T_2$
or $T = (N + \beta) T_s$; T_s is the sampling time, N is a non negative integer and $0 < \beta < 1$.

Expression for C_k^2 is given by

$$C_k^2 = B C_{k-1}^2 + B k_p \left(\frac{1}{C} - 1\right) u_{k-N-2} + k_p (1-B/C) u_{k-N-1} \tag{42}$$

If no delay is present i.e. T=0, the expression for the undelayed model takes the form

$$C_k^1 = k_p (1-B) u_{k-1} + B C_{k-1}^1 \tag{43}$$

where $B = Exp(-T_s/\tau_p)$ and $C = Exp[-\beta T_s/\tau_p]$

From Fig.2, the error signal e_k can be expressed as

$$e_k = r_k - C_k - (C_k^1 - C_k^2) \tag{44}$$

If a conventional PI digital control algorithm is employed, u_k takes the form

$$u_k = k_c [e_k + \frac{\tau_s}{\tau_I} \sum_{i=1}^{k} e_i] \qquad (45)$$

where k_c is the controller gain and τ_I is the reset time.

3.1.10 *State feedback controllers.* In the last three decades considerable work has been done in design and implementation of state variable feedback controllers [14, 15, 16, 17]. Various design methods have been employed e.g., pole placement [14] or minimization of quadratic performance index (linear optimal control [17]). These controllers can accommodate external measurable or unmeasurable disturbances [18].

We consider the system

$$\dot{x} = A x + B u + E w \qquad (46)$$

$$y = C x + D u + F w \qquad (47)$$

Where x is n dimensional state vector, u and y are m-dimensional control, and output vectors, w is r dimensional disturbance vector. Defining the augmented state vector,

$$z = \begin{bmatrix} x \\ y \end{bmatrix}$$

a new state equation can be derived as

$$\dot{z} = A_1 z + B_1 \vartheta \qquad (48)$$

Where

$$A_1 = \begin{bmatrix} A & o \\ C & o \end{bmatrix}$$

$$B_1 = \begin{bmatrix} B \\ D \end{bmatrix} , \quad \vartheta = \dot{u}$$

A state feedback control law of the form $\vartheta = -kz$ can be designed either for placement of closed-loop poles of system (48) in desired locations or for minimising an index of the form

$$J = \int_0^{\alpha} (y^T Q y + \dot{u}^T R \dot{u}) \ dt \qquad (49)$$

The resulting controller takes the form of a proportional plus integral control. The approach can be used for designing feedforward control. A more general treatment is discussed in section 3.2.2(c).

Since the state feedback controllers need all the state variables for feedback, it is necessary to reconstruct the state variables using a Luenberger observer or a Kalman filter [7, 9, 14, 17]. We will consider a general form of these state estimators in the multi input section.

Auslander et. al. [19] have discussed optimal state feedback control for a single-input single-output system based on unit step response of the system.

92

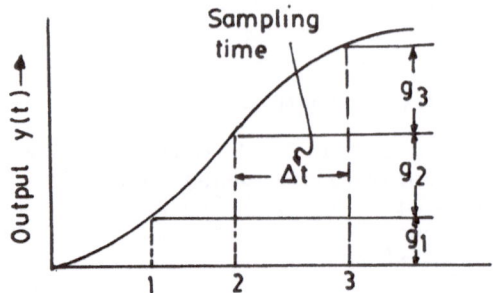

Fig.4 Unit step response of industrial plant (typical)

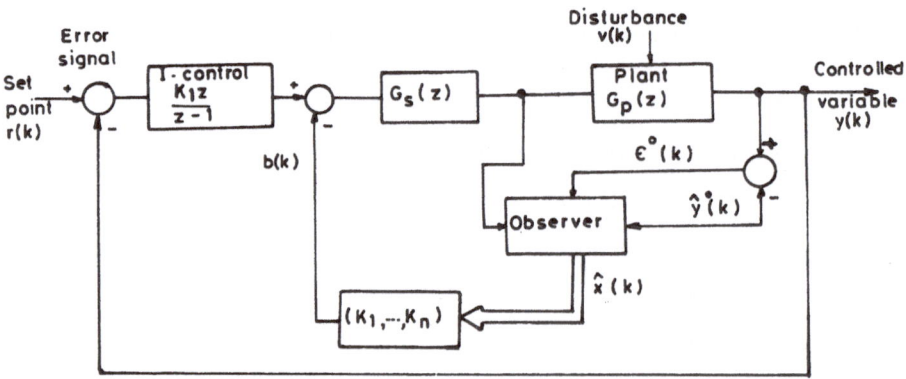

Fig. 5 State vector feedback with observer

Unit step response of an industrial process is shown in Fig.4 as the sigmoid shaped response curve which is typical in many process control problems. It's impulse transfer function is:

$$G_p(z^{-1}) = g_1 z^{-1} + g_2 z^{-2} + \cdots + g_n z^{-n} / (1-p z^{-1})$$
$$= (b_1 z^{-1} + b_2 z^{-2} + \cdots + b_n z^{-n}) / (1-p z^{-1}) \tag{50}$$

where $b_1 = g_1$, $b_i = g_i - p\, g_{i-1}$ (i = 2 --- n)

$$p = 1 - \frac{g_n}{k_p - \sum_{i=1}^{n-1} g_i}$$

and k_p is the plant gain.

The input output relation (50) can be equivalently described by

$$x(k+1) = P\, x(k) + q\, u(k) \tag{51}$$

$$y(k) = cx(k) \tag{52}$$

P, q and c are given by

$$P = \begin{bmatrix} 0 & 1 & 0\text{--------}0 \\ 0 & 0 & 1\text{--------}0 \\ \hline \text{----------------------------} \\ 0 & 0 & 0\text{--------}p \end{bmatrix} \qquad q = \begin{bmatrix} g_1 \\ g_2 \\ | \\ | \\ g_n \end{bmatrix}$$

$$c = [1\ 0\text{------}0]$$

The state feedback control law can be expressed as

$$u(k) = k_I \sum_{i=0}^{k} [r(i) - y(i)] \quad - \quad \sum_{j=1}^{n} k_j\, x_j\,(k) \tag{53}$$

Where $k_I = \dfrac{1}{b_1 + b_2 + \cdots\, b_n}$ is for the main I action,

$$k_1 = 0,\ k_2 = k_3 \cdots = k_{n-1} = k_I$$

$$k_n = \frac{1}{g_n}\ [\,(1+p) - (g_1 + g_2 + \cdots + g_{n-1})\, k_I\,]$$

The state feedback control law is shown is fig.5.

The observer block [14] is for estimating the state of the system for feedback. The observer algorithm for the discrete plant (51) is given by [19].

$$\hat{x}\,(k+1) = \hat{x}^\circ\,(k+1) + f\,[\,y\,(k+1) - c\,\hat{x}^\circ\,(k+1)\,] \tag{54}$$

$$\hat{x}^\circ\,(k+1) = P\hat{x}(k) + q\,u(k) \tag{55}$$

$$\hat{x}^\circ\,(o) = o \tag{56}$$

If there is no unmeasurable disturbance input then f^T is given by

$$f^T = [\,1\ p\ p^2\ p^3 \text{-----}p^{n-1}\,] \tag{57}$$

CEGB, UK has reported implementation of a state estimator based controller for superheater steam temperature control in power boiler. This scheme is shown in fig. 6.

94

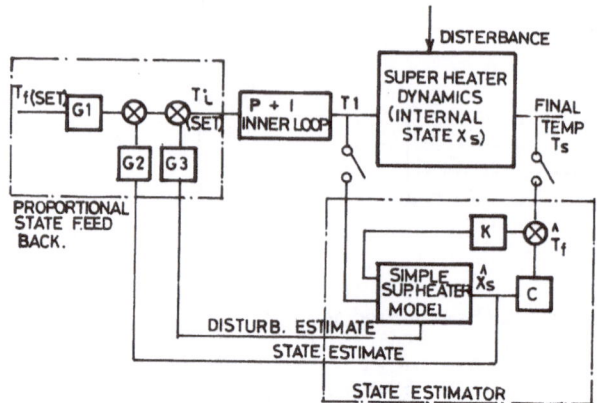

FIG-6 STATE ESTIMATOR BASED CONTROLLER

3.1.11 _Predictive Control._ There has been increasing interest in predictive control methods. Following pioneering work of Richalet and co-workers [20, 21] many workers have proposed different control strategies. These are classified as Long Range Predictive Control [22], Generalised Predictive Control [23], Extended Horizon Adaptive Control [25], etc. A discussion on the Long Range Predictive Control algorithm of De Keyser has been included in a separate chapter by him in this book. We will cover here a methodology of predictive control using impulse response method. This will be followed by Generalised Predictive Control of Clarke [23]. For comparison various single input predictive control strategies e.g. Dynamic Matrix Control (DMC) of Cutler [24] and Extended Horizon Adaptive Control (EHAC) of Ydstie [25] will be covered here. In the multi input section we will discuss the multi input form of controller of Richalet [21] and other variants proposed by industry e.g. Dynamic Matrix Control (DMC) and Quadratic Dynamic Matrix Control (QDMC).

a) Predictive control using impulse response model
We will first discuss a predictive control methodology in which impulse response of the process is used to predict the output of the process for a certain period of time. The predicted trajectory is compared with a desired trajectory and an algorithm is derived to calculate the reference of future input values which minimise the difference between both trajectories. The impulse response of the process is shown in fig. 7 and the prediction horizon in fig. 8.

The output of a sampled linear continuous system H(P) can be described as

$$y(kT) = \sum_{h=0}^{\alpha} h(nT) \; u(kT - nT) \tag{58}$$

The assumption $h(o) = o$ and $h(NPAR + i) = o$ for $i > o$ gives the following simplification

$$y(kT) = \sum_{n=1}^{NPAR} h(nT) \; u(kT - nT) = \underline{h}^T \underline{u}_{k-1} \tag{59}$$

where $\underline{h}^T = [h_1 \; h_2 \; h_3 \qquad \qquad h_{NPAR}]$

We consider the prediction horizon counts NPRED = 4 sample periods. The model is used to predict the behaviour of the process over the prediction horizon.

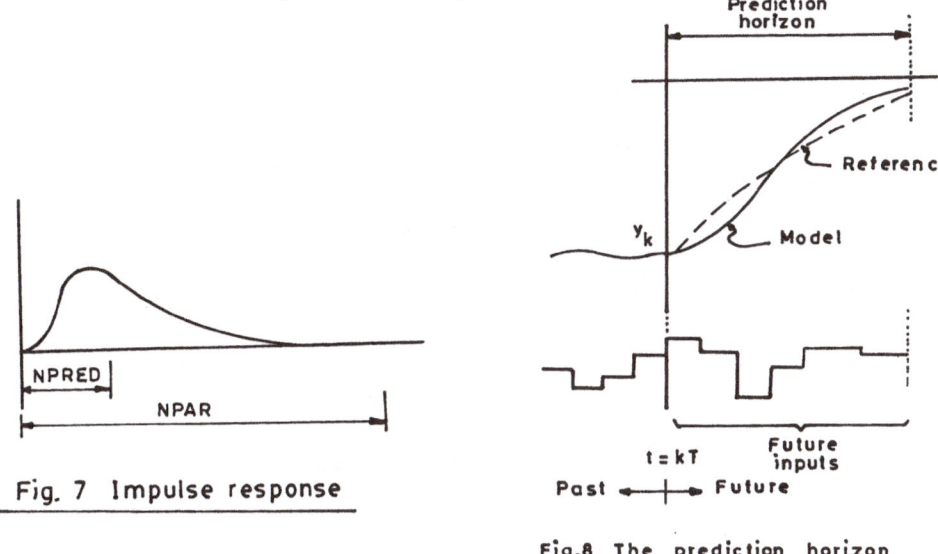

Fig. 7 Impulse response

Fig.8 The prediction horizon

The model trajectory $\underline{YG} = [\, yG(1), yG(2), yG(3), yG(4)\,]^T$ shown in Fig.9 is based on an assumed input sequence of future inputs $\underline{UG}$, here an extension of the last applied system input u(k). $\underline{YG}$ is given by

$$
\begin{bmatrix} yG(1) \\ yG(2) \\ yG(3) \\ yG(4) \end{bmatrix} =
\begin{bmatrix}
UG & u(k) & u(k-1) & u(k-2)\text{- - - -}u(k-NPAR+2) \\
UG & UG & u(k) & u(k-1)\text{- - - -}u(k-NPAR+3) \\
UG & UG & UG & u(k)\text{- - - -}u(k-NPAR+4) \\
UG & UG & UG & UG\text{- - - -}u(k-NPAR+5)
\end{bmatrix}
\begin{bmatrix} h_1 \\ h_2 \\ h_3 \\ | \\ | \\ h_{NPAR} \end{bmatrix}
$$

(60)

The algorithm calculates the input value to be applied at the process at the next sampling period. In this way a delay of one period is introduced which is taken into account by computing yM (k+1). $yM(k+1) = \underline{h}^T\underline{u}_k$ is the model output at t = k+1 and is used as starting value for reference trajectory. The reference trajectory $\underline{yR}$ is derived from a first order model

$$yR(1) = \alpha\, yM(k+1) + (1-\alpha)\, R$$
$$yR(n+1) = \alpha\, yR(n) + (1-\alpha)\, R$$

To minimise the difference between model trajectory and reference trajectory, the input sequence is to be adjusted. This will be done by adding an extra vector $\underline{\delta}$ to the assumpttion $\underline{UG}$. When u = $\underline{UG} + \underline{\delta}$ is used as a future input sequence the error vector $\underline{e}$ may be written as

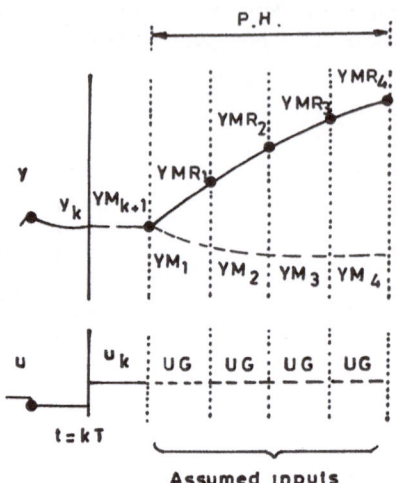

P.H.

YMR$_4$

YMR$_3$

YMR$_2$

YMR$_1$

y

y$_k$ YM$_{k+1}$

YM$_1$ YM$_2$ YM$_3$ YM$_4$

u   u$_k$ UG UG UG UG

t = kT

Assumed inputs

Trajectories for the prediction horizon

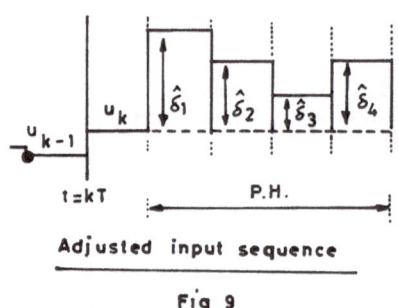

u$_k$ $\hat{\delta}_1$ $\hat{\delta}_2$ $\hat{\delta}_3$ $\hat{\delta}_4$

u$_{k-1}$

t = kT   P.H.

Adjusted input sequence

Fig. 9

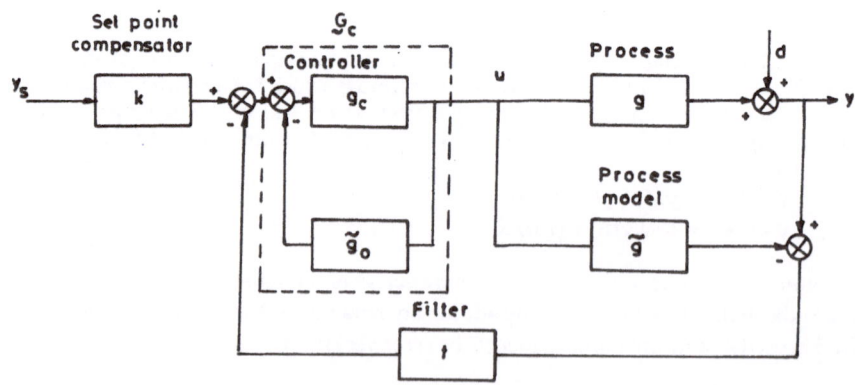

Set point
compensator $\underline{g}_c$

Controller   Process   d

y$_s$ k   g$_c$ u   g   y

$\tilde{g}_0$

Process
model

$\tilde{g}$

Filter

f

Fig. 10 Internal model control

$$\underline{e} = \underline{y}R - \underline{y}G \quad - \quad \begin{bmatrix} \delta_1 & & & \\ \delta_2 & \delta_1 & 0 & \\ \delta_3 & \delta_2 & \delta_1 & \\ \delta_4 & \delta_3 & \delta_2 & \delta_1 \end{bmatrix} \begin{bmatrix} h_1 \\ h_2 \\ h_3 \\ h_4 \end{bmatrix} \tag{61}$$

We can rewrite (61) as

$$\underline{e} = \underline{y}R - \underline{y}G \quad - \quad \begin{bmatrix} h_1 & & & \\ h_2 & h_1 & o & \\ h_3 & h_2 & h_1 & \\ h_4 & h_3 & h_2 & h_1 \end{bmatrix} \begin{bmatrix} \delta_1 \\ \delta_2 \\ \delta_3 \\ \delta_4 \end{bmatrix} \tag{62}$$

$$= \underline{y}R - \underline{y}G - H\underline{\delta}$$

The optimal $\hat{\underline{\delta}}$ to minimise $\underline{e}$ can be found from minimisation of quadratic cost function V

$$V = \underline{e}^T . \underline{e} = \sum_{i=1}^{NPRED} e_i^2 \tag{63}$$

The optimal $\hat{\underline{\delta}}$ is the well known least square solution

$$\hat{\underline{\delta}} = [H^T H]^{-1} H^T (\underline{y}R - \underline{y}G) \tag{64}$$

When the cost function is extended with $\beta^2 \underline{\delta}^T \underline{\delta}$ to smooth $\hat{\underline{\delta}}$ solution, $\hat{\underline{\delta}}$ is given by

$$\hat{\underline{\delta}} = [H^T H + \beta^2 I]^{-1} H^T (\underline{y}R - \underline{y}G) \tag{65}$$

The optimal input sequence is given by

$$\hat{\underline{u}} = \underline{U}G + \hat{\underline{\delta}} \tag{66}$$

and is shown is fig. 9.

b) Generalised Predictive Control
The model adopted for the plant is called CARIMA (controlled auto-regressive and integrated moving average) and has the form [23].

$$A (q^{-1}) y (t) = B(q^{-1}) u(t-1) + c(q^{-1}) \in (t) / \Delta \tag{67}$$

Here Δ is the operator $(1-q^{-1})$ so that $\Delta x(t)$ equals $x(t) - x(t-1)$. In this model the dead time k is absorbed in B so that its leading (k-1) elements are zero.

A long range predictive design projects outputs given currently available data [$y(j)$, $u(j-1)$, $j \le t$] upto a prediction horizon, based on a set of assumptions about present and future controls [$u(j)$, $j \ge t$] (or specifically about increments in control $\Delta u(j)$, $j \ge t$). The predicted output can be decomposed into two components:

1. $y_1 (t + j/t)$ being the free response of the plant assuming future controls equal u (t-1) or $\Delta u (t + i) = o$

2. $y_2 (t + j)$, being the part of response due to controls
   $[\Delta u(t + i), i \geq o]$

The predicted output vector can be written as

$$y = G\tilde{u} + p + \epsilon \qquad (68)$$

Where y is the vector of future outputs $y(t + j)$

$\tilde{u}$ is the vector of future controls
p is the vector of predictions with elements
$y_1 (t + j/t)$ and
ϵ is the vector of errors due to future noise terms

The matix G is of dimension N x N, where N is the prediction horizon

A set of future system errors $e (t + j) = w(t + j) - y(t + j)$ may be defined as a vector

$$e = [w (t+1) - y(t + 1), - - - - - - - - - - - -, w(t+N) - y(t+N)]^T \qquad (69)$$

where $\{w (t + j)\}$ is the future set point sequence
(can be constant in process control, i.e., $w(t + j)=w$)
The crucial step in GPC is to define a cost function and to find a minimising solution

$$J (N_1, N_2, NU, \lambda) = \sum_{j=N1}^{N2} e^2 (t + j) + \lambda \sum_{j=1}^{NU} \Delta u^2 (t + j-1) \qquad (70)$$

Where N_1 is the minimum costing horizon
         N_2 is the maximum costing horizon (=N)
         NU is the control horizon
subject to the constraint that $\Delta u (t + j) = o$ for $j \geq NU$

For a deterministic plant with $\epsilon (t + j) = o$, minimising solution gives
$$\tilde{u} = (G^T G + \lambda I)^{-1} G^T (w-p) \qquad (71)$$

c) Dynamic Matrix Control (DMC)
In the late 1970's workers at Shell (Cutler and Ramaker [24]) developed a predictive control strategy named 'Dynamic Matrix Control'. This technique has been successfully applied in Shell organisation. The DMC algorithm models the process output $y(t)$ at discrete time instants by means of a discrete step response $a(j)$. The prediction horizon L for the DMC algorithm equals the number N of step response co-efficients (finite number of terms) taken into account. The DMC technique further allows for l_u consecutive changes in the input variable $(l_u \leq N)$, l_u being called the control horizon. Changes in the model output over the prediction horizon due to consecutive changes in the input variable over the control horizon can be expressed in vector notation form as

$$\Delta y = A. \Delta u \qquad (72)$$

Where $\Delta y^T = [\Delta y(t+1) \ \Delta y(t+2) \ - - - - \qquad \Delta y(t+N)]$
         $\Delta u^T = [\Delta u(t) \ \Delta u(t+1) \qquad - - - - \qquad \Delta u(t+l_u-1)]$

and

$$A = \begin{bmatrix} a(1) & & & O \\ a(2) & a(1) & & \\ \text{-----------------} \\ a(l_u) & a(l_u-1)\text{-------}a(1) \\ a(N) & a(N\text{-}1)\text{-------}a(N\text{-}l_u\text{+}1) \end{bmatrix} \qquad (73)$$

$a(j)$: The co-efficients of the step response.

As each sampling instant the predicted values of the model output over the prediction horizon are computed taking into account all the previously calculated and applied input variable moves.

These predictions are corrected using the measured actual output $y(t)$.

$$y_{im}^* (t+k) = y_{im} (t+k) + (y(t) - y_{im} (t)) \qquad (74)$$
$$k = 1, 2 --- N$$

Difference between the setpoint w and the sequence of corrected predictions of process output over the prediction horizon results in an error vector e

$$e = \begin{bmatrix} w \\ w \\ \cdot \\ \cdot \\ \cdot \\ w \end{bmatrix} - \begin{bmatrix} y_{im}^* (t+1) \\ y_{im}^* (t+2) \\ \cdot \\ \cdot \\ \cdot \\ y_{im}^* (t+N) \end{bmatrix} = w - y_{im}^* \qquad (75)$$

Allowing for new consecutive input variable moves over the control horizon, the error vector can be minimised over the prediction horizon. A quadratic cost criterion is used

$$v = e^T . e \qquad (76)$$
and $e = w - y_{im}^* - A \Delta u$

Solution for Δu is obtained as

$$\Delta u = (A^T . A)^{-1} A^T (w - y_{im}^*) \qquad (77)$$

Using a penalization factor β to multiply the diagonal elements of $(A^T . A)$ smooth solution for Δu is obtained. This factor is used as a tuning parameter for the algorithm.

d) Extended Horizon Adaptive Control (EHAC)
Ydstie has proposed this method [25].
Process can be described by a mathematical input/output model

$$A (z^{-1}) y(t) = B(z^{-1}) u(t-d) \qquad (78)$$

The basic idea of EHAC is to compute at each sampling instant a sequence of inputs [$u(t), u(t+1)$ ------ $u(t+l-d)$] so as to be satisfy the identity

$$E \{ y(t + l) - \omega(t + l) \} = 0 \text{ with } l \ge d \qquad (79)$$

Only the first element of the computed sequence is applied to the process and the procedure is repeated at each sampling instant. Obviously solution of the identity (79) is not unique (unless l=d).

Possible approaches are, for instance, to assume that the control will be constant over the interval [l, t+l-d]

$$u(t) = u(t+1) - - - - - - = u(t+l - d)$$

or to choose a strategy where the control effort is minimised ie to compute $u(t) - - - - u(t+l - d)$ so as to minimise

$$J_1 = \sum_{i=0}^{l-d} u^2(t+i) \tag{80}$$

under the constraint (79).

An incremental equivalent of this approach can deal with load disturbances.

e) Internal Model Control (IMC)
So far we have discussed a number of predictive control strategies. Other forms of predictive controllers are dead beat controller and self tuning regulators. Brasilow [26] has given an interpretation to Smith Predictor as an inferential controller. Motivated by the work Morari [27] has unified all these into a frame work named 'Internal Model Control'. The general structure of an Internal Model Control of a single input single output system is shown in fig. 10.

Here g_c is some controller, $\tilde{g}$ is the process model, $\tilde{g}_0$ some function of the process model, f a filter for robustness and k is a setpoint compensator. It has been shown by Garcia and Morari that following types of control easily fit into the framework of IMC.

1. PID Control : $f = k = 1$, $\tilde{g}_0 = \tilde{g}$, $g_c = PID$
2. Time Delay Compensator (Smith Predictor): $f = k = 1$
    $$\tilde{g}_0 = e^{-Ts} \tilde{g}, g_c = PID$$
3. Deterministic Linear Quadratic Optimal Feedback Control : f = specified filter, K = setpoint compensator, $\tilde{g}_0 = \tilde{g}$, $g_c = p$.
4. Model Algorithmic Control: f=k= reference trajectory
    $$G_c = \tilde{g}^{-1} z^{-1} \text{ (one step ahead predictive controller).}$$
5. Dynamic Matrix Control: $f = k = 1$, $G_c \cong \tilde{g}^{-1} z^{-1}$ (approximate realization).

Internal model control techniques have been applied on a lot of processes in the past 20 years, e.g., distillation column (level 1 and 2), catalytic reactors, Hydrocrackers-Hydrotreaters, Boilers, Heat furnaces and heat exchangers.

3.1.12 _Adaptive Control._ Adaptive Control can be described as the changing of controller parameters based on the changes in system operating conditions. Knowledge of the input and output variations of the system being controlled allows the parameters of a dynamic model of the plant to be estimated and from these the controller parameter values required to give a specified closed-loop performance can be calculated. Three types of adaptive control schemes are popular, namely, 'Gain Scheduling', 'Model Reference Adaptive Control (MRAC)' and Self Tuning Control (STC)'. The predictive control algorithms discussed in section 3.1.11 are used together with on line identification procedure so that they are also adaptive. A Gain Schduling scheme is shown in fig. 11.

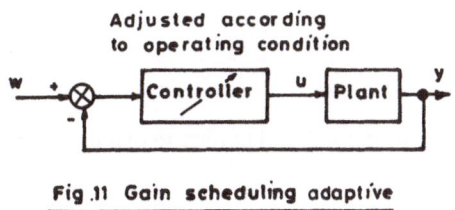

Fig.11 Gain scheduling adaptive
controller

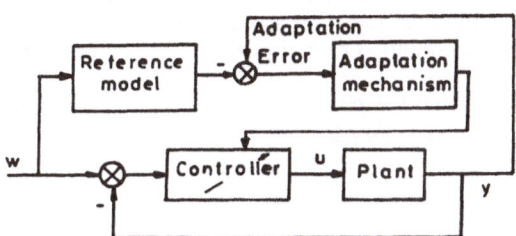

Fig.12 Model reference adaptive control system

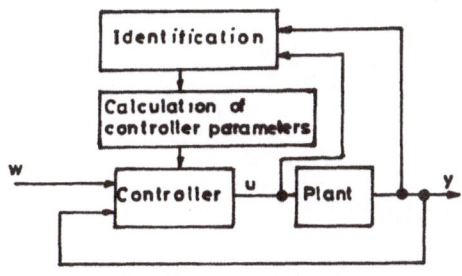

Fig.13 Self tuning controller

If the way in which the dynamic characterstics change with operating condition are known apriori then this scheme can be adopted. Gain Scheduling Controllers are often used to compensate for nonlinearity in control valve transfer function [7]. In 'Model Reference Adaptive Control' a model representing the desired behaviour of the closed-loop system is chosen to specify a desired performance of the system and the actual performance is measured against this. The regulator parameters are adjusted depending upon the error between the system output and the reference model output. The scheme is shown in fig. 12. Let the plant $x_p(t)$ and the reference model $x_m(t)$ behaviours be given by the nth order state equations.

$$\dot{x}_p\ (t) = A_p\ (t)\ x_p\ (t) + B_p\ (t)\ u\ (t) \tag{81a}$$

$$\dot{x}_m(t) = A_m(t)\ x_m\ (t) + B_m\ (t)\ u\ (t) \tag{81b}$$

x, u, A, and B have the usual meaning. Subscripts p and m refer to plant and model respectively.

An error function can be formed by writing

$$e(t) = x_m\ (t) - x_p(t) \tag{81c}$$

The aim of the control is to force this error function to zero [28].

Landau and Lozano [65] has reported an evaluation of various designs for discrete time explicit Model Reference Adaptive Control (MRAC) based on a unified stability point of view both in tracking and regulation. We discuss a simplified procedure for a SISO plant described by:

$$A\ (q^{-1})\ y(k)\ =\ q^{-d}\ B(q^{-1})\ u\ (k) + e(k) \tag{82}$$
$$d > 0,\ y(0) \neq 0$$

where
$$A(q^{-1}) = 1 + a_1\ q^{-1} + \text{-------} + a_{nA} + q^{-nA}$$
$$B(q^{-1}) = b_0 + b_1\ q^{-1} + \text{---------} + b_{nB}\ q^{-nB},\ b_o \neq 0$$

$\{q^{-1}\}$ is the backward shift operator, $\{d\}$ represents the plant time delay, $\{u(k)\}$ and $\{y(k)\}$ are the plant input and output and e(k) is a bounded disturbance.

The control should be such that the regulation $[u^M\ (k) = 0]$ an initial disturbanmce $[y(0) \neq 0]$ is eliminated with the dynamics defined by

$$C_2\ (q^{-1})\ y\ (k+d)\ =\ 0 \qquad\qquad k \geq 0 \tag{83a}$$

where
$$C_2\ (q^{-1}) = 1 + C_1^2 q^{-1} + \text{-------} + C^2_{n_{C2}}\ q^{-nC2}$$
is an asymtotically stable polynomial.

The control law is given by

$$p^T\ (k)\ \phi\ (k)\ =\ C_2\ (q^{-1})\ y^M\ (k+d) \tag{83b}$$

where
$$\phi\ (k)\ =\ [\ u(k), u(k\text{-}1) \text{- - - -} u(k\text{-}d\text{-}n_B + 1), y(k), \text{- - - - - -}\ y(k\text{-}n_R)]$$

$y^M\ (k)$ is the output of the reference model and $n_R\ =\ \max\ (n_A\text{-}1, p\text{-}k)$
The controller parameter is adjusted by the algorithm

$$P\ (k) = P(k\text{-}1) + F_k\ \phi\ (k\text{-}d)\ \epsilon\ (k) \tag{83c}$$

$$F_{k+1} = \frac{1}{\lambda_1(k)}\ [F_k - \frac{F_k\ \phi(k\text{-}d)\ \phi^T(k\text{-}d)\ F_k}{\lambda_1(k)/\lambda_2(k) + \phi^T(k\text{-}d)\ F_k\ \phi(k\text{-}d)}] \tag{83d}$$

with

$$\epsilon\ (k) = \frac{C_2\ (q^{-1})\ y(k) - P^T\ (k\text{-}1)\ \phi(k\text{-}d)}{1 + \phi^T\ (k\text{-}d)\ F_k\ \phi(k\text{-}d)} \tag{83e}$$

when
$$O < \lambda 1\,(k) \le 1, \quad O < \lambda 2\,(k) < 2, \quad Fo > O$$

The algorithm was applied to a catalytic fluidized bed reactor for ammoxidation of propylene to produce acrylonitrile using an Apple II microcomputer by Koutchoukali and co workers [64]. The host program used in the control was written in PASCAL. It calls external assembly language routines related to real time control, data acquisition routine, DA converter routine and clock reading routine.

In recent years the most popular adaptive control strategy has been the self tuning approach [1,29,30]. Parameters in a dynamic model of the process are updated on line from input-output data. Controller settings are adjusted based on the new parameter estimate. A block diagram of self tuning control is shown in fig. 13.

We consider a single input single output plant and its ARMAX model (auto regressive moving average model with exogenous input).

$$A\,(q^{-1})\,y(t)= B(q^{-1})\,u(t-k) + C(q^{-1})\,S(t) + d(t) \tag{84}$$

Here y is the output, u is the input, S is a stochastic noise variable (random variable with normal distribution and zero mean), d is the load disturbance (unmeasurable), k is the known time delay expressed as integer multiple of sampling time. A, B and C polynomials are defined by

$$A\,(q^{-1}) = 1 + \sum_{i=1}^{n} a_i q^{-i}$$

$$B\,(q^{-1}) = \sum_{i=0}^{m} b_i q^{-i}$$

$$C\,(q^{-1}) = \sum_{i=0}^{n} c_i q^{-i}$$

Assuming the noise model parameters in (84) as zero and the disturbance d as constant, the model (84) can be rewritten as

$$y(t)= \varphi^{T}\,(t\text{-}1)\,\Theta + \epsilon\,(t) \tag{85}$$

where φ and Θ are defined as

$$\varphi^{T}\,(t-1) = [\ y(t-1)\ y(t-2) - - - - - - u(t-n)\ u(t-k-1) - - - - - -u(t-k-m-1), 1\ ]$$
$$\Theta^{T} = [\ a1\ a_2 - - - - -a_n,\ b_1\ b_2 - - - - - -b_m,\ d\ ] \tag{86}$$

The recursive least square algorithm can be written as

$$\hat{\Theta}\,(t) = \hat{\Theta}\,(t-1) + P(t)\,\varphi(t-1)\,[\ y(t) - \varphi^{T}(t-1)\,\hat{\Theta}\,(t-1)\ ] \tag{87}$$

$$P(t) = \frac{1}{\lambda}\{\,P(t-1) - P(t-1)\,\varphi(t-1)\,[\ \varphi^{T}(t-1)\,P(t-1)\,\varphi(t-1) + \lambda\,]^{-1}\,\varphi^{T}(t-1)\,P(t-1)\ \} \tag{88}$$

Where P(t) is the covariance matrix of the estimation error $y(t)-\hat{y}(t)$, $\hat{y}(t)$ being given by

$$\hat{y}(t) = \varphi^T(t-1) \, \hat{\Theta} \, (t-1) \text{ (setting } \epsilon = o) \tag{89}$$

In eqn. (88), λ is the exponential forgetting factor $o < \lambda \leq 1$. When $\lambda < 1$, recent data are weighed more heavily than past data during the recursive calculation.

Self tuning algorithm has two activities—parameter estimation and controller design. In the Clarke Gawthrop [1] procedure the controller is designed by minimising a quadratic cost function of the form

$$J = E \, \{[\, P(q^{-1}) \, y(t+k) - R(q^{-1}) \, y_r \, (t) \,]^2 + [\, Q(q^{-1}) \, u(t) \,]^2 \,\} \tag{90}$$

where y_r is the setpoint, E is the expectation operator, $P(q^{-1})$, $Q(q^{-1})$ are polynomials with $P_o=1$ and $R(q^{-1})$ is a rational transfer function. P, Q and R are user specified design parameters.

The optimal control law which minimises cost function J can be written as

$$u(t) = \frac{-Fy(t) + CR \, y_r(t)}{EB + CQ} \tag{91}$$

where E and F are polynomials in q^{-1} which satisfy the identity

$$CP = EA + q^{-K} F \tag{92}$$

In principle the self tuning version (91) could be obtained by estimating the A, B and C polynomials on line and solving the identity in (92).

From the foregoing discussion it is clear that STC controller design procedure requires following design parameters.

Model Parameter
  Sampling period, t
  model order n and m
  time delay k

Estimation Parameter
  Initial parameter estimate $\hat{\Theta}$ (o)
  Initial covariance matrix, P(o)
  forgetting factor, λ
  diagnostic parameter, β

Control parameters
  output weighting, $P(q^{-1})$
  input weighting, $Q(q^{-1})$
  setpoint weighting, $R(q^{-1})$

The diagnostic parameter β is used to monitor the performance of the estimator.
A number of guidelines for the selection of STC design parameters are available in [1,29,30]

3.1.13 *Fuzzy Control Algorithm.* Several investigators reported that incorporating human intelligence into automatic control would be a more efficient solution and this has lead to the development of fuzzy control algorithms [3,31]. The fuzzy algorithm is based on intuition and experience and can be regarded as a set of heuristic decision rules. Typical structure of linguistic rules of fuzzy logic controller may be :

If < process --- state > then < control --- Action > where <---> represents some fuzzy proposition. This kind of rules is no more than a map with nonlinear gains from process states to control actions. A control block diagram of fuzzy controller is shown in figs. 14 and 15.

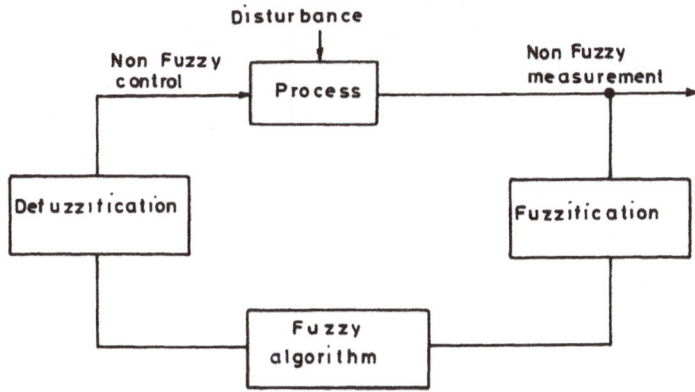

Fig.14 Concept of the Fuzzy control

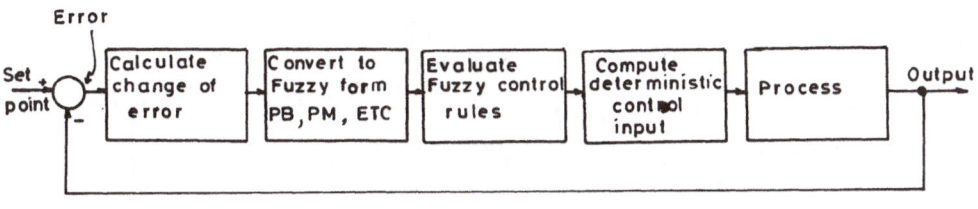

Fig. 15 General representation of Fuzzy control system

In order to quantize the qualitative statements the following linguistic sets are assigned
Large Positive (LP)
Medium Positive (MP)
Small Positive (ZP)
Zero (ZE)
Small Negative (SN)
Medium Negative (MN)
Large Negative (LN)
The statement of control rule can be:

If the error (set point minus process output) is large positive and error change (error from process output minus error from the last process output) is small positive, then the input to the system is large positive.

A SISO fuzzy control loop is shown in figs. 16 and 17.

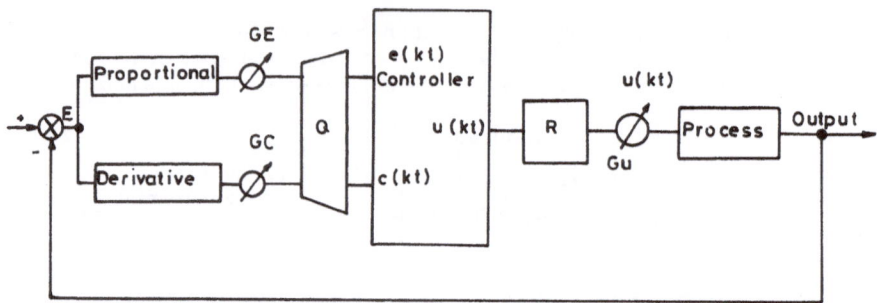

Fig.16 Basic SISO Fuzzy control loop

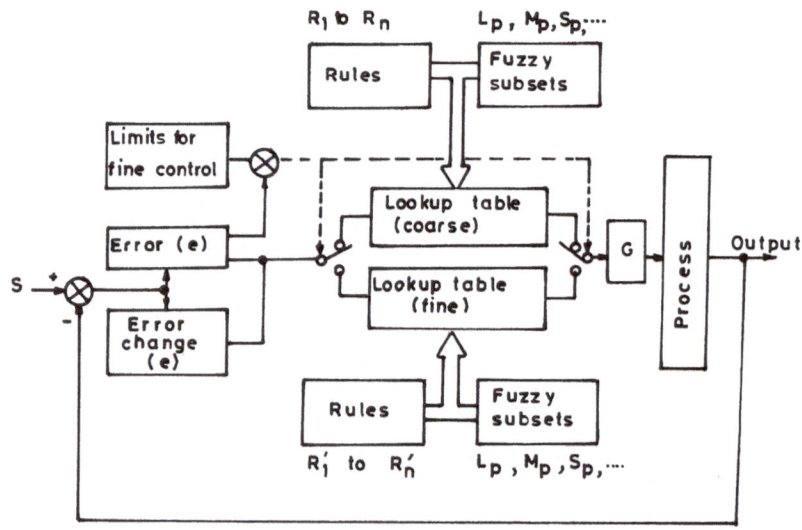

Fig. 17 Block diagram of Fuzzy controller

The actual values of error and the change of error are scaled by the scaling factor (gains) GE and GC and the resulting fuzzy variables e(KT) and C(KT) after quantization procedure are given by [32].

$$e(kT) = Q\{ (s - x(kT)) \times GE\} \tag{93}$$

$$C(kT) = Q\{ (x(kT) - x (k-1)T) \times GC \} \tag{94}$$

These fuzzy variables result in a fuzzy subset U(kT) which by the decision rule R { } employed gives the control element u(kT) required i.e.,

$$u(kT) = R \{ U(kT) \} \tag{95}$$

The value is again scaled by GU and provides the actual change Δi (kT) in the process input i(kT)

$$i((k+1)T) — i(kT) = GU \times u(kT) \tag{96}$$

Various forms of relational matrices (or decision tables) R can be found in literature.

Expert fuzzy controllers are also in use. These controllers have more knowledge of process control. They can have higher level rules to decide which low-level rules to apply, e.g.,

If < Process	----	State 1 >	then < use ---	Rule set 1>
If < Process	----	State 2 >	then < use ---	Rule set 2>
If < Process	----	State N >	then < use ---	Rule set N>

The low level rules may be fuzzy linguistic rules for design of fuzzy logic controller or non fuzzy control policies.

Fuzzy control algorithms have been used in automatic kiln control in cement industry [71]. Ihara discussed one application in Train Control in his chapter.

3.2 Controllers for Multiple Input and Multiple Output Systems

Having noted various control strategies for single input single output systems we will cover in this section popularly used control strategies for multiple input multiple output systems. We will discuss some new topics typical of multi input systms e.g. decoupling and will also include multi input version of various control strategies discussed in the previous section (i.e. optimal, predictive and adaptive control).

3.2.1 *Decouplers*. An excellent account for decoupling theory of multi variable process control systems is available in reference [33]. Decoupling is concerned with making the transfer function matrix of a multi variable process diagonal. We discuss the decoupler design considering a two input two output plant shown in fig. 18. The transfer function matrix of the plant is given by

$$\begin{bmatrix} y_1(s) \\ y_2(s) \end{bmatrix} = \begin{bmatrix} G_{11}'(s) & G_{12}'(s) \\ G_{21}'(s) & G_{22}'(s) \end{bmatrix} \begin{bmatrix} u_1(s) \\ u_2(s) \end{bmatrix} \tag{97}$$

Following matrix relationship holds

$$\begin{bmatrix} u_1(s) \\ u_2(s) \end{bmatrix} = \begin{bmatrix} G_{c1} & D_1 G_{c2} \\ D_2 G_{c1} & G_{c2} \end{bmatrix} \begin{bmatrix} e_1(s) \\ e_2(s) \end{bmatrix} \tag{98}$$

Using equations (97) and 98) the following relation can be obtained

$$\begin{bmatrix} y_1 \\ y_2 \end{bmatrix} = \begin{bmatrix} G_{11}' & G_{12}' \\ G_{21}' & G_{22}' \end{bmatrix} \begin{bmatrix} G_{c1} & D_1 G_{c2} \\ D_2 G_{c1} & G_{c2} \end{bmatrix} \begin{bmatrix} e_1 \\ e_2 \end{bmatrix} \tag{99}$$

$$= \begin{bmatrix} A_{11} & A_{12} \\ A_{21} & A_{22} \end{bmatrix} \begin{bmatrix} e_1 \\ e_2 \end{bmatrix} \tag{99a}$$

Let A_{12} and A_{21} be zero i.e.,

$$G_{11}' D_1 G_{c2} + G_{12}' G_{c2} = 0 \tag{100}$$

and

$$G_{21}' G_{c1} + G_{22}' G_{c1} D_2 = 0 \tag{101}$$

It follows that

$$D1 = -\frac{G_{12}'}{G_{11}'} \tag{102}$$

$$D2 = -\frac{G_{21}'}{G_{22}'} \tag{103}$$

Equations (102) and (103) are basic formulae for decouplers.

With these decouplers the open-loop transfer function matrix becomes

$$H'(s) = \frac{G_{11}' G_{22}' - G_{12}' G_{21}'}{G_{11}' G_{22}'} \begin{bmatrix} G_{11}' G_{c1} & O \\ O & G_{22}' G_{c2} \end{bmatrix} \tag{104}$$

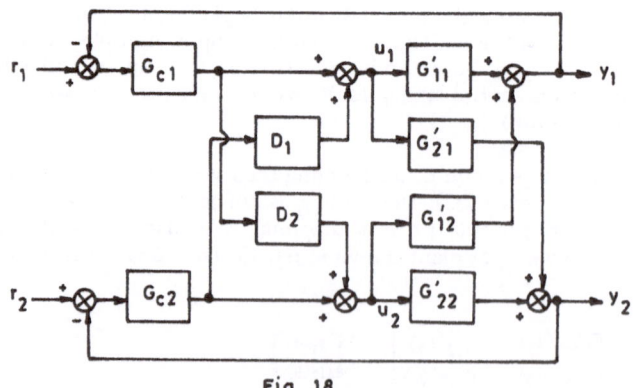

Fig. 18

It is obvious that the system is now decoupled at a particular operating point and the controllers G_{c1} and G_{c2} will not influence the decouplers D_1 and D_2. The decouplers make it possible to adjust G_{c1} and G_{c2} independently to regulate the two loops.

Some of the multi variable frequency domain methods developed in Great Britain show some similarities (such as diagonal dominance) with the decoupling approach. Amongst the various methods Inverse and Direct Nyquist Array, Characteristic Loci and Sequential Return Difference [34] are popular.

Traditional industrial control strategy for multi variable control problems is to use a multiloop control scheme consisting of several PI or PID controllers connected in SISO loop. Pairing of variables is done based on relative gain array [35].

3.2.2 _State Feedback Controllers_. We will discuss in this section multi input version of state feedback controllers which have been used for multi variable plants. Optimal control design methods result in a state feedback controller. We will cover three formulations as used by industry. We will also cover multivariable control based on internal model principle.

a) Stochastic Optimal Control:
System model is given by

$$x(k+1) = A\,x(k) + B\,u(k) + v(k) \tag{105}$$

$$y(k) = C\,x(k) + w(k) \tag{106}$$

Where $v(k)$ and $w(k)$ are both white gaussian with zero mean and of known covariance.

$$E\,[\,v(k)\,v^T(j)\,] = Q_e\,\delta_{kj} \tag{107}$$

$$E\,[\,w(k)\,w^T(j)\,] = R_e\,\delta_{kj} \tag{108}$$

We choose $Q_e = I$, $R_e = I$, δ_{kj} is Dirac-function. We consider the quadratic performance index of the form

$$J1 = \sum_{k=0}^{n-1} x^T(k)\,Q_k\,x(k) + u^T(k)\,R_k\,u(k) \tag{109}$$

According to separation theorem, the design steps are given below:

Step 1.   Optimal Control Problem:

$$u(k) = -\,K(k)\,\hat{x}(k) \tag{110}$$

$$K(k) = (R_k + B^T\,S(k+1)\,B)^{-1}\,B^T\,S(k+1)A \tag{111}$$

$$S(k) = A^T\,S(k+1)A - A^T\,S(k+1)\,B\,(B^T\,S(k+1)\,B + R_k)^{-1}\,B^T\,S(k+1)A + Q_k \tag{112}$$

Where $K(k)$ is the feedback gain matrix. $S(k)$ is the Riccati gain matrix. When $k=n$, $S(n)=0$.

Step 2.   Optimal Filtering Problem:

$$\hat{x}(k) = \hat{x}(k\,|\,k-1) + k_e(k)\,[\,y(k) - C\hat{x}(k\,|\,k-1))] \tag{113}$$

$$\hat{x}\,(k\,|\,k{-}1) = A\hat{x}\,(k{-}1) + B\,u(k{-}1) \tag{114}$$

$$k_e\,(k) = P(k\,|\,k-1)\,C^T\,(C\,P(k\,|\,k-1)\,C^T + R_e)^{-1} \tag{115}$$

$$P\,(k{+}1\,|\,k) = A\,P(k\,|\,k{-}1)\,A^T - A\,P(k\,|\,k{-}1)\,C^T$$
$$(C\,P(k\,|\,k{-}1)\,C^T + R_e)\,CP(k\,|\,k{-}1)\,A^T + Q_e \tag{116}$$

$$P\,(1\,|\,o) = A\,P_0\,A^T + Q_e \tag{117}$$

When $K_e\,(k)$ is the filter gain matrix, $P(k)$ is the Riccati gain matrix. When $k=o$, $P(o)=$ var $x(o)$.

The above controller is being implemented for 200MW units of North East Electricity Generating Board, People's Republic of China [36].

b) Optimal controller with plant model and state estimates:
CEGB, UK has developed an optimal controller [2] with plant model and state estimates. Very few parameters in boiler model are contant which can be calculated apriori. Even if the parameters can be measured, it is often better to rely upon good estimates rather than a multiplicity of measurements. They have used bias estimates to update the model parameters. The state and output equations are as defined in Eqn. (105) and (106). Let the estimated values of A, B and C matrices be $\hat{A}(k)$, $\hat{B}(k)$ and $\hat{C}(k)$. The parameter errors can be compensated by addition of bias states b(k). The augmented state equations become

$$x(k{+}1) = \hat{A}(k)\,x(k) + \hat{B}(k)\,u(k) + v(k) + M(k)\,b(k) \tag{118}$$

$$b(k{+}1) = b(k) + n(k) \tag{119}$$

$$y(k) = \hat{C}\,(k)\,x(k) + N(k)\,b(k) + w(k) \tag{120}$$

Where n(k) is an independent white noise vector. The bias states can then be related to the parameter errors by the equations

$$-\,M(k)\,b(k) = \tilde{A}(k)\,x\,(k) + \tilde{B}(k)\,u(k) \tag{121}$$

$$\tilde{A}(k) = \hat{A}(k) - A(k) \tag{122}$$

$$\tilde{B}(k) = \hat{B}(k) - B(k) \tag{123}$$

Both the state x(k) and bias state b(k) can be estimated by applying the normal Kalman filter. Considering an approximate state equation

$$\tilde{x}(k{+}1) = \hat{A}(k)\,\tilde{x}(k) + \hat{B}(k)\,u(k) + v(k) \tag{124}$$

and the error term

$$e(k) = \tilde{x}(k) - x(k) \tag{125}$$

$$e(k{+}1) = \hat{A}(k)\,e(k) + \tilde{A}(k)\,x(k) + \tilde{B}(k)\,u(k) \tag{126}$$

If a new error function is defined as

$$\tilde{e}(k) = e(k+1) - \hat{A}(k)\ e(k) \tag{127}$$

and observing that

$$-\tilde{e}(k) = M(k)\ b(k) = \tilde{A}(k)\ x(k) + \tilde{B}(k)\ u(k) \tag{128}$$

parameter estimates can then be updated by the following algorithm

$$a_{ij}(k+1) = a_{ij}(k) + \frac{\alpha}{\hat{x}^T(k)\ \hat{x}(k)}\ [M(k)\ \hat{b}(k)]^T\ \hat{x}_{ij}(k) \tag{129}$$

$$b_{ij}(k+1) = b_{ij}(k) + \frac{\alpha}{\hat{x}^T(k)\ \hat{x}(k)}\ [M(k)\ \hat{b}(k)]^T\ u_{ij}(k) \tag{130}$$

a_{ij} and b_{ij}'s are elements of matrices A and B

This parameter update algorithm can be used in conjunction with eqn. (121) to produce a time update for the bias estimates

$$\hat{b} = \hat{b}(k) - (M^T M)^{-1}\ M^T\ [\Delta \hat{A}(k)\ x(k) + \Delta \hat{B}(k)\ u(k)] \tag{131}$$

Thus an extended Kalman filtering algorithm is developed to carry out the following tasks.
  i)   Time update of state and bias estimates;
  ii) Measurement update of state and bias estimates;
  iii) Recalculation of explicitly dependent model parameters;
  iv) Estimation of nonmeasurable model parameters;
  v)   Model update.

Optimal Controller:

If the two stage estimator is used, then only the states need be considered in the controller design. Following performance index was considered

$$J = \sum_{k=0}^{\alpha} [x^T(k)\ Q\ x(k) + u^T(k)\ R\ u(k) + \Delta u^T\ S\ \Delta u(k)] \tag{132}$$

Where $\Delta u(k) = u(k) - u(k-1)$.

By defining an augmented state variable

$$z(k) = \begin{bmatrix} x(k) \\ u(k-1) \end{bmatrix} \tag{133}$$

the performance index (132) can be transformed to

$$J = \sum_{k=0}^{\alpha} [z^T(k)\ Q_A\ z(k) + u^T(k)\ R_A\ u(k) + 2\ u^T(k)\ M_A\ z(k)] \tag{134}$$

where the augmented state equation becomes

$$z(k+1) = \begin{bmatrix} A(k) & 0 \\ 0 & 0 \end{bmatrix} z(k) + \begin{bmatrix} B(k) \\ I \end{bmatrix} u(k) \tag{135}$$

and matrices Q_A, R_A and M_A are defined as

$$Q_A = \begin{bmatrix} Q & 0 \\ 0 & S \end{bmatrix} \qquad R_A = [R+S] \tag{136}$$

$$M_A = [0 \mid -S]$$

The minimisation of index (134) leads to a control law

$$u(k) = [\, K_1(k) \mid K_2(k)\,] \begin{bmatrix} x(k) \\ u(k-1) \end{bmatrix} \tag{137}$$

Where $K_1(k)$ and $K_2(k)$ are obtained from a partitioned version of the Riccati equation.

If we consider a performance index dropping the third term in the right hand side of performance index (132) the control law takes the form

$$u(k) = K(k) x(k) + L(k) y(k) \text{ set} \tag{138}$$

Where $K(k)$ is obtained from optimal control theory and $L(k)$ can be choosen to produce a correct steady state response.

$L(k)$ can be modified due to additional lag for the previous case as [70] as

$$L(k) = [C(k) (I-A(k) - B(k) (1-K_2(k))^{-1} K_1(k))^{-1} B(k)]^{-1} \tag{139}$$

This algorithm with adaptive Kalman filter for model parameter estimation belongs to the class of adaptive optimal controller of Hitachi Ltd. discussed in section. 3.2.4 (b).

c) State feedback industrial regulator design via optimal control theory:
We discuss in this section two algorithms for solving industrial regulators design. First one is proposed by Davison and Smith [18] and has been introduced in SISO section. The second one is a block noninteracting optimal controller design proposed by Hitachi Ltd Japan [37].

(i) Formulation of Davison and Smith:
The state and error equations of the system are defined by

$$\dot{x} = Ax + Bu + Ew \tag{140}$$

$$e = Cx + Du + Fw \tag{141}$$

Where x is n dimensional state vector, u is r dimensional input vector and e is m dimensional error vector, w is a q dimensional vector of unmeasurable constant disturbance, measurable disturbance and reference inputs to the plant.

$$\begin{bmatrix} E \\ F \end{bmatrix} \quad w = \begin{bmatrix} Eu & Em & o \\ Fu & Fm & -M \end{bmatrix} \begin{bmatrix} w_u \\ w_m \\ y_{ref} \end{bmatrix} \tag{142}$$

The system equations (140) and (141) can be rewritten in terms of an augmented state vector

$$z = \begin{bmatrix} \dot{x} \\ e \end{bmatrix}$$

$$\dot{z} = \begin{bmatrix} A & o \\ C & o \end{bmatrix} z + \begin{bmatrix} B \\ o \end{bmatrix} \vartheta \tag{143}$$

$$\vartheta = \dot{u} \tag{144}$$

We can design a state feedback control law of the form

$$v = - Kz = - [K_1 \, K_2] \, z \tag{145}$$

for satisfying any of the three criteria:

i) for placing closed-loop poles in desired locations [14, 17].
ii) for minimising a quadratic index of the form [17, 18]

$$J = \int_0^\alpha [z^T Q z + \dot{u}^T R \dot{u}] \, dt \tag{146}$$

iii) for placing closed-loop poles in desired locations and at the same time minimising an index of the form (146) [17, 41].

The control law (145) can be expressed as

$$u = K_1 x + K_2 \int_0^t e(t) \, dt + (K_1 - I) G^* \begin{bmatrix} E_m \\ F_m \end{bmatrix} w_m + (-K_1 - I)G^* \begin{bmatrix} o \\ M \end{bmatrix} y_{ref} \tag{147}$$

with $\quad G^* = G^T [G G^T]^{-1}$

$$G = \begin{bmatrix} A & B \\ C & D \end{bmatrix}$$

An implementation of modal control as in criterion (i) above has been reported for a reactor flasher systeem [38]. The control study retains a cascade control system structure with model controller as its subsystem. The system model however did not have the w terms.

An implementation of optimal control as in criterion (ii) above has been reported for a drum type boiler in an Italian power station by Mafezzoni and cori [39]. They have modified the basic optimal regulator to accommodate some practical considerations and have retained PID controllers in the local level and have designed two level optimal controller similar in concept to ADC of Nakamura and Akaike [44].

Mulholland [40] has reported implementation of multi variable optimal controller in Ammonia plant of AECI, South Africa. They have utilised critenion (ii) above and have implicitly utilised criterion (iii) for acceptable closed-loop response. The system model however did not include w term.

(ii) Blocked non-interacting optimal controller:

This technique has been employed by Hitachi Ltd., Japan [37]. The process model is represented by system equation (140) without the disturbance term w. B matrix in eqn. (140) is assumed to be diagonal. In order to reduce influence of the terms A_{ij} $(i \neq j)$ in the system equation (140) an additional feedback compensation D. X is added. The state equation of ith block can be written as

$$\dot{x}_i = A_{ii}\, x_i + A_{ij}\, x_i + B_{ii}\, u_i + B_{ii}\, D_{ij}\, x_i \tag{148}$$

If $A_{ij} + B_{ii}\, D_{ij} = o$, the influence of j the block on ith block is eliminated.

The system equation (148) is reduced to

$$\dot{x}_i = A_{ii}\, x_i + B_{ii}\, u_i \tag{149}$$

Using a standard quadratic index

$$J = \int_{o}^{\alpha} (\, x_i{}^T Q_1\, x_i + u_i{}^T R u_i\,)\, dt \tag{150}$$

a proportional state feedback law

$$u_i = -k_i\, x_i \tag{151}$$

is obtained. The expression for k_i is

$$k_i = R^{-1}\, B_{ii}\, P_i \tag{151a}$$

P_i being the solution of Riccati equation

$$A_{ii}{}^T P_i + P_i\, A_{ii} - P_i\, B_{ii}\, R^{-1}\, B_{ii}\, P_i + Q_1 = o \tag{152}$$

This has been applied to steel rolling mills with parallel processing. By this technique the thickness control accuracy has been improved from an actual of $\pm 0.8\%$ upto $\pm 0.4\%$.

d) Dynamic Multivariable Constraint Control:
This method has been developed by scientists at Laboratoire d' Automatique de Grenoble, France [42].

The main objectives of the dynamic control are

   i) Tracking of set points variation.
  ii) Compensation of input measured disturbances.
 iii) Compensation of non measured disturbances (regulation).

Moreover the control system has to satisfy following constraints e.g., inequality constraint on the inputs, on line changing of dynamic specification etc.

A multivariable control structure suggested by them is shown in Fig.19.

All the models considered are described by state equation (105) and (106) as given below:

$$x(k + 1) = Ax(k) + B u(k)$$
$$y(k) = Cx(k)$$

Suffixes c, p, r_a, rr are used to obtain the state equation of system model for control input, system model for disturbance input, the reference model for tracking and the reference model for regulation.

Let us choose a criterion

$$J = \sum_{k=0}^{\alpha} (y(k) - y_d(k))^T Q(y(k) - y_d(k)) + (u(k) - u_d(k))^T R(u(k) - u_d(k)) \qquad (153)$$

in which $Q > o$, $R > o$, and y_d is the wanted output for the model.

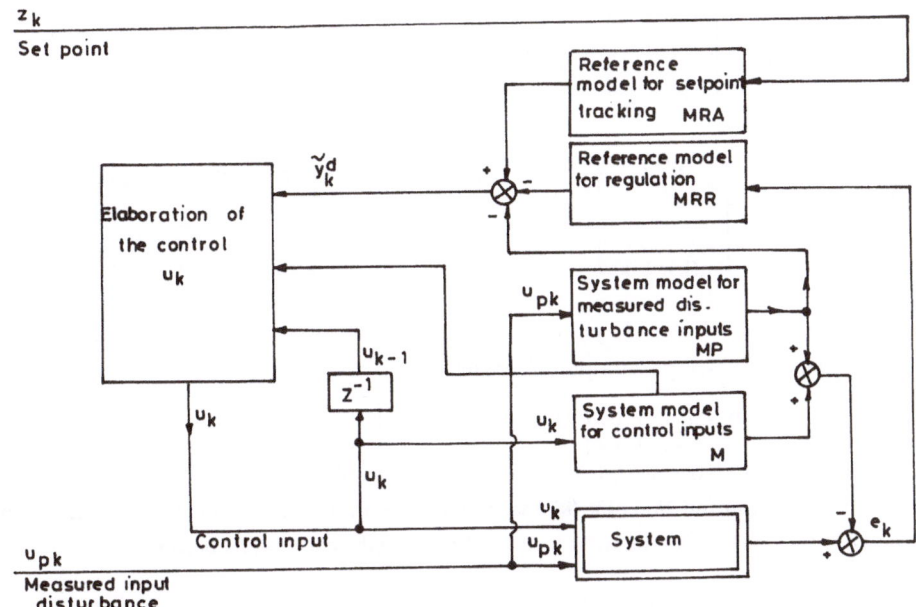

Fig.19 Multivariable dynamic control structure

If the reference z and the measured disturbance 'up' remain constant for the future, the solution of a Riccati equation leads to a control law of the form:

$$u(k) = K_x x(k) + K_{ra} x_{ra}(k) + K_z z(k) - K_{rr} x_{rr}(k)$$
$$+ K_e e(k) - K_{up} up(k) + K_{xp} x_p(k)$$
$$+ K_{ud} u_d(k) \qquad (154)$$

u_d is chosen to obtain a null static error.

On Line Reference Model Modification:

For being able to adjust on line reference staying in a linear context, these authors have proposed a finite moving horizon critenion.

$$J = \sum_{k}^{k+N-1} (y(i+1) - y_d(i+1))^T Q(y(i+1) - y_d(i+1))$$
$$+ (u(i) - u(i-1))^T R (u(i) - u(i-1))$$

(155)

The modification on the u term gives the possibility to avoid the explicit solution of the static error problem.

The above criterion can also be expressed as

$$J = (\tilde{y}(k) - \tilde{y}_d(k))^T \tilde{Q} (\tilde{y}(k) - \tilde{y}_d(k)) + (\tilde{u}(k) - \tilde{u}_d(k))^T \tilde{R} (\tilde{u}(k) - \tilde{u}_d(k))$$

(156)

with $\quad \tilde{y}(k) = \begin{bmatrix} y(k+1) \\ \vdots \\ y(k+N) \end{bmatrix}$, $\quad \tilde{y}_d(k) = \begin{bmatrix} y_d(k+1) \\ \vdots \\ y_d(k+N) \end{bmatrix}$ $\quad \tilde{Q} = \begin{bmatrix} Q & O \\ O & Q \end{bmatrix}$

$\tilde{u}_k = \begin{bmatrix} u(k) \\ \vdots \\ u(k+N-1) \end{bmatrix}$, $\quad \tilde{u}_d(k) = \begin{bmatrix} u(k-1) \\ 0 \\ \vdots \\ 0 \end{bmatrix}$, $\quad R = \begin{bmatrix} 2R & -R & - - - - - & O \\ -R & 2R & & \\ & & 2R & -R \\ O & & -R & 2R \end{bmatrix}$

For the solution at instant k, $\tilde{y}_d(k)$ is used and the state vectors x_p, x_{ra}, x_{rr} are not used. To compute $\tilde{y}_d(k)$ following assumpttions are made

$$up(i) = up(k), \; i = k+1 \text{ -----------------} \; k+N-1$$
$$z(i) = z(k), \; i=k+1\text{----------------------} \; k+N-1$$
$$e(i) = e(k), \; i=k+1 \text{ ---------------------} \; k+N-1$$

$\tilde{y}(k)$ being a linear function of $\tilde{u}(k)$ and $x(k)$, the solution for $\tilde{u}(k)$ minimising J can be obtained analytically. Considering the first term u(k) of $\tilde{u}(k)$ the control law becomes

$$u(k) = - K_k x(k) - K_y \tilde{y}_d(k) + K_u u(k-1)$$

(157)

The control matrices are no more dependent on reference models which can then be modified on line. The control law is linear and requires computation of $\tilde{y}_d(k)$ at every sampling instant.

Input Constraints:

The problem is set like in the preceeding paragraph with the following input constraints.

$$v(i) \le u(i) \le w(i), \; i = k\text{- - - - -}k + N-1$$

(158)

These authors have solved the problem by relaxation method and modified gradient method with off line computation.

A comparative feature of the algorithmic version for number of parameters to be stored and number of multiplications has been given in [42], alongwith simulation results on Vinylchloride unit of Rhone Poulene Company and a simulated model of a refinery debutanizer.

e) AR Modelling and Optimal Control:

AR modelling of dynamic systems and then using quadratic index to design an optimal controller have become very popular. Many industrial implementations have been reported. Uchida has described the technique in his chapter on thermal power plants. Kawai and Koike have discussed the implementation of this technique in cement plant in this book, Miyasugi [69] has discussed application of the technique for steam reformer and reformed gas temperature control system design. He has employed optimal controller with integral compensation. The basic approach has been described by Akaike [43] and details of application in the context of thermal power plant has been described by Nakamura and Akaike [44]. Pseudo random binary sequence (PRBS) are applied at inputs for identification of the plant and the signals are chosen to be independent. Offline LWR (Levingson-Wiggins-Robinson) method is used to identify the plants. While discussing their approach Nakamura and Akaike [44] have drawn a comparison with identification and control method (model algorithmic control) developed by Richalet and co-workers [21] (see section 3.1.11 and 3.2.3). Richalets method has been successfully used in thermal power plant [45] apart from being extremely popular in petroleum industries. The fundamental idea of Richalet's approach is close to this AR modelling. Their basic model is based on a direct expression of the impulse responses which require a large memory of past histories. In this approach [43,44] this difficulty is avoided by the use of an AR model which leads to an efficient identification of the necessary state space representation.

Although the approach has been covered by Uchida, Kawai and Koike in their respective chapters in this book, we discuss in brief AR modelling method and optimal control design.

The autoregressive representation of the (r+l) dimensional vector $\dot{x}(n)$ at time n is

$$\dot{x}(n) = \begin{Bmatrix} x(n) \\ u(n) \end{Bmatrix} = \sum_{m-1}^{M} A_m \, x \, (n-m) + w(n) \tag{159}$$

Where x(n) is r-dimensional output variable,
　　　u(n) is l-dimensional input variable
　　　w(n) is a (rxl) x 1 random vector satisfying specific relations, M is order of the model and A_m is an (rxl) x (rxl) matrix.

Eqn. (159), can be transformed into a state space representation with respect to r-dimensional vector $x_0(n)$, $x_1(n)$, -- x_{M-1} (n) given by the expression

$$x_0 \, (n) = x(n)$$

$$x_k \, (n) = \sum_{m=k+1}^{M} (a_m \, x(n+k-m) + b_m \, u(n+k-m)) \tag{160}$$

$$k = 1,2 \text{-----} M - 1$$

a_m and b_m are obtained from the co-efficient matrices.

A_m in Eqn.(159) is given by

$$A_m = \begin{bmatrix} \overset{r}{a_m} & \overset{l}{b_m} \\ {}^* & {}^* \end{bmatrix} \begin{matrix} |r \\ |l \end{matrix}$$

* denotes entries that are irrelevant for designing controller. The state space representation i given by

$$z(n) = \Phi z(n-1) + Fu(n-1) + v(n) \tag{161}$$

$$x(n) = H z(n-1) \tag{162}$$

Where

$$z(n) = \begin{bmatrix} x_0(n) \\ x_1(n) \\ \vdots \\ x_{M-1}(n) \end{bmatrix} \begin{matrix} |r \\ \\ M_r \end{matrix}$$

$$\Phi = \begin{bmatrix} a_1 & I & 0 - - -0 \\ a_2 & 0 & I - - -o \\ \vdots & & \\ a_{M-1} & o & o - - - -I \\ a_M & o & o - - - -o \end{bmatrix} |r \qquad F = \begin{bmatrix} b_1 \\ b_2 \\ \vdots \\ b_{M-1} \\ b_M \end{bmatrix} \begin{matrix} |r \\ \\ M_r \end{matrix} \qquad H = \underbrace{[\ I \ o - - \ \ - -o]}_{M_r} | r$$

and v(n) is given by the relation

$$w(n) = \begin{bmatrix} v(n) \\ x \end{bmatrix} \begin{matrix} |r \\ | \ 1 \end{matrix} \tag{163}$$

The optimal state feedback gain matrix is determined by minimising the quadratic index

$$J = E \sum_{i=1}^{k} (z^T (i) \, Qz(i) + u^T (i-1) \, R \, u(i-1)) \tag{163b}$$

A flow chart to design the optimal controller using dynamic programming method is given in fig. 20.

f) Multivariable Control based on Internal Model Principle:
Wonham and co workers [66] have proposed solution of regulation problem of linear time invariant systems with deterministic disturbance and reference signals. Control action is generated by a compensator and is aimed at providing closed-loop stability and output regulation in the face of small variations in certain system parameters. It has been shown that a structurally stable synthesis must utilize feedback of the regulated variable and incorporate in the feedback path a reduplicated model of the dynamic structure of the disturbance and reference signals. The necessity of this structure constitutes the 'internal model Principle'. An observer based multivariable form of this principle has been in commercial operation in Sumitomo Light Metal Works, Japan since 1986. This is likely to be inplemented in all other rolling mills of this company [67]. The synthesis procedure is composed of three major steps:

i)   The first one is to improve the response characteristics by state feedback.
ii) The second one is to compute the feedforward control input to counter balance the disturbance.

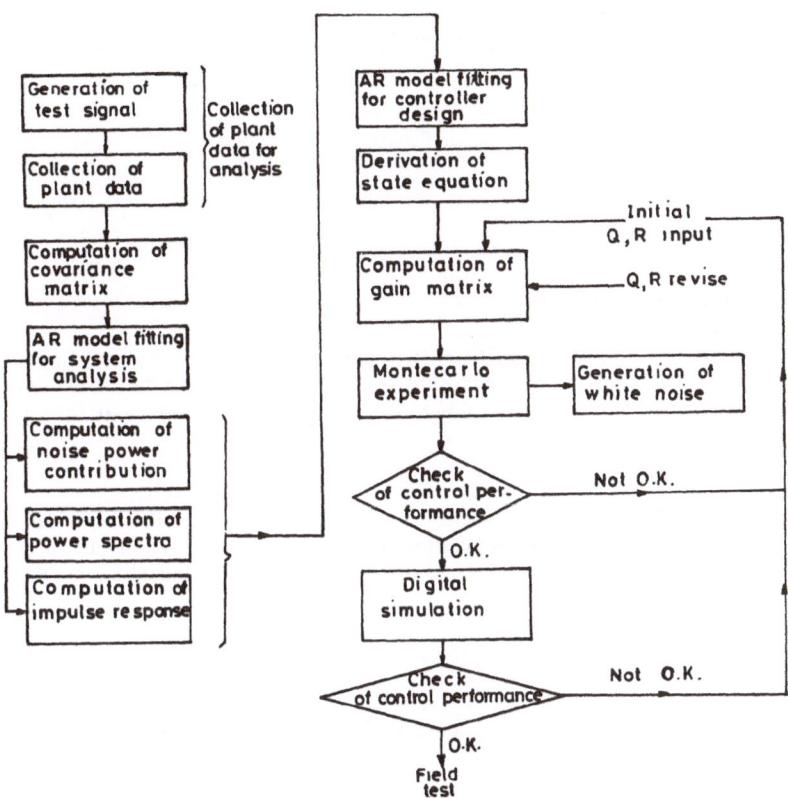

Fig. 20 Procedure for system identification and controller implementation

iii) The third one is to implement the feedforward control by feedback control based on the estimation of the disturbance by observer.

The state and output equations of the plant

$$\dot{x} = Ax + Bu + D_1d \tag{164a}$$

$$y_1 = C_1x + D_2d \tag{164b}$$

$$y_2 = C_2x + D_3d \tag{164c}$$

$$\dot{d} = A_1d \tag{164d}$$

Here the x, u, d are n-dimensional state, m-dimensional input and r-dimensional disturbance respectively. y_1, and y_2 are m_1 and m_2 dimensional outputs. The output regulation problem is to drive y—> 0 as t—> α. Class of exogeneous signals to be processed is expressed as in eqn. (164d). This could be step and ramp.

The controller structure derived in [67] is given by

$$\dot{z} = Fz + Gy_1 + Hy_2 \tag{164e}$$

$$u = pz + Qy_1 + Ry_2 \tag{164f}$$

Where z is the state of the controller.

The algorithm was implemented in a 16 bit process computer by Sumitomo [67] and the thickness variations caused by acceleration is reduced much more by the new system than the conventional control. For 0.32 mm delivery thickness the strip length without tolerance ($\pm$ 0,004 mm) is reduced 59% approximately.

3.2.3. *Predictive Control.* In this section we will discuss the multi input version of model algorithmic control, dynamic matrix control and its variant Quadratic Dynamic Matrix Control (QDMC). These techniques have been successfully implemented mostly in chemical industry.

a) Predictive control using impulse response model—
We have discussed this algorithm in the previous section for a SISO system. In this section we derive the algorithm for multivariable system. Consider a two input two output system [20]. Notations have the same meaning as in section 3.1.11 except that subscripts 1 and 2 are added to indicate first and second channel respectively. The model outputs can be written as

$$yM1 = h_{11}^T u_1 + h_{12}^T u_2 \tag{165}$$

$$yM2 = h_{21}^T u_1 + h_{22}^T u_2 \tag{166}$$

The desired behaviour of the outputs is given by two reference models, one for each output: yR1 and yR2. For a prediction horizon of NPRED samples, two reference trajectories yR1 and yR2 are derived from the two reference models. Again the assumption of two input vectors UG1 = { $u_1(k)$ } and UG2 = { $u_2(k)$ } will allow the calculation of two model trajectories YG1 and YG2 using above equations (165-166).

To minimise the difference between model trajectories and the corresponding references, we add δ_1, to UG_1 and δ_2 to UG_2. Thus we get

$$u_1 = UG_1 + \delta_1 \tag{167}$$

$$u_2 = UG_2 + \delta_2 \tag{168}$$

The error vector e_1 for the first output

$$e_1 = YR1 - YG1 - [H_{11} : H_{12}] \begin{bmatrix} \delta_1 \\ \delta_2 \end{bmatrix} \tag{169}$$

$$[H_{11} : H_{12}] = \begin{bmatrix} h_{11}^{(1)} & o & h_{12}^{(1)} & o \\ h_{11}^{(2)} & h_{11}^{(1)} & h_{12}^{(2)} & h_{12}^{(1)} \end{bmatrix}$$

To reduce matrices NPRED = 2, error vector e_2 can be found in similar way. Thus we can write

$$\begin{bmatrix} e_1 \\ e_2 \end{bmatrix} = \begin{bmatrix} yR_1 \\ yR_2 \end{bmatrix} - \begin{bmatrix} yG_1 \\ yG_2 \end{bmatrix} - \begin{bmatrix} H_{11} & H_{12} \\ H_{21} & H_{22} \end{bmatrix} \begin{bmatrix} \delta_1 \\ \delta_2 \end{bmatrix} \tag{170}$$

or

$$E = yR - yG - H\Delta$$

Least square minimisation of E gives

$$\underset{\Delta}{\text{Min}} \quad E^T E + \beta^2 \Delta^T \Delta \tag{171}$$

$$\hat{\Delta} = \begin{bmatrix} \hat{\delta}_1 \\ \hat{\delta}_2 \end{bmatrix} = [H^T. H + \beta^2. I]^{-1} H^T [yR - yG] \tag{172}$$

The optimal input sequence for the prediction horizon

$$\hat{u}_1 = UG_1 + \hat{\delta}_1 \tag{173}$$

$$\hat{u}_2 = UG_2 + \hat{\delta}_2 \tag{174}$$

We can express the generalised version of predictive control discussed above. The technique has become popular and has been named model algorithmic control [21].

In the generalised form constraints on the controls are introduced

$$u_{mm} < u < u_{max} \tag{175}$$

$$|\, u(n) - u(n-1)\, | < V_{max} \tag{176}$$

Constraints on internal variables or secondary outputs (e.g., $y_s(n) = G^T. u$) can be introduced.

Under these constraints many approaches can be used for solution:
a) Linear programming
b) Least square minimization
c) Quadratic programming
d) Dynamic programming

Identification and control algorithm can be written as below:

$$\text{Identification} \qquad \underset{\substack{| \\ \text{Measure}}}{y(N)} = \underset{\substack{| \\ A_M=A_o}}{A^T} \underset{\substack{| \\ \text{Measure}}}{U} \tag{177}$$

$$\text{Control} \qquad y(N) = A^T . U = U^T . A \tag{178}$$

Past Future   Past Future

Measure Reference   Measure Reference
Model   Model

Identification

In the identification scheme y(N) and U are given. Problem is to find A. In the control problem y(N) is known in the past from collected data and in the future by the reference model trajectory. A is given by previous identification. U is given in the past from stored computed controls and in the future by the same type of algorithm.

A flow chart of the control algorithm is shown in fig. 21 which gives the outline of the control computation.

The above control algorithm can be employed in direct digital control and transparent control mode. In the first case the internal model used is the process model and the controller directly acts on the process. In the transparent control mode the identification is performed on the closed-loop system with their PID analog controls. The outputs of this control algorithm are the setpoints of these conventional controllers. Thus transparent mode can be seen as a dynamic setpoint control. This mode has similarity to the approach adopted by Kyushu Power Company for optimal control design of thermal power plant (see Uchida in this book) and Mafezzoni and Cori [39] for practical optimal controller for a 160 MW Italian power plant.

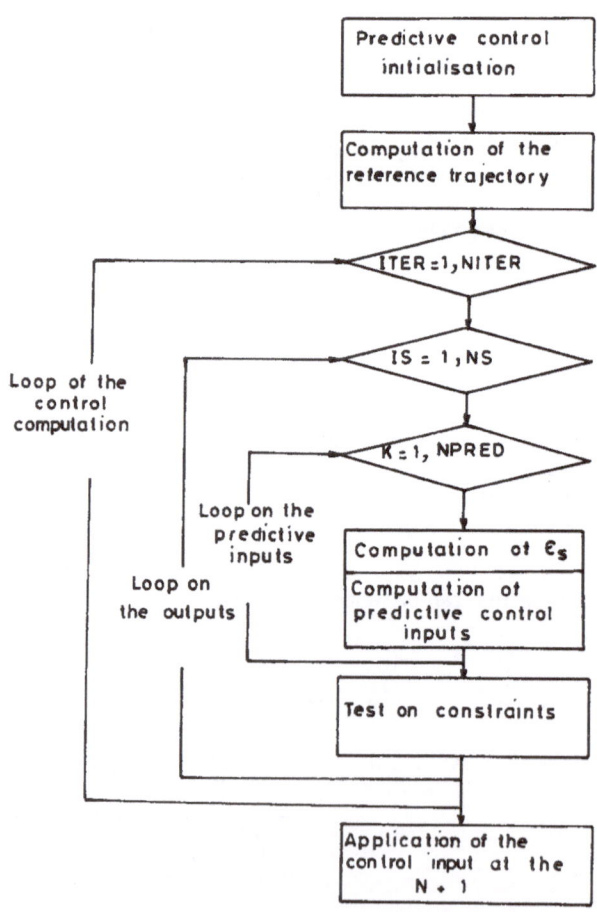

Fig. 21 Flow chart of MAC

b) Dynamic Matrix Control:

In section 3.1.11 we have discussed SISO version of DMC algorithm of Cutler and Ramaker. We derive here the DMC equations for multivariable system [46,47]. For a r output s input system a linear dynamic representation is given by

$$\Delta \underline{y} = \underline{A}\, \Delta \underline{u} \qquad (179)$$

Here the output projection vector

$$\Delta y(t+1) = [\, \Delta y_1^T (t+1)\ \Delta y_2 (t+1) \ldots \Delta y_r^T (t+1)\,]^T$$

and the corresponding vector of input variable move

$$\Delta u = [\, \Delta u_1^T (t)\ \Delta u_2^T (t) \ldots \Delta u_s^T (t)\,]^T$$

$$\underline{A} = \begin{bmatrix} A_{11} & A_{12}\text{-----}A_{1s} \\ A_{21} & A_{22}\text{----- } A_{2s} \\ \text{---------------} \\ A_{r1} & A_{r2} \qquad A_{rs} \end{bmatrix}$$

Where $\underline{A}$ is composed of blocks of step response co-efficients as in eqn. (73) relating the ith output to the jth input. Notations are same as used in section 3.1.11.

In DMC it is usually necessary to restrict or suppress the amplitude of the input moves. Thus DMC equations are generally formulated as

$$\begin{bmatrix} \Delta \underline{y} \\ O \end{bmatrix} = \begin{bmatrix} \underline{A} \\ \underline{A}_1 \end{bmatrix} \Delta \underline{u} \qquad (180)$$

Where for multivariable systems

$$\underline{A}_1 = \mathrm{Diag}\ [\underbrace{\lambda_1\ \lambda_1\text{---}\lambda_1}_{l_u}\ \lambda_2\ \lambda_2\text{--}\lambda_2\text{---}\lambda_s\ \lambda_s\ \text{---}\lambda_s] \qquad (181)$$

and $\lambda_i > 0$ in the ith input move suppression co-efficient.

It is possible in DMC to give tighter control to particular controlled variable(s) by increasing the relative weight on the corresponding least square residual. This is achieved by pre-multiplying DMC equations with the matrix weight $\gamma_i > 0$.

$$\Gamma = \mathrm{Diag}\ [\underbrace{\gamma_1\ \gamma_1\text{---}\ \gamma_1\ \gamma_2\ \gamma_2\text{---}\ \gamma_2\text{---}\gamma_r\ \gamma_r\ \text{---}\gamma_r}_{N}] \qquad (182)$$

Including this weighting matrix the solution of the DMC equations become

$$\Delta u(t) = (\underline{A}^T \Gamma^T \Gamma\ \underline{A} + \underline{A}_1^T \underline{A}_1)^{-1}\ \underline{A}^T\ \underline{A}_1^T\ \underline{A}_1\ \Delta y(t+1) \qquad (183)$$

c) **Quadratic Dynamic Matrix Control (QDMC):**
In on line applications, the moves computed in eqn. (183) may not be implementable due to process operating limit violations. There could be constraints on manipulated variable, control variable and key process variables. The QDMC algorithm developed by Shell Oil Company combines the proven model based control structure of dynamic matrix control with the constraint handling capabilities of quadratic programming. The algorithm is described by the following objective function.

$$\min_{\Delta u} \; 1/2 \, \Delta u^T \, H \Delta u - g^T \, \Delta u \tag{184}$$

subject to the constraint

$$c \, \Delta u \geq c \tag{185}$$

and $\quad -\Delta u_{max} \leq \Delta u \leq \Delta u_{max} \tag{186}$

H is the Hessian matrix defined by

$$H = A^T \Gamma^T \, \Gamma \, A + A_1^T \, A_1 \tag{187}$$

and g is the gradient vector of QP, given by

$$g = A^T \, \Gamma^T \, \Gamma \Delta y \tag{188}$$

Solution of the QP algorithm at each sampling interval produces an optimal set of moves Δu which satisfies the constraints. Any commercially available quadratic programming algorithm could be used for on line computation.

Note: i) A state space formulation of model predictive control has been reported [48]. The step (impulse) response model can be put into state space form and a Kalman filter can be designed to predict the future output trajectory of the process and the control action is calculated using the usual techniques in model predictive conrol.

ii) Internal model control realization of a Linear Quadratic Gaussian design has been shown to be possible by Grimble [49].

3.2.4 *Adaptive control.* We will cover in this section adaptive control methods for multi variable system. There are numerous text books on the subject [51, 52, 53]. A multivariable form of self tuning controller and an adaptive optimal controller will be described.

a) **Multivariable Self Tuning Controller:**
A multivariable self tuning regulator applied successfully to a drum boiler in a power plant [50] is described. The formulation of the problem is applicable for any multivariable process. The multivariable process model is given by

$$y(k) = \sum_{i=1}^{N} A_i \, y(k-1) + \sum_{i=0}^{N} B_i \, u(k-i) + e(k) = P \, d(k) + e(k) \tag{189}$$

$y_c(k)$ and $u(k)$ are n dimensional input and output respectively. $v(k)$ is r-dimensional measurable disturbance. $e(k)$ is $(n+r)$ vector noise term (stochastic, normal distribution with zero mean and covariance R), $y^T(k)$ is an extended vector, $y^T(k) = [\, y_c^T(k) \; v^T(k) \,]$. A and B matrices are of dimensions $(n+r) \times (n+r)$ and $(n+r) \times n$. P is an unknown parameter matrix of dimension $(n \times r) \times (N \times (2n \times r) \times n)$ and is given by

$$P = [B_0, A_1 - - - -A_N, B_N]$$
The vector for real time measured process data

$$d(k) = [\, u^T(k),\, y_c^T(k-1),\, v^T(k-1)- - - -y_c^T(k-N)\, v^T(k-N),\, u^T(k-N)] = [\, u^T(k)\, x^T(k-1)\,] \quad (190)$$

When x(k-1) denotes formally 'state vector'.

An exponential forgetting factor is applied to past data d(k-j), j=o,1 .... Data weight is chosen as f^j, f = const <1 for identification. Unknown parameters are estimated by minimising the predicted error.

$$\hat{e}(k) = y(k) - P(k-1)\, d(k)$$

$$J1 = \sum_{j=k_o}^{K} \hat{e}^T(j)\, f^{2(k-j)}\, \hat{e}(j),\, k_o \ge 1 \quad (191)$$

A multi step control criterion is used as the criterion for the stochastic optimal control

$$J = E\,[\sum_{i=1}^{M} x^T(k+i)\, Q(i)\, x(k+i)\, |\, d(k)\,] \quad (192)$$

M denotes the number of control steps and x(k) is the state vector.

The criterion generally requires the solution of a Riccati equation and an optimal minimum variance control is theoretically obtained. Choice on the selection of Q (i) and results of application of self tuning feedforward and feedback control for a drum boiler plant are available in references [50,54].

b) Adaptive optimal controller—
In section 3.2.2 we have discussed AR modelling and optimal controller design technqiue proposed by Nakamura and Akaike [44]. This technique employs off line identification method of AR model. The technique has been extended for on line identification and online changes for control gains by Nomura and Sato [55] of Hitachi Ltd., Japan. These authors have developed the technique for thermal power plant. But this is applicable for any general multivariable process.

The process model is considered as

$$x(k) = \sum_{i=1}^{L} A(i)\, x(k-i) + \sum_{i=1}^{L} B(i)\, u(k-i) + w(k) \quad (193)$$

Where x(k) is output variable, u(k) is input variable, L is model dimension, w(k) is white noise and

$$x^T(k-1) = [\, x_1(k-1)\, x_2(k-1)- - - - - -x_n(k-1)\,]$$
$$u^T(k-1) = [\, u_1(k-1)\, u_2(k-1)- - - - -u_m(k-1)\,]$$
$$w^T(k) = [\, w_1(k)\, w_2(k)- - - - -w_n(k)\,]$$

A(1) and B(1) are matrices of dimension (nxn) and (nxm) respectively.

A Kalman filter is employed to identify the model parameters A(1) and B(1) on line.

Equation (193) can be written as

$$x(k) = H(k)\, \Phi + w(k) \tag{194}$$

Where
$$H(k) = \begin{bmatrix} y^T(k) & o & o \\ o & y^T(k) & o \\ o & o & y^T(k) \end{bmatrix}$$

$$y^T(k) = [\, x^T(k{-}1)\ x^T(k{-}2) \,\text{-}\,\text{-}\,\text{-}\,\text{-}\, x^T(k{-}L)\ u^T(k{-}1)\ u^T(k{-}2)\ \text{-}\,\text{-}\,\text{-}\,\text{-}\, u^T(k{-}L)\,]$$

$$\Phi^T = [\, \Phi^T_1\, \Phi_2^T \text{-}\text{-}\text{-}\text{-}\text{-} \Phi_n{}^T\,]$$

$$\Phi i^T = [\, Ai\,(1)\ Ai\,(2) \text{-}\text{-}\text{-} Ai\,(L)\ Bi\,(1)\ Bi\,(2) \text{-}\text{-} Bi\,(L)\,]$$

$$Ai\,(1) = [ai_1\,(1)\ ai_2\,(1)\ \text{-}\text{-}\text{-} ai_n\,(1)\,]$$

$$Bi\,(1) = [bi_1\,(1)\ bi_2\,(1)\ \text{-}\text{-}\text{-} bi_m\,(1)\,]$$

The model parameter Φ is assumed to change with time, Φ is given by the state transition equation

$$\Phi\,(k) = \Phi\,(k{-}1) \tag{195}$$

Eqn.(194) can be considered as observation equation for model parameter Φ.

Using equations (194) and (195) an estimate of model parameter Φ can be obtained using a Kalman filter.

$$\hat{\Phi}(k) = \hat{\Phi}\,(k{-}1) + P(k)\, H^T(k)\, W^{-1}(k)\,[\,x(k) - H(k)\,\hat{\Phi}\,(k{-}1)\,] \tag{196}$$

$$P(k) = [\, P^{-1}(k{-}1) + H^T(k)\, W^{-1}(k)\, H(k)\,]^{-1} \tag{197}$$

When
$\hat{\Phi}\,(k)$ is estimate of $\Phi\,(k)$
$W\,(k)$ is covariance of $w(k)$
$P\,(k)$ is errors of $\Phi\,(k)$

Eqn. (193) is written is state space form as

$$z(k) = C\, z(k{-}1) + D u(k{-}1) + v(k) \tag{198}$$
$$x(k) = [I\ o\ \text{-}\,\text{-}\,\text{-}\,\text{-}\,\text{-}o]\, z(k)$$

where
$$z^T(k) = [\, z_0{}^T(k)\ z_1{}^T(k) \text{-}\text{-}\text{-}\text{-}\text{-} z^T_{L{-}1}(k)\,]$$

$$v^T(k) = [w^T(k)\ o\ \text{-}\,\text{-}\,\text{-}\,\text{-}\,\text{-}o]$$

$$C = \begin{bmatrix} A(1) & I & o\text{-}\,\text{-}\,\text{-}\,\text{-}\,\text{-}\,\text{-}o \\ A(2) & o & I\text{-}\,\text{-}\,\text{-}\,\text{-}\,\text{-}\,\text{-}o \\ A(L{-}1) & o & o\text{-}\,\text{-}\,\text{-}\,\text{-}\,\text{-}\,\text{-}I \\ A(L) & o & o\text{-}\,\text{-}\,\text{-}\,\text{-}\,\text{-}\,\text{-}\,o \end{bmatrix}$$

$$D = [\, B^T(1)\, B^T(2) - - - B^T(L-1)\, B^T(L)\,]$$

A quadratic criterion is used to obtain control law

$$J = E\,[\,\sum_{i=1}^{N} \{z^T(k+i)\,Q\,z(k+i) + u^T(k+i-1)\,R\,u(k+i-1)\}\,\,]\tag{199}$$

Q and R are nonnegative definite and positive definite matrices, respectively, of dimentions Ln x Ln and m x m.

Applying dynamic programming methods the optimal control u° (k) can be obtained

$$u^\bullet(k+N-i) = -[\,R+D^T\tilde{Q}(I-i+1)\,D]^{-1}\,D^T\tilde{Q}\,(I-i+1)\,C\,z(k+I-i)\tag{200}$$

$$S(I-1) = \tilde{Q}(I-i+1) - \tilde{Q}(I-i+1)\,D\,[R-D^T\tilde{Q}(I-i+1)\,D]^{-1}\,D^T\,\tilde{Q}(I-i+1)\tag{201}$$

$$\tilde{Q}(I-i) = C^T S(I-i)\,C\,Q,\ \ \tilde{Q}(I) = \tilde{Q}\,(i=1, 2 - - I)\tag{202}$$

Finally u° (k) is calculated using

$$u^\bullet(k) = G(1)\,z(k)\tag{203}$$

$$G(1) = -[R+D^T\tilde{Q}(1)\,D]^{-1}\,D^T\tilde{Q}(1)\,C\tag{204}$$

For large values of N, control gain G(1) converges to a constant value [17].

Hitachi has already implemented a state prediction based controller for superheater temperature control in boiler. In this a 2x2 process dynamic model is used to predict future trends in main steam temperature. The control algorithm implemented is a velocity type PI control law.

$$\Delta F_f(i) = k_1(i)\,e(i, n) + k_p(i)\,\{\,e\,(i,n) - e(i-1, n)\,\}$$

Where $\Delta F_f(i)$, control output at ith sampling period. $k_1(i)$, $k_p(i)$ are integral and proportional gain respectively

e(i, n) = Main steam temperature prediction error after nth sampling at ith sampling period = R (i, n) - x(i, n).
R(i, n) = Target value after nth sampling instant at ith sampling instant.

In this control system a predicted value of x(i, n) is used instead of the measured values x(i).

The block diagram of the advanced control system in fig. 22.

Since the above method requires an accurate physical model of the plant, the technique may not prove suitable for large multi-input multi output plant. For this purpose the adaptive technique described in this section may prove more suitable.

3.3 Auto Tuning

In recent years there has been considerable progress in on line tuning of PID regulators. In fact the commercial adaptive controllers can be divided into

- Tuners for PID controllers
- Adaptive tool boxes
- Adaptive controllers for specific applications.

To the first category belong Foxboro's Exact controller [56] and Satt control's auto tuner [57]. These are controllers that tune the parameters in conventional PID controllers. Foxboro's Exact is based on pattern recognition techniques and Satt control auto-tuner is based on relay tuning and the controller is tuned by pressing a button. Uchida in his chapter on thermal power plant has discussed about other commercially available auto tuning controllers. Kawai and Koike in their chapter on cement plant have discussed an auto tuning algorithm for PID controller. Other commercially available controllers are Turnbull control system TCS 6355 and ASEA NOVATUNE, Recently Dumont and Zervos have proposed a novel method of automatic Tuning of PID controllers. Here the closed-loop system is represented by a series of Laguerre functions and the PID settings are found by optimization of an ISE criterion.

We briefly describe here a fuzzy mathematics based PID auto tuning method developed by Hitachi Ltd., Japan [58]. The auto tuning function has been decomposed into two major operations, namely, initial tuning and on line tuning. The initial tuning unit carries out identification of the process by sending a signal step wise to the process at start up. It estimates the process characteristics referring the step response from the control system and sets the P, I and D constants initially. The on line tuning unit monitors the controlled variable after the completion of the initial tuning. The on line tuning uses expert system methodology and has rejection capability for signal disturbances. The unit consists of process variable waveform recognizer, fuzzy reasoning unit and adjustment rule sections. Fig. 23 illustrates the operations of the units. The waveform recognizer determines the performance index of control characteristics as overshoot, damping and ratio of period. As shown in fig. 23 the fuzzy reasoning unit determines the rate of increase and decrease of the correction co-efficients of P, I and D constants according to actual performance indices by aid of the adjustment rules. The constants are changed everytime the response varies. This procedure is repeated several times until the P, I and D constants are tuned properly to the desired values. The method has been incorporated in a single loop controller.

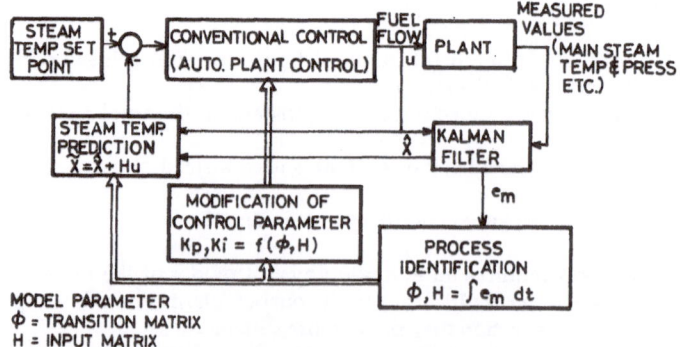

FIG-22 STATE PREDICTION BASED CONTROLLER

3.4 Expert Control

The purpose of an expert system tool is to facilitate the building of a knowledge based system which can be substituted for expert operators in process control. Components of knowledge in control [59] problems are shown in Fig.24. In control context the system can have several algorithms for control and estimation e.g. PID regulator, a relay tuner, a least square recursive estimator etc. The system also has other algorithms for supervision and analysis and signal generation algorithms to improve identifiability. All the algorithms are co-ordinated by an expert system which decides when to use a particular algorithm [60]. Uchida in his chapter on thermal power plant has discussed use of expert systems and Saito in his chapter on steel industry has covered an application of knowledge based control. We have discussed fuzzy logic controller grown from the work of Zadeh in section 3.1.13.

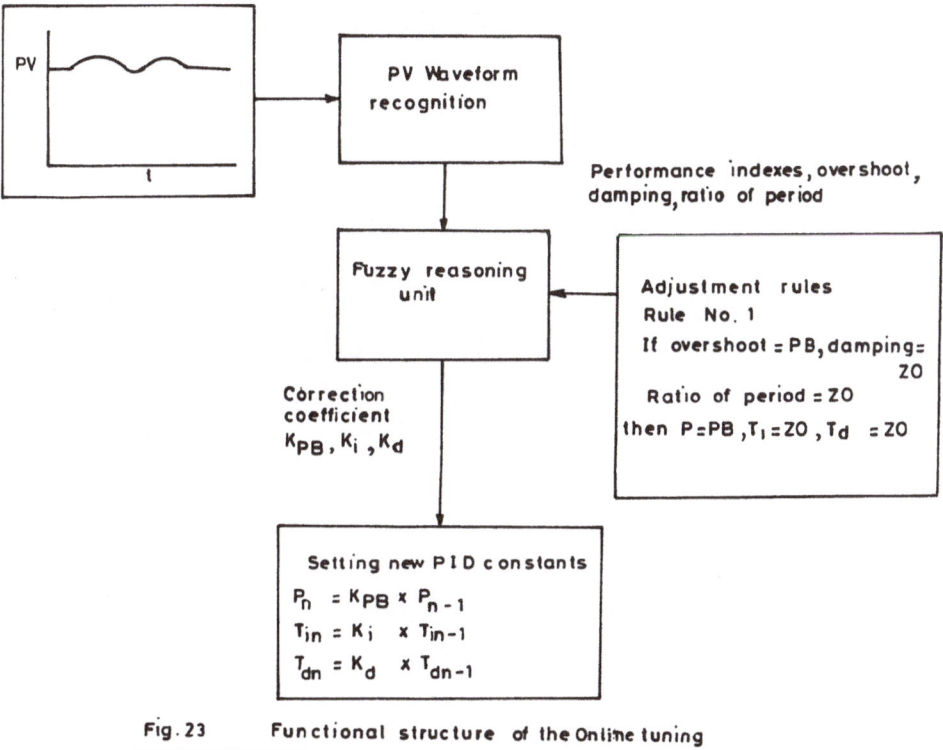

Fig. 23   Functional structure of the Online tuning

Amongst on-line process industry examples, Exxon and Texaco, USA, use two processors in parallel, a Lisp Processor to handle expert system and a 68010 processor for real time data acquisition. The 68010 uses an external MUTIBUS to communicate with the distributed process control system. Current on line application include:
- Alarm management
- operator aid on difficulty to control process (start up/shut down)
- Adaptive control strategies
- Process and equipment trouble shooting.

130

3.5 Implementation of control algorithms

We will discuss in this section implementation of control algorithms in plant control systems. In commercially available distributed digital control systems a variety of elementary functions or operations are available which can be flexibly configured or combined to yield a custom control system at the first level.

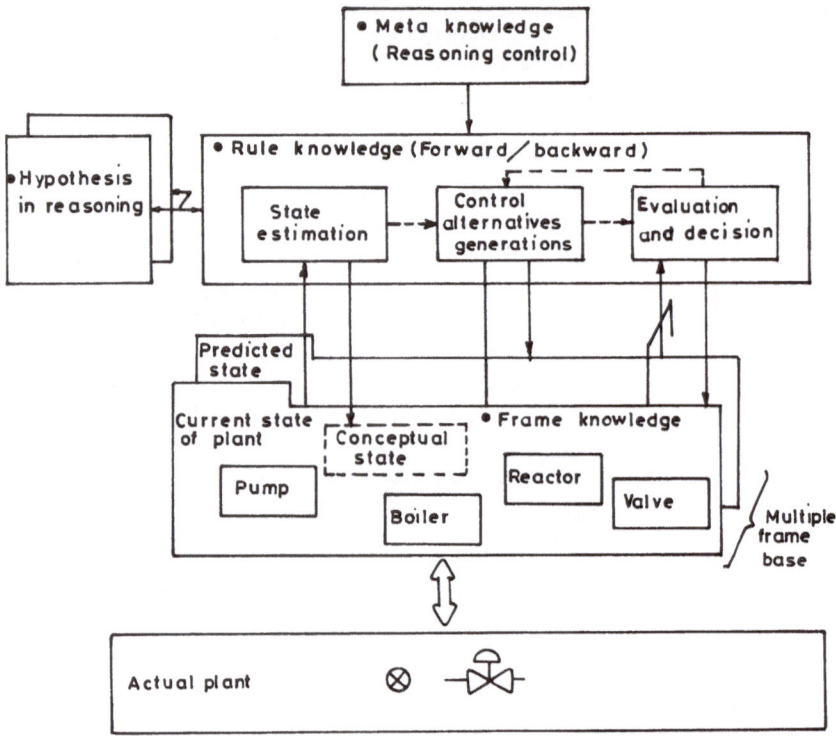

Fig.24 Components of knowledge in control problems

Examples of function blocks available at various modules are [61]:

a) Arithmatic operations involving multiple inputs and including multiply and divide
b) Signal processing including lead/lag filtering, high/low limiting, square root function, function generation and rate limiting.
c) Binany logic operations including 'and', 'or' 'not', set/reset memory, comparator, compare inputs against alarm points, validity checks and digital time delay.
d) Control operations including PID action, pulse positioning, adaption of tuning parameters and interface to other control modules.
e) Station operations including basic setpoint, cascade, ratio and manual transfer.
f) Communication operations including input/output to other modules in the same node and modules in remote node.

131

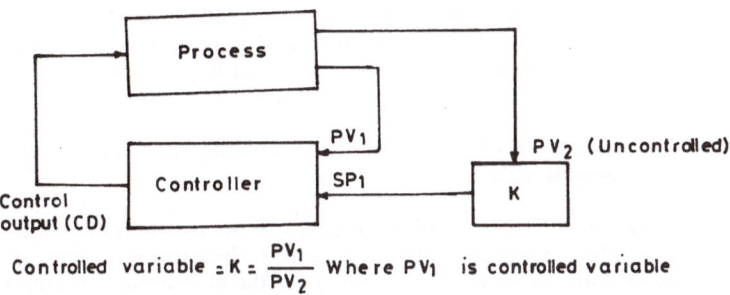

Controlled variable $= K = \dfrac{PV_1}{PV_2}$ Where PV_1 is controlled variable

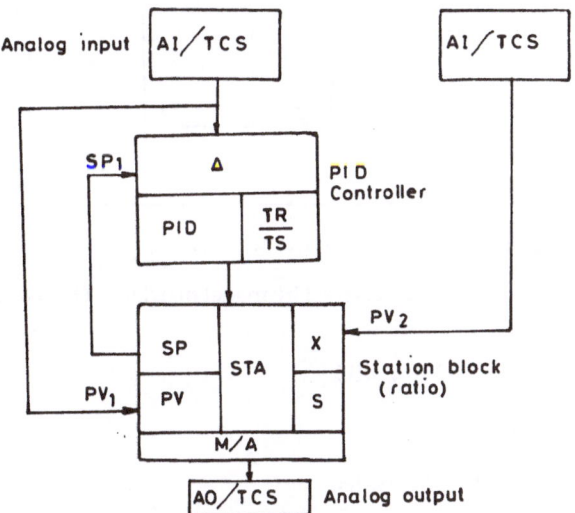

Fig. 25 Ratio control structure

With the help of operator communication console or plug in configuration and tuning module the user simply selects the function blocks required for his control system and indicates them in the form of block type, block location and specification list (inputs and tuning parameters). Thus the control algorithm can be implemented without the need of actual programming.

Fig. 25 shows implementation of ratio control structure using function blocks provided in control system of Bailey Controls. USA.

We also cover here the implementation of a two input two output process by use of both classical PI controllers together with an adaptive decoupling precompensator. We first describe the control algorithm and then show how the algorithm including the Kalman filter parameter estimator is implemented in Micro-z control system [62] of Controle Bailey, France. The linear control structure is shown in fig. 26.

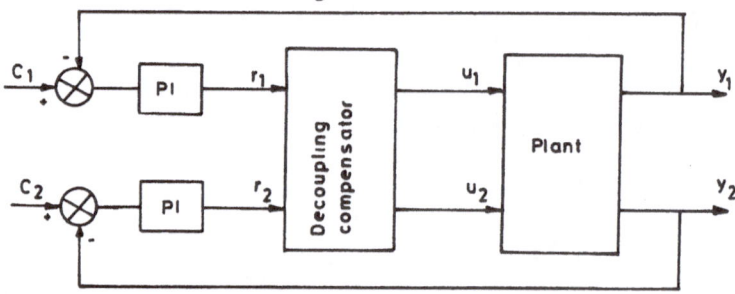

Fig. 26   Adaptive control with decoupling

The decoupling pre compensator is designed by model matching. The reference model considered is.

$$y_m (k+1) = A_m y_m(k) + B_m u(k) \tag{205}$$

A_m and B_m are 2x2 matrices.

The plant to be controlled has the model

$$y (k+1) = A y(k) + B u(k) \tag{206}$$

An adaptive solution for decoupling control law u(k) is given by

$$u (k) = \hat{B}^{-1} (k) [B_m r(k) + (A_m - \hat{A}(k)) y(k)] \tag{207}$$

$\hat{A} (k)$ and $\hat{B}(k)$ are estimated on line.

Considering

$$\in (k) = y_m (k) - y(k) \tag{208}$$
$$\in_s (k) = \in (k) - A_m \in (k-1) \tag{209}$$
$$\in_s (k+1) = B_m r(k) + \Phi^T(k) \theta \tag{210}$$

with

$$\phi^T (k) = [y_1 (k) \ y_2 (k) \ u_1(k) \ u_2 (k)]$$

$$\Phi(k) = \begin{bmatrix} \varphi^T(k) & 0 \\ 0 & \varphi^T(k) \end{bmatrix}$$

$$\theta^T = [\,\Delta a_{11}\ \Delta a_{12}\ b_{11}\ b_{12}\ \Delta a_{21}\ \Delta a_{22}\ b_{21}\ b_{22}\,]$$

The estimate of $\hat{\theta}(k)$ of θ is given by Kalman filter

$$\hat{\theta}(k) = \hat{\theta}(k-1) + K(k)\,\epsilon_s(k) \tag{211a}$$

$$D(k+1) = [\,R(k+1) + \Phi^T(k)\,P(k)\,\Phi(k)\,] \tag{211b}$$

$$P(k+1) = P(k) - P(k)\,\Phi(k)\,D^{-1}(k+1)\,\Phi^T(k)\,P(k) \tag{211c}$$

$$K(k+1) = P(k)\,\Phi(k)\,D^{-1}(k+1) \tag{211d}$$

with $P(o) > o$, and $R(k)$ is covariance matrix of $\epsilon(k)$

Some modifications in updating of $P(k)$ have been incorporated. Trace of $P(k)$ is kept within $(\underline{t_r}, \overline{t_r})$.

```
IF ϵ_s(k) > ϵ_s) then
compute D(k), P(k), K(k), (eqn. 211(b-c))

IF ( | ϵ_s(k) | – | ϵ_s(k-1) | < v ) then
IF (trace P(k) < t_r) then
P(k) = 1.15 P(k)
ENDIF                                                    (212)
ELSE IF (trace P(k) < t_r) then
P(k) = 1.15 [ P(k) + 0.05 Diag (P(k)) ]
ENDIF
ENDIF
```

The above algorithm has been implemented using three cards of Micro Z system. It has been observed that implementation of 8x8 $P(k)$ computation needs more than 96 additions and 64 summation blocks for its own sake while upto 99 operations can be configured in one card. Accordingly the updating of 8x8 $P(k)$ matrix is done by two (4x4) matrices $P_1(k)$ and $P_2(k)$. This is inferred from both the choice of a diagonal $P(o)$ matrix and hypothesis that $R(k)$ be diagonal. The decoupled version of Kalman filter equation can be written as $i = 1, 2$.

$$\hat{\theta}(k) = [\,\hat{\theta}_1{}^T(k)\ \hat{\theta}_2(k)\,],\ \theta_1, \theta_2 \in R^4 \tag{213a}$$

$$\hat{\theta}_i(k) = \hat{\theta}_i(k-1) + K_i(k)\,(\epsilon_s(k))_i \tag{213b}$$

$$\delta_i(k+1) = R_{ii}(k+1) + \varphi^T(k)\,P_i(k)\,\varphi(k) \tag{213c}$$

$$P_i(k+1) = P_i(k) - (P_i(k)\,\varphi(k)\,\varphi^T(k),\,P_i(k))/\delta_i(k+1) \tag{213d}$$

$$K_i(k+1) = P_i(k)\,\varphi(k)\,/\,\delta_i(k), \tag{213e}$$
$$P_i(o) = I_4$$

134

For robustness conditions the basic updating of $P_i(k)$ is modified by use of the trace control algorithm.

The implementation architecture is shown in Fig. 27. The first card contains the linear control structure (Eqn.207) with the reference model (Eqn.205) and the evaluation of ϵ_s (Eqns.208-210). The remaining cards 2 and 3 implement respectively the updating of the model parameter estimates $\hat{\theta}_1(k)$ and $\hat{\theta}_2(k)$ [Eqns.211 (b) - (c) and 213 (a) - (e)]. The transmission of ϵ_s, y, u_1 $\hat{\theta}_1$, $\hat{\theta}_2$ between cards (1,2) and (1,3) are realized by analog transmission as per Micro-Z architecture.

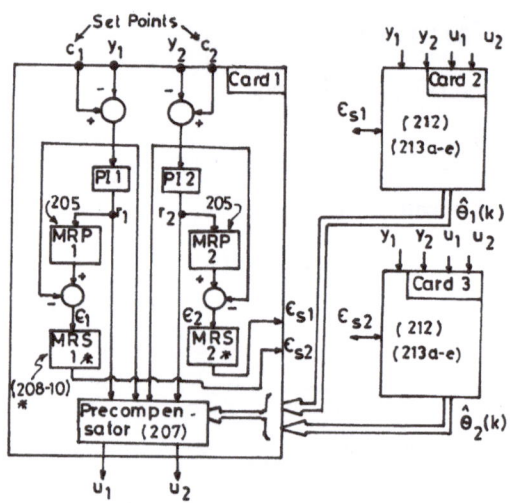

Fig. 27 Implementation architecture on micro-Z

However some advanced algorithms will need programming for their implementation. A functional segregation for implementation of blocked non-interacting optimal controller discussed in section 3.2.2 (c) in Hitachi control system is shown in Fig.28. The adaptive optimal control discussed in section 3.2.4 (b) has been implemented in Hitachi control computer V90. As per Nomura and Sato [55] this can also be implemented in Hitachi DDC system controller.

Self tuning controllers have been successfully applied in the Czechoslovak paper industry for the paper machine control [68] DEC's PDP 11/10, PDP 11/03 machines have been used among others. Fortran language is maximally utilized for the writing of source programme module.

As discussed earlier Uchida, Saito, Kawai and Koike have discussed implementation of several advanced control and PID auto tuning strategies in thermal power, steel and cement plants in their respective chapters. These authors have also described control system architectures used by them for these implementations.

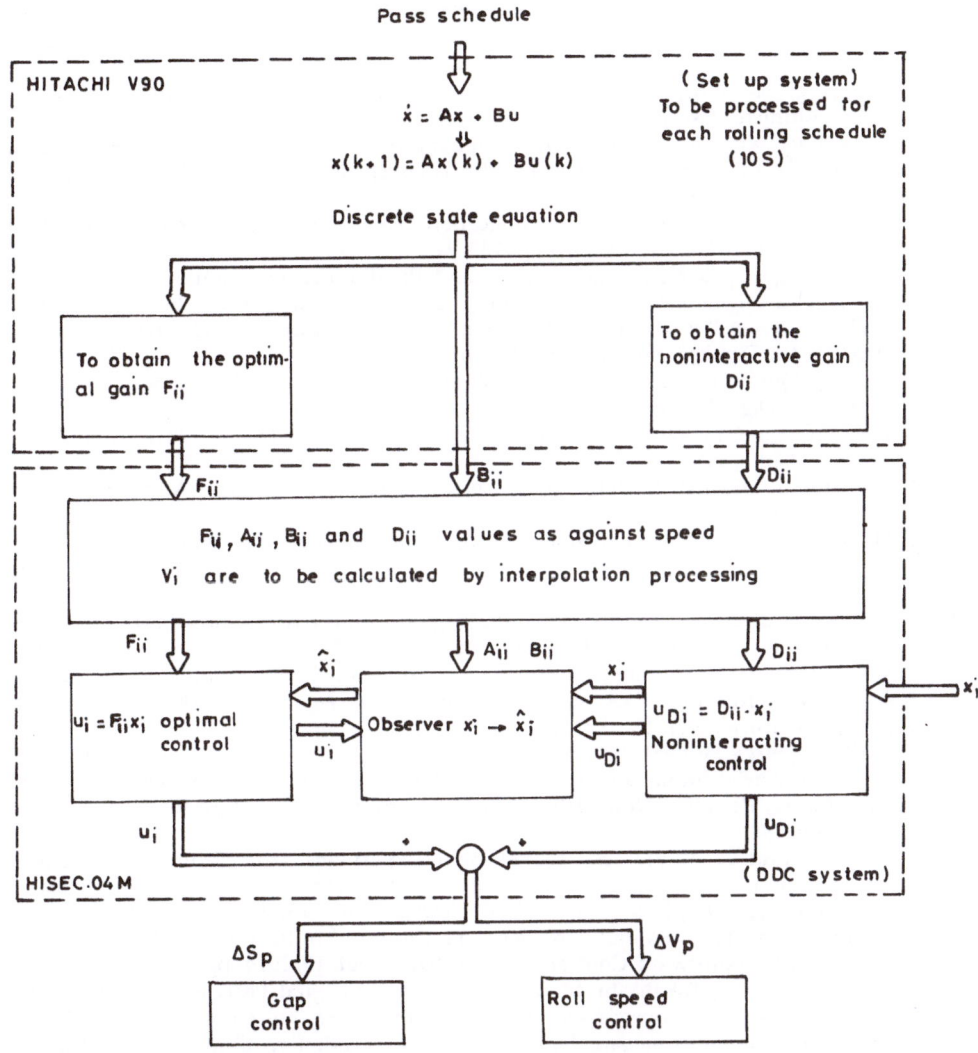

Fig. 28 Configuration of blocked non-interaction optimal control system

136

REFERENCES

1. D.E. Seborg, 'The Prospects of Advanced Process Control', *Proceedings, 10th IFAC World Congress*, Munich, 1987, pp. 281-289.
2. J.N. Wallace and R. Clarke, 'The Application of Kalman Filtering Estimation Techniques in Power Station Control', *IEEE Transactions on Automatic Control*, vol. AC–28, No. 3, 1983, pp. 416–427.
3. P.J. King and E.H. Mamdani, 'The Application of Fuzzy Control Systems to Industrial Processes', *Automatica*, vol. 13, 1976 pp. 235–242.
4. H.B. Verbruggen, 'Basics, Fundamentals and Possibilities for Digital Control', *IFAC proceedings on Digital Computer Applications to Process Control*, No. 6, 1986, pp. 95–103.
5. P. Eykhoff, 'New Trends in Identification', *IFAC proceedings on Control Science and Technology for Development*, Beijing, 1985, pp. 27–36.
6. B.C. Kuo, *Digital Control Systems*, Holt, Rhinehart and Winston, Inc., 1980.
7. R. Isermann, *Digital Control Systems*, Springer verlag, Berlin (1981).
8. K.J. Ästrom and B. Wittenmark, *Computer Controlled Systems*, Prentice Hall, Inc., Englewood Cliffs, 1984.
9. G.F. Framklin and J.D. Powell, *Digital Control of Dynamic Systems*, Addison-Wesley, Reading, 1980.
10. G.H. Hostetter, *Digital Control System Design*, Holt, Rinehart and Winston, 1988.
11. D.W. Clarke, 'PID Controllers and their Computer Implementation', *Oxford University report no. O.U.E.L. 1482/83*.
12. K.C. Chiu, A.B. Corripio and C.L. Smith, 'Digital Control Algorithms Parts I, II & III', *Instrument and Control Systems*, Oct, Nov. and Dec., 1973.
13. B.T. Condon and C.A. Smith, 'A Comparison of Control Algorithms as Applied to a Stirred Tank Reactor,' *ISA symposium on Instrumentation in the Chemical and Petroleum Industries*, vol. 13, 1977, pp. 11–19.
14. T. Kailath, *Linear Systems*, Prentice Hall, 1980.
15. J.K. Pal, 'Industrial Implementations of State Feedback Controllers', *Recent Advances in System Theory*, Papyrus, New Delhi 1988 (Proceedings, National Systems Conference, Coimbatore, 1988).
16. H. Krause and R. Vahldieck, 'Operational Experience with Progressive Control Concepts Demonstrated on a State Controller', *VDI Beriehte*, No. 589, 1986, pp. 73–86.
17. B.D.O. Anderson and J.B. Moore, *Linear Optimal Control*, Prentice Hall, 1971.
18. E.J. Davison and H.W. Smith, 'A Note on the Design of Industrial Regulators: Integral Feedback and Feedforward Controllers', *Automatica*, vol. 10, 1974, pp. 329–332.
19. D.M. Auslander, Y. Takahashi, M. Tomizuka, 'Direct Digital Process Control: Practice and Algorithms for Microprocessor Application', *Proc. IEEE*, vol. 66, 1978, pp. 199–208.
20. P.M. Brujin, L.J. Bootsma and H.B. Verbruggen, 'Predictive Control using Impulse Response Models', *IFAC Proceedings on Digital Computer Application to Process Control*, Düsseldorf, 1980, pp. 315-320.
21. J. Richalet, A. Rault, J.L. Testud and J. Papon, 'Model Predictive Heuristic Control: Applications to Industrial Processes', *Automatica*, vol. 14, 1978, pp. 413–428.
22. R.M.C. De Keyser, P.G.A. Van de Velde and F.A.G. Dumortier, 'A Comparative Study of Self Adaptive Long Range Predictive Control Methods', *Automatica*, vol. 24, 1988, pp. 149–163.
23. D.W. Clarke, 'Application of Generalized Predictive Control', *IFAC Proceedings on Adaptive Control of Chemical Processes*, No. 6, 1989, pp. 1–8.
24. C.R. Cutler and B.L. Ramaker, 'Dynamic Matrix Control—A Computer Control Algorithm', paper no. 51b, *AICHE, 86th National Meeting*, 1979.
25. B. Ydstie, 'Extended Horizon Adaptive Control', Proceedings, *IFAC 9th World Congress*, Budapest, 1984.

26. C. Brosilow, 'The Structure and Design of Smith Predictors from the Point of View of Inferential Control', *Proceedings, JACC*, 1979.

27. C.E. Gracia and M. Morari, 'Internal Model Control, 1.A. Unifying Review and Some New Results', *I&EC Process Design and Development*, vol. 21, 1982, pp. 308–323.

28. I.D. Landau, 'A Survey of Model Reference Adaptive Techniques—Theory and Applications', *Automatica*, vol. 10, 1974, pp. 353–379.

29. D.E. Seborg, T.F. Edgar and S.L. Shah, 'Adaptive Control Strategies for Process Control: a Survey', *AICHE Journal*, No. 32, 1986, pp. 881–913.

30. K.J. Ästrom, 'Adaptive Feedback Control', *Proc. IEEE*, No. 75, 1987, pp. 185.

31. J.J. Buckley and H. Ying, 'Fuzzy Controller Theory: Limit Theorems for Linear Fuzzy Control Rules', *Automatica*, vol. 25, 1989, pp. 469–472.

32. S.G. Tzafestas, 'Knowledge Engineering Approach to System Modelling, Diagnosis, Supervision and Control', *IFAC Proceedings on Simulation of Control Systems*, No. 13, 1987, pp. 15–22.

33. L.C. Hui, *General Decoupling Theory of Multivariable Process Control System*, Springer-verlag (Lecture notes in control and information sciences No. 53), 1983.

34. P.K. Sinha, *Multivariable Control —An Introduction*, Marcel Dekker, 1984.

35. T.J. Mcavoy, *Interaction Analysis*, ISA, 1983.

36. W.Y. Xu and H.G. Zhang, 'Modern Control Theory Applied to 200 MW Boiler Turbine Unit Control', *IFAC Proceedings on Power Systems and Power Plant Control*, No. 8, 1987, pp. 361–366.

37. Y. Saito, S. Hattori and Y. Katayama, 'Application of Blocked Non-interacting Optimal Control to Rolling Mill', *Hitachi Review*, vol. 37, No. 4, 1988, pp. 221–228.

38. S. Chotai, 'Modal Control of a Chemical Reactor Flasher System', *Proceedings IFAC World Congress*, Munich, 1987, pp. 315–320.

39. R. Cori and C. Mafezzoni, 'Practical Optimal Control of a Drum Boiler Plant', *Automatica*, 10, 1984.

40. M. Mulholland, 'Multivariable Control of an Ammonia Plant—Modelling and Control Theory', *IFAC Proceedings on Digital Computer Applications to Process Control*, No. 6, 1986, pp. 105-110.

41. A.K. Mahalanabis and J.K. Pal, 'Optimal Control of Linear Multivariable Systems with Prescribed Pole Locations in the Presence of System Disturbances', *Proceedings IEE, PT.D*, Sept. 1985.

42. Y. Yem, G. Bornard, J.P. Gauthier and R. Perret, 'Dynamic Multivariable Constraint Control—Applications to Petrochemistry' *Proc. 1st AL-FATEH/IFAC Workshop*, Tripoli, May 1980, pp. 39-47.

43. H. Akaike, 'Autoregressive Model Fitting for Control', *Ann. Inst. Statist Math*, No. 23, 1971 pp. 163-180.

44. H. Nakamura and H. Akaike, 'Statistical Identification for Optimal Control of Supercritical Thermal Power Plants', *Automatica*, vol. 7 No. 1, 1981, pp. 143-155.

45. M. Lecrique, A. Rault, M. Tessier and J.L. Testud, 'Multivariable Regulation of a Thermal Power Plant Steam Generator', *Proceedings IFAC World Congress*, Helsinki, 1978.

46. C.E. Gracia and A.M. Morshedi, 'Quadratic Programming Solution of Dynamic Matrix Control', *Chemical Engg. Communication*, vol. 46, 1986, pp. 73-87.

47. S.J. Kelly, M.D. Rogers and D.W. Hoffman, 'Quadratic Dynamic Matrix Control of Hydrocracking Reactors', *Proceedings American Control Conference*, Atlanta, 1988.

48. S. Li, K.Y. Lim and D.G. Fisher, 'A State Space Formulation for Model Predictive Control', *AICHE Journal*, vol. 35, 1989, pp. 241-249.

49. M.J. Grimble, 'Relationship between Internal Model Control and LQG Controller Structure', *Automatica*, vol. 25, 1989, pp. 41-53.

50. J. Fessel, 'An Application of Multivariable Self Tuning Regulators to Drum Boiler Control', *Automatica*, vol. 22, No. 5, 1986 pp. 581–585.

138

51. C.J. Harris and S.A. Billings, *Self Tuning and Adaptive Control: Theory and Application*, Peter Peregrinus Ltd., 1981.
52. H. Unbehauen, *Methods and Applications in Adaptive Control*, Springer-Verlag, 1980.
53. K.J. Ästrom and B. Wittenmark, *Adaptive Control*, Addison-Wesley, 1988.
54. J. Fessel and J. Jarkovsky, 'Steam Superheater Control via Self Tuning Regulator', *IFAC Proceedings on Digital Computer Application to Process Control*, No. 6, 1986, pp. 235–239.
55. M. Nomura and Y. Sato, 'Adaptive Optimal Control of Steam Temperatures for Thermal Power Plants, *IEEE Transactions on Energy Conservation*, vol. 4, No. 1, 1989, pp. 25–33.
56. E.H. Higham, 'A Self Tuning Controller Based on Expert System and Artificial Intelligence', *Proceedings IEE Conference 'Control '85'*, pp. 110-115.
57. K.J. Ästrom and T. Hagglund, *Automatic Tuning of PID Regulators*, ISA, Triangle Research Parks, NC, 1988.
58. K. Tachibana, T. Suehiro and T. Saito, 'A Single Loop Controller with Auto-Tuning System using the Expert Method', *Hitachi Review*, vol. 36, No. 6, 1987, pp. 357-362.
59. M. Funabashi and K. Mori, 'Knowledge-based Control Systems and Software for Building Expert Systems "EUREKA–II" ', *Hitachi Review*, vol. 37, No. 4, 1988, pp. 267-274.
60. K.J. Ästrom, 'Toward Intelligent Control', *IEEE Control Systems Magazine*, vol. 9, No. 3, 1989, pp. 60–64.
61. J.D. Schoeffler, 'The Incorporation of Advanced Control Techniques in Process Control', *Proceedings Seventh Annual Control Conference*, Control Engineering, 1981, pp. 17–46.
62. K.L. King, 'Using a Building Block Approach to Implement Advanced Control Techniques', *Proceedings Seventh Annual Control Conference*, Control Engineering, 1981, pp. 149–167.
63. Hanselmann, 'Implementation of Digital Controllers—A Survey', *Automatica*, vol. 23, 1987, pp. 7–32.
64. M.S. Koutchoukali, C. Laguerie and K. Najim, 'Model Reference Adaptive Control System of a Fludized Bed Reactor', *Automatica*, Vol. 22, No. 1, 1986, pp. 101-104.
65. ID Landau and R. Lozano, 'Unification of Discrete Time Explicit Model Reference Adaptive Control Design, *Automatica*, 17, No. 4, 1981, pp. 593-611.
66. B.A. Francis and W.M. Wonham, 'The Internal Model Principle of Control Theory', *Automatica*, vol. 12, 1976, pp. 457-465.
67. I. Hoshino et. al, 'Observer-based Multivariable Control of the Aluminium Cold Tandem Mill', *Automatica*, vol. 24, 1988, pp. 741-754.
68. A. Lixr et. al, 'Application of Adaptive Control Algorithms in the Czechoslavak Paper Industry', *IFAC Proceedings on Instrumentation and Automation in the Paper, Rubber, Plastics, Polymerization Industries*, Belgium, No. 8, 1984, pp. 213-217.
69. T. Miyasugi, 'Identification of Dynamics of a Steam Reformer and Subsequent Design of Reformed Gas Temperature Control System', *Proceedings 1983 Summer Computer Simulation Conference*.
70. R. Clarke, 'The Application of Multivariable Control to a 500 MW Oil Fired Boiler/Turbine Unit', *CEGB, UK, SSD/SW/82/NS1*, 1982.
71. S. E. Sheridan and P. Skjoth, 'Automatic Kiln Control at Oregon Portland Cement Company's Durkee Plant Utilizing Fuzzy Logic', *IEEE Trans. Industry Applications*, 1984, pp. 562-568.

Chapter 4

ESTIMATION AND SIGNAL PROCESSING ALGORITHMS

A. K. Mahalanabis
Department of Electrical Engineering
The Pennsylvania State University
University Park, PA   16802, U.S.A.

ABSTRACT. This chapter presents the essential elements of the minimum variance filtering and prediction algorithms derived on the basis of the state space and the CARMA models. After a brief review of the two forms of stochastic signal models, the estimation algorithms are first derived for systems with known models. This is followed by the introduction of a sequential parameter estimation algorithm for systems with unknown models. A solution of the adaptive filtering and prediction problem for systems with unknown models is then discussed. Finally, a numerical example illustrating the application of the adaptive prediction algorithm for solving a simple control problem is also presented.

1. INTRODUCTION

Minimum variance (MV) prediction and filtering of stochastic signals from their noise corrupted observations play an important role in the implementation of the LQG and the MV optimal control laws for specified stochastic systems (see, e.g., Goodwin and Sin [1] and Astrom and Wittenmark [2]). Recursive solutions of these estimation problems have been developed by many researchers following the pioneering work of Kalman [3]. Detailed descriptions of the various results available in this area may be found, e.g., in [1]-[8]. The earlier texts emphasize both the continuous time and discrete time results. However, in view of our interest in their digital implementation, we consider here only the discrete time filtering and prediction algorithms. Two different forms of the model based filtering and prediction algorithms will be discussed. The first form makes use of the stochastic state space model of the given signal process while the second form is derived on the basis of the stochastic time series model.

In the first part of our discussion, a-priori knowledge of the model parameters will be assumed and solutions of the signal estimation problems will be derived using the minimum mean squared error criterion. For most practical problems, however, the model

139

parameters are seldom known a-priori and it is necessary to estimate them from the available input-output data. A sequential least squares algorithm that permits an on-line estimation of the parameters of the time series model will be presented and its application for the parameter estimation of a canonical state space model will also be discussed. Combining the signal and parameter estimation algorithms, simple adaptive but sub-optimal filtering and prediction algorithms will then be developed.

We shall start in Section 2 with a brief look at the state space and the time series models needed for our subsequent developments. In Section 3, the Kalman filtering and prediction algorithms based on the state space model will then be derived and will be followed in Section 4 by the development of the linear prediction algorithm based on the time series model. The problem of sequential estimation of the parameters of an assumed signal model will be considered in Section 5 and the development of an adaptive prediction algorithm will be briefly indicated in Section 6.

2. STOCHASTIC SIGNAL MODELS

We assume that the signal of interest {y(k)} represents the output of a dynamic system driven simultaneously by a deterministic input (the control) and a stochastic input (the disturbance) which is further corrupted by an additive noise process. Two forms of mathematical models of such a process have been widely studied. The first form corresponds to the use of a higher order input-output difference equation and is known as the "time series" model. The second form, known as the "state space" model, makes use of a first order difference equation in a suitably defined state vector along with an algebraic relationship between the state vector and the output process. We consider here both these categories of the stochastic signal models for our subsequent use.

2.1. The State Space Model

A state space description of the system may be given in terms of the pair of equations

$$x(k+1) = A \ x(k) + B \ u(k) + w(k) \qquad (1a)$$

$$y(k) = C \ x(k) + v(k) \qquad (1b)$$

where $k = 0, 1, 2, \ldots$ is the discrete time index, $x(k)$ represents the n dimensional state vector, $u(k)$ is the m dimensional control input vector, $y(k)$ is the p dimensional output vector and A, B and C are the shift invariant model matrices of appropriate dimensions. {w(k)} and {v(k)} represent respectively the system distrubance and the observation error, assumed to be zero mean independent white Gaussian sequences with the respective covariance $P_w(k)$ and $P_v(k)$. The following additional assumptions are also made:

(a) The initial state x(0) is a Gaussian rv with known mean $m_x(0)$ and covariance $P_x(0)$.

(b) The rv x(0) and the noise processes {w(k)} and {v(k)} are mutually independent for all $k \geq 0$.

(c) The matrices A, B and C constitute a stable observable and controllable triple.

A solution of the state equation is easily derived using the method of recursion to get the result [4]

$$x(k) = A^k x(0) + \sum_{i=1}^{k} A^{k-1} B u(i-1) + \sum_{i=1}^{k} A^{k-1} w(i-1) \qquad (2a)$$

This equation relates x(k) to the initial state x(0) and to the past set of inputs. Sometimes we may also relate x(k) to the state x(j) for some $j \leq k$. We may manipulate the right hand side of Eq. (2a) to get the relation

$$x(k) = A^{k-j} x(j) + \sum_{i=j+1}^{k} A^{k-1} B u(i-1) + \sum_{i=j+1}^{k} A^{k-1} w(i-1) \qquad (2b)$$

Eq. (2a) shows that the solution for x(k) is Gaussian whose mean and covariance may be determined from the assumed statistics of x(0) and {w(k)}. Alternatively, we may derive simple difference equations for the statistics of the solution process from the given state equation. For example, taking the expectation of the two sides of Eq. (1a), we get the following equation for the mean $m_x(k) = E\{x(k)\}$:

$$m_x(k+1) = A m_x(k) + B u(k) \qquad (3a)$$

Subtracting Eqs. (3a) from Eq. (1a) and using the defining relation $P_x(k) = E\{[x(k) - m_x(k)][x(k) - m_x(k)]^T\}$ for the covariance, we may obtain the relation

$$P_x(k+1) = A P_x(k) A^T + P_w \qquad (3b)$$

where P_w is the covariance of w(k). We may also obtain the following expression for the covariance kernel $P_x(k, j) = E\{[x(k) - m_x(k)][x(j) - m_x(j)]^T\}$ with the help of Eq. (2b):

$$P_x(k, j) = A^{k-j} P_x(j), \quad k \geq j \qquad (3c)$$

Once the solution of the state equation has been obtained, it is simple to write the solution for the observed process {y(k)} using Eq. (1b). Note that the process {x(k)} is not only Gaussian but is also Markov in view of the form of Eq. (1a). The observed process is, however, only Gaussian but not Markovian. The statistics of {y(k)} are obtained easily from its relationship with the state process {x(k)} [4].

REMARKS. (a) The solution process {x(k)} is, in general, non-stationary since the mean $m_x(k)$ and the covariance $P_x(k)$ are shift variant even if the matrices A, B and P_w are constant. However, in view of the assumption of the stability of the A matrix, the limiting forms of $m_x(k)$ and $P_x(k)$, as the index $k \rightarrow \infty$, become shift invariant. This implies that the solution process of the state equation asymptotically behaves as a stationary process. With the matrix C and the covariance matrix P_v of the noise term {v(k)} also shift invariant, the observation processes will be asymptotically stationary too.

(b) The model equations are easily generalized to the case of general non-stationary .signals by letting the model matrices to be shift variant. All the relations derived above for the shift invariant model, however, remain valid for the shift variant model.

2.2 The CARMA Model

The second form of the stochastic signal model of interest to us is the "controlled auto-regressive moving-average" (CARMA) model which corresponds to the equation

$$A(q^{-1}) \, y(k) = B(q^{-1})u(k) + D(q^{-1}) \, v(k) \qquad (4)$$

As before, {u(k)} is the control input sequence, {y(k)} is the output sequence and {v(k)} is the disturbance sequence which is assumed to be a zero mean white noise sequence with covariance $P_v = \sigma^2 I$. $A(q^{-1})$, $B(q^{-1})$ and $D(q^{-1})$ are matrix polynomials of the forms

$$A(q^{-1}) = I + A_1 \, q^{-1} + \ldots + A_n \, q^{-n}$$

$$B(q^{-1}) = B_1 q^{-1} + B_2 \, q^{-2} + \ldots + B_m \, q^{-m}$$

$$D(q^{-1}) = I + D_1 \, q^{-1} + \ldots + D_n \, q^{-n}$$

where q^{-1} is the backward shift operator. The following additional assumptions are usually made:

(a) The zeros of the polynomials $A(q^{-1})$ and $D(q^{-1})$ are strictly inside the unit circle.

(b) The polynomials $A(q^{-1})$ and $B(q^{-1})$ are relatively prime.

Note that the CARMA model involves the input and the output sequences only and is thus simpler. A solution of the difference Eq. (4) may be obtained, if necessary, by making use of standard techniques. For example, for the special case of a single-input single-output system, we may use the method of long division to express the solution in the form of an infinite series [9]

$$y(k) = \sum_{i=0}^{\infty} \alpha_i \, u(k-i) + \sum_{i=0}^{\infty} \beta_i \, v(k-i) \qquad (5)$$

where the coefficients α_i and β_i may be determined for given $A(q^{-1})$, $B(q^{-1})$ and $D(q^{-1})$. This form of solution has the advantage of expressing $\{y(k)\}$ as a linear function of $\{u(k\}$ and $\{v(k)\}$ and of yielding the mean and covariance of the solution process in the form of infinite series. More specificially, this shows that the mean and covariance of $\{y(k)\}$ may be expressed by the relations below:

$$m_y(k) = \sum_{i=0}^{\infty} \alpha_i \, u(k-i) \qquad (6a)$$

$$P_y(k) = \sum_{i=0}^{\infty} \beta_i^2 \, \sigma^2 \qquad (6b)$$

It can be shown that under the assumption that all the zeros of $A(q^{-1})$ lie inside the input circle, the mean of $\{y(k)\}$ is bounded if the sum of the input sequence $\{u(k)\}$ is bounded. Under the same condition, the output variance is bounded if the input variance σ^2 is bounded.

2.3 A Canonical State Space Model

It is interesting to note that the CARMA model (also called the ARMAX model [1]) discussed above is equivalent to a state space model if the matrices A, B and C are selected to have particular canonical forms. To demonstrate this, consider the single-input single-output state space model

$$x(k+1) = A \, x(k) + B \, u(k) + w(k) \qquad (7a)$$

$$y(k) = C \, x(k) + v(k) \qquad (7b)$$

with the matrices A, B and C having the following forms:

$$A = \begin{bmatrix} -a_1 & 1 & 0 & \cdots & 0 \\ -a_2 & 0 & 1 & \cdots & 0 \\ \cdot & \cdot & \cdot & & \cdot \\ \cdot & \cdot & \cdot & & \cdot \\ -a_{n-1} & 0 & 0 & \cdots & 1 \\ -a_n & 0 & 0 & \cdots & 0 \end{bmatrix}$$

$$B = [b_1 \quad b_2 \cdots \quad b_n]^T, \quad C = [1 \quad 0 \quad 0 \quad \cdots \quad 0]$$

Let us also assume that $w(k) = F v(k)$ with F having the form

$$F = [f_1 \quad f_2 \ldots \quad f_n]^T$$

This form of the state space model is called the "observer form" and may be shown to be equivalent to the CARMA model with the polynomials $A(q^{-1})$, $B(q^{-1})$ and $D(q^{-1})$ identified from the elements of the matrices A, B and F. To determine the CARMA model parameters, we note first that the observer form of the matrix C implies the relation

$$y(k) = x_1(k) + v(k)$$

Using the first row of the state equation, $x_1(k)$ may be replaced to get the expression

$$y(k) = -a_1 y(k-1) + x_2(k-1) + b_1 u(k-1) + [f_1 - a_1] v(k-1)$$

We can now replace the term $x_2(k-1)$ using the second row of the state equation and so on. Continuing this process, we may rewrite the model equations (7a)-(7b) in the CARMA form

$$A(q^{-1})y(k) = B(q^{-1}) u(k) + D(q^{-1}) v(k)$$

where the polynomials $A(q^{-1})$, $B(q^{-1})$ and $D(q^{-1})$ have the following explicit forms:

$$A(q^{-1}) = 1 + a_1 q^{-1} + \ldots + a_n q^{-n}$$

$$B(q^{-1}) = b_1 q^{-1} + \ldots + b_n q^{-n} \qquad (8)$$

$$D(q^{-1}) = 1 + [f_1 - a_1] q^{-1} + \ldots + [f_n - a_n]q^{-n}$$

Clearly, then, the observer form of the state space model with the state noise $w(k) = F v(k)$ is equivalent to the CARMA model if the polynomials are selected as in Eq. (8).

It is possible to generalize this form to the case of systems with more than one input and output. In this case, the matrix A may be expressed as below:

$$A = \begin{bmatrix} A_{11} & A_{12} & \ldots & A_{1p} \\ A_{22} & A_{22} & \ldots & A_{2p} \\ \cdot & \cdot & & \cdot \\ \cdot & \cdot & & \cdot \\ \cdot & \cdot & & \cdot \\ A_{p1} & A_{p2} & \ldots & A_{pp} \end{bmatrix}$$

where the block matrices have the specific structures

$$
\underset{(n_i \times n_i)}{A_{ii}} =
\begin{bmatrix}
x & 1 & 0 & \cdots & 0 \\
x & 0 & 1 & \cdots & 0 \\
\cdot & \cdot & \cdot & & \cdot \\
\cdot & \cdot & \cdot & & \cdot \\
\cdot & \cdot & \cdot & & \cdot \\
x & 0 & 0 & \cdots & 1 \\
x & 0 & 0 & \cdots & 0
\end{bmatrix}
$$

$$
\underset{(n_i \times n_j)}{A_{ij}} =
\begin{bmatrix}
x \\
x \\
\cdot \\
\cdot \\
\cdot \\
x \\
x
\end{bmatrix}
\quad [0]
$$

where x represents a possibly non-zero element and n_i, $i = 1, 2, \ldots p$ represents the observability sub-indices of the system. The matrix C, on the other hand, has its rows given by

$$
C_i = [0 \quad 0 \ldots 0 \quad 1 \quad 0 \ldots \quad 0]
$$

where the unit element appears in the $1 + n_1 + \ldots + n_{i-1}$ column.

3. THE KALMAN FILTERING AND PREDICTION TECHNIQUES

In this section, we shall introduce the mathematical statements of the filtering and prediction problems, give a brief derivation of the Kalman algroithms for their solution and note some of the important properties of these algorithms.

3.1 The MV Estimation Problems

Consider the system modeled by Eqs. (1a) and (1b) under the assumptions stated earlier. A general minimum variance state estimation problem for this system may be introduced as below:

Given the system (1a)-(1b), find the estimate $\hat{x}(j/k)$ of the state $x(j)$ at the discrete time point j given the data for the set of observations $Y(k):\{y(0), y(1), \ldots y(k)\}$ so as to minimize the performance index

$$
J(\hat{x}) = \tilde{E}\{x^T(j/k) \ \tilde{x}(j/k)]/Y(k)\} \tag{9}
$$

where $\tilde{x}(j/k) = x(j) - \hat{x}(j/k)$ is the estimation error and $E\{[.]/Y(k)$ indicates the expectation of [.] conditioned on the data for $Y(k)$.

Three special cases of this general problem, which arise depending on the relative values of the indices j and k, may now be introduced:

(a) The filtering problem: In this case, we have j = k so that the estimate is denoted as $\hat{x}(k/k)$ and is defined to be the "filtered estimate" of x(k). Note that, the index k is considered to be the current time point so that the filtered estimate represents the estimate of the present state obtained by processing the data for all the past as well as the present observation.

(b) The prediction problem: In this case, j > k which implies that data for the past and present observations are used to estimate the value of the state at some future point of time.

(c) The smoothing problem: In this case, j < k which implies that the data for the past and the present observations are used in order to estimate the value of the state at a past point of time.

It is apparent that, of the three forms of estimates, the smoothed estimate employs the most amount of data which is thus expected to be the most accurate form of estimates possible. However, this not only requires more computational effort but there is a definite time delay between the points of observations and estimation which makes this form of estimate generally unsuitable for feedback control. We shall, accordingly, pay further attention to the filtering and the prediction problems only.

Before attempting to derive the desired solutions for these two problems, let us examine briefly the solution of the general MV state estimation problem. The performance index may be explicitly written as

$$J(\hat{x}) = E\{[x^T(k) \; x(k)]-2 \; \hat{x}^T(j/k)x(k)$$
$$+\hat{x}^T(j/k)\hat{x}(j/k)/Y(k)\}$$

To find the optimum value of the estimate, we use the first order optimality condition

$$\text{grad } J = \underline{0} \tag{10}$$

where the gradient is taken with respect to the components of the estimate $\hat{x}(j/k)$. It is easy to check that the following result is then obtained:

$$\hat{x}(j/k) = E\{x(j)/Y(k)\} \tag{11}$$

In other words, the minimum variance estimate of the state x(j) given the data for Y(k) is the same as the conditional mean of x(j) given Y(k).

REMARKS. Note that taking the conditional expectation of the two sides of Eq. (1b), we can get the result

$$\hat{y}(j/k) = C \hat{x}(j/k) \qquad (12a)$$

Thus, the filtered and the predicted estimates of the signal process {y(k)} may be easily obtained from the corresponding estimates of the state process {x(k)}. Alternatively, we may determine the estimate $\hat{y}(j/k)$ directly by minimizing the performance index $\tilde{F}\{y^T(j/k)$ $\tilde{y}(j/k)\}$. This yields the result

$$\hat{y}(j/k) = E\{y(j)/Y(k)\} \qquad (12b)$$

The two relations, however, yield the same result as should be expected.

3.2 Derivation of the Kalman Filter

The Kalman filtering algorithm provides with an attractive recursive form of the solution of the filtering problem. The filter estimated $\hat{x}(k/k)$ is obtained as a linear combination of the predicted estimated $\hat{x}(k/k-1)$ and a correction term that depends on the innovations $\tilde{y}(k/k-1) = y(k) - \hat{y}(k/k-1)$. To derive this algorithm, we note first that, with j = k, Eq. (11) yields the following relation

$$\hat{x}(k/k) = E\{x(k)/y(0), y(1),...y(k)\}$$

$$= E\{x(k)/Y(k-1), y(k)\} \qquad (13)$$

Let us assume that the data for Y(k-1) have been processed already to generate the estimates $\hat{x}(k/k-1)$ and $\hat{y}(k/k-1)$. Since the estimate $\hat{y}(k/k-1)$ represents the part of y(k) that can be predicted from the past data Y(k-1), the new information in y(k) is the same as that in the innovations $\tilde{y}(k/k-1)$. Eq. (13) may thus be rewritten as

$$\hat{x}(k/k) = E\{x(k)/Y(k-1), \tilde{y}(k/k-1)\} \qquad (14)$$

It is also easy to show that the variable $\tilde{y}(k/k-1)$ is orthogonal to the elements of Y(k-1) [4] so that Eq. (14) reduces to

$$\hat{x}(k/k) = E\{x(k)/Y(k-1)\} + E\{x(k)/\tilde{y}(k/k-1)\} \qquad (15)$$

The first term on the right hand side is recognized as the one step ahead prediction $\hat{x}(k/k-1)$ of the state x(k). To evaluate the second term, we recall that x(k) and y(k) (and so $\tilde{y}(k/k-1)$) are Gaussian random variables (rv). Using the expression for the conditional expectation of the Gaussian rv $x(k)$ given the value of the rv $\tilde{y}(k/k-1)$, we may rewrite Eq. (15) in the form [4]

$$\hat{x}(k/k) = \hat{x}(k/k-1) + P_{xy}(k) \, P_{yy}^{-1}(k) \, \tilde{y}(k/k-1) \qquad (16)$$

where $P_{xy}(k)$ is the cross-covariance of $x(k)$ and $\tilde{y}(k/k-1)$ and $P_{yy}(k)$ is the covariance of $\tilde{y}(k/k-1)$. To determine the two covariances explicitly, we note that the innovations may be expressed by the relation

$$\tilde{y}(k/k-1) = C \, \tilde{x}(k/k-1) + v(k) \qquad (17)$$

We are then led to the following expressions for the covariances:

$$P_{xy}(k) = C \, P_x(k/k-1) \qquad (18)$$

$$P_{yy}(k) = C \, P_x(k/k-1) \, C^T + P_v \qquad (19)$$

where $P_x(k/k-1) = E\{\tilde{x}(k/k-1) \, \tilde{x}^T(k/k-1)\}$ is the state prediction error covariance.

It is customary to rewrite Eq. (16) in the form

$$\hat{x}(k/k) = \hat{x}(k/k-1) + K(k) \, \tilde{y}(k/k-1) \qquad (20a)$$

To make use of Eq. (20a), we must also determine the prediction $\hat{x}(k/k-1)$. A relation for this quantity is easily obtained if we take the expectation of the two sides of Eq. (1a) given the data for the set $Y(k-1)$. This gives

$$\hat{x}(k/k-1) = A \, \hat{x}(k-1/k-1) + B \, u(k-1) \qquad (20c)$$

Note that the filter gain $K(k)$ may be determined if the prediction error covariance $P_x(k/k-1)$ is known. An equation for this quantity is obtained by noting first the equation for the prediction error $\tilde{x}(k/k-1)$ which is obtained by subtracting Eq. (20b) from Eq. (1a) (rewritten with k replaced by k-1):

$$\tilde{x}(k/k-1) = A \, \tilde{x}(k-1/k-1) + w(k)$$

The prediction error covariance is then obtained from the relation

$$P_x(k/k-1) = A \, P_x(k-1/k-1) \, A^T + P_w \qquad (20d)$$

Clearly, the prediction error covariance may be computed if we have the value of the previous filtering error covariance $P_x(k-1/k-1)$. To get an equation for the filtering error covariance, we note that the filtering error $\tilde{x}(k/k)$ may be expressed as

$$\tilde{x}(k/k) = x(k) - \hat{x}(k/k)$$

$$= \tilde{x}(k/k-1) - K(k)C \ \tilde{x}(k/k-1) - K(k)v(k)$$

The covariance of $\hat{x}(k/k)$ is thus given by the relation

$$P_x(k/k) = [I-K(k)C] \ P_x(k/k-1) \ [I - K(k)C]^T + K(k) \ P_v kT(k).$$

If the optimal value of $K(k)$ given in Eq. (20b) is substituted, this reduces to

$$P_x(k/k) = [I - K(k)C] \ P_x(k/k-1) \qquad (20e)$$

Eqs. (20a)-(20e) constitute a recursive chain which may be initialized by selecting $\hat{x}(0) = \hat{x}(0/-1)$ and $P_x(0) = Px(0/-1)$ where $\hat{x}(0)$ and $P_x(0)$ are respectively the mean and the covariance of the initial state.

3.3 The Kalman Predictor

It is possible to combine the equations for the filtered and the predicted estimates and covariances to obtain a dynamic form of the Kalman predictor. It is easy to check that the one-step-ahead predictor is specified by the following three equations:

$$\hat{x}(k+1/k) = A \ \hat{x}(k/k-1) + B \ u(k)$$
$$+ K*(k) \ \tilde{y}(k/k-1) \qquad (21a)$$

$$K*(k) = A \ P_x(k/k-1)C^T[C \ P_x(k/k-1) \ C^T + P_v]^{-1} \qquad (21b)$$

$$P_x(k+1/k) = AP_x(k/k-1)\{I - C^T[CP_x(k/k-1) \ C^T$$
$$+ P_v]^{-1}C \ P_x(k/k-1)\}A^T + P_w \qquad (21c)$$

Eq. (21c) serves as the recursive equation for computing the prediction error covariance starting with the initialization indicated earlier. Note that the right hand side of this equation is a nonlinear function of the covariance $P_x(k/k-1)$ and is called the matrix Riccati difference equation. The need to solve this equation causes the main computational burden of the Kalman filtering and prediction algorithms. Once the covariance matrix is updated, the gain matrix $K*$ may be evaluated using Eq. (21b) and then the estimate may be updated using Eq. (21a).

It may some times be required to predict the state d steps ahead with $d > 1$. The concerned equation may be obtained from Eq. (11) by setting j = k+d which yields the result

$$\hat{x}(k+d/k) = E\{x(k+d)/Y(k)\} \qquad (22a)$$

To evaluate the right hand side explicitly, we first imitate Eq. (2b) to write

$$x(k+d) = A^{d-1} x(k+1) + \sum_{i=k+2}^{k+d} A^{k-1} B u(i-1)$$

$$\sum_{i=k+2}^{k+d} A^{k-1} w(i-1) \qquad (22b)$$

Taking the expectation of the two sides of this equation conditioned on Y(k), we get the relation

$$\hat{x}(k+d/k) = A^{d-1} \hat{x}(k+1/k) + \sum_{i=k+2}^{k+d} A^{k-1} B u(i-1) \qquad (22c)$$

Thus, the d-step-ahead prediction of the state may be obtained recursively if we process Eqs. (21a)-(21c) and (22c) starting with the initial choices of the estimate and the covariance indicated earlier.

3.4 Some Properties

The derivation ensures that the estimates above minimize the respective error variances. Some other interesting properties of the Kalman filter are now noted:

UNBIASEDNESS. The MV estimate $\hat{x}(j/k)$ is an "unbiased" estimate in the sense that the following relation holds:

$$E\{\tilde{x}(j/k)\} = E\{[x(j) - \hat{x}(j/k)]\} = 0 \qquad (23)$$

To check that this is true, note that we may express the left hand side as

$$E\{\tilde{x}(j/k)\} = E\{E\{\tilde{x}(j/k)/Y(k)\}\}$$

where the inner expectation is conditioned on Y(k). An explicit evaluation of this expectation may be made as below:

$$E\{\tilde{x}(j/k)/Y(k)\} = E\{[x(j) - \hat{x}(j/k)]/Y(k)\}$$

$$= E\{x(j)/Y(k)\} - E\{\hat{x}(j/k)/Y(k)\}$$

But, the quantity $\hat{x}(j/k)$ is already the expected value of x(j) conditioned on Y(k) so that there is no effect of the second expectation operation on this term. Accordingly, we get

$$E(\tilde{x}(j/k)) = \hat{x}(j/k) - \hat{x}(j/k) = 0$$

STABILITY OF THE ASYMPTOTIC FILTER

The filtering and prediction equations presented earlier are valid for both shift variant and shift invariant models. Note that the Kalman filter and predictor are shift variant systems in either case. This follows from the fact that the covariance matrix $P_x(k/k-1)$ is shift variant and so the gain matrices $K(k)$ and $K^*(k)$ are also shift variant. The implementation of the Kalman filter may be considerably simplified if the gain matrix turns out to be a constant matrix. This may happen for the stationary signal case if the solution of the Riccati equation approaches a constant value P_x as $k \to \infty$. Assuming that P_x exists, this matrix must satisfy the algebraic Riccati equation (ARE)

$$P_x = A\, P_x\{I - C^T[CP_xC^T + P_v]^{-1}CP_x\}A^T + P_w \qquad (24)$$

This equation is obtained easily from Eq. (21c) by noting that we must have $P_x(k+1/k) = P_x(k/k-1) = P_x$ for $k \to \infty$.

Eq. (24) plays an important role on the existence of the asymptotic Kalman filter and on its stability. Since this is a nonlinear algebraic equation, there are more than one possible solution. A necessary and sufficient condition for the asymptotic Kalman filter to exist and to be asymptotically stable is that a positive definite solution for P_x of Eq. (24) exists. It can be shown that a sufficient condition for this to happen is that the pair (C, A) is detectable and the pair (A, Q), where Q represents the square root of the matrix P_w, is stabilizable (see, e.g., [1]). Recall that for a stationary signal process, the state transition matrix A is stable so that the detectability and the stabilizability conditions are automatically satisfied. However, even for some non-stationary signal models, the asymptotic Kalman filter may be stable.

ROBUSTNESS. The Kalman filtering algorithm obviously requires the availability of a suitable state space model for the signal process before it may be employed. Fortunately, the stable Kalman filter is relatively insensitive to the initial conditions $\hat{x}(0)$, $P(0)$ and also to the noise covariances P_w and P_v which are needed for the implementation of the filter. This ensures that the filter may often be implemented with approximately evaluated initial statistics and noise covariances. It is also possible to apply the Kalman filtering algorithm in order to obtain the best linear estimate under situations where the noise processes are not Gaussian white, though the results are no longer optimal. Further details of these and other useful properties of the Kalman filter may be found, for example, in [3] – [8].

4. LINEAR PREDICTION BASED ON THE CARMA MODEL

It is relatively simple to employ the CARMA model equation for setting up a steady state predictor for the signal process $\{y(k)\}$. To demonstrate this, we shall consider the scalar version of Eq. (4) under the two assumptions stated earlier and will find the equation for the single step prediction $\hat{y}(k+1/k) = E\{y(k+1)/Y(k)\}$. Let us introduce the new polynomial $G(q^{-1})$ by the relation

$$[D(q^{-1}) - A(q^{-1})] = q^{-1}G(q^{-1}) \qquad (25)$$

Since both $D(q^{-1})$ and $A(q^{-1})$ are assumed to be order n, the polynomial $G(q^{-1})$ would be of the order n-1. We may thus, express the new polynomial in the form

$$G(q^{-1}) = g_0 + g_1 q^{-1} + \ldots + g_n q^{-n+1} \qquad (26)$$

with $g_{i-1} = d_i - a$, $i = 1, 2, \ldots n$. Let us consider now the model equation

$$A(q^{-1})y(k) = B(q^{-1})u(k) + D(q^{-1})v(k)$$

Using Eq. (26), we may rewrite this as

$$[D(q^{-1})-q^{-1}G(q^{-1})]y(k) = B(q^{-1})u(k) + D(q^{-1})v(k)$$

which, in turn, may be rewritten as

$$[D(q^{-1})[y(k) - v(k)] = q^{-1}G(q^{-1})y(k) + B(q^{-1})u(k) \qquad (27)$$

Defining the prediction $\hat{y}(k/k-1)$ by the relation

$$\hat{y}(k/k-1) = y(k) \quad - v(k) \qquad (28a)$$

we may rewrite Eq. (27) in the form

$$D(q^{-1})\hat{y}(k+1/k) = G(q^{-1})y(k) + qB(q^{-1})u(k) \qquad (28b)$$

This is a difference equation for the one-step-ahead prediction of the output process. To establish the optimality of $\hat{y}(k+1/k)$, note that the right hand side of Eq. (28b) is completely determined from the knowledge $y(k),\ldots y(k-n)$, $u(k),\ldots u(k-m)$ since the polynomials $G(q^{-1})$ and $B(q^{-1})$ are known. The estimate $\hat{y}(k+1/k)$ thus satisfies the relation

$$E\{\hat{y}(k+1/k)/Y(k)\} = \hat{y}(k+1/k)$$

However, Eq. (28a) implies that the left hand side is equal to $E\{y(k+1)/Y(k) + E\{v(k+1)/Y(k)$. Since the second term is identically zero, we see that $\hat{y}(k+1/k) = E\{y(k+1)/Y(k)\}$ and is thus the optimal prediction of $y(k+1)$.

REMARKS. Note that in order to generate the optimal prediction, it is necessary to solve Eq. (28) with the correct initial conditions. Since these are seldom known, the predictor equation is usually solved with arbitrary initial conditions. As long as the zeros of $D(q^{-1}$ are inside the unit circle, the transients generated by the (wrong) initial data will decay to zero so that the solution approaches the correct value of the prediction in the steady state.

5. SEQUENTIAL PARAMETER ESTIMATION

In many practical cases, the parameters of the process model may not be known a-priori and need to be estimated on-line from the available input and output data before the signal estimation algorithms may be employed. The resulting solutions are then referred to as the "adaptive" or the "self-tuned" filtering and prediction algorithms. The adaptive algorithms are also required in situations where the model parameters may be known initially but are expected to change during the operation of the system. There are many possible forms of on-line parameter estimation algorithms for both the state space and the time series models (see, e.g., [10]). We shall briefly describe here a sequential least squares algorithm for the estimation fo the parameters of the CARMA and a canonical stochastic state space model of a single-input single-output system.

5.1 Estimation of the CARMA Model Parameters

The CARMA model of a single-input single-output system has the form

$$y(k) = -\sum_{i=1}^{n} a_i(y(k-i) + \sum_{j=1}^{m} b_j u(k-j)$$

$$+ \sum_{i=1}^{n} d_i\, v(k-i) + v(k) \qquad (29a)$$

Since the disturbances $\{v(k)\}$ are not measurable, we consider the relation

$$y(k) = \hat{y}(k/k-1) + e(k) \qquad (29b)$$

where $e(k) = y(k) - \hat{y}(k/k-1)$ is the innovations and $\hat{y}(k/k-1)$ is expressed (using Eq. (28a)) as

$$\hat{y}(k/k-1) = G(q^{-1})y(k-1) + qB(q^{-1})u(k-1)$$
$$+ [1-D(q^{-1})\hat{y}(k/k-1)] \qquad (30a)$$

Introducing the regression vector $\mathbf{\Phi}(k-1)$

$$\mathbf{\Phi}(k-1) = [y(k-1)...y(k-n)\ u(k-1)\ ...\ u(k-m)$$
$$-\hat{y}(k-1/k-2)\ ...\ -\hat{y}(k-n/k-n-1)]^T$$

and the parameter vector θ

$$\theta = [g_0 \cdots g_{n-1} \ b_1 \cdots b_m \ d_1 \cdots d_n]^T$$

we may rewrite the model Eq. (29b) in the regression form

$$y(k) = \phi^T(k) \ \theta + e(k) \qquad (30b)$$

Assuming that the regression vector is completely known, the least squares estimation theory [1], [10] may be employed to develop a sequential algorithm for the estimation of the parameter vector θ. The estimation index $J(\theta)$ is defined by the relation

$$J(\theta) = (1/2)\{ \sum_{i=1}^{k} [y(i) - \phi^T(i-1)\theta]^2$$

$$+ [\theta-\theta(0)]^T P_0^{-1}[\theta-\theta(0)]\} \qquad (31)$$

where $\theta(0)$ is the assumed initial estimate of θ and p_0 is a positive definite matrix that indicates our confidence on this estimate. The optimal estimate of the vector θ is defined to be the value that minimizes the index $J(\theta)$. Using the first order optimality condition

$$\text{grad } J(\theta) = \underline{0}$$

where the gradient is taken with respect to the components of θ, we get the desired least squares parameter estimate as

$$\theta(k) = P(k-1)[P_0^{-1}\theta(0) + \Phi^T(k-1)Y(k)] \qquad (32a)$$

We have used the notation $\theta(k)$ to indicate the estimate of θ based on k observations $Y(k) = [y(1) \ y(2) \ \cdots \ y(k)]^T$ and have defined the regression matrix $\Phi(k-1)$ and the matrix $P(k-1)$ by the relations

$$\Phi(k-1) = [\phi(0) \ \phi(1) \cdots \phi(k-1)]^T$$

$$P^{-1}(k-1) = \Phi^T(k-1)\Phi(k-1) + P_0^{-1} \qquad (32b)$$

Note that Eq. (32b) may be rewritten as

$$P^{-1}(k-1) = P^{-1}(k-2)+\phi(k-1)\phi^T(k-1), \ P^{-1}(-1) = P_0^{-1} \qquad (33a)$$

Also, we may rewrite Eq. (32a) as

$$\theta(k) = P(k-1)[P_0^{-1}\theta(0) + \Phi^T(k-2)Y(k-1) + \phi(k-1)y(k)]$$

$$= P(k-1)[P^{-1}(k-2)\theta(k-1) + \phi(k-1)y(k)]$$

The right hand side may be rewritten with the help of Eq. (33a) in the intuitively appealing form

$$\theta(k) = P(k-1)[P^{-1}(k-1)-\phi(k-1)\phi^T(k-1)\theta(k-1)$$
$$+ P(k-1)\phi(k-1)y(k)$$

$$= \theta(k-1)+P(k-1)\phi(k-1)[y(k)-\phi^T(k-1)\theta(k-1)] \quad (33b)$$

Eqs. (33a) and (33b) specify a sequential least squares algorithm for the estimation of the parameters of a system with known past inputs, past outputs and past predictions.

In this form, the matrix P needs to be computed in the inverse form and then inverted in order to determine the gain term $P(k-1)\phi(k-1)$. A better alternative is to make use of the matrix inversion lemma [1], [10]) in order to rewrite Eq. (33a) in the form

$$P(k-1) = P(k-2) -P(k-2)\phi(k-1)\phi^T(k-1)P(k-2)$$
$$[1 + \phi^T(k-1)P(k-2)\phi(k-1)]^{-1} \quad (33c)$$

REMARKS. (i) Note that the constant parameter vector θ satisfies the model equation

$$\theta(k+1) = \theta(k)$$

If this is considered to be the state equation and if the regression model Eq. (30b) is considered to be the corresponding observation equation, a recursive algorithm for the estimate $\theta(k)$ may be easily obtained from the Kalman filtering algorithm. It is easy to check that this yeilds Eqs. (33b) and (33c) provided we assume $\{e(k)\}$ to have unit variance.

(ii) The algorithm above is often called the "pseudo linear regression algorithm" since the regression vector, being a function of $\hat{y}(1/1-1)$, $1 = k-1,...k-n$, is actually a non-linear. The solution obtained above is necessarily approximate in view of the dependence of the regression vector on the parameter estimates.

5.2 Estimation of the State Space Model Parameters

The original state space model specified by Eqs. (1a) and (1b) has more parameters than we may efficiently estimate from the information contained in the input-output data. It has been shown that an appropriate state space model, whose parameters may be estimated from these data, corresponds to the steady state "innovations model."

$$\hat{x}(k+1/k) = A \hat{x}(k/k-1) + B u(k) + K^* e(k) \quad (34a)$$

$$y(k) = C \hat{x}(k/k-1) + e(k) \quad (34b)$$

where K^* is the steady state value of the Kalman predictor gain and A and C have the observer canonical form. Since the innovation process $\{e(k)\}$ is white, the results of sub-section 2.3 may be utilized in

156

order to replace the innovations model by an equivalent CARMA model. Equations (33a)-(33c) are then directly applicable to this case. More details may be found in [1].

5.3 The Convergence Issue

An important question about any sequential parameter estimation algorithm is that of its convergence to the correct parameter starting with an arbitrary initial estimate. This issue is discussed at length in standard texts on parameter estimation (e.g., |1|, [10]). An interesting conclusion of the convergence analysis is that under reasonable assumptions on the disturbance sequence, such as

$$\sup_{k} (1/k) \sum_{i=1}^{k} v^2(i) < \infty$$

and on the system, such as

$$[1/D(z) - 1/2] \text{ is strictly passive}$$

the pseudo linear regression algorithm produces parameter estimates with bounded norm of the estimation error. For more details of this and other convergence behaviors, the reader may refer to the above references.

6. ADAPTIVE PREDICTION

For signals with unknown models, an adaptive signal estimation algorithm may be obtained if we combine the parameter estimation algorithm of the preceding section with the signal estimation algorithms of Sections 3 and 4. For example, an adaptive one-step-ahead prediction algorithm would consist of the following equations:

$$\hat{y}(k+1/k) = \phi^T(k)\theta(k) \tag{35a}$$

$$\theta(k) = \theta(k-1)+P(k-2)\phi(k-1)[1+\phi^T(k-1)P(k-2)\phi(k-1]^{-1}$$
$$[y(k)-\phi^T(k-1)\theta(k-1)] \tag{35b}$$

$$P(k-1) = P(k-2) -P(k-2)\phi(k-1)\phi^T(k-1)P(k-2)$$
$$[1+\phi^T(k-1)P(k-2)\phi(k-1)]^{-1} \tag{35c}$$

A Numerical Example

It is possible to combine the algorithms of Sections 3 and 5 in order to derive an adaptive Kalman filtering and prediction algorithm. In order to demonstrate the results of application of the adaptive Kalman predictor, we consider a simple second order system

$$x(k+1) = \begin{bmatrix} -a_1 & 1 \\ -a_2 & 0 \end{bmatrix} x(k) + \begin{bmatrix} b_1 \\ b_2 \end{bmatrix} u(k) + \begin{bmatrix} f_1 \\ f_2 \end{bmatrix} v(k)$$

$$y(k) = [1 \quad 0] \, x(k) + v(k)$$

The system has been simulated using P_v = .25, a_1 = -1.8, a_2 = .8, b_1 = 1, b_2 = -.4, f_1 = 1 and f_2 = 1. The control input $u(k)$ has been generated as the state feedback law

$$u(k) = -M(k) \, \hat{x}(k/k-1)$$

with $M(k)$ computed as

$$M(k) \, M(k) = [B^T(k)B(k) + .5]^{-1} \, B^T \, A(k)$$

where the hat indicates the estimated values of the model matrices. This relation may be obtained by using the minimum state error variance control theory [11]. The hat indicates the estimated parameters obtained using the pseudo linear regression algorithm. Figure 1 shows the parameter estimates and the output of the system obtained starting from the indicated initial estimates. Note that K_1 and K_2 have the values $f_1 - a_1$ = 0.8 and $f_2 - a_2$ = 0.2.

7. REFERENCES

1. G. C. Goodwin and K. S. Sin, "Adaptive filtering, prediction and control," Prentice Hall, NJ, 1984.

2. K. J. Astrom and B. Wittenmark, "Computer controlled systems: Theory and design," Prentice Hall, NJ, 1984.

3. R. E. Kalman, "A new approach to linear filtering and prediction problems," J. Basic Eng., Trans. ASME, Series D, Vol. 82, pp. 35-45, 1960.

4. J. S. Meditch, "Stochastic optimal linear estimation and control," McGraw Hill, NY, 1969.

5. A. P. Sage and J. L. Melsa, "Estimation theory with applications to communication and control," McGraw Hill, NY, 1971.

6. T. Kailath, "Lectures on linear least squares estimation," Sprinter Verlag, NY, 1976.

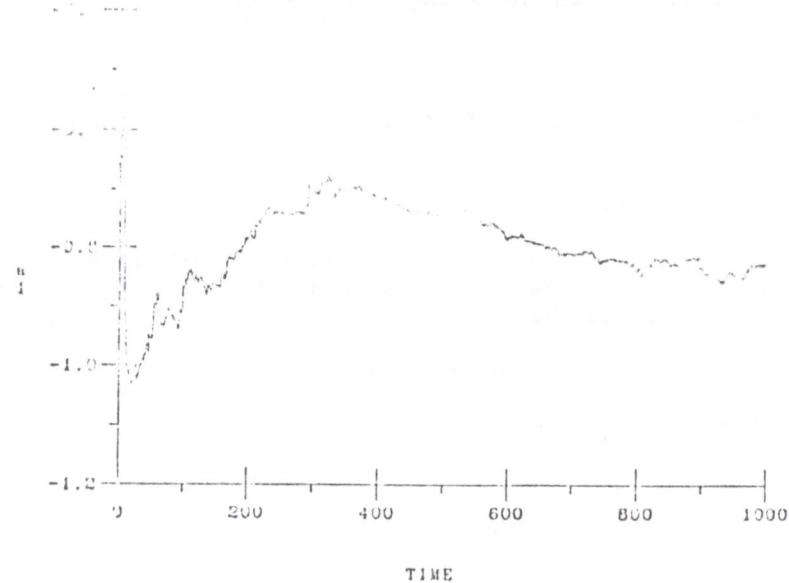

TIME

Fig. 1(a)   Estimate of a_1

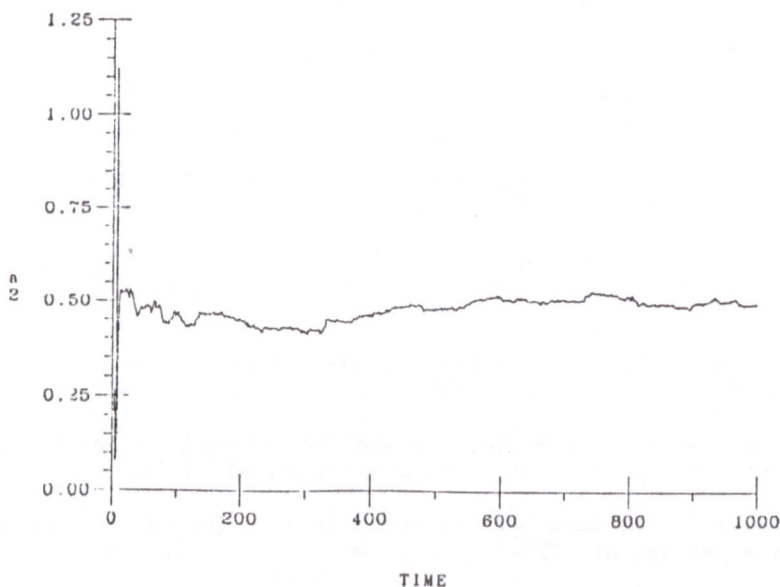

TIME

Fig. 1(b)   Estimate of a_2

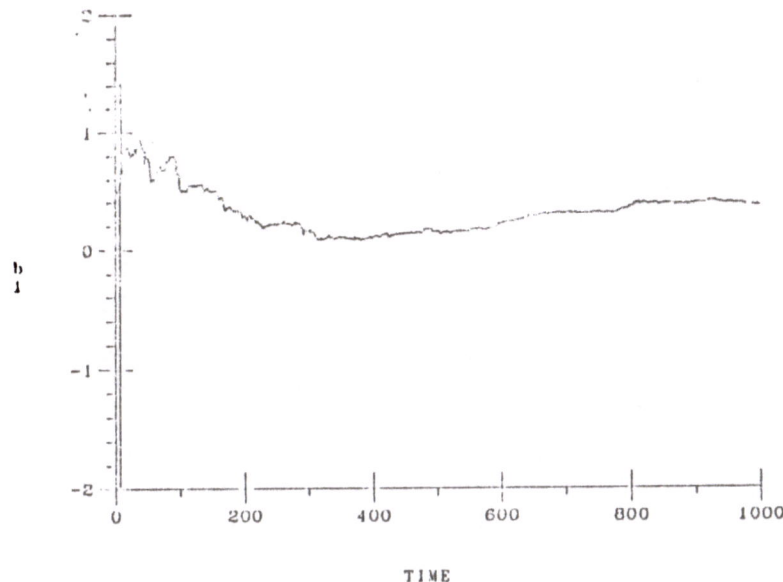

TIME

Fig. 1(c)   Estimate of b_1

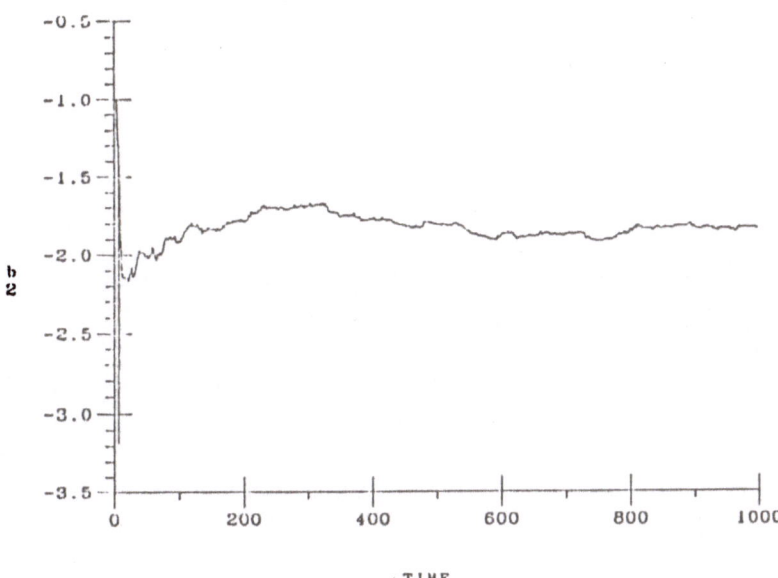

TIME

Fig. 1(d)   Estimate of b_2

160

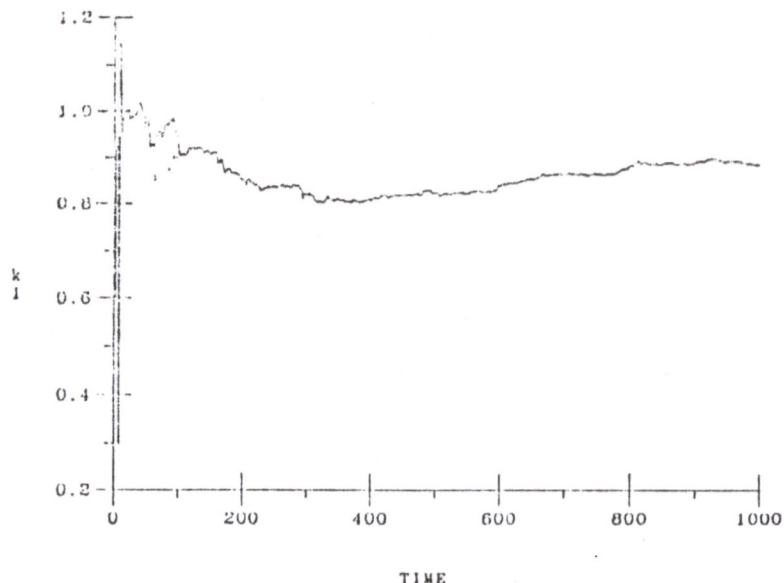

TIME

Fig. 1(e) Estimate of k_1

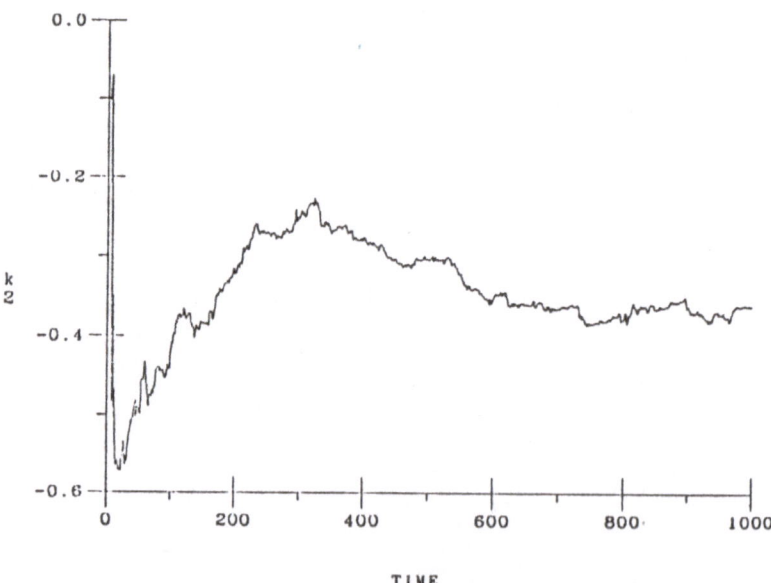

TIME

Fig. 1(f) Estimate of k_2

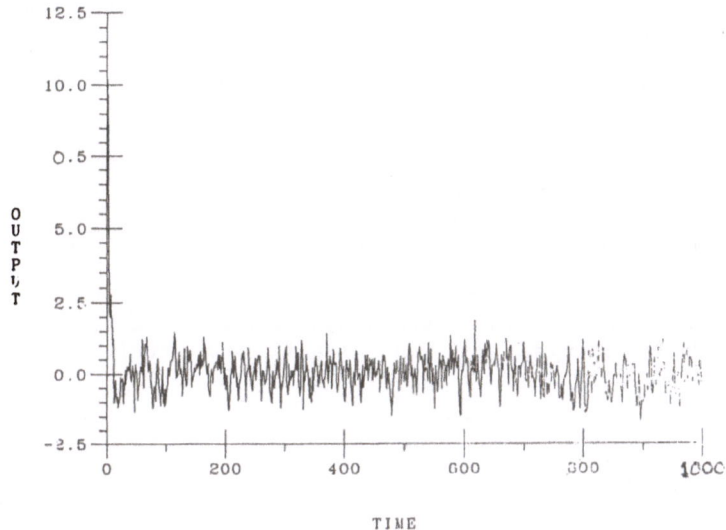

Fig. 1(g)   Output of the system under adaptive control.

162

7. G. J. Bierman, "Factorization methods for discrete sequential estimation," Academic Press, NY, 1977.

8. B.D.O. Anderson and J. B. Moore, "Optimal filtering," Prentice Hall, NJ, 1979.

9. E. A. Robinson and M. T. Silvia, "Digital foundations of time series analysis," Holden Day, San Francisco, 1979.

10. L. Ljung, "System identification: Theory for the user," Prentice Hall, NJ, 1987.

11. P. Desai, "A study of minimum state error variance self-tuning controllers," Ph.D. Thesis, The Pennsylvania State University, May 1988.

Chapter 5

CONFIGURATION OF DISTRIBUTED CONTROL SYSTEMS

P.Purkayastha
Desein (New Delhi) Pvt Ltd
Consulting Engineers
GK II New Delhi 110048
India

ABSTRACT. In this chapter, we trace the evolution of the current generation of distributed systems from two origins -the digital computers and the process control systems based on pneumatic and electronic controllers. The various generations of distributed systems are discussed briefly. Based on present system capabilities, the control system architecture is discussed in relation to the plant structure. Simplified models of the process are developed and a method of impact analysis is used to evaluate control system architecture. The use of such techniques can provide a firmer basis of design than the intuitive methods used in current system designs.

1. INTRODUCTION

The advent of microprocessors saw their wide spread use in plant controls in the form of distributed control systems. Such systems were able to introduce the advantage of digital technology without sacrificing the reliability and fault tolerance of the earlier analog systems. The major problems of catastrophic failure associated with the earlier digital controls based on centralized computer control architectures were largely overcome. Though distributed systems were to see widespread use in the industry, design principles on which the architecture of such systems were based were rarely articulated except in very general terms. As the distributed systems were themselves being pulled in differing directions by vendors offering competitive conceptual premises, the practicing control system designer was not well placed to define the design requirements of such systems. With more than a decade of experience of distributed systems, it is now possible to see the broad contours of developments taking place as well as to formulate design considerations for the configuration of plant control systems.

The major consideration in control system architecture is its relation to the plant and the nature of the control problem. The consideration for control system

163

S. G. Tzafestas and J. K. Pal (eds.), Real Time Microcomputer Control of Industrial Processes, 163–186.
© 1990 Kluwer Academic Publishers.

architecture for plants which have hard real time constraints or where the consequence of failure of the control system could be catastrophic can be quite different from that of systems where the process can be easily controlled manually on the failure of the system. Similarly, the nature of the process, the subsystem boundaries and the relation between the subsystems in structural terms are important considerations for defining the control system architecture (Purkayastha and Pal:1985). Here we will examine various considerations for system architecture and relate them to the process and the plant. We will attempt to evolve certain general principles for control system design architecture.

2. CONTROL SYSTEM EVOLUTION

The distributed systems of today have originated from two different streams. One stream is digital computers and the other is process control systems starting from the pneumatic systems to the electronic controllers.

2.1 Pneumatic and Analog Systems

The pneumatic instruments and controls of the 1930's were distributed in nature. The standardization of signal to the 3-15 psi level occurred with the further development of pneumatic controls and made possible the centralized operation of the plant. The centralized control room where the operator had an overview of the plant as well as had control of all remote devices is the legacy of the late 40's and 50's, and retains its validity even today.

The electronic controllers made their appearance almost simultaneously with the digital computers. Though in functional terms the electronic controllers were virtually analogues of the pneumatic systems, the type of signal transmission changed to the more convenient electrical form. The pneumatic systems had been characterized by long transmission lags in the pneumatic lines which were eliminated by the speed of transmission of the electrical signal. The electrical signal was prone to the problem of being sensitive to "noise" which could be dealt with by extra circuitry. (Kompass: 1981)

The electronic controllers soon came to be separated from the indicators and the operator's interface , namely the hand auto stations. This was the "split concept", which is still commonly used in a number of industries. The ease of maintenance, the modularity of the systems, the relative ease of change and the miniaturization of the systems all contributed to the popularity of such systems.

2.2 Digital Control with Centralised Systems

The chemical industry was the first to introduce digital computers for process control. The use of such systems was at the supervisory level where the computer generated the set point for a set of analog controllers. The high cost of such computers could only be justified if the process was complex and required such changes of set points based on changing process conditions.

Supervisory control was first tried by Monsanto for it Ammonia Plant at Luling in 1959 (Suski and Rodd:1985). The first application of a process computer for control applications was by Texaco (Astrom:1985). Supervisory control was generally restricted to static optimization of a process to decide the best operating point and derive from that the set points of the lower level analog controllers. The use of the lower level controllers of the conventional type ensured that the availability of the system was not affected by the use of a computer, while the benefits of optimization etc. could be derived from it.

The use of computers encountered certain problems, which in some measure are still present today. The use of static optimization required plant models which were generally not available (Astrom:1985). Optimization for a quick changing process required dynamic models which were even more difficult. To compound the problem, chemical processes were closely guarded secrets of the concerned companies who were loath to part with this knowledge to the control system vendors. The initial attempts to use supervisory control resulted in intense activity to develop plant models based on the physics and chemical kinetics of the process. Even today, this remains an area which requires far more work.

This period also saw the introduction of process computers for monitoring, for working out plant balances and efficiency calculations, and for generating plant logs and management reports. On-line monitoring was restricted by the slow speed of the system and peripherals, though logging, report generation and plant balance calculations provided useful tools for overall supervision of the process.

The attempt to use direct on-line digital control was made in the 60's (Suski and Rodd:1986). The Imperial Chemical Industry was the first to introduce direct digital control in one of its plants. The use of direct digital control met with a mixed success. In plants where the process dynamics were fast and the control loops tightly coupled, the application was not popular. The digital computer of those days was an expensive device of low computational power. Further, it was failure prone requiring specialized maintenance personnel not easily available in the industrial environment. In order to justify the cost of such systems, the designers tried to pack more and more functions into the systems resulting in further increase of the already heavy computational load. The speed of response of such systems degraded during large plant upsets as the amount of information required to be processed under such conditions was beyond the computational resources of the system. Further, the catastrophic failures, in the sense of the entire plant controls being lost on the failure of the computer, also ensured the unpopularity of direct digital controls for tightly coupled plants.

The philosophy of direct digital control was to implement the controls on a small computer if the plant was small, and on a large system if the plant was large. Figure 1 shows the centralized control structure adopted for such systems.

Many measures were evolved to take into consideration the problems encountered in actual applications. Direct digital control may have had serious limitations as outlined above. It, however, also provided insights into the control software, evolved algorithms which went beyond the conventional PID algorithms then available on analog systems, and developed various redundancy and back-up schemes relevant even today.

Coupled with the use of direct digital control, an attempt was made to automate the plants (Adam, Kinney, Hamm and Robichau:1967). Computers could be softwared to handle a number of modes of plant start-ups and shut-downs. Here again, though valuable insight was gained in the automation of the process, the attempts did not prove very successful. Sensors and final control elements, along with the inherent problems of reliability and low computational power of the systems, were to severely restrict such applications.

As computers became cheaper and more reliable, there was a gradual increase of computer applications for control purposes. The mini-computer developed for process applications made its appearance in the late 60's. By 1970, there were more than 5,000 process control computers. This number was to rise to 50,000 by 1975 (Suski and Rodd:1986). This, of course, includes all process control applications including supervisory, monitoring and direct digital controls.

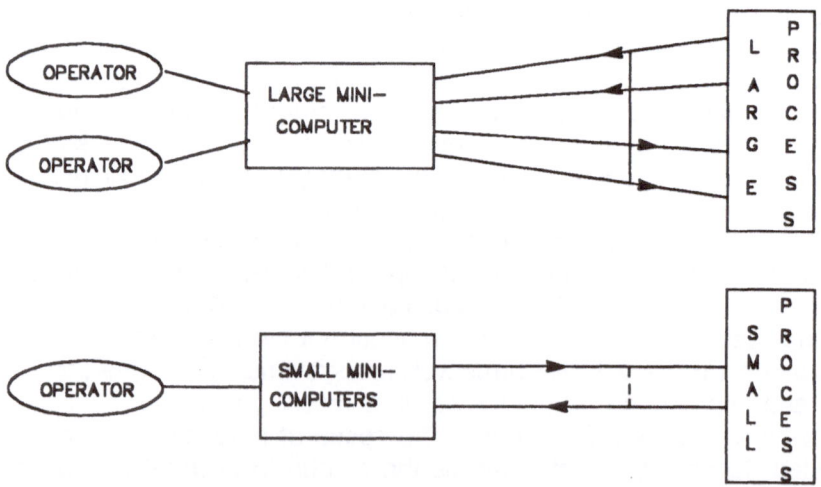

FIGURE-1   Direct Digital Control

2.3. Distributed Systems

The pneumatic and analog systems had functionally distributed the control functions in hardware even if they were geographically centralized. The use of direct
digital control was to accentuate the centralized control concept even more. The
distributed system introduced by Honeywell was to provide a break in the
continuing centralization that was taking place since the 40's.

TDC 2000, the model T of the distributed system of Honeywell, was
the first to geographically distribute the controllers in the plant while retaining
centralized operator functions. Further, the advantage of digital electronics was
retained along with functional distribution similar to the analog controllers. Though
more than one loop could be resident on the same controller, the catastrophic
failure associated with the centralized computer was avoided. Soon after Honeywell, a number of other control system manufacturer's introduced similar
systems. In plants which are geographically distributed over a large area, the
advantage of reduced cabling along with the flexibility of digital systems, made
distributed systems quite popular. Soon distributed systems were rapidly
entering into other areas like power stations, metallurgical industries etc.

2.3.1. Hierarchic Distributed Systems. The distributed systems of the earlier variety
consisted of a number of local controllers communicating with each other, as well
as a large central computer for operator communication. The controllers were
mostly 8 bit with small amounts of memory and low communication speeds.
The majority of distributed systems in use today are such systems. They have been
called second generation systems, second generation in terms of digital control
systems (Suski and Rodd:1986). Such systems have a hierarchic structure and
have different processors for different functions. Figure 2 gives the general structure
of such systems.

The use of a number of processors for implementing the control functions
of the plant meant that there had to be a means for the processors to talk to each
other. Further, there was a need to communicate with the host computer where
a number of operator communication functions were resident. Sharing of
information and maintaining coherence in a distributed system within the
constraints of real time was the key task of the system. The communications
network in a distributed environment was the key to system performance.

The amount of information transmitted in such a system depended on the
amount of processing that could be performed in the controllers themselves. The
more the processing that was carried out on raw process data, the less the
information that was required to be transmitted. However, as the operator
communicates to all the nodes, there had to be a minimum throughput of the
system in order to provide a quick system response to operator demand. This
might be an occasional demand, but the system had to be able to meet this worst
case "scenario".

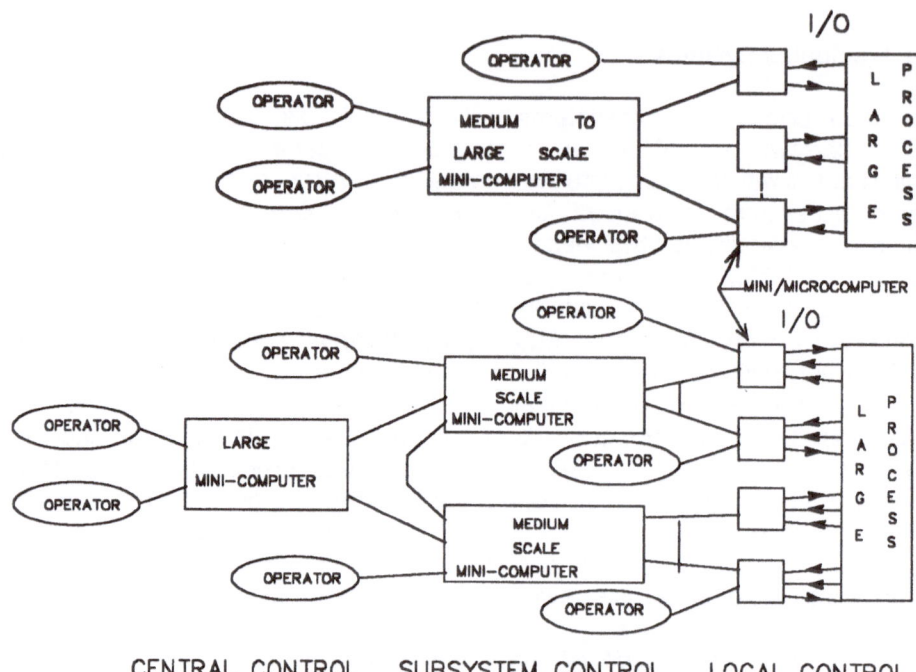

CENTRAL CONTROL   SUBSYSTEM CONTROL   LOCAL CONTROL

FIGURE-2   Hierarchic Distributed Control System

Various network topologies were worked out to meet the requirements of distributed systems. Figure 3,4 and 5 show the star, point to point and data highway structures. The star topology used a large central computer to "talk" to all the distributed processors. The central processor had to participate in all the communications and obviously the system was limited by the capacity of the central processor. The failure of this processor would lead to the failure of communications of the entire system. The point to point communication was faster and less prone to such failure of communications. It had, however, the problem of being very expensive, as all nodes had to have as many ports as the number of nodes. The expansion of such systems was rather difficult, as for each extra node added, every node would have to be modified to add an extra port.

The data highway concept where all the nodes talk to each other through a shared resource - the high speed bus - was a major conceptual advance in distributed controls. A serial data bus could be utilized by a geographically and functionally distributed system with a number of nodes. The Honeywell's TDC 2000 was the first to introduce this concept followed by other leading manufacturers. The bus speeds and the capacity of individual processors were limited in such systems.

To overcome these problems, data concentrators were used along with separation of the plant areas. These systems also allowed for local operators, who could communicate to a centralised plant operator through the plant data highway.

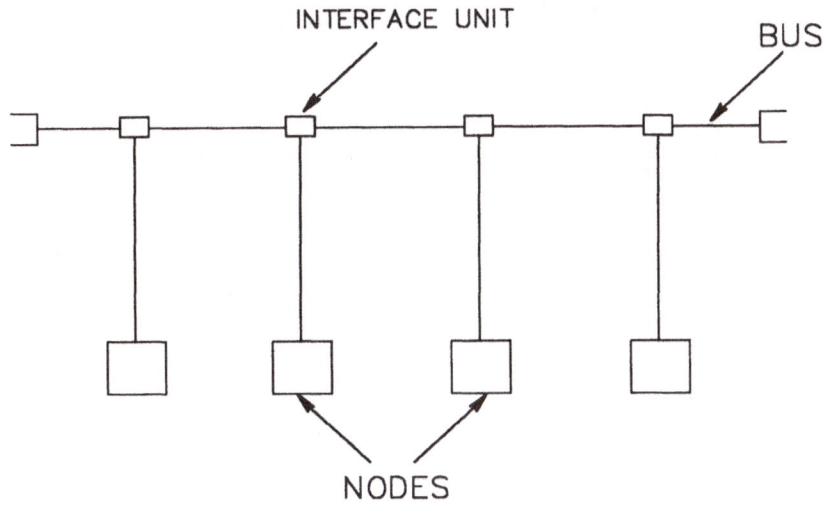

FIGURE-3   Bus Network

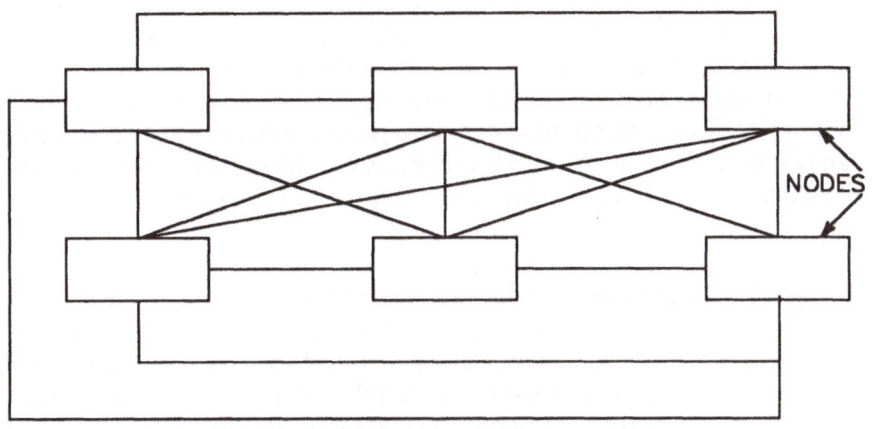

FIGURE-4   Point to Point Connection

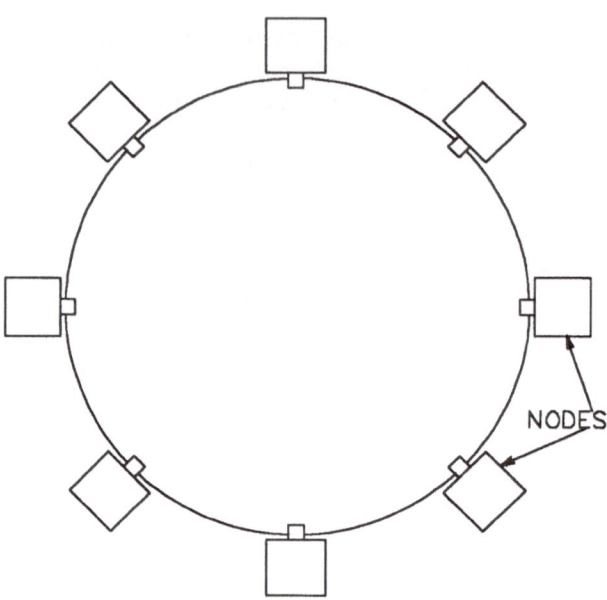

FIGURE-5   Ring Network

The major concern in the highway based systems was how a number of processors could share the same resource such that all processors could communicate without some being frozen out of the network or one hogging the network (Ibe:1987). Catastrophic failures of communications had also to be engineered against. In conventional data processing applications, where real time constraints are not involved, bus contentions are allowed. If collisions are detected, the nodes which are transmitting "back off" using different rates, till a collision free transmission takes place. This is the basis of the popular "ethernet" protocol. The ethernet types of protocols, have however, the problem of being indeterministic in terms of a guaranteed minimum time period within which a communication will take place. Though increased bus speeds can be used to virtually guarantee communications, the inderministic protocols have not proved so popular for control purposes.

There have been two ways by which bus contentions have been avoided in distributed control systems. In one method a bus master mediates between nodes and decides who has access to the bus. The other method is to have a token which moves across the network with the holder of the token having access to the bus. This method has gained a lot of popularity and more and more systems are using this.

To protect against catastrophic failures of the communication system, generally redundant communications are used such that a single point failure does not lead to the   failure of all communications.

The distributed systems that emerged in this period were really digital counterparts of analog systems in terms of control strategies and functions. They did not use the full power of digital computations, but provided the flexibility of digital systems. The systems could be modified or configurations changed quite simply, the digital electronics were relatively drift free, systems were easily expand- able, the use of CRT graphics gave the operator a valuable insight in to the process.

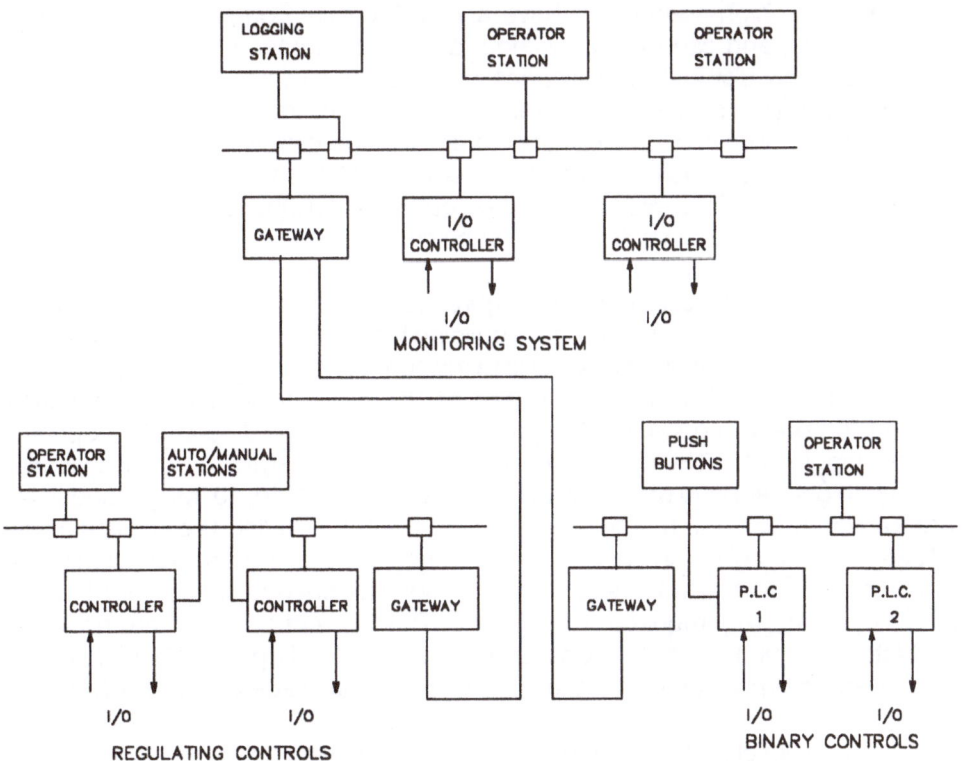

FIGURE-6 Interconnection between various systems

There were, however, certain problems with the systems. As can be seen from Figure 6, a number of separate systems were in coexistence for the same plant. Quite often, PLCs would be used for open loop functions, a distributed system for closed loop functions and a separate computer based system for data acquisition and other computational tasks. Tying up the systems was a major problem in such an approach. Normally, gateways existed in the different systems to communicate to each other. In practice, this proved a bottleneck if large volumes of data had to be moved, apart from the inherent communication difficulties in a mixture of protocols. The response time for operator communications was also slow in such systems, as the centralized computer could not really meet the computational load under all plant conditions.

2.3.2. Unilevel Systems. In the mid 80's, third generation systems based on more powerful processors - 16/32 bit micros - made their appearance. These systems are characterized by a unilevel structure (Imamichi and Inamoto:1985). These systems use a high speed bus, have almost identical computational resources at each mode, and can function without a host computer in the network. This permits a highly distributed intelligence in the system with fewer building blocks, greatly simplifying system structure. The removal of the host computer makes such systems more acceptable to the industrial plant operating staff who are still wary of computers. Such systems also have the ability to combine closed loop and open loop functions, along with data acquisition and other computational tasks. Figure 7a,b shows the architecture of such systems.

The unilevel structure has the simplicity of a near identical node configuration. The specific node is configured to meet a certain number of functional tasks, but is basically the same set of building blocks arranged in a different way. The software for each node is specific to the functions the node performs, which again can be drawn from a common pool of library software. The peripherals for the node of course depends on the specific functions involved - archiving, operator station, logging etc. The data highway structure in such systems is also simpler. The much greater baud rates available - in the range of 1 to 10 Megabauds - is a great advance over the much slower speeds available earlier. The earlier highway speeds required either the use of data concentrators or the segmentation of the plant areas and highways such that the highway speeds would not become bottlenecks to the required throughput. The current systems are no longer restricted by such considerations and can permit a large number of nodes/ controllers communicating over the same highway with no deterioration of performance.

The unilevel systems have the advantage of having only a few types of cards for the entire system. A lower inventory of spares and ease of maintenance are obvious benefits to such an approach. Some of the systems which have adopted a unilevel approach are those of Westinghouse, Taylor, Hitachi, Honeywell etc.

Some of these systems have combined with the above unilevel structure, compatibility to their earlier systems. The consideration here has been to provide upgrades from their existing installed base of such systems, so as not to render them obsolete. The TDC-3000 seeks to incorporate the TDC-2000 in order to upgrade the existing installations of TDC-2000. Others like Hitachi, have used more than one network in order to transfer information from one control area to another. However, conceptually the trend is towards a more unilevel approach.

The current generation of systems all offer modulating controls, binary controls and monitoring functions within the same system. Unlike the earlier situation, where there existed a difficult problem of matching the vendors of programmable logic controllers, with modulating controllers and with data acquisition systems, the current generation of systems offers all these functions as a part of the same system. Fast loop responses like 10 to 50 milliseconds, required for certain interlocks, configuration of complex loops and large computations can now be met with the same hardware.

In the next sections we will first examine the plant control problem and see how it can be distributed over a number of processing elements. The control system architecture can then be examined in relation to the plant structure.

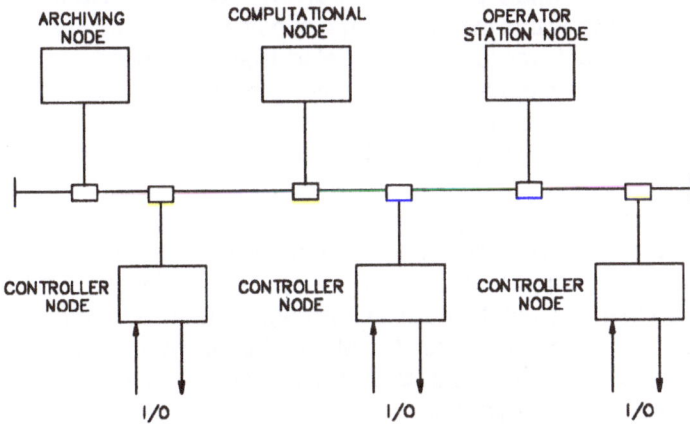

FIGURE - 7a   Unilevel Structure - Bus

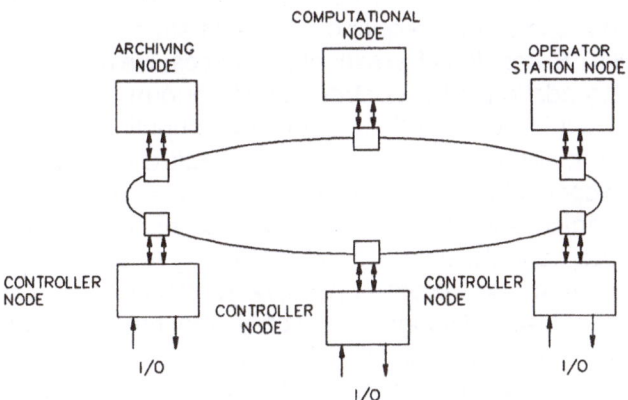

FIGURE - 7b   Unilevel Structure - Ring

3.0 DISTRIBUTED CONTROL PROBLEM

A control problem for a complex process or a complex of processes can be decomposed into a set of control problems which require some information exchange. Normally, the decomposition of the control problem corresponds to the natural boundaries of the process. While this is true for process control plants, there are certain industrial control problems where large computational resources are required in real time. Here, the computing problem has to be broken up for its solution within the given time constraints leading to a distributed configuration.

The distributed problems in robotics, avionics etc. are problems which have hard real time constraints - in the range of milliseconds (Kopetz:1986). The control problem is also quite often one large problem split up due to the requirements of computing needs within the given time constraint. In other words, the control problem here is much more tightly coupled than a process control problem. Further, the control systems of aircraft cannot tolerate mid flight failures and fault tolerance here is an important consideration for system architecture.

The process control problem is characterized by slower real time constraints, smaller data throughput from the plant sensors, and natural boundaries of the plant subsystems. Though the subsystems can be tightly or weakly coupled, the failure of one control loop can be tolerated by the operator taking over manually the failed loop. Only in certain loops, their critical nature and their fast dynamics may lead to a plant outage in the case of manual control. Generally this is more of an exception. Apart from the above, the distributed control problem has another characteristic. Past data states can be used even if current states are not available as the process dynamics are not fast. This results in much greater autonomy to the distributed computing problem where each processor need not wait for the other processors to furnish the latest computed values but can go ahead with the past value available. In most industrial control problems, the computational speed and sampling frequency available are an order higher than the process dynamics. Except for a few exceptional cases, the use of past states does not lead to the degradation of the control system performance.

From above it is clear that the distributed computing problem is not similar for all systems. Normally, the control systems available do permit a range of design options within the constraints of their systems. However, a specific system may be well suited for a particular application but not well suited for another.

The process control problem has also differing functions embedded in the control system. Generally in a process plant, both closed loop controls and open loop controls are involved. Though intrinsically, the protection and sequential control problems are subsumed under open loop controls, there are fundamental differences which have a bearing on system architecture. Protection requirements have the highest priority in the system, while the sequential control and regulating controls have a lower order of priority. This introduces a hierarchy in the control functions (Fig 8).

As has already been stated, the control system can function based on past states as the sampling speeds and control computations are an order higher than the process dynamics. It must, however, be kept in mind that the speed with which protection must act necessitates a faster sampling cycle than for regulating or sequential control tasks. The combining of all binary functions can blur this distinction and lead to wrong selection of cycle times. The allocation of control tasks to controllers and "prioritizing" the control schemes are the key problems in control system design. The control systems available, however, permit a wide flexibility to the designer in terms of such configuration.

From the foregoing, it may be seen that the system architecture is determined primarily by process considerations. It is the nature of the plant and its processes which must be clearly understood for configuring the system.

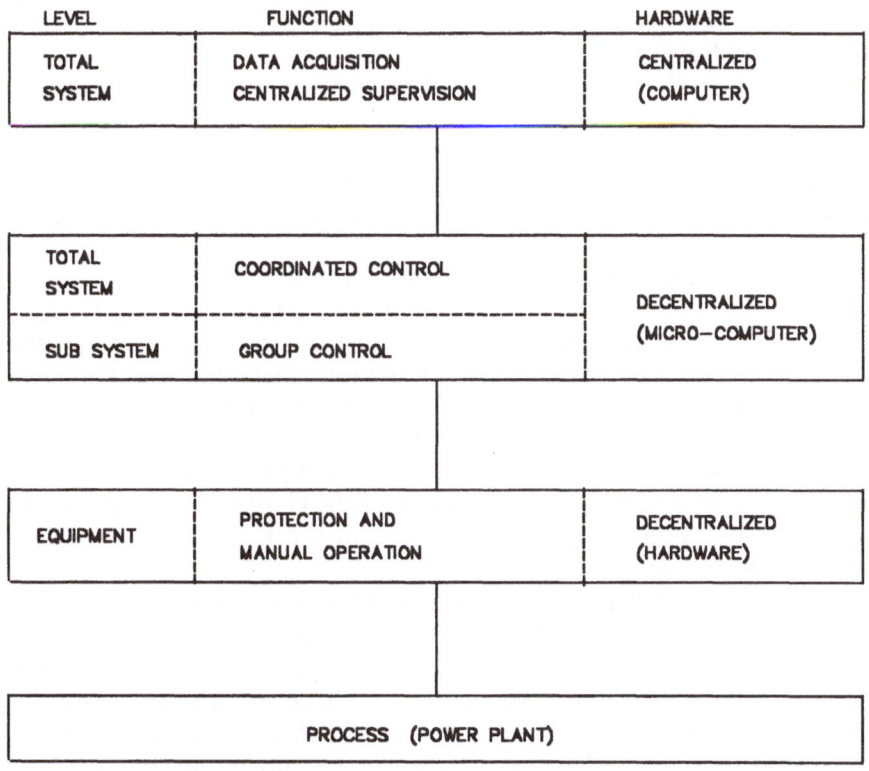

FIGURE - 8   Concept of System Structure

4. CRITERIA FOR SYSTEM SEGMENTATION

In the earlier type of control system, the segmentation was based on different functions being incorporated in different hardware. Thus, the relay based or solid state systems performed open loop functions, while the regulating functions were performed on different hardware. Though this is no longer a necessity as the current microprocessor based systems can incorporate both functions, the practice amongst a large number of practitioners is still to keep the functions of regulation and binary controls separate.

The present generation of systems allows both the regulating and the logic tasks to be built into the same hardware. Therefore, we see that the programmable logic controller (PLC) manufacturers offer regulating software blocks as a part of their standard software. Similarly, more and more control system manufacturers are also offering logic functions also as a part of their repertoire. Differences of approach and hardware exist based on past practices, though it is increasingly becoming clear that an evolution of systems in a common direction is taking place.

Apart from the availability of hardware and software to do both functions, there may still be reason for system segmentation to be based on control system tasks. In fact, two clearly discernible trends are visible regarding system segmentation. One scheme is to segment the system in terms of control functions. The other is to define the control system by the subsystem boundaries of the process.

Two common fallacies exist regarding control system design. Given the power of today's processors, it is possible to load a large number of functions on to a few nodes. This does not take into account that a control system is embedded in the plant, and therefore, structurally should not violate the basis of plant design. There is no point in building up a large number of parallel streams and making all their controls resident on a single piece of hardware. Unfortunately, the process designer and the control system designer seem to inhabit different domains presupposing an autonomy which does not exist in reality. The control system designer should structurally take into account the plant structure and not be designed autonomously.

The other fallacy is the multiplication of hardware with the belief that this ensures a better system. The plant structure replicated so faithfully may result in an amount of hardware which reduces overall reliability rather than increasing it. The correct approach is the optimal loading of the controllers with an appropriate segmentation of the system based on plant subsystem boundaries.

We are now in a position to identify the major system design goals (Purkayastha:1988). They are as follows:-
   a) Embed the control system structure in the process structure
   b) Reduce information transfer between nodes so that each node can function nearly autonomously

c) Optimally load each node after taking into account the above
d) Provide the necessary redundancy depending on plant criticality and increase system reliability
e) Minimise system costs

While b) and c) above have been recognised as a problem of optimal design (Imamichi and Inamoto:85), the relation of the control system to the plant has been generally underplayed. The practicing engineer, on the other hand, has used intuitive techniques to partition the system based on his feel for such systems. A good design under such circumstances is more of an art and not easily replicable.

After the basic system is designed, standard reliability tools are used to evaluate the availability of the system so designed (Rooney:1983). Again, this fails to take into account that the impact of failure on the plant is not the same for all loops or functions. While for non critical loops or loops with slower dynamics, a failure can easily be tolerated, the plant may require a shut down if a certain other portion of the system fails. In a hazardous process, the plant may have to provide for a safe shut down under conditions of partial system failure. The standard tools of reliability analysis have to be modified somewhat to take into account the plant structure and the criticality of certain functions.

In the next section, we will discuss the approaches used to take into account some of the above considerations. We believe, that this, combined with standard techniques of optimizing processor loading provides a better approach to the now practiced hit and try methods.

5. TASK ALLOCATION IN A DISTRIBUTED ENVIRONMENT

In this section, we will develop a method of impact analysis which will seek to quantify the impact of failure of a node or processor on the physical plant. This is based on the loss of load principles (Billinton and Allan:1984) which are used in power system studies. For the use of such techniques, we will first develop certain simplified models of the plant.

5.1 CONTROL TASK ALLOCATION PROBLEM

The distributed system design problem is that given a set of controllers and a set of tasks, the tasks have to be allocated to various controllers satisfying the basic design objectives as identified in the previous section. Given a set of control tasks N, and a set of "m" controllers, N is to be divided into a set of "n" tasks such that $m*n = N$. This segmentation of the control tasks necessitates a set of information exchange "p" between the "m" controllers. Various methodologies have been used to allocate the control tasks. Some have minimized the information flow "p" between the controllers, while distributing the "n" tasks in such a way that individual controller upper bounds regarding loading and memory capacities are met. Others have tried to meet response time requirements of the system while minimizing the number of controllers. Mixed integer programming techniques (Purkayastha and Singh:89) as well as dynamic programming techniques (Ma, Lee and Tsuchiya:1982; Imamichi and Inamoto:1985) have been used to solve the optimal allocation of tasks . While the optimisation problem can be solved using any of the standard techniques, the more important consideration is to incorporate the design objectives into the constraints and the minimising function.

With the advent of high speed data highways, minimizing information exchange between controllers has ceased to be an important criteria by itself. The necessity to reduce information flows stems from the need to reduce the impact of failure of one controller on another, which is an independent design objective. Similarly, memory restrictions are no longer serious, given that the current generation of processors can readily access large amounts of memory.

The major consideration in control loop allocation is that the impact of failure of one controller on the plant as a whole has to be minimized. Before we examine this, we would like to introduce a simple method which can model the plant topology in a meaningful way.

It can be shown that any process plant can be modelled for our purpose by using two types of system blocks and two types of connection between the blocks. The two types of system blocks are active blocks and passive blocks while the two types of connections are parallel connections and the series connections. Though more complex models can be built, for the purpose of impact analysis, the topology of the process is captured sufficiently by this.

An active block is a plant subsystem within which a particular process is taking place - there is always a process flow through an active block which is acted upon in some way within the block. In a passive block, there is a flow in or a flow out of the block, but no transformation takes place within the block. An active block, is therefore, a unit of the process which has a process flow and a set of associated control tasks. The active block transforms the input in some way to produce a set of outputs. For instance, reactors, pumps, compressors, heaters etc. are all active blocks by the above definition.

Passive blocks are used in any plant to provide a built-in capacitance in the system. They act as storage elements of the process fluid and can decouple a set of active blocks from each other. Sometimes, the storage provided is so small that the block acts only as a temporary "flywheel" in the system and cannot really be called a passive block. If the storage provided in a process block can sustain the process for a time T, then it will be considered a passive block if $T >> t$, where t is the largest of the time constant of the active blocks of the process. However, if T and t are of the same order, then the storage is not sufficient for it to be called a passive block.

We can examine a small plant topology using the above. Figure 9 and Figure 10 shows two active blocks A and B connected in series and in parallel. Figure 11 shows how a storage block can be used to decouple the series process and convert it to a parallel process.

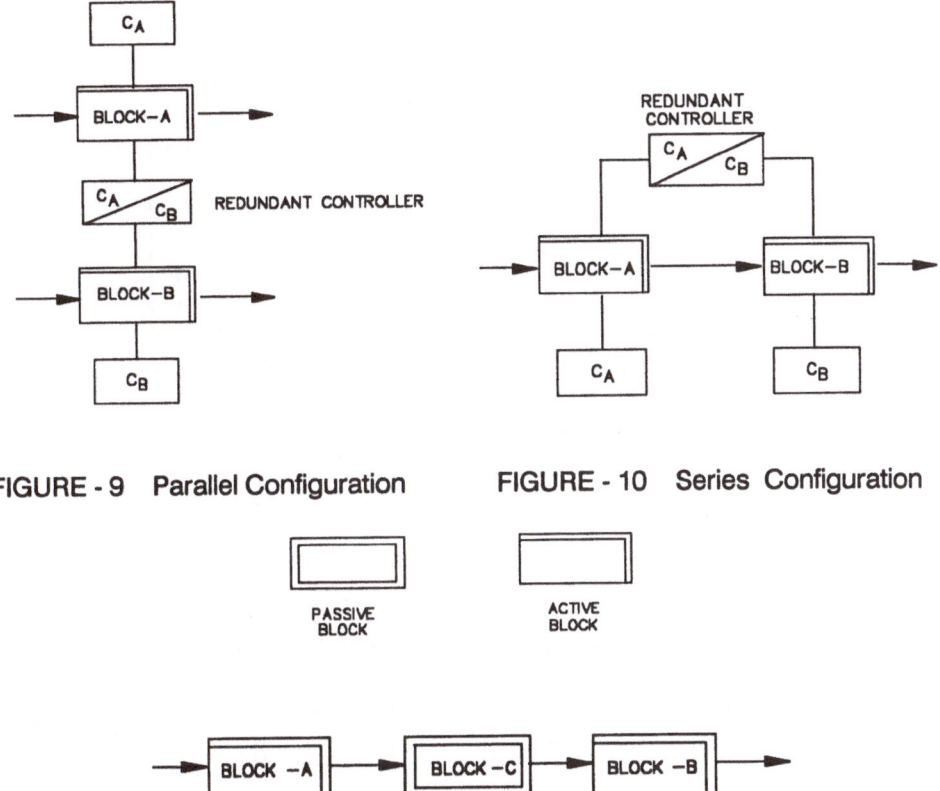

FIGURE - 9   Parallel Configuration   FIGURE - 10   Series Configuration

FIGURE-11   Conversion of Series Process to Nearly Parallel - One

Consider a flow diagram as given in Fig.12a and its topological model. In this plant there are two streams, each stream having a set of control tasks and control elements. In a conventional task allocation problem, either the data transfers between controllers or the execution times of control tasks are minimized. The decoupling of the control tasks of two streams is not considered in the allocation problem. However, in case such a coupling occurs and the control tasks of two different streams are allocated to one controller, the failure of this controller can seriously upset both the process streams. The allocation of control tasks of the two streams to one controller, then, leads to upsetting both streams simultaneously, due to failure of a single element - the controller. In effect, the parallelism of the process in terms of the two streams has been destroyed by inadvertently coupling them through the control task allocation.

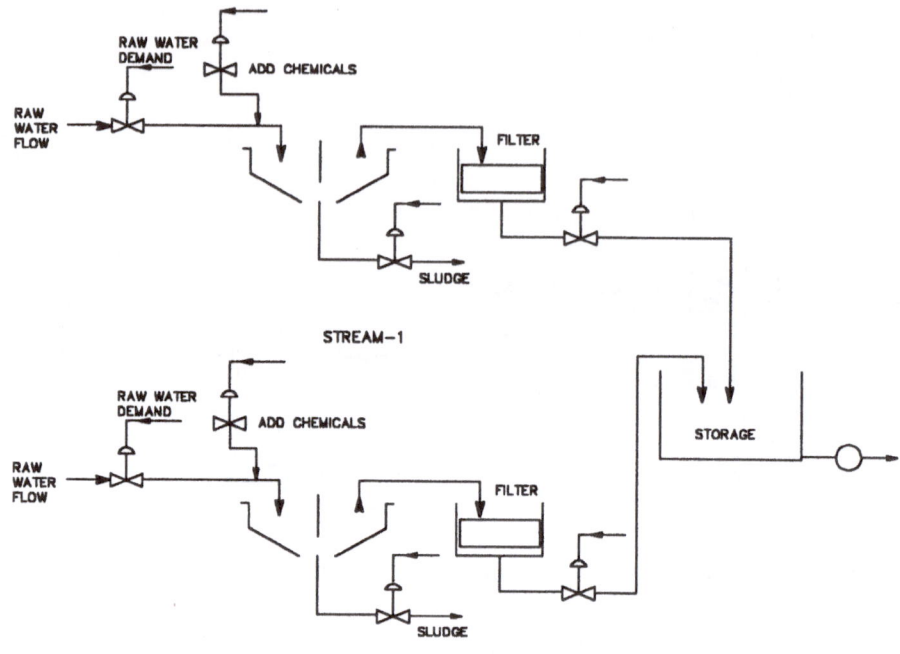

STREAM-1

STREAM-2

FIGURE - 12a   Process Configuration for a Water Treatment Plant

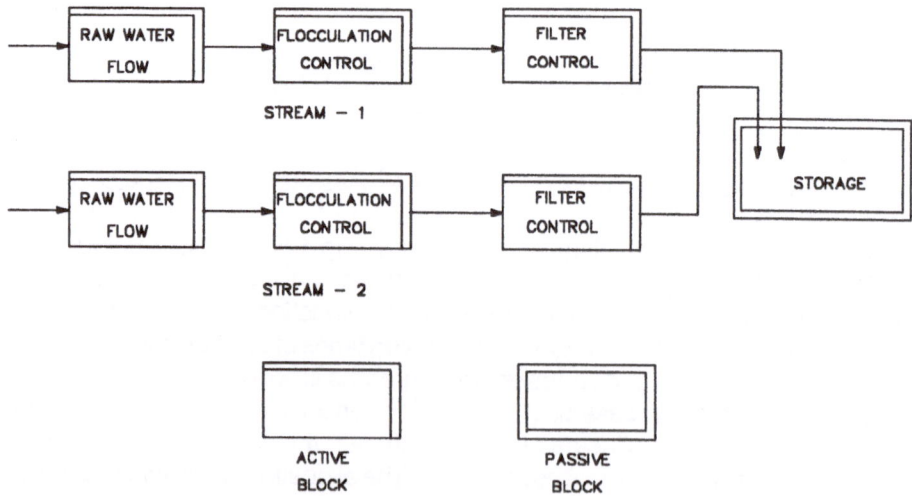

STREAM - 1

STREAM - 2

ACTIVE
BLOCK

PASSIVE
BLOCK

FIGURE - 12b   Process Topological Model of the Water Treatment Plant

5.2. Process Models and Reliability Relations

For a given control system, we have discussed the use of simple reliability relations for evaluating total system reliability. Here, we will discuss the control system reliability as a part of the process itself.

There are many votaries in the industrial control environment for single loop integrity. In fact, this appears more as a conditioned reflex of dealing with analog systems which necessarily had single loop integrity.

The hardware in analog systems did not allow resource sharing. Single loop integrity was the result of such an inability. The development of digital systems permit resource sharing. This allows the flexibility of choosing the specific loops or control tasks which will share a processor or controller.

Simple reliability relations can be used to work out the availability of the various controller configurations. We give below the various parameters used for evaluation of reliability.

$$A = \frac{MTBF}{MTBF + MTTR}$$

Where MTBF = Mean time before failure

MTTR = Mean time to repair

MTBF = $1/\lambda$

A = Availability

U = Unavailability

λ = is the failure rate of the modules

MTTR = $1/\mu$

μ = is the number of repairs carried out per unit of time or the repair rate of modules.

As λ is much smaller than μ, the unavailability U is given by

$$U = \frac{\lambda}{\lambda + \mu} = \frac{\lambda}{\mu} = \lambda * (MTTR)$$

For a series chain connection

$$A_{SYS} = A_1 * A_2 * A_3 ... A_n$$

Where A_{SYS} is the availability of the system and A_1, A_2...A_n are the individual availabilities of the modules in the chain. For two modules in parallel, the combined availability is

$$A_{SYS} = A_1 + A_2 - A_1 * A_2$$

This provides availability for a one to one back up. However, availability computations can also be made for a one is to n backup, where one controller backs up n controllers (Moore and Herb:1987). A more detailed exposition of reliability relations is available in the literature (Rooney:1983).

Given the failure rates and repair rates of modules, it is possible to work out the overall reliability of the systems and also the loss probabilities associated with each failed state. Capacity utilization tables identify such probabilities and can be used to compare alternate configurations. Obviously, configurations which are less likely to be unavailable are better configurations from this stand point.

Let us examine the serial process modelled in Figure 9. In this process model, if we use individual controllers with availability of 0.9975, and assume that the failure of the controller leads to the loss of the process, the availability of the process using single loop integrity is

$$A = 0.9975 * 0.9975 = 0.995$$

Obviously, this is lower than the use of one controller for both the processes. Using redundant controllers, availability increases dramatically.

$$A = (1 - .0025^2) = 0.99999375$$

The above results show that by simply increasing hardware in a serial process model, one does not gain in terms of availability. On the contrary, the overall availability of the system is lowered due to the nature of the reliability relations. The more this chain of controllers is built up in the serial process, the more fragile the overall system is to failure. However, in case the plants have parallel structures, this would not hold good. In parallel processes, extra hardware increases overall availability. As each process is autonomous of others, obviously, independent controllers preserve this autonomy.

Chemical plants generally have a decoupled structure with large amounts of intermediate storage (Stephanopoulous:1983). The oil price shock and the consequent high cost of energy, has resulted in reduction of intermediate storage thereby coupling the plants more tightly.

In decoupled plants, the process dynamics are generally slower and the control loop structures simpler. This permits the operator to take on a number of loops on manual, without impairing greatly plant performance.

The quantitative method which has been used here, attempts to evaluate alternate control structures by using impact analysis. In such a scheme, the impact of the failure of a controller on plant capacity is used to evaluate alternate partitioning of the system. The method is equivalent to the loss of load in power system studies.

If Z is the derated plant capacity on loss of a controller, then the impact factor α is defined as follows :

$Z = \alpha$ C, where C is the capacity of the plant. In other words, the factor α embodies the impact of the failure of the controller on the plant. As manual back-up is available, the loss of a controller does not lead to the loss of the plant. α can vary from 0 to 1 depending on the criticality of the loop.

A simple example will illustrate the use of this method. Consider two plants with two sub-systems A and B. In case I, the two sub-systems are in series so that the loss of any sub-system leads to the loss of the plant (Fig.9). In case II, the sub-systems work in parallel (Fig.10), and the loss of any sub-system leads to loss of 50% of the output. For the above two cases, the impact of loss of controllers on the plant have been worked out to evaluate alternate control strategies. For the purpose of this evaluation, the availability of the mechanical system is taken to be one. This results in showing up plant availability only in terms of control system availability and is therefore relevant to our specific problem.

Based on certain numerical values of the impact factor, the failure rate and the mean time to repair the capacity utilization tables have been constructed for two differing control structures. In one case, single loop integrity has been preserved and in another redundant multi-loop controllers have been used. The results are tabulated below :

Table I : Capacity Utilization
Series Plant Configuration with $\alpha = 0.0$

Derated Capacity (Z)	Individual Probability	
	Single loop Integrity	Redundant Multifunction Controllers
0.0	0.0197	0.0067
0.25	0.0	0.0
0.50	0.0	0.0
1.0	0.9802098	0.9933

$\lambda = 5000$ f/10^6

Table 2 : Capacity Utilization
Series Plant Configuration with $\alpha = 0.5$

Derated Capacity (Z)	Individual Probability	
	Single loop Integrity	Redundant Multifunction Controllers
0.0	0.00	0.0
0.25	0.0081	0.0067
0.50	0.0116	0.0
1.00	0.980298	0.9933

$\lambda = 5000$ f/10^6

184

Table 3   :   Capacity Utilization
                 Parallel Plant Configuration with α= 0.0

| Derated Capacity (Z) | Individual Probability | |
	Single loop Integrity	Redundant Multifunction Controllers
0.0	0.0009801	0.0067
0.25	0.0	0.0
0.50	0.0178417	0.0
0.75	0.0	0.0
1.00	0.9811781	0.9933

λ = 5000 f/10^6

Table 4   :   Capacity Utilization
                 Parallel Plant Configuration with α = 0.5

| Derated Capacity (Z) | Individual Probability | |
	Single loop Integrity	Redundant Multifunction Controllers
0.0	0.0	0.0
0.25	0.0009801	0.0067
0.50	0.0	0.0
0.75	0.0178417	0.0
1.00	0.9811781	0.9933

λ = 5000 f/10^6

For the case of the series plant configuration, the Tables 1 and 2 make clear that irrespective of the impact factor and system failure parameters, the probability of being in each derated state is always lower for the redundant controllers. For the case of parallel plant configuration, the issue is not so clear. In fact, with α = 0.5, it can be seen that there is a strong argument in favour of single loop integrity. The impact of controller loss as above, can be coupled with the loss of production and a straight forward analysis made of the cost of outage (Kuusisto:1982). A trade off between the cost of outage and cost of extra hardware can be made if desired.

The above analysis required the formulation of the impact factor for each controller. As this is a hypothetical figure based on the subjective judgement of the designer, there may be questions regarding this. However, it may be noted that subjective weighting factors are already used when a designer computes availability for his system. Instead of giving a uniform weight for each controller, a differential weighing, corresponding to the importance of each loop to the overall plant, appears to be a more relevant approach. A detailed examination has been made regarding the relative importance of each control loop for a thermal power plant elsewhere (Purkayastha and Pal:1985). It is possible to generate similar tables for any process plant and then use various partitioning and levels of redundancy to arrive at a control structure. Though this would still have certain subjective elements built into it, it would generate more appropriate structures than those based on pure intuition of the designers.

An alternate way of attempting the task allocation problem would be to dissociate certain loops and associate certain loops based on plant structure. Associative and dissociative constraints can then be introduced in the optimization problem. This would ensure that loops belonging to one process stream are dissociated from other such process streams in case the streams are parallel.

The key design problems in a distributed system are the segmentation of the plant and the control task allocation. The levels of redundancy required are a function of the critical nature of the control tasks. No scheme can achieve all design goals. A designer can at best optimize a conflicting set of design objectives based on actual plant topology and functional requirements.

REFERENCES

Adams B, W F Kinney, C R Hamm and J O Robichau (1967) Extending Power Plant Automation with a Process Computer, Bailey Meter Company Reprint E-19

Astrom K J (1985), Process Control - Past, Present and Future, Control Systems Magazine, Vol 5 No.3 PP 3-9

Billinton R and R N Allan (1984), Reliability Evaluation of Power Systems, Plenum Press.

Fisher D G (1983), Computer Control of Decentralised Process Systems, Journal of Computers and Chemical Engineering, PP 395-421

Ibe, Oliver C (1987), Introduction to Local Area Networks for Manufacturing and Office Systems, IEEE Control Systems, Vol 7 No.3, PP 36

Imamichi C and A Inamoto (1985), Unilevel Homogenous Distributed Computer Control Systems and Optimal System Design, 6th IFAC Distributed Computer Control Workshop, Monterey, California

Martinovic, A. (1983) Architecture of Distributed Control System, Chemical Engineering Progress, February, PP67-77.

Kompass Edward J. (1981), A Long Perspective on Integrated Process Control Systems,Control Engineering,Vol 28 No.9 PP4-9.

Kopetz H (1986), Time Rigid Scheduling in Distributed Real Time System, 7th IFAC Distributed Computer Control Workshop, Mayschoss, West Germany.

Kuusisto, T H (1982) Redundancy in Process Automation System from Viewpoint of Running Economy. IFAC Conference on Direct Digital Control of Power Plants, London PP 49-56.

Moore J A and S M Herb (1987), Understanding Distributed Process Control, ISA.

Purkayastha P (1988), Distributed Control Systems - Implementation StrategiesProceedings of the International Seminar on Microprocessor Applications for Productivity Improvement New Delhi. (Ed) V P Bhatkar and Krishna Kant, Tata Mcgraw Hill PP 329-340.

Purkayastha P and S Singh(1989), Task Allocation and Optimal Design for Distributed Systems, 9th IFAC Distributed Computer Workshop, Tokyo, Japan.

Purkayastha P (1988), Communication Network for Distributed Systems, COMMEX 2000, New Delhi.

Purkayastha P and J K Pal (1985), Design Considerations of Distributed Control Architecture for a Thermal Power Plant, 6th IFAC Distributed Computer Control Workshop, Monterey, California.

Rooney J P (1983), System Reliability Assessment Techniques - a Comparison, Proc International Reliability, Availability and Maintainability Conference IEEE.

Stephanopolous G (1983), Synthesis of a Control System for Chemical Plants - a Challenge to Creativity, Journal of Computers and Chemical Engineering, Vol 7 no.4, PP 331-365.

Suski G J and M G Rodd (1986), Current Issues in Design, Analysis and Implementation of Distributed Computer Based Control Systems, 6th IFAC Distributed Computer Control Workshop,Mayschoss, West Germany.

Chapter 6

MODELLING AND SIMULATION

J. K. Pal
Engineering Technology and Development Division
Engineers India Limited
1, Bhikaiji Cama Place
New Delhi 110066, India

and

S. G. Tzafestas
Control and Robotics Group
Computer Engineering Division
National Technical University
Zografou, Athens 15773, Greece

ABSTRACT. The chapter presents a state of the art review of modelling and simulation for industrial process control. Introduction to mathematical modelling, validation of models, and numerical methods for solution are included. Both lumped and distributed parameter systems are covered. Various popular simulation languages, and industrially available simulation softwares are discussed. Cases for common industrial equipment and plant, i.e. boiler, turbo-generator, distillation column, pipeline, and digital control systems are considered. Application of multiprocessors and expert-systems as future of simulation are indicated. An industrial case study of a combined cycle power plant contributed by Chiyoda Corporation is presented in the Appendix.

1. INTRODUCTION

The rapid development of highly capable and cheap microprocessors has greatly changed the prospects of implementing sophisticated control strategies in industrial plants. In the last three decades there have been important developments in control theory. Developments in microprocessor based control systems and control theory have helped in the realisation of advanced control and information systems in industrial equipment and plants. In the case of industrial plants the steady state design stage preceeds the control system design stage. The steady state design is particularly amended only when a control system with sufficient dynamic performance cannot be designed due to poor steady state behaviour of the process. The controller design procedure

187

S. G. Tzafestas and J. K. Pal (eds.), Real Time Microcomputer Control of Industrial Processes, 187–248.
© 1990 *Kluwer Academic Publishers.*

188

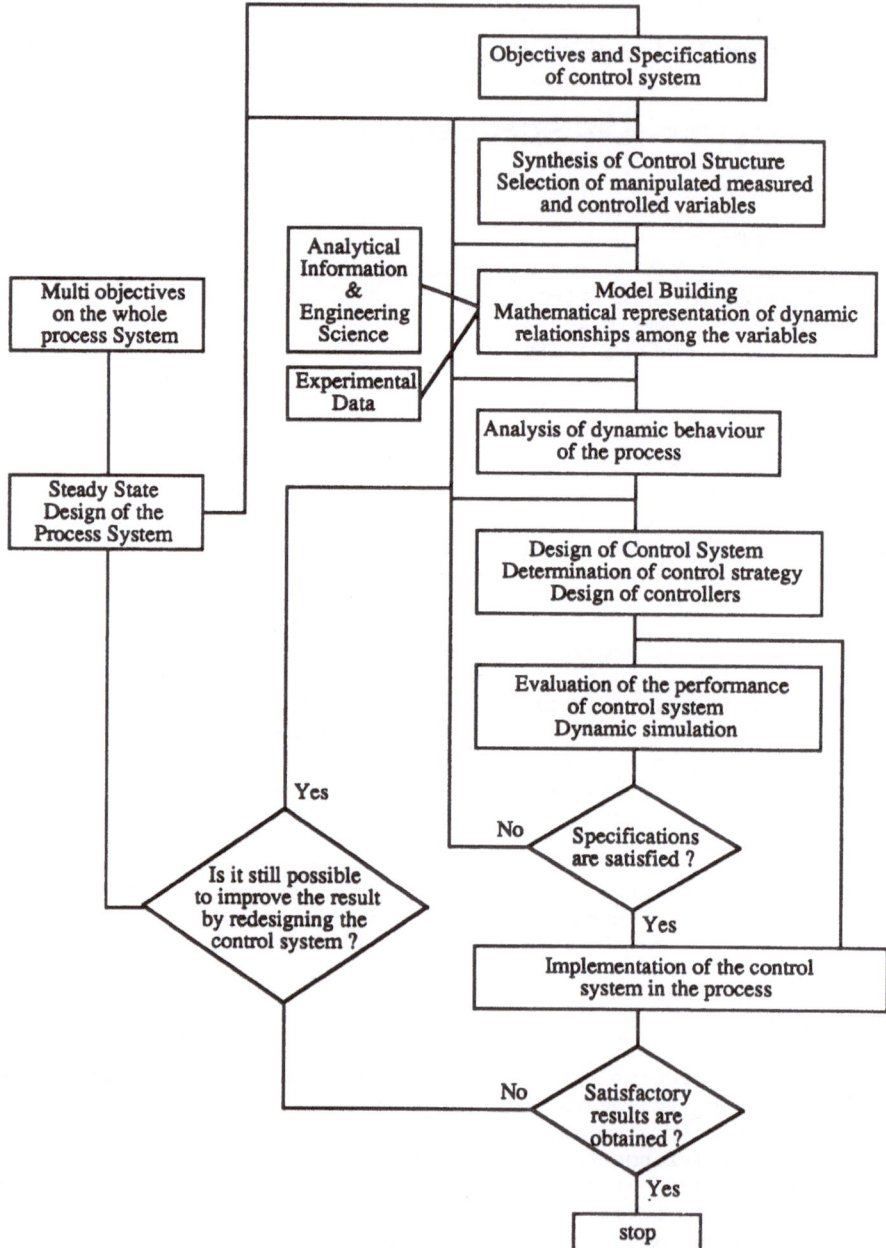

Fig. 1 Design Procedure for a control system

of a plant involves verification of the selected strategy by simulation. The procedure has been best depicted by Hashimoto [1] and is shown in Fig.1. This figure explains the role of dynamic simulation in finalising a control scheme of an individual equipment or a plant complex.

Simulation is a modern methodology. It provides an efficient and economic way to analyse a system. It has been in use in control system engineering for more than 40 years [2]. It has been an established procedure to get proper feedback based on simulation results about the effectiveness of a control scheme before its implementation. Modern process monitoring and control methods are generally based on good plant simulation results. Implementation of state estimation, multivariable control, dynamic and static optimization, and predictive control are based on simulation results. Many digital control techniques include plant modelling and/or simulation inherent in the algorithm.

Simulation [3] can be seen as a multistage problem solving process as shown in Fig.2. To start with, a detailed analysis of the original system and the given problem has to be done to arrive at an appropriate mathematical model. This model can be implemented in computer. For the implementation of this model, programming requirements, numerical methods and availability of simulation hardware and software have to be checked. The model should also be validated with respect to the actual behaviour of the original system. After all these stages are completed the simulation process can be used for application.

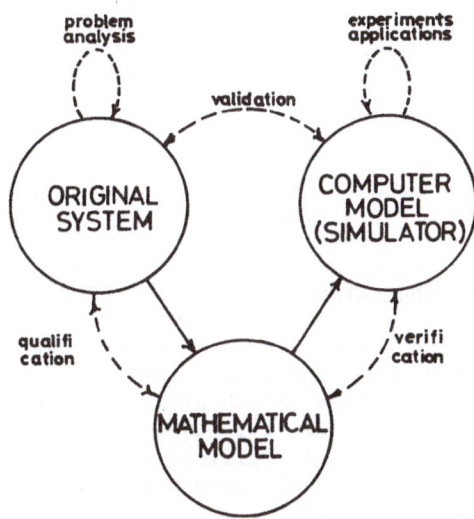

Fig 2 Simulation as multistage
problem solving process

Our discussion in this chapter will be concentrated on simulation of process control systems which may involve a controller dedicated to

190

an equipment or part of a plant (i.e. a low dimensional problem) or a
controller for the integrated plant control system (i.e. a large-scale
system) consisting of many interconnected subsystems like boiler,
turbine, generator and power system or batch reactors, storage units,
distillation columns and also the distributed process monitoring and
control system. Thus modern simulation techniques deal with continuous
and discrete dynamic systems together. A different representation of
Fig.2 shown in Fig.3 indicates the various steps involved in conducting

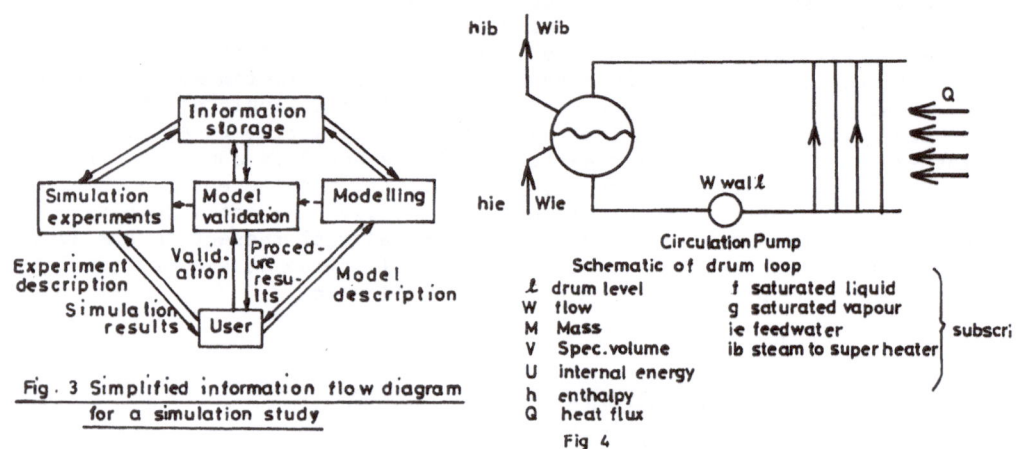

Fig. 3 Simplified information flow diagram
for a simulation study

Schematic of drum loop

ℓ drum level   f saturated liquid
W flow   g saturated vapour
M Mass   ie feedwater   } subscri
V Spec.volume   ib steam to super heater
U internal energy
h enthalpy
Q heat flux

Fig 4

a dynamic simulation study to design a control system [4]. Extensive
test runs are required to validate the model for simulation.

2.   MATHEMATICAL MODELLING

Mathematical modelling is concerned with the mapping of
relationships between the physical variables in the system into
mathematical structures like algebraic equations and differential
equations or systems of differential equations [5]. There are two basic
classes of models [6,7]. (a) Empirical or black box models and (b)
Deductive models, where the model structure and parameters are obtained
from physical conservation laws.

The black box model is derived from experiments. Many simulation
studies are conducted before the plant or system is actually
constructed. To this end, a deductive model derived from the physical
laws governing the plant process is used. Simulation enables the
validation of the model by qualitative study of the dynamic response,
better understanding of the inner functioning of the system,

representation of nonlinearities, simplification of the geometry of the system physical construction, and choice of a suitable reference coordinate system.

Deductive models can be deterministic or stochastic in nature. Our discussion will be devoted mostly on the deterministic model. Models can further be classified into two classes, viz distributed parameter model and lumped parameter model. Distributed parameter models use partial differential equations of the parabolic, hyperbolic or elliptic type to describe the dynamic system, whereas lumped parameter models use ordinary linear or nonlinear differential equations. Both models have associated algebraic equations linear or nonlinear to describe the boundary conditions or other relevant constraints. Distributed parameter models can be reduced to lumped parameter models by spatial discretization or modal expansion. Lumped parameter models can also be discretized to get discrete time models described by difference equations. In the above discussion we have considered only continuous dynamic models where variables change their value continuously with time. However, some discrete changes in the model variables may occur due to discrete events [3]. In process control situations these are due to static nonlinearities e.g., On-off switches, and constraints on internal variables of the model. Thus in real plant or process simulation a more accurate mathematical representation of the process will involve a continuous model with imbedded discrete event submodels.

The mathematical modelling will be discussed in a later section. Here we note that lumped and distributed parameter models have the following forms:

Lumped parameter model

$$\dot{x} = f(x,u), \qquad x(o) = 0 \qquad \qquad (1)$$
$$g(x,u) = 0 \qquad \qquad (2)$$

Where x and u are the state and input vector respectively, and f(.) and g(.) are nonlinear vector-valued functions.

Distributed parameter model (single dimensional)

$$\frac{\partial x}{\partial t} = f(x, \quad \frac{\partial x}{\partial z}, \frac{\partial^2 x}{\partial z^2}, u) \qquad \qquad (3)$$

$$x(z,o) = x^o(z) \qquad \qquad (4)$$

$$g_1(x, \frac{\partial x}{\partial z}, u) = 0 \text{ for } z=0 \qquad \qquad (5)$$

$$g_2(x, \frac{\partial x}{\partial z}, u) = 0 \text{ for } z=1 \qquad \qquad (6)$$

where x and u are time and space dependant, $z \epsilon (0,1)$ is the space coordinate in this one-dimensional system. The vector non-linear functions g_1 and g_2 represent boundary conditions in general form.

Having noted the forms of lumped and distributed parameters models the basics of mathematical modelling techniques are discussed with the help of some examples. There are two widely used techniques for mathematical modelling [7], which are based on the variational approach and the use of conservation laws. We will discuss the use of conservation laws. All systems have certain entities which are conserved, e.g. mass, energy and momentum. The conservation of mass equation can be written as

$$\left|\begin{array}{l}\text{Time rate of}\\\text{change of mass}\\\text{in the system}\end{array}\right| = \left|\begin{array}{l}\text{Mass flow}\\\text{into the}\\\text{system.}\end{array}\right| - \left|\begin{array}{l}\text{Mass flow}\\\text{out of the}\\\text{system}\end{array}\right|$$

The energy conservation equation may be written as :

$$\left|\begin{array}{l}\text{Time rate of}\\\text{change of}\\\text{internal kinetic}\\\text{and potential}\\\text{energy inside}\\\text{the system}\end{array}\right| = \left|\begin{array}{l}\text{Flow of internal}\\\text{kinetic and}\\\text{potential energy}\\\text{into the system}\\\text{by convection}\\\text{or diffusion}\end{array}\right| - \left|\begin{array}{l}\text{Flow of internal}\\\text{kinetic and}\\\text{potential energy}\\\text{out of the system}\\\text{by convection}\\\text{or diffusion}\end{array}\right|$$

The equation of conservation of momentum can be written as :

$$\left|\begin{array}{l}\text{Time rate of}\\\text{change of momentum}\\\text{in i direction}\end{array}\right| = \left|\begin{array}{l}\text{Net sum of}\\\text{forces in i}\\\text{direction.}\end{array}\right|$$

2.1   Conservation Model

Using the above conservation laws one can derive the mathematical model of a system. Some examples of deductive and black box models are discussed below.

2.1.1 <u>Boiler model</u>: We take the case of a simplified boiler model [8]. The schematic of the drum loop is shown in Fig.4. Applying mass conservation for the liquid and vapour states in the drum loop (drum, downcomers, water walls etc) gives

$$\frac{dM_f}{dt} = W_{ie} - W_{wall} \tag{7}$$

$$\frac{dM_g}{dt} = W_{wall} - W_{ib} \tag{8}$$

since the drum loop contains only liquid and vapour

$$M_f V_f + M_g V_g = V \cdot \tag{9}$$

Likewise the conservation of energy can be written as

$$Q + W_{ie}\, h_{ie} - W_{ib}\, h_{ib} = d(M_f\, u_f + M_g\, u_g)/dt \tag{10}$$

Using u=h-pv, the above equation can be manipulated to yield

$$W_{wall} = \frac{Q - (M_f \dfrac{dh_f}{dp} + M_g \dfrac{dh_g}{d_g} - V) \dfrac{dp}{dt} - W_{ie} (h_f - h_{ie})}{(h_g - h_f)}$$ (11)

Equation (11) states that the steam produced is equal to the heat absorbed less the heat required to maintain drum contents at saturation, less the heat required to raise incoming feedwater to saturation, all divided by the enthalpy difference between vapour and liquid. Once Once W_{wall} is known it can be used to predict drum level using the formula :

$$\text{drum level} = 1 = \frac{M_f V_f + W_{wall} (V_g - V_f) T_r - C_1}{C_2}$$ (12)

where C_1 is the volume of downcomers and risers, and C_2 converts the volume of water in the drum to height. T_r is the residence time of bubbles in the risers.

2.1.2 Water to air cross flow heat exchanger model: Mathematical modelling of heat exchangers have been studied extensively in the literature [9]. Here first we discuss the development of nonlinear models of the dynamics of flow and heat processes in total volume in the explicit form of the mass flow, unit energy and pressure. This approach is utilised to develop the model of primary to secondary cross flow heat exchanger, and a case of water to air cross flow heat exchanger is then discussed:

The equations of mass, energy and momentum balance are :

$$\frac{d\rho}{dt} = 1/V (M_1 - M_2)$$ (13)

$$\frac{de}{dt} = \frac{1}{V.\rho} [(M_2 - M_1) e + E_1 - E_2 + Q]$$ (14)

$$\frac{dw}{dt} = \frac{w_2 - w_1}{L} .w - \frac{A}{V}.\rho [P_2 - P_1 + f.L + g\rho \quad (h_2 - h_1)]$$ (15)

where V and L are constants, and various symbols, subscripts and indices have been used as under:

e–unit energy, u–internal energy, w–velocity, h–gravitational spatial coordinate, P–pressure, v–specific volume, M–Mass flow, E–total energy flow, z–spatial coordinate.

A–cross–sectional area,	k–friction constant
f–frictional coefficient	x–state vector
g–gravitational constant	U–input vector
ρ–density	y–output vector
q–heat flux	F_{cu}–heating surface (Primar side)

C–specific thermal capacity,

$\emptyset$–temperature

i–enthalpy

V–total volume

L–total length.

F_{cv}–heating surface (secondar side)

$\propto_{1,2}$– heat transfer coefficient (primar side), (secondar side)

S–entropy

Q–total heat flux

Subscripts

1–inlet   p–primar   C–set of walls of HE tubes (transfer)

2–outlet   s–secondar   pc–primar to set of walls of HE tubes

cs–Set of walls of HE tubes to Secondar

indices

HE–Heat exchanger (s)   PDP–Partial differential equations

DP– distributed parameter   CP–Concentrated parameters

Following the procedure described in [9], these equations can be written in the state space form

$$\dot{x} = f(x) + G(x)\overline{U} + B(x)\underline{U} \tag{16}$$

where the state and input vectors are defined as

$$x = [x_1\, x_2\, x_3]^T, \overline{U} = [u_1\, u_2\, u_3]^T, \underline{U} = [u_4\, u_5\, u_6\, u_7]^T, x_1 = M,$$

$$x_2 = e, x_3 = P, u_1 = M_1 - M_2, u_2 = (M_1 / \rho_1) - (M_2 / \rho_2), u_3 = (e_1 + \frac{P_1}{\rho_1})M_1$$

$$u_4 = h_2 - h_1, u_5 = P_2 - P_1, u_6 = (e_2 + \frac{P_2}{\rho_2})M_2, u_7 = Q.$$

A nonlinear model of primary to secondary cross flow heat exchanger can be developed using the subscript "p" and "s" respectively. In this case the following equations are obtained:

$$\dot{x}_p = f_p(x_p) + G_p(x_p)\overline{U}_p + B_p(x_p)\underline{U}_p \tag{17}$$

$$\dot{x}_s = f_s(x_s) + G_s(x_s)\overline{U}_s + B_s(x_s)\underline{U}_s \tag{18}$$

Where $u_{7p} = Q_{pc} \in \underline{U}_p$ and $u_{7s} = Q_{cs} \in \underline{U}_s$ are the total heat flux primary to set of walls of the ribbed or unribbed heat exchanger tubes, and the total heat flux from these walls to secondary, respectively. The energy accumulation process in the set of walls of heat exchanger tubes is given by

$$\frac{de_c}{dt} = \frac{1}{V_c\, \rho_c}[Q_{pc} - Q_{cs}] \tag{19}$$

Let us now consider a water to air cross flow heat exchanger as shown in Fig.5. Neglecting the dynamics of the mass flow (M) and pressure (P) the model can be described as

$$\frac{de_p}{dt} = \frac{M_p}{V_p \, \rho_p}(e_{1p} - e_{2p}) - \frac{1}{V_p \, \rho_p} Q_{pc} \qquad (20)$$

$$\frac{de_c}{dt} = \frac{1}{V_c \, \rho_c}(Q_{pc} - Q_{cs}) \qquad (21)$$

$$\frac{de_s}{dt} = \frac{M_s}{V_s \, \rho_s}(e_{1s} - e_{2s}) - \frac{1}{V_s \, \rho_s} Q_{cs} \qquad (22)$$

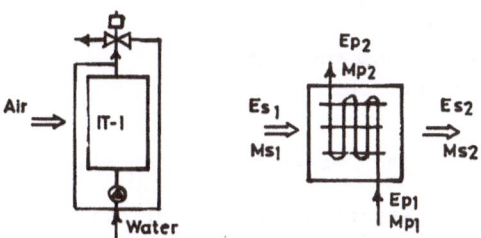

Fig 5 Scheme of water to Air cross flow HE

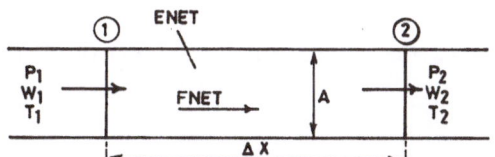

Fig 6. Generalised one dimensional flow element

Defining the state input and output variables as $x1 = \emptyset_p$, $x2 = \emptyset_c$, $x3 = \emptyset_s$, $u_1 = \emptyset_{1p}$, $u_2 = \emptyset_{1s}$, $y_1 = \emptyset_{2p}$, $y_2 = \emptyset_{2s}$ and supposing that $e \approx u \approx c\emptyset$ the following state space model can be obtained as

$$\dot{x} = Ax + B\underline{U} \qquad (23)$$

where $X = [x_1 \, x_2 \, x_3]^T$, $\underline{U} = [u_1 \, u_2 \, y_1 \, y_2]^T$. Various elements in A and B have been defined in [9].

2.1.3 Hydro power equipment model : Fasol [10] has described computer simulation methods to improve hydro power control. We describe below the models of the hydraulic conduit and hydro turbine used for this purpose.

Hydraulic conduit : Assuming one-dimensional flow through a pipe, balance of mass flow yields the equation of continuity as

$$L_1 = \frac{\partial H}{\partial t} + V \frac{\partial H}{\partial X} + \frac{a^2}{g} \frac{\partial V}{\partial X} = 0 \qquad (24)$$

and balance of forces, pressure and inertia results in the equation of motion

$$L_2 = \frac{\partial V}{\partial t} + V \frac{\partial V}{\partial x} + g \frac{\partial H}{\partial X} + \frac{\lambda}{2D} \, |V| V - g \sin \alpha = 0 \qquad (25)$$

where the different variables are defined as under:

$H = p/\rho g$ – Head of Pressure, p – pressure
$V = Q/A$, Mean velocity of flow, Q – flow rate
a – Pressure wave propagation velocity
λ – wall-friction coefficient, A-cross-sectional area of the pipe
D – Diameter, $\qquad\qquad$ α - Slope of the pipe

Equations (24) and (25) are called "water hammer" equations.

Turbine : Since the transient performance of a hydraulic turbine is very fast in relation to conduit, the dynamics of the turbine is generally not taken into consideration for simulation. In the time domain the turbine is described by the following model:

$$\begin{pmatrix} q_T\,(t) \\ P_T\,(t) \end{pmatrix} = \begin{pmatrix} f_{qu}\,(u) & f_{qh}\,(u) & f_{gy}\,(u) \\ 0 & f_{nu}\,(u)\,q_T\,(t) & 0 \end{pmatrix} \begin{pmatrix} 1 \\ h_T\,(t) \\ y\,(t) \end{pmatrix} \qquad (26)$$

where q_T, h_T, P_T, y are normalised deviations from rated values of flow, head, power and rotating speed, u is given guide vane position, and $f_{qu}\,(u)$, $f_{qh}\,(u)$ etc., are nonlinear functions received from the static performance graph of the turbine.

2.1.4 Compressor model : A generalised one dimensional flow element is shown in Fig.6 [11]. The momentum, continuity and energy equations applied to this element are :

Momentum

$$\int_v \frac{\partial}{\partial t}(\rho_s \, \bar{C}) \, dv = - \int_s P_s \, d\bar{S} - \int_s \rho_S \, \bar{C} \, (\bar{C}.d\bar{S}) + F_{NET} \qquad (27)$$

Continuity

$$\int_v \frac{\partial \rho_s}{\partial t} \, dv = - \int_s \rho_s \bar{C}.d\bar{S} \qquad (28)$$

Energy

$$\int_v \frac{\partial}{\partial t}[\,\rho_s\,(U_i + KE)]dv = - \int_s \,(h+KE)\,\rho_s\,\bar{C}.d\bar{S} + E_{NET} \qquad (29)$$

The flow is assumed to be one-dimensional in the axial direction with a mean flow area, and Mach numbers are assumed relatively low (<0.3) at the element boundaries. The lumped parameter distribution is assumed such that integration of (27) to (29) gives.

Momentum

$$\frac{dw_1}{dt} = \frac{1}{\Delta x}[A_m (P_{t1} - P_{t2}) + F_{NET}] \tag{30}$$

Continuity

$$\frac{d(\rho_s)_2}{dt} = \frac{1}{A_m \Delta x} (w_1 - w_2) \tag{31}$$

Energy

$$A_m \Delta x \frac{d}{dt}[\rho_s (C_v T_s + \frac{c^2}{2}] = (\rho_s HC_x A)_1 - (\rho_s HC_x A)_2 + E_{NET} \tag{32}$$

It has been shown that the dynamics of energy equation add little to the accuracy of the model and thus the energy equation (32) reduces to

$$H_2 = H_1 + \frac{E_{NET}}{w_1} \tag{33}$$

The thermodynamic model uses a simple equation of state, namely

$$\frac{P_t}{\rho} = ZRT \tag{34}$$

together with the Schulz relationship to define the polytropic compression paths where

$$W_P = \frac{C_p T_1}{(1/e + x)} \left[\left(\frac{P_{t2}}{P_{t1}}\right)^m - 1\right], m = \frac{ZR}{c_p} (\frac{1}{e} + X), Ws = \frac{W_P}{e} + \Delta KE$$

For an adiabatic process $P_s \rho_s - \frac{1}{(1-m)} = constant$, and so

$$\frac{d\rho_s}{dt} = (1-m) \frac{\rho_s}{P_s} \frac{dP_s}{dt}$$

Substituting $P_s = Z\rho RT_s$ and assuming low Mach numbers yields

$$\frac{d\rho_s}{dt} = \frac{(1-m)}{ZRT_t} \frac{dP_t}{dt}$$

Using this relation in the continuity equation (31), one gets

$$\frac{dP_{t2}}{dt} = \frac{ZRT_{t2}}{(1-m)v} \ (W_1-W_2) \tag{35}$$

where

A – cross sectional area,	C – velocity
Cp – specific heat at constant pressure	
Cv – specific heat at constant volume	
e – polytropic efficiency,	KE – kinetic energy
E NET –Rate of change of energy,	F NET – Body force
f – frequency	h – Enthalpy
H – stagnation enthalpy	m – polytropic index
M – Mach number,	P – Pressure
R – gas constant	S – Surface area
T – temperature	U_i – Internal energy
V – Volume	W – Mass flow rate
Wp – Polytropic head	Ws – Shaftwork
x – Axial length	X – Compressibility function
Y – ratio of specific heats	ρ – density
Δ – change in variables	

Equations (30), (33) and (35) form the basis of the dynamic model. In order to simulate a system it is necessary to divide it into discrete elements to which equations (30), (33) and (35) can be applied. In a system, different types of elements will be present e.g., compressor, valve and elements of ducting. The maximum interaction between system elements is achieved by discretization into elements of approximately equal volumes. The manner in which each element is modelled varies. The force and energy input terms F_{NET} and E_{NET} will be different for various types of elements. It is also assumed that steady state information can be used to calculate these terms for a given instantaneous element inlet condition. Values of polytropic head and efficiency are determined from instantaneous volume flow rate and compressor characteristics. Upstream and downstream ductwork can be suitably represented and also throttle valve elements are represented using isentropic nozzle flow relationship. Thus this simple model can be used in different types of compressor elements. In order to solve these equations, one needs the system geometry distribution, the element characteristics, the inlet conditions and the working fluid data.

2.1.5 Pipeline network model : The basic equations are the continuity equation, the momentum equation and the energy equation and an equation of state [12].

Continuity equation:

$$\frac{d}{dt} \ (\rho A) + \frac{d}{dx} \ (\rho AV) = o \tag{36}$$

Momentum equation :

$$\frac{dv}{dt} + v\frac{dv}{dx} + \frac{1}{\rho}\frac{dp}{dx} + g\frac{dH}{dx} + \frac{fv|v|}{2D} = 0 \qquad (37)$$

Energy equation:

$$\frac{dT}{dt} + V\frac{dT}{dx} + \frac{T}{\rho c}\frac{dP}{dt}\frac{dv}{dx} - \frac{f|v^3|}{2\,cD} + \frac{4U}{\rho cD}\,(T-T_g) = 0 \qquad (38)$$

where

where various symbols are defined as

A – cross sectional area,	V – velocity
U – Heat transfer coefficient	ρ – density
C – specific heat at constant volume,	p – Pressure
D – Diameter,	f – Moody friction factor
T – Temperature,	Tg – ground temperature
x – incremental distance along the pipeline	
t – Time,	g – gravitational constant.

Equations (36) to (38) represent the basic single dimensional pipeflow equations and are used to model transient pipe flows. A relationship between pressure, density and temperature for the fluid is used to solve them. This equation is called "equation of state". For liquids that can be regarded as incompressible the following equation is used

$$\rho = \rho_0\left[1 + \frac{P-P_0}{B} + \alpha\,(T - T_0)\right] \qquad (39)$$

where B is the bulk modulus, and α the thermal expansion coefficient. For light hydrocarbon gases the equation of state $P = \rho.RzT$ is appropriate, where z (the compressibility factor) is a known function of temperature and pressure. For lines carrying ethylene, butane, propane, and LNG/LPG products BWRS (Benedict Webb-Rubin correlation as modified by Starling [13]) can be used. This equation of state has the form :

$$P = \rho\,RT + (B_0RT - A_0 - C_0\,|\,T^2 + D_0\,/\,T^3 - E_0\,/\,T^4)\,\rho^2 + (bRT - a - d/T)\,\rho^3 +$$
$$6a\,(a + d\,/\,T)\,\rho^6 + (C\rho^3/R^2)\,(1 + \gamma\rho^2)\,Exp\,(-\gamma\rho^2) \qquad (40)$$

The various parameters have been defined in [13].

2.1.6 Distillation model : A nonideal multicomponent tray column model has been developed in [14]. The following assumptions apply.

(1) The vapour and liquid of each stage are in equilibrium,
(2) The liquid in each stage is perfectly mixed, and
(3) The vapour hold up is negligible.

The overall mass balance gives

$$\frac{dM_i}{dt} = L_{i+1} + V_{i-1} - V_i - L_i + F_{1i} + F_{vi} - S_i \qquad (41)$$

The component mass balance gives

$$\frac{dM_iX_{ij}}{dt} = L_{i+1}\ X_{i+1,j} + V_{i-1}Y_{i-1,j} - V_iY_{ij} - L_iX_{ij} + F_{1i}X_{fi,j} + F_{vi}Y_{fi,j} - S_iX_{i,j} \qquad (42)$$

The energy balance gives

$$\frac{dM_iV_i}{dt} = L_{i+1} + h_{i+1} + V_{i-1}H_{i-1} - V_iH_i - L_ih_i + F_{1i}\ h_{fi} + F_{vi}H_{fi} - S_ih_i \qquad (43)$$

The tray equilibrium gives

$$K_j = Y_j \,/\, X_j = \Upsilon_i f_j^{01} / \varnothing_j P \qquad (44)$$

The tray efficiency gives

$$E_{ij} = (Y_{i,\,j} - Y_{i-1,\,j}) \,/\, (Y^*_{\,i,j} - Y_{\,i-1,j}) \qquad (45)$$

The tray dynamics gives

$$L_i = KL_w\,h_{ow}^{\,3/2}$$

The pressure drop gives

$$h_T = h_1 + h_v + h_\sigma$$

where

E - Murphree vapour efficiency
F_1 - liquid feed,
f^{01} - standard-state fugacity
H - vapour enthalpy
h_{ow} - height over weir,
h_T - total pressure drop,
h_σ - Pressure drop due to surface tension
K - Equilibrium constant,
$\varnothing$-fugacity coefficient,
f – Feed (subscript)
j - component index (subscript)
L - Liquid flow rate
M - liquid hold up
S - side stream,
V - vapour flow rate
y - actual vapour composition

F_v – vapour feed
h - liquid enthalpy
h_1 - height of the liquid inventory
on the tray
h_v - Pressure drop due to resistance of
passage of gas through Tray.
γ - activity coefficient

i - Tray number (subscript)
k - constant
l_w - length of weir
P - pressure
U - internal energy
X - liquid composition
y^* - vapour composition in phase equilibrium with x.

2.1.7 <u>Synchronous generator model</u> : Considering a symmetric rotor case a synchronous generator model can be written as [15].

Rotor circuit equations:

$$\varphi_{ad} = X''_{ad} (-I_d + \varphi_{fd}/X_{fd} + \varphi_{1d}/X_{1d}) \tag{46}$$

$$\varphi_{aq} = X''_{aq} (-I_q + \varphi_{1q}/X_{1q}) \tag{47}$$

$$I_{fd} = (\varphi_{fd} - \varphi_{ad}) / X_{fd} \tag{48}$$

$$I_{1d} = (\varphi_{1d} - \varphi_{ad}) / X_{1d} \tag{49}$$

$$I_{1q} = (\varphi_{1q} - \varphi_{aq}) / X_{1q} \tag{50}$$

$$\frac{d}{dt} (\varphi_{fd}) = R_{fd} (E_{fd} / X_{ad} - I_{fd}) \tag{51}$$

$$\frac{d}{dt} (\varphi_{1d}) = -R_{1d} \cdot I_{1d} \tag{52}$$

$$\frac{d}{dt} (\varphi_{1q}) = -R_{1q} \cdot I_{1q} \tag{53}$$

$$E''_d = -X''_{aq} (\varphi_{1q}/X_{1q}) \tag{54}$$

$$E''_q = X''_{ad} (\varphi_{fd}/X_{fd} + \varphi_{1d}/X_{1d}) \tag{55}$$

$$E'' = \sqrt{E''_d \cdot E''_d + E''_q \cdot E''_q} \tag{56}$$

where I_d, I_q = d and q axis armature current, I_{fd}, I_{1d}, I_{1q} = field, d and q axis damper circuit current, E_{fd}, E''_d, E''_q = field voltage, d and q axis component of E", R_{fd}, R_{1d}, R_{1q} = field, d and q axis damper resistance, X_{ad}, X_{fd}, X_{1d}, X_{1q} = air gap flux, field, d and q axis damper reactance. X''_{ad}, X''_{aq} = d and q axis air gap flux subtransient reactance, φ_{ad}, φ_{aq} = d and q axis air gap flux. φ_{fd}, φ_{1d}, φ_{1q} = field, d and q axis damper flux.

Motion equations:

$$\frac{M}{\omega_o} \frac{d}{dt} (\omega) = T_m - T_e - P_d.\omega \tag{57}$$

$$\frac{d\delta}{dt} = \omega$$

where

M = Moment of inertia, ω_o = rated angular velocity
T_m = Mechanical torque, T_e = Electromagnetic torque
δ = phase angle, P_d = Mechanical damping constant.

The excitation control system AVR and speed control system Gov can be represented as per IEEE model, namely :

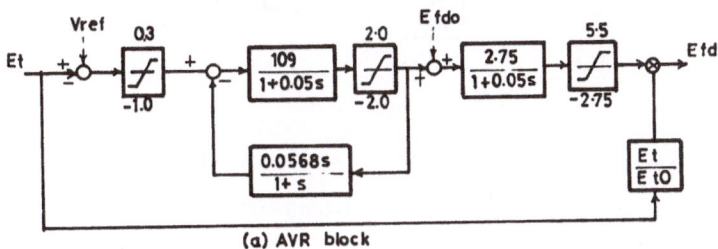

(a) AVR block

202

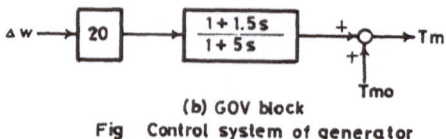

(b) GOV block

Fig   Control system of generator

2.2 Model of Connecting Network

In this section, we describe the model of the connecting network of plant process or system.

2.2.1 <u>Plant piping system model</u> : Almost all dynamic simulations associated with power plants and chemical processes must include a mathematical representation of plant piping. That is the pipes, valves, pumps and compressors that interconnect the major elements in a simulated plant (e.g. tanks, heat exchangers etc.). From a simulation modelling view point, plant-piping can be represented in a simplified form by a set of algebraic material balance equations. Piping model simplification is limited by the tight coupling between the piping and plant controls (control system transmitters and control valves are imbedded in the piping [16]). This is described below with the help of Fig.7.

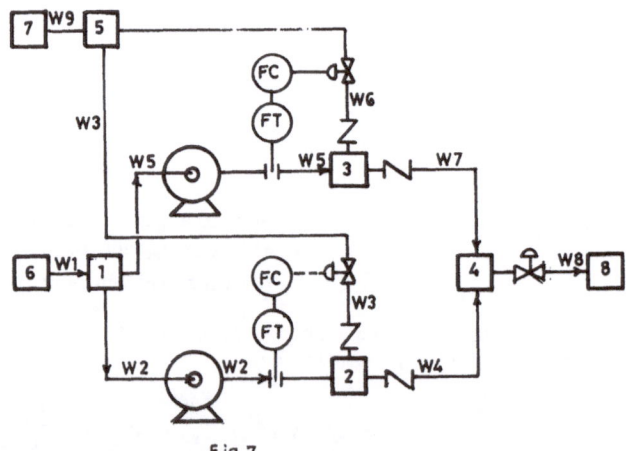

Fig 7

The algebraic material balance equations are:

$$w_1 = w_2 + w_5, \qquad w_2 = w_3 + w_4, \ w_5 = w_6 + w_7, \ w_8 = w_4 + w_7, \ w_9 = w_3 + w_6 \qquad (58a)$$

The node mass balance error equation is simply

$$E_n = \Sigma w_i^n - \Sigma w_o^n \qquad (58b)$$

where w_i^n and w_o^n are the mass flow rates of the streams feeding and leaving the nth mode respectively.

2.2.2 Electrical network model : A large power system can be subdivided into areas. The areas may be obtained, for example, by breaking the system along boundaries of different power companies or regions in the same company. Each area of a power system will consist of : (a) a generating unit with respective regulators (Exciter and governor) each connected to the transmission network through a node; (b) static and dynamic loads (c) a transmission network, consisting of transmission lines, transformers, generating, load and interconnection node. Here we are concerned with the transmission network. If the transmission network has n generators and m load nodes, it can be completely defined by (n+m) equations in terms of nodal voltages and currents [17]. The linearised form of this model has 2(n+m) real equations:

$$\begin{pmatrix} \Delta I \\ \Delta I_c \end{pmatrix} = Y \begin{pmatrix} \Delta V \\ \Delta V_c \end{pmatrix} \qquad (59a)$$

where
$$\begin{aligned}
\Delta I &= [\, \Delta I_1 \quad \Delta I_2 \ldots \ldots \quad \Delta I_i \ldots \ldots \quad \Delta I_n \,] & (2n \times 1) \\
\Delta V &= [\, \Delta V_1 \quad \Delta V_2 \ldots \ldots \quad \Delta V_i \ldots \ldots \quad \Delta V_n \,] & (2n \times 1) \\
\Delta I_c &= [\, \Delta I_{c1} \quad \Delta I_{c2} \ldots \ldots \quad \Delta I_{ci} \ldots \ldots \quad \Delta I_{cm} \,] & (2m \times 1) \\
\Delta V_c &= [\, \Delta V_{c1} \quad \Delta V_{c2} \ldots \ldots \quad \Delta V_{ci} \ldots \ldots \quad \Delta V_{cm} \,] & (2m \times 1)
\end{aligned}$$

The hth component of the vector ΔI_c represents load current at the hth node i.e.

$$\Delta I_{ch} = \Delta I_{Lh} + \Delta I_{ah} \qquad (59b)$$

where I_{Lh} and I_{ah} indicate static and asynchronous dynamic loads, and Δ is a perturbation variable. The matrix Y is symmetric and is composed of (2 x 2) subblocks given by

$$Y_{\mu v} = \begin{pmatrix} g_{\mu v} & - & b_{\mu v} \\ b_{\mu v} & & g_{\mu v} \end{pmatrix} \qquad (59c)$$

with $\mu, v = 1, 2, \ldots n+m$

2.3   Function and Controller Models.

The steady state performance is modelled by algebraic functions [18]. The very fast responses are often treated using the pseudostatic approach. In other words, if the inertia is negligible, it can be neglected. For example internal power rate delivered by steam in a turbine is directly calculated without any delay as

$$w = \sum D\,(h_i - h_o) \tag{60}$$

where D is the flow rate of steam in turbine.

Algebraic equations arise from thermodynamic equilibrium relations in chemical processes. The simulation of controllers is very important. Controllers are intelligent part of model complexes. An example was given in section 2.1.7. The controllers could be of any advanced type like multivariable, adaptive or optimal controllers.

2.4   Event and Logic Model.

Event variables [18] are included in the model so that by state change of these variables start/stop, open/close, and on/off conditions can be simulated. The interlock between the event variables contributes most of the logic models. The event variable set (A) at least is a sum of four subsets, i.e.

$$(A) = (C) \cup (D) \cup (E) \cup (F) \tag{61}$$

where   (C) = subset of operation modes
        (D) = subset of internal malfunctions
        (E) = subset of manual malfunctions
        (F) = subset of overshoot signals

To execute the interlock function between the event variables a set of rules in the form of "If-Then" statements can be built in the model. This logic structure guarantees safety of operation.

3.   VALIDATION OF MODELS

As discussed in section 2, a mathematical model usually consists of equations which describe the physical process of interest according to the laws of physics. In order to test the validity of a mathematical model, the predictions of the model are compared with the corresponding experimental results for a range of conditions. In practice the range of experimental tests is usually limited and model validation is restricted to the available experimental tests. A model possesses a higher degree of credibility if it has been validated against a large number of experimental tests under a wide range of conditions.

Butterfield and Thomas [19] discussed a number of techniques for assessing and improving dynamic simulation models. They discussed a method for selecting "best fit parameters" for models to be used in the design and/or analysis of control systems. They have also developed the concept of "model distortion" to assess quantitatively the extent of time variation or distortion needed in the model's nominally constant parameters to match perfectly a recorded transient. They have successfully applied the distortion technique for model validation in a nuclear power plant.

We have described in section 2.1.4 a compressor model. To validate the model Elder et al [11] carried out experimental tests on a small and a large single stage centrifugal compressor. Fast transients were introduced into the test system by means of a pneumatically operated blowoff valve/butterfly valve respectively. High performance transducers were mounted on the test compressor to provide transient measurements of compressor inlet flow, inlet pressure, outlet pressure and blowoff valve station pressure. Three transients were considered. The compressor was taken to a predetermined point on the characteristic. At this setting the blow-off valve was opened. After steady state was reached, the valve was closed so that the operating point returned to the original point on the characteristic. Finally from this predetermined point, the valve was closed and the compressor was taken into surge. Experimental and simulated results for the surge cycle is shown in Fig.8. The authors have attributed to slight difference in time to surge between the simulated and experimental results to the inaccuracy in the blow-off valve representation. In both cases the model was largely able to reconstruct the incidents.

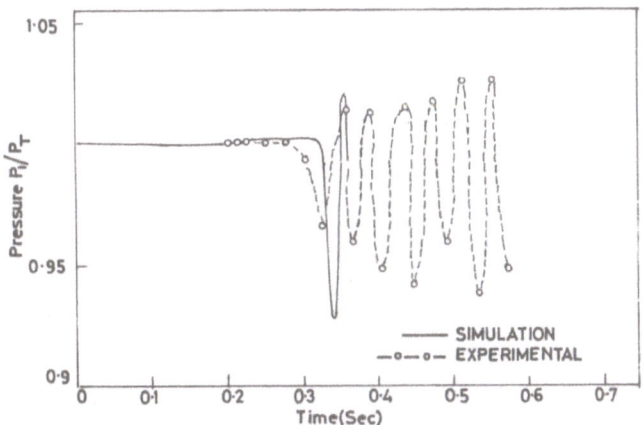

Fig 8 Simulations for the first test compressor:surge
cycle simulation

Turner and Simpson [20] have developed a compressor station model for gas pipeline network shown in Fig.9. The model has been developed considering mass and energy conservation at inlet and outlet header, heat transfer in cooler, flow resistance, gas compression equation, fuel flow, recycle and control system. To validate the model aganist operating data, a simple network was considered comprising a 200 meter long pipeline with representation of compressor station at the halfway point. The measured suction temperature and flow were used as the boundary conditions at the pipeline inlet, and the measured discharge pressure as the boundary condition at the pipeline outlet. Operaing modes and set points in the simulation were identical to those in the station during measurements. Unit A was in discharge pressure control mode with set point at 7.2 megapascals, and unit B was in shut down

mode.　Fig.10 shows the measured and model calculated compressor power,
speed and recycle flow.

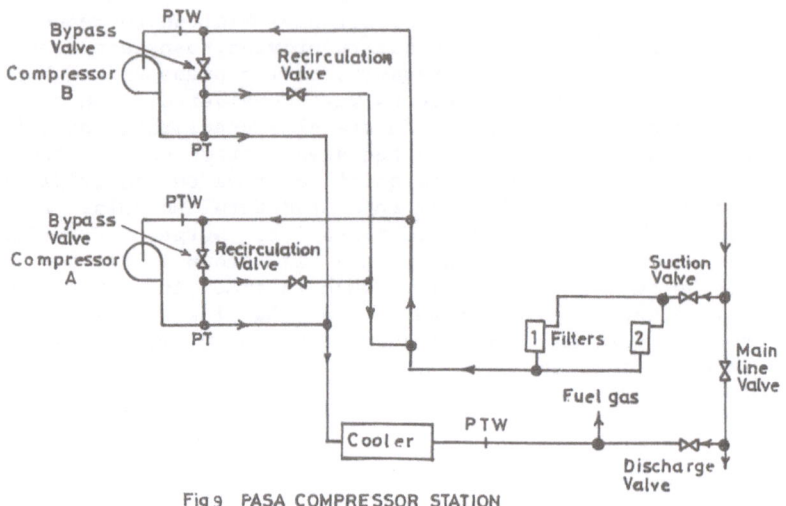

Fig 9 PASA COMPRESSOR STATION

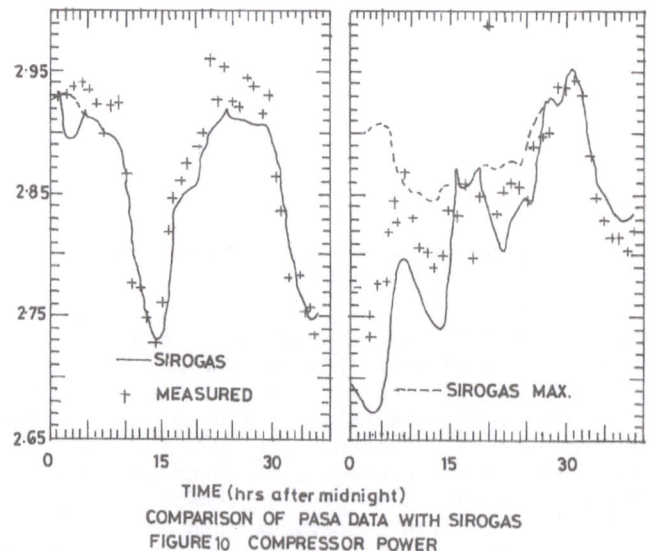

TIME (hrs after midnight)
COMPARISON OF PASA DATA WITH SIROGAS
FIGURE 10 COMPRESSOR POWER

The existence of scatter in the measured power is to be expected
since it is quite sensitive to small errors in the two temperature
measurements.　The measured power has been obtained by using the

equation of state and measured pressures and temperatures to calculate the change in the enthalpy of gas, on passage through the compressor.

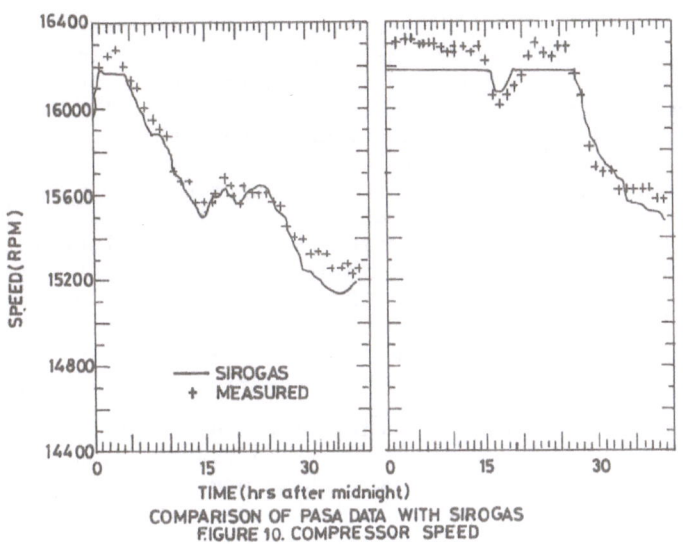

COMPARISON OF PASA DATA WITH SIROGAS
FIGURE 10. COMPRESSOR SPEED

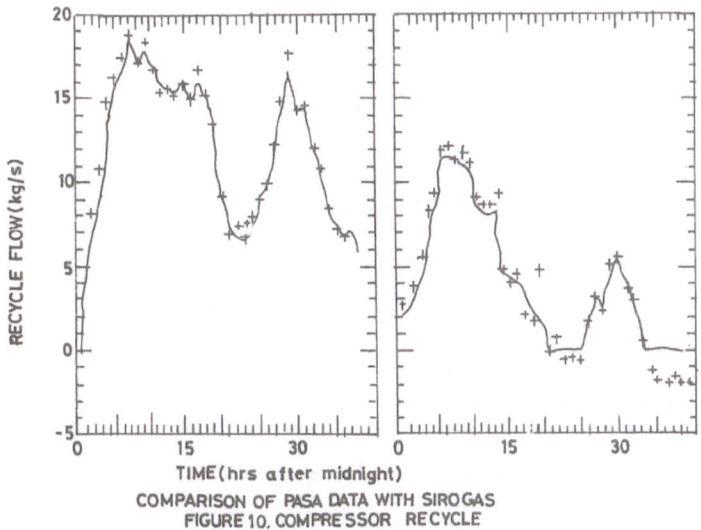

COMPARISON OF PASA DATA WITH SIROGAS
FIGURE 10. COMPRESSOR RECYCLE

The compressor temperatures and pressures in the simulation and the experiment were in excellent agreement. With one station operating data model was found valid and experiments with other stations and operating conditions are recommended to check the model fully.

4.   SOLUTION TECHNIQUES

In general, the simulation problem is concerned with the solution of ordinary and partial differential equation defining the dynamics of the system, and nonlinear/linear algebraic equations defining the boundary conditions. The algebraic equations arise from the assumption of quasistationarity and also from thermodynamic equilibrium relations. For ordinary differential equations, the 4th order Runge-Kutta algorithm, Predictor-corrector rules and Gear's method are popular. For partial differential equations the basic approach is to transform them into a set of ordinary differential equations and then integrate numerically these differential equations. For the solution, the finite difference and finite element methods, the characteristic method, and the spectral (modal) method are popular. There are numerous text books discussing these methods [3] and interested readers are referred to them. New techniques named multigrid methods are aimed at providing fast numerical solution procedures for easy implementation on multiprocessor systems.

In simulation, standard codes for numerical integration and algebraic equation solution are used. It is far beyond the scope of this chapter to discuss them. Below we discuss one technique for solution of differential equations and another technique for simultaneous solution of differential-algebraic equations. Our problem first is to solve a set of coupled nonlinear differential equations [21] :

$$\dot{x} = f(x, t) \tag{62}$$

The simplest implicit integration method may be written as

$$x_{n+1} = x_n + \Delta t f(x_{n+1}, t_{n+1}) \tag{63}$$

where $$x_n = x(t_n) \tag{64}$$

Linearising equation (62) at $x = x_n$ and $t = t_n$ one obtains :

$$\dot{x} = f(x_n, t_n) + J_n (x - x_n) \tag{65}$$
$$= J_n x + f(x_n, t_n) - J_n x_n$$

where the Jacobian is obtained as

$$J_n = \frac{\partial f(x_n, t)}{\partial x_n} \bigg| t = t_n \tag{66}$$

Applying the implicit method to (65) one gets

$$x_{n+1} = x_n + \Delta t (J_n x_{n+1} + f(x_n, t_n) - J_n x_n) \tag{67}$$

This may be written as

$$x_{n+1} = x_n + \left(\frac{1}{\Delta t} I - J_n\right)^{-1} f(x_n, t_n) \tag{68}$$

Thus the problem of solving equation (62) is reduced to the problem of inverting the matrix

$$\emptyset = \frac{1}{\Delta t} I - J_n (x_n, t_n) \tag{68}$$

For solution of a large problem the inversion problem becomes easier if J_n takes the form of a sparse matrix with diagonal structure. We now discuss the solution of coupled differential and algebraic equations which appear in a dynamic simulation problem [22]. The dynamic plant model can be written as :

$$\dot{x} = F(x, z, v, p, t), \quad x(t_0) = x_0 \tag{69}$$
$$0 = G(x, z, v, p, t) \tag{70}$$
$$W = H(x, z, p, t) \tag{71}$$

where x involves the dynamic state variables, and z involves additional process varibles evaluated by algebraic equations. The input and output variables of the whole plant are contained in v and W, and p is the process parameter. Equation (70) can be solved iteratively to get

$$z = z(x, v, p, t) \tag{72}$$

This can be used to rewrite (69) in the form

$$\dot{x} = F(x, z(x, v, p, t), v, p, t) \tag{73}$$

Equation (73) can be solved by a standard ODE solver. At each call of the right hand side F the implicit equation (72) is to be solved for z. We will discuss here a more convenient way of simultaneous solution of differential and algebraic equations. Generally a multistep integration scheme of order p can be written in the following form:

$$x_{n+1} = \sum_{i=1}^{p} a_i \, x_{n+1-i} + \Delta t \sum_{i=0}^{p-1} b_i \, f_{n+1-i} \tag{74}$$

If the coefficient b_0 is not zero, we have to solve an implicit equation for x_{n+1}. Equation (74) can be rewritten as

$$0 = \sum_{i=0}^{p} a_i \, x_{n+1-i} + \Delta t \sum_{i=0}^{p-1} b_i \, f_{n+1-i} \tag{75}$$

The algebraic equation (70) of the plant is solved together with equation (75). Alternatively, Gear's approach [22] could be adopted.

5.   SIMULATION LANGUAGES AND SIMULATION METHODOLOGY:

In this section we will discuss the simulation languages and structures
of some dynamic simulators. We will start with the simulation languages
CSMP III, ACSL and EASY5, and then we shall discuss the dynamic process
simulators DYNSYL, DPS and DIVA [22].

5.1   Simulation Languages

The simulation of a dynamic system can be done using higher level
programming languages (like FORTRAN, PASCAL), and simulation programming
systems or languages like CSSL, DARE, ASCL, CSMP III, GPSS, GASPIV, SLAM
II, SIMSCRIPT II, DPS, DESIRE etc. A comprehensive survey of available
simulation programming systems is available in references [23]. The
essential advantages of simulation languages are that they help the
inexperienced user to perform simulation studies and also experienced
user is relieved from standard tasks. A good simulation language
provides four features which make it´s use attractive: i) Translation
ii) input/output iii) integration algorithms and (iv) linear analysis
routines. These features are briefly discussed below.

Translation: The simulation languge acts as a pre-compiler to translate
a user´s source code into a FORTRAN program. This allows the programmer
to take advantage of functions and routines provided by the language,
like tables, limits, deadbands, random number generators, switches etc.
The translator also sorts the source program statements into an
executable order providing great flexibility.
Input/Output: The simulation language provides an executive routine
which includes many user-convenience features for printing and plotting
the output.
Integration algorithm : The simulation language includes integration
algorithms.
Linear analysis routines : While transient response analysis is the
major use of a plant model, it is not the only use. Particularly in the
area of control system design and analysis, a great deal of valuable
information is available from analysis of the linear model. The linear
model can be used to generate root locus, Bode plot, Nyquist plot,
stability margins, and optimal control design.

        In order to discuss the impact of simulation programming systems we
consider the case of CSMP III developed by the IBM corporation. This
would be followed by a brief discussion on Advanced Continous Simulation
Language (ACSL) developed by Mitchell Gauthier associates, Inc., and the
Engineering Analysis System 5 (EASYS) developed by Boeing Computer
Services Company.

a) CSMP III : CSMP III is a general purpose digital computer program
which can be used for the solution of algebraic and differential
equations describing a dynamic system. The program provides an
application oriented language that allows models to be prepared directly
and simply from either a block diagram representation or from a set of

ordinary differential equations. CSMP III accepts FORTRAN statements.
CSMP functional blocks [24] can be explained with the aid of the
following figure.

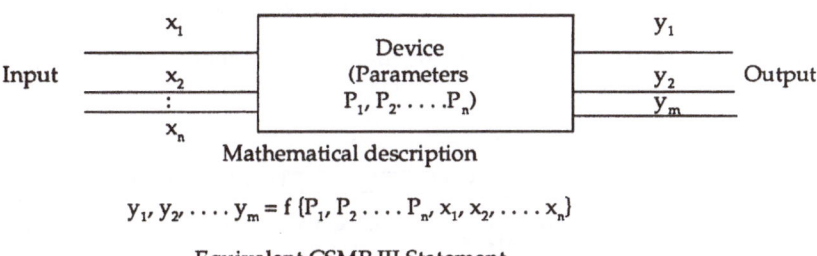

Mathematical description

$$y_1, y_2, \ldots y_m = f \{P_1, P_2 \ldots P_n, x_1, x_2, \ldots x_n\}$$

Equivalent CSMP III Statement

$$y_1, y_2, \ldots y_m = \text{DEVICE} (P_1, P_2 \ldots P_n, x_1, x_2, \ldots x_n)$$

Fig. CSMP   Block

A library of CSMP III functional blocks is available in [24–25].
This program offers the capability of modelling each system component
independently as macro-functions, each basically defined by the above
input-output relationships. This flexibility allows the user to create
a library of CSMP III system component macro functions and these
macrofunctions can be interconnected as required by the system
configuration. In order to illustrate this we use the power system
simulation problem of Prasad and Dunlop [26]. The power system single
line diagram and CSMP III macro functions are shown in Fig.11.
General outline of simulation program is shown below:

```
CONSTANT
          — Machine data
          — Saturation data
          — Network data
          — Switching data
          — Initialize variables
          — Compute necessary equivalent circuit
            parameters.
          — Call initial conditions subprograms.
DYNAMIC
          — Call Generator Macro
          PROCEDURE
          — Correction for Generator saturation
          END PROCEDURE
          — Call step-up transformer Macro
          PROCEDURE
          — Correction for transformer saturation
          END PROCEDURE
          — Call Transmission line Macro
```

```
                ~ Call external system~infinite bus macro
TIMER
                ~ Specify simulation time, integration interval and desired outp
                  interval.
OUTPUT ~ Plotted/printed output of desired values.
END
SUBPROGRAMS ~ Initial conditions of network currents and voltages machin
                  fluxes etc.
```

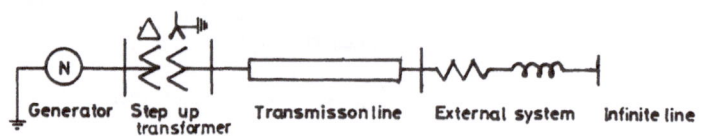

Generator   Step up   Transmisson line   External system   Infinite line
            transformer

a) Single line diagram

| Generator MACRO | Step up transformer MACRO | Transmission line MACRO | External system infinite bus MACRO |

b) C S M P III Macro functional description

Fig11

The "CONSTANT" segment allows the calculation of the necessary initial conditions required to specify the initial integrator outputs. In this one time calculation step, various fixed parameters of machies, network and other switching constants may be stored. Althogh CSMP III permits the use of standard FORTRAN IV, calculations involving complex algebra have to be done within subprograms of CSMP III. After the initialisation problem is over, the simulation enters the "DYNAMIC" segment. Within this segment various power system MACRO function blocks are called and input and output variables properly arranged to form the interconnections as per the configuration of the system (Fig.11(a)). Where FORTRAN logic is necessary, e.g. for correction of saturation effects, circuit breaker operation etc., appropriate ´PROCEDURE´ segments have to be provided within the "DYNAMIC" segment. The "DYNAMIC" segment is generally terminated by the "TIMER" statement which includes total time of simulation (FINTIM), the integration interval (DELT) and the output interval (OUTDEL).

The authors [26] have used a synchronous generator model, an excitation system and governor~turbine model, a network model and a machine~network interface to study the dynamic behaviour. They found that the variable step Runge~Kutta method is a numerically stable integration technique for simulation, though CSMP III has several integrating techniques. The authors have also shown an example of how to develop a CSMP III Macrofunction taking the case of a shunt reactor.

The Electric Power Research institute (EPRI) has developed the modular modelling system (MMS) [27], a dynamic analysis code for simulation of fossil and nuclear power plants or subsystems. EPRI has developed MMS in a simulation language framework. ACSL and EASY5 were used for this purpose. We discuss below the basic features of these languages.

b) ACSL: ACSL is block oriented. There are three major blocks : INITIAL, DYNAMIC and TERMINAL. INITIAL block defines a set of calculations which may be performed prior to simulation. DYNAMIC block defines a set of calculations which is performed during simulation, and TERMINAL block does the same after simulation. Any one or all of the parameters are defined in the model program. ACSL program statements are quite similar to Fortran Statements except that many additional internal functions are available, like dead time, transfer function, step and ramp functions etc. ACSL programs are more structured than Fortran and can be used to model systems defined by time dependent, non-linear differential equations and/or transfer functions. The model consists of free format FORTRAN - like statements and FORTRAN subroutine-like invocations of any of 112 ACSL MACROS which in many cases have exact FORTRAN equivalent. FORTRAN subroutines may be appended at the end of the program. ACSL does not generate a model schematic. It allows the user to generate MACROS within the program.

c) EASY-5 : EASY-5 is component oriented. The model consists of definitions of locations and interconnections for any of 48 standard EASY5 components and imbedded FORTRAN statements. EASY 5 standard components are mainly for control system hardware. It provides eleven matrix operations e.g., addition, multiplication, inversion. EASY5 automatically generates a schematic representation of the model from user specified component locations showing all interconnections. It allows to define MACROS within the program. Once defined, MACROS are used in the same context as standard components.

d) DPS: The Dynamic Process Simulator (DPS) [28] was originally developed by the Japanese Union of Scientists and Engineers (JUSE) and now, jointly by JUSE and CAD Centre, U.K. Simulation development in DPS follows the approach employed in steady-state process flow sheeting programs, in which the user defines the problem by linking previously defined models through material and information flows. Models in DPS may be one of two types :

(1) Element: A model defined in terms of a set of algebraic and ordinary differential equations. Some examples from the standard library are heat exchanger; flash unit; control valves; PID controller etc.

(2) Unit : A model defined by a collection of elements with its own internal connections. A typical example would be a distillation column model consisting of a tower, a condenser, an accumulator, a reboiler, control valves and controllers each of which is modelled by an element.

Many common items of standard models of process equipment exist in the DPS standard model library. If a model is required that is not available in the library it can be created by writing the model equations in a simple non-procedural language.

Although DPS has been developed for the study of continuous time system behaviour, the program allows for a wide variety of discrete events i.e. switching of flows, starting/stopping of pumps, closing/opening of valves etc. Such events may be specified to occur at a particular time or as a result of conditions that arise during the simulation. DPS consists of two programs: The model processor and the problem processor. The model processor is used only for defining and processing the model equations and storing the models in the model library. Problem specification and calculation is handled by the problem processor, which accesses the necessary models from the model library.

Standard DPS functions include the mathematical functions available in a language such as FORTRAN (e.g. SQRT, ABS, EXP etc), as well as a collection of special functions which are useful particularly in the modelling of multicomponent processes and process control systems. Some of these functions are: ´IP´ ~ Inner product, ´S´ ~ Sum, ´IP/´ ~ Inner product and Division, ´AVR´ ~ Average, ´SWCH´ ~ Logic Switch, ´DEDZ´ ~ Dead zone. The sum function $y = 's'(x1, x2, xn)$ performs the mathematical function

$$y = \sum_{j=1}^{n} . x_j$$

Use of this function in calculating total molar flow rate of a stream, TMFLO can be made as:

TMFLO = ´s´(IN3.C(1 to NC))

where IN3.C(k) = the molar flow rate of component number k in stream IN3.

The program provides a choice of Euler, Runge~Kutta and backward Euler methods for integration of the differential equations, and the direct substitution and Newton~Raphson methods for solving the system of algebraic equations.

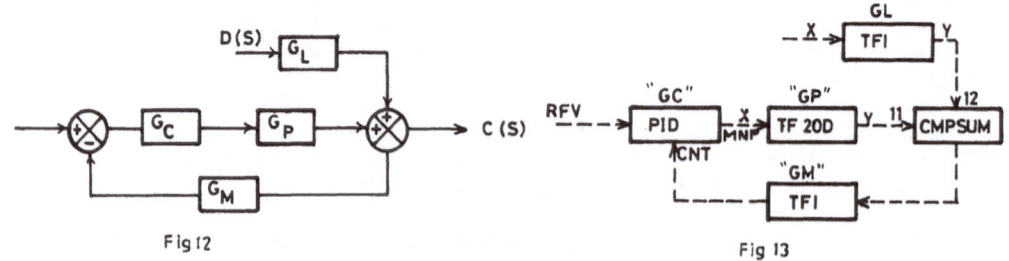

Fig 12   Fig 13

We consider a single variable feedback control system for simulation as shown in fig.12. The DPS block diagram of the feedback control system is as shown in Fig.13.

The DPS elements used are TF1, PID, TF20D, and CMPSUM are user selected equipment codes of "GL", "GC", "GP", "SUM" etc. As it can be easily seen, the elements are connected only by information flows.

5.2   Simulation Methodology.

As discussed earlier the structure of a computer program for dynamic simulation will be as shown in Fig.14. In order to discuss the simulation methodology in detail, we include here the detailed flow diagram of DYNSYL with its important subroutines. This program was

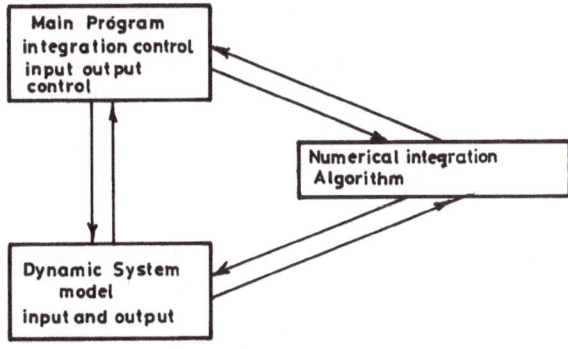

Fig 14

developed by Lawrence Livermore Laboratory for process unit simulation. The flow diagram, as shown in Fig.15, is self explanatory. We also include a simulation flow chart for a distillation column simulation shown in Fig.16 and Fig.17 respectively. The models that describe the dynamics of the tower and its related equipment have been programmed in the SEL32/7780 computer. Fig.17 shows the program execution flow chart. The programs were developed and modularised by using a structured programming technique which reduced the program complexity and made the program easy to modify and understand. The programs were implemented either to perform on-line simulation or on-line real-time simulation.

6.   SIMULATION SOFTWARE FOR INDUSTRY

Dynamic simulation in industry is used as a tool for process analysis, control design and operator training. Hashimoto [1] has briefly discussed five dynamic simulation packages developed in Japan. Most of the packages allow the user to build the model of the overall process by connecting "blocks" according to the topological structure of the process. Here "block" means a set of "modules" connected in advance in order to represent the mathematical model of a processing unit, such as a reactor, heat exchanger, turbogenerator, boiler, distillation column or controller. The input data required are: (i) The number of components and streams, (ii) Mathematical models of processing units and equipment parameters, (iii) The topological structure of the process, (iv) Initial value and (v) Conditions to be attended in the course of integration e.g., opening and closing of valves, and switches. In this section some details of three powerful simulation software packages will be given.

(a) Dynamic Analysis Program (DAP): Bechtel [29] developed the dynamic analysis program (DAP) using Fortran IV. They proposed to adopt a standarised approach for simulation studies. The decision to standardise led to three important conclusions:

i) Simulation standardisation should include process industry oriented software which has user interactive communication and plotting.

ii) Software library should contain basic "off the shelf" programs or modules that would be limited to basic items such as controllers, valves pumps, compressors, etc. For example a module for multicomponent vapour—liquid equilibrium would be available but a module to simulate distillation column dynamics would have to be developed by the user.

iii) The objective of standardisatin is to improve the efficiency and productivity of experienced simulation engineer. The DAP development project produced a simulation methodology covering :

- ⁓ The notation used on simulation piping and instrumentation diagrams (P&IDs).
- ⁓ The generalised form of the user interactive inputs to the software editor or command processor.
- ⁓ The Fortran names used for process simulation variable parameters in predefined Fortran commons.
- ⁓ The format of both printed and graphical simulation outputs and the associated notation.

The DAP functional diagram is shown in Fig.18. DAP has the following features to solve user—written model programs:

- ⁓ A command processor for overall model control, enabling the use to read or modify model parameters (e.g. controller setting) and initialise test runs,
- ⁓ Choice of several integration routines (e.g. Euler RK2, RK4, PC2)
- ⁓ Comprehensive data storage and retrieval,
- ⁓ Time history plotting capability of all model variables
- ⁓ Tabular summaries of data and results,
- ⁓ A library of frequently used programs (e.g. controllers, valve dynamics, pressure—flow relations),
- ⁓ Operation in interactive batch environment.

A typical simulation P&ID for a compressor station with notation for simulation P&ID are shown in Fig.19. DAP has a three part program library to assist the user in generating the simulation model program.

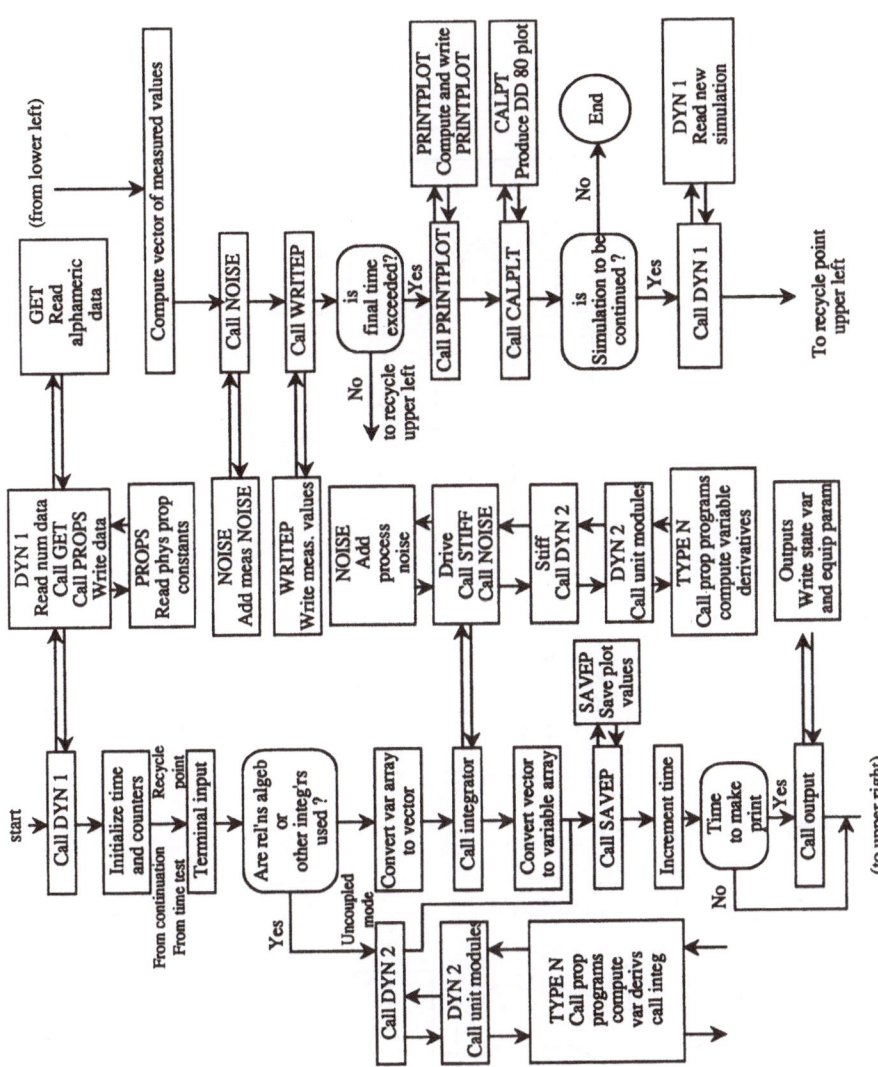

Fig.15 Detailed flow diagram for DYNSYL showing the important subroutines

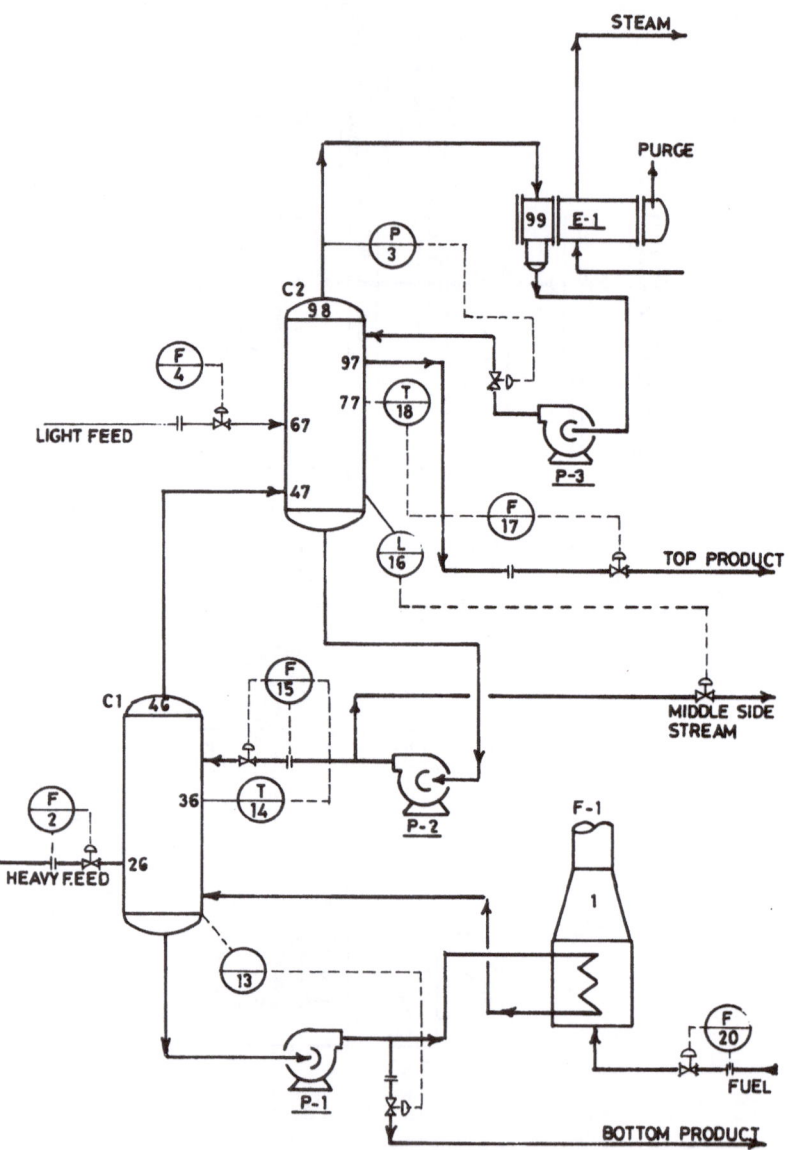

Fig16.Simulation diagram of multicomponent distillation columns

219

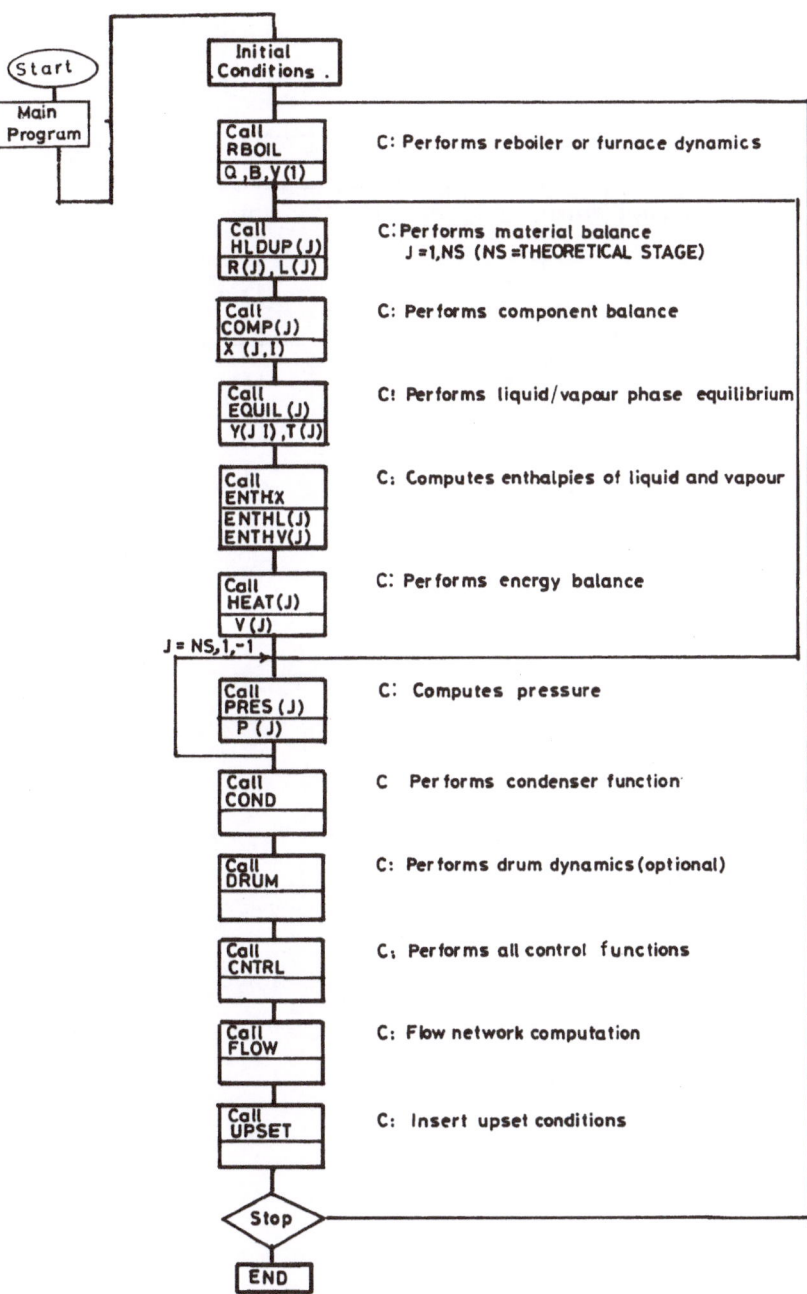

Fig 17 Flow chart for executing tower model

220

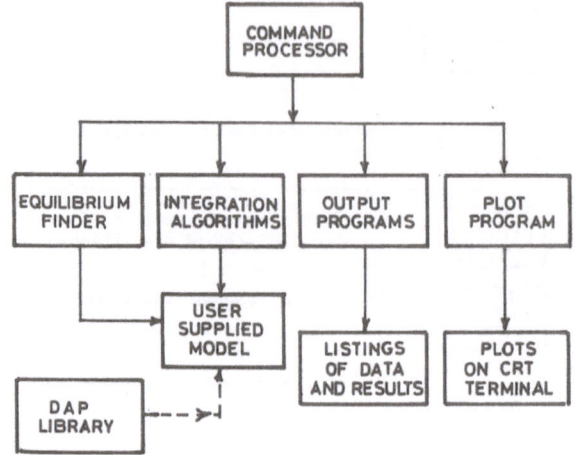

Fig18 DAP FUNCTIONAL DIAGRAM

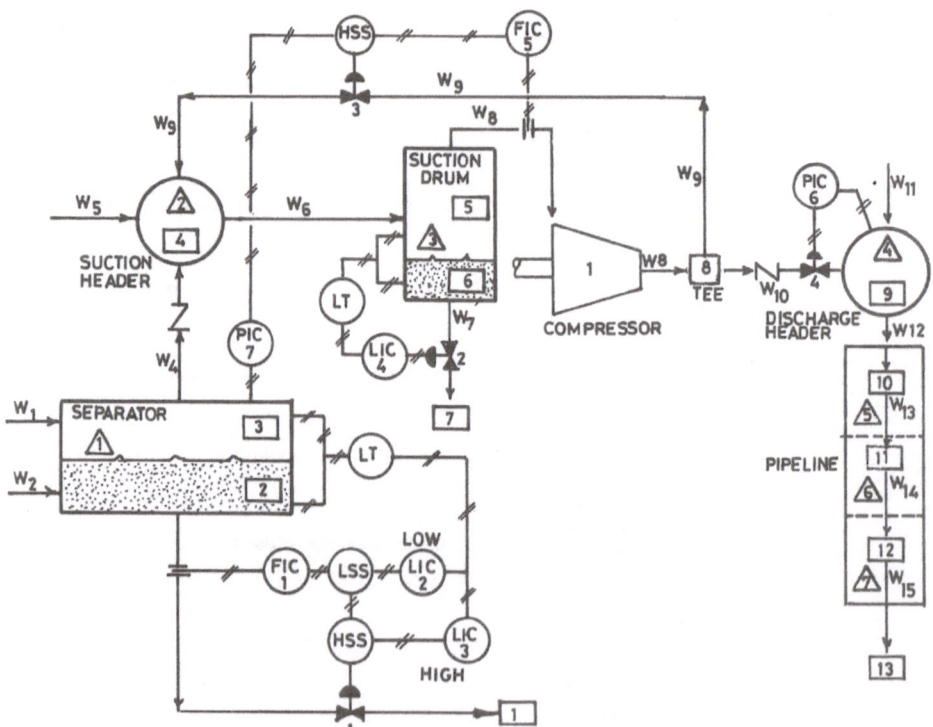

Fig 19   TYPICAL SIMULATION P & ID

This library includes :

i) A set of steam table programs valid over a 01 ~ 1500 psia range.
ii) A set of approximately 60 basic programs to simulate controllers and control system elements, valves, a variety of pressure~flow mechanisms, vapour~liquid equilibrium, etc.
iii) A set of eleven program skeleton, including two mandatory programs to be completed by the user. The simulation model program and a program supplying the names and numerical values for a ten variable simulation monitoring output must be supplied to obtain an executable DAP simulation. The other nine optional programs provide the user with the capability of inserting commands, storing and recalling user supplied data from mass storage, etc.

The interested reader can refer to a simulation study done for an Ethylene vaporization system [30] and for a compressor station using DAP [31].

(b) DIVA~Dynamische Simulation Verfahrentechnisches Anlagen : A new flow sheet oriented dynamic simulator has been developed by researchers at University of Stuttgart [22]. A block diagram of DIVA is shown in Fig.20. There are three groups of program modules :

~ Interface routines for user and process interaction,
~ The simulator core which sets up and solves the plant equations,
~ Various Libraries and data bases.

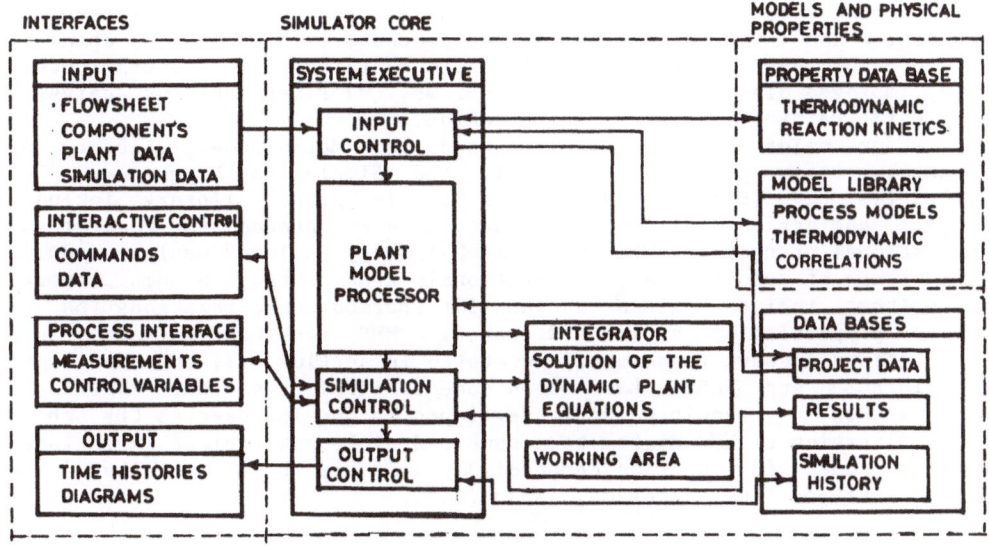

Fig.20 Block diagram of the dynamic simulator

System executive is the most essential part of the simulator. Input and output control, generation of a structured problem description, automatic generation of a plant model for the specified simulation problem and the control of the simulation itself are done by this. Plant model processor does automatic generation of a calculation instruction to solve a specified simulation problem. User friendly input interface has been provided. In DIVA the user sketches the flowchart on the screen of a graphic terminal, employing a light pen. This operation is supported by a graphic editor which permits an interactive design of the flowsheet. In addition to the usual off line simulations, on line simulations can be carried out in DIVA. In this case measurements of process variables are fed into the simulator and control variables are returned at fixed time intervals. DIVA also supports model based measurement techniques such as Kalman filters and Luenberger observers. DIVA has been used in modelling and simulation of industrial production plants consisting of reactors and coupled distillation columns. Simulation has also been carried out for development of modern control systems, and online estimation of temperature and concentration profiles in multicomponent distillation. Future developments in DIVA are expected to be directed towards enlarging the model library, improving the property data base access, embedding expert systems and optimising the solution strategy and the numerical methods.

(c) Modular Modelling System (MMS): As stated eearlier in section 5.1 modular modelling system [27] has been developed by EPRI, USA for simulation studies in both fossil and nuclear power plants. It consists of a library of pre-engineered models (modules) of power plant componets which can be interconnected in a physically realistic arrangement. MMS is intended to provide an improved basis for : (i) Specification, selection and integration of plant components, (ii) Design and checkout of control systems, (iii) Simulation to expedite plant commissioning, (iv) Improved diagnostic of plant problems, (v) Plant accident analysis, and (vi) Training simulator qualification.

Four major elements of the code are: Methodology, Library of modules, Steam property algorithm, and Simulation Languages. Two Libraries have been developed. The "Single phase" library includes fossil, nuclear and balance of plant modules intended for control and operational transients. The "two phase" library includes nuclear modules for loss of coolant and operational transients. Steam property algorithms include FORTRAN routines for thermodynamic relationships to define properties of steam and water. MMS has been developed in a simulation language framework. As stated in section 5.1, two languages namely ACSL and EASY5 have been used. EPRI also have developed translators to translate models developed in one language to the other. An application of MMS system to a combined cycle power plant is included in the companion section (Appendix). This has been contributed by Chiyoda Corporation, Yokohama, Japan [45].

7. SIMULATION HARDWARE.

Minis, superminis and mainframes are traditionally used for simulation work. Carlson [29] reported the use of medium size digital computer (EAI 3200) for process control simulation studies. DIVA has been implemented in DEC VAX-11/750 and MicroVAX-II and DPS has been released for PRIME 50 series (PRIMOS), DECVAX (VMS) and IBM (MVS/TSO) computers. SEL32/77 has been used in a number of simulation studies. Significant effort has been made in recent years to use personal computers in simulation. They can be used to build dedicated simulation workstations with graphic capabilities. More and more simulation software originally developed for Minis and Mainframes are being transferred to PCs. Multiprocessor-based systems are being developed for simulation, especially for real-time simulation. We will discuss this in a separate section. Use of computing facility at plant control system itself for simulation is also very popular. Three such implementations are cited here [32,33,34]. An excellent account for verificatiion of advanced control strategies using this facility has been discussed in [35].

Hitachi [47] has verified the behaviour of an adaptive optimal control strategy for a power plant using simulator and a distributed control system of Hitachi where the adaptive optimal control strategy has been implemented. This adaptive optimal control strategy has been described in the chapter "Digital Control Algorithms".

8.   MULTIPROCESSORS IN SIMULATION.

One of the drawbacks of digital computer simulation lies in the poor cost performance, if simulation is to be carried out in real-time. In order to overcome these difficulties the use of mini or micro-computers in parallel has been studied extensively. The multipurpose simulator HOSS developed at Hokkaido University employed 34 parallel processors each one being equivalent to a PDP-11/34 with a VAX-11/780 as a host computer. The configuration of the parallel processor array [36] is shown in Fig.21.

Kasahara et al [37] have developed a parallel processing scheme for simulation of dynamic systems by decomposing the simulation problem at the equation level or at the operation element level (e.g. addition, substraction, multiplication and division, trigonometric functions etc.). The simulator designed and constructed by the authors to demonstrate the proposed parallel processing scheme on the basis of the multiprocessor scheduling algorithm is shown in Fig.22.

De Keyser, Coen and Verdiere [38] have designed a multiprocessor-based real-time simulator for a cutter suction dredging ship. The multiprocessor simulator has been discussed in a separate chapter in this book by De Keyser and Van Ostaeyen. Kokai et al. [15] have discussed a multiprocessor-based real-time simulator. The configuratiion of the multiprocessor system is shown in Fig.23. The flow chart of the generator simulation program is shown in Fig.24. The parallel processing on four processors and the task assignment of each processor are shown in Fig.25 and 26. The most difficult problem to run the four processors in parallel was synchronisation of processors. Semaphores (integer type variables) have been used to synchronise the processors. Semaphores are located in a common memory and are accessed by each processor. Three

Semaphores used as shown in Fig.26 are:

    Semaphore S1:   Synchronising the master and slave processor 1

    Semaphore S2:   Synchronising the master and slave processor 2

    Semaphore S3:   Synchronising the master and slave processor 3

The model used has been described in section 2.1.7.

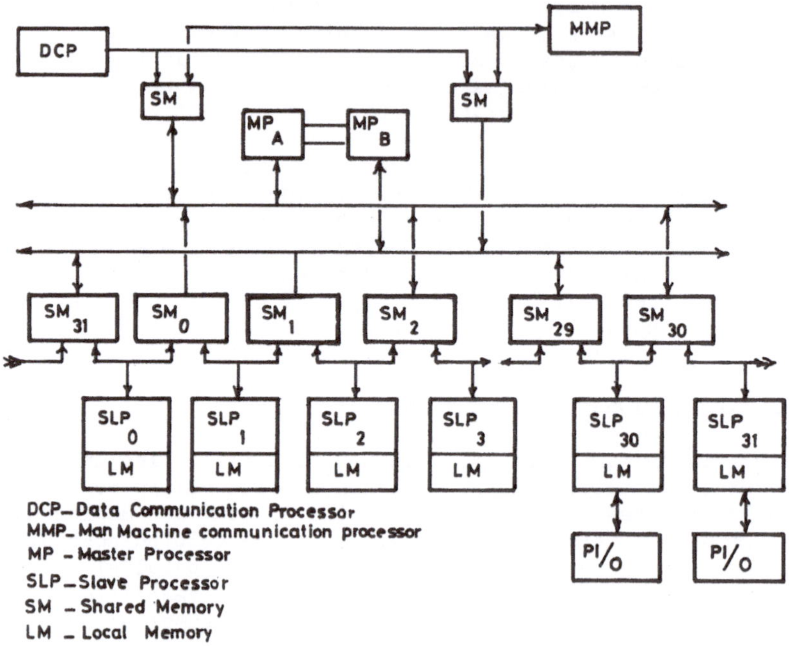

DCP_ Data Communication Processor
MMP_ Man Machine communication processor
MP _ Master Processor
SLP_ Slave Processor
SM _ Shared Memory
LM _ Local Memory

Fig 21.

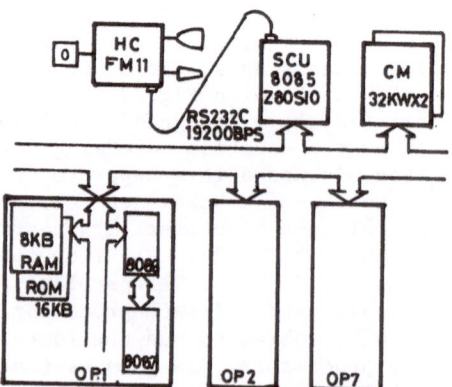

Fig 22 System configuration of WAMUX

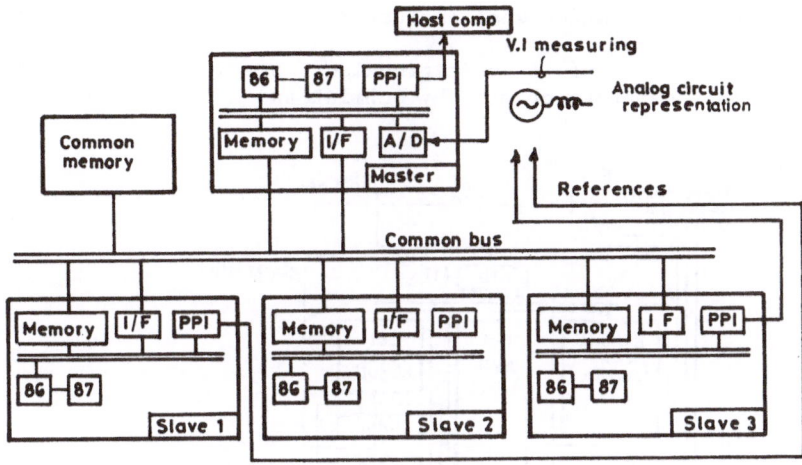

86: Intel 8086   I/F: Bus interface
87: Intel 8087   PPI: Peripheral interface

Fig23 Configuration of multiprocessor system

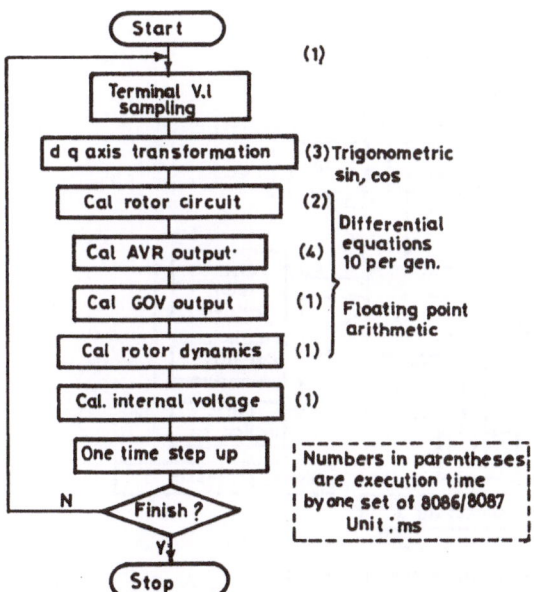

Fig24 Flow chart of generator simulation program

The simulator performance was evaluated with offline simulation results using EMTP program.

Another case of a multiprocessor based simulation has been discussed in a separate chapter by Schmidt and Lappus in this book.

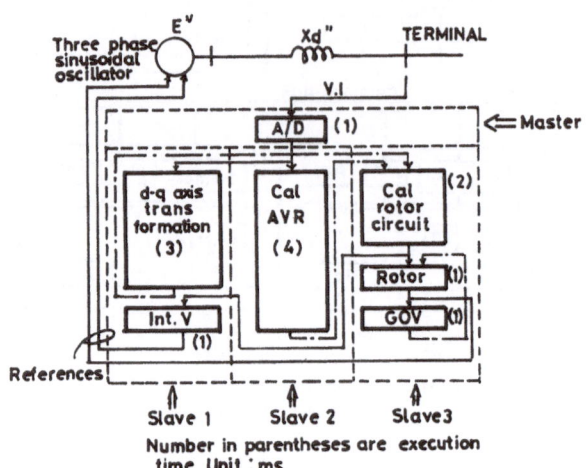

Number in parentheses are execution time Unit : ms

Fig25 Parallel processing on four processors

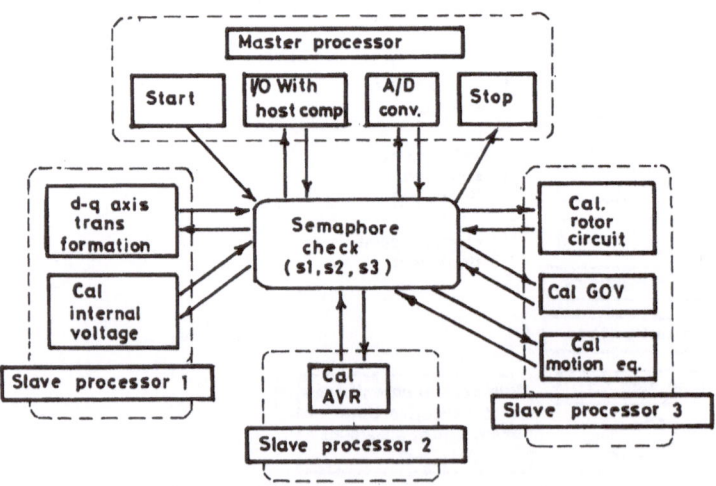

Fig26 Task assignment of each processor

9. SIMULATION OF DIGITAL CONTROL SYSTEM PERFORMANCE

There has been some interest to simulate the performance of distributed computer control systems for the purpose of design and evaluation of alternative architectures [39,40]. These simulations cover
a) communication system performance evaluation
b) system reliability evaluation
c) control performance evaluation
Functional requirement and evaluation criteria of industrial LAN include:

i) Responsiveness (real-time capabilities),
ii) Fairness (n to n communication),
iii) High data transmission rate,
iv) Expandibility, and
v) High reliability and maintainability.

Various evaluation indices have been attributed to industrial LANs e.g. message transmission time, waiting time and queue length of message, system throughput and channel utilisation.

Davidson and Houle [40] have developed a program SIMORD to simulate hierarchical configurations of distributed process control computers. SIMORD is run interactively and begins the deterministic simulation with the help of a questionnaire. In this stage the execution time for the cyclical control programs of each computer is established and CPU utilization rate is calculated. The number of alarms that the CPU will be able to accommodate per scanning period for a given utilisation rate is calculated. During the second stage, the impact due to random arrival of alarms on the control computer operation is simulated. Depending on the alarms interarrival time distribution (exponential, general etc) queing theory can be used to establish the memory buffer size necessary to store the alarms and the average amount of time spent by an alarm on the queue.

For the deterministic simulation the user should specify the configuration of the computer system i.e. number of hierarchical levels (maximum three), number of computers at each level, the type of CPU, the operating system, the quantity of memory for each computer and it´s disk type. Further information concerning the input and output processing is required. This comprises of total number of analog and digital inputs and outputs, scanning period of digital and analog points, average number of digital and analog points per scanning period and average number of variables calculated and messages per scanning period.

Information concerning number of alphanumeric and graphic terminals, total number of control loops, average number of variables displayed on a graphic page and admissible CPU utilization rate is required. Based on the received data, the simulation program calculates the execution time of various fill in the blank process control programs and displays this time. The deterministic simulation program also calculates the response times to an alarm (impulse response) generated by one of the first level computer. This time has two components:

i) Time necessary for an alarm, generated by a first level computer, to arrive at the main computer.

ii) Time necessary for a command, generated by the main computer, to arrive at the control element of the process variable which generated the alarm.

The program carries out alarm simulation in two ways:

i) alarm simulation for a server utilisation SU < 1
ii) alarm simulation for a server utilisation SU > 1

The value of server utilisation is determined by

$$SU = b/a$$

where b = average alarm service time, and
a = average alarm interarrival time.

SU can also be expressed as:

$$SU = (N \times T)/P$$

where N = number of alarms per scanning period,
T = average alarm service time, and
P = scanning period.

For case-i (i.e. SU<1) queing theory has been employed and for case-ii (ie SU>1) when alarms are received in bursts their absorption time has been calculated.

The results of simulation of a hierarchical configuration of computers (Digital PDP-11/70 (1 No), PDP-11/44 (4 Nos), and PDP-11/24 (20 Nos)) controlling a waste water treatment plant have been presented in reference [40]. These authors have carried out the simulation on a VAX-11/750. Program execution time measurements and their validation have been carried out on a PDP-11/44 computer.

A simulation case study for a distributed process control system based on utilisation of data highway and process control unit nodes has been described by Schoeffler [41]. An industrially available (Bailey Controls, USA) control system architechture has been considered by this author and a case study of a paper mill has been described. The problem has been solved by considering sub system performance rather than using an integrated queuing theory model of the overall system.

Performcnce evaluation of telesupervisory/telecontrol systems can also be done by simulation. This class of systems is used for remote supervision and control of geographically distributed plants/processes (ie pipelines, power systems). These systems comprise of complex digital computer based data acquisition and control systems and run advanced level applications softwares including real-time simulation of plant process to aid plant operation. Evaluation and selection process of different configurations offered by suppliers to meet purchaser's specification can be aided by simulation. Configuration features, characteristics, processor loads and requirements guide the selection process. Different solutions offered by vendors can be compared in terms of meeting requirements and performance. Performance data is estimated through simulation. Case studies for two large Brazilian electric utility companies for selection of control centre energy management system configurations have been reported [42]. System requirement specification include descriptions of power network, EMS functions (applications software functions for power system control), input/output data (including scan time for each type of data), RTU, communication link and Master Station configuration. A typical performance requirement specification is described below:

Specified execution time of various programs
      Automatic generation Control (AGC) – 4 secs.
      Data acquisition – 2 sec.
      State estimation – 180 sec.
      Contingency evaluation – 600 sec.
      Display update – 5 sec.
Response times should be
      Network analysis – 90 sec. (for 100 buses, 100 lines, 20 contingencies)
      Display – 2 sec.
      Status change annunciation – 2 sec.
      Power flow – 30 sec. (for 500 buses, 800 lines and 260 TCUL)
      Failover – 5 sec.
Other requirements
      Load on each processor – 50 %
      1 limit viloation / UTR, minute.
      4 display requests / console, minute.
      Full alarm logging
      No alarm loss or degradation of response for alarm bursts of upto 200.

Simulation performance of one chosen configuration has been obtained as:
      Processor load at SCADA/AGC level – 32.26%, most of it is due to MMI function 11.5%.
      Communications – 5%
      Operating system – 5%
      Processor load at application software level – 22.12% comprising of
          State estimation
          Bus load forecasting and   5%
          Network solution
      Contingency analysis – 6.9 %
      Study mode function – 1.05 %
      Configuration management – 1.17 %
      Operating system – 3 %
      Other programs – 5 %

230

10. THE FUTURE OF SIMULATION

An wider use of dynamic simulation and analysis in process industry is shown in Fig.27. A computer-aided control system package is linked to

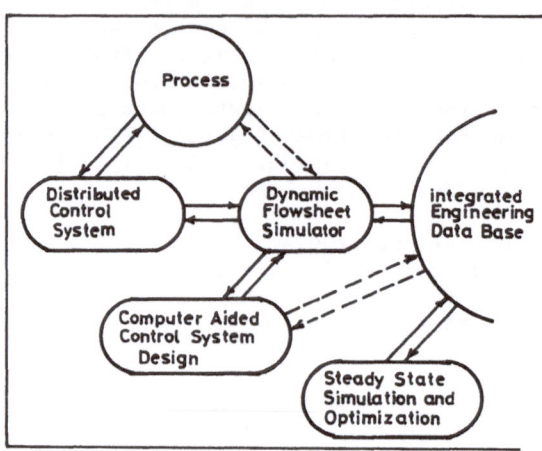

Fig 27 CAD environment for process dynamics investigation

the simulator. The simulator and a distributed control system are coupled to the real plant. Such a high-level simulation activity will be based on detailed submodels and a bunch of coordinated subsimulators. These simulations will be used for integrated decision making for design, engineering, early man-machine problems and operator training. Support from dedicated expert systems for the preparation of such simulators is expected. Expert systems integrated in the user interface may be used in future to advise the user on the choice of the model description and the simulation method itself. A possible scheme for model construction by expert systems has been discussed by Schmidt and Lappus [3]. Use of artificial intelligence for network simulation in gas pipeline has been discussed in [43] and for the glass industry in [44].

11. REFERENCES

1. T.F.Edgar and D.E. Seborg (Ed) <u>Chemical Process Control 2</u>, Proceedings of the Engineering foundation Conference 1981.
2. F.E. Cellier, <u>Progress in Modelling and Simulation</u>, Academic Press, 1982.
3. G.Schmidt and G.Lappus, ʹDigital simulation methods – A Surveyʹ <u>IFAC Proceedings Digital Computer Applications to Process Control, No. 6, 1986</u>, pp 83-93.
4. T.G.Hicks, <u>Power Plant Evaluation and Design Reference Guide</u>, McGraw-Hill.

5. K.H.Fasol and H.P.Jorge, ´Principles of model building and identification´, <u>Automatica</u>, 16, 1980 pp 505–518.
6. W.J.Karplus, ´The spectrum of mathematical modelling and system simulation´, <u>Simulation of Control Systems</u>, North Holland Publishing Company, 1976.
7. D. Maudsley, ´An approach to the mathematical modelling of dynamic systems´, <u>Measurements and Control</u>, 1978, pp 181–189.
8. N.W.Rees, P.T.Nicholson and D. H. Mee, <u>IFAC Proceedings on Modelling, and Control of Electric Power Plants</u>, Italy, 1983, pp 51–58.
9. B.M.Novakovic, ´Design of the dynamics of heat exchanger´, <u>IFAC Proceedings on Digital Computer Applications to Process Control,</u> No. 6, 1986, pp 333–340.
10. K.H.Fasol, ´Computer simulation to improve hydro power control, <u>IFAC proceedings on Electric Energy Systems</u>, 3, 1986, pp 241–247.
11. R.L. Elder, M.E.Gill and A.M.Y. Razak, ´Validation of a compressor model´, <u>Trans. Inst. Measurement and Control</u>, vol.8, No. 4, Oct.1986, pp 171–181.
12. Scientific Software Intercomp, Inc., Houston, <u>Real Time Model</u> (Company Technical Literature).
13. K.E. Starling, <u>Fluid Thermodynamic Properties for Light Petroleum Systems</u>, Gulf Pulbishing Co. Houston, 1973.
14. T.C.Chen et al, ´Maryland University dynamic simulator for chemical processes´, <u>Proceedings Summer Computer Simulation</u>, 1984, pp 636–641.
15. Y.Kokai, I.Matori and J.Kawakami, ´Multiprocessor based generator module for a real-time power system simulator´, <u>IEEE Trans on Power Systems</u>, Nov. 1988, pp 1633–1639.
16. A.M.Carlson, ´Some techniques for simulating plant piping´, <u>Proceedings Summer Simulation Conference</u>, Vancouver, July 1983, pp 872–877.
17. M.Brucoli, F. Torelli and M. Trovato, ´An alternate model of an interconnected power system in polynomial matrix form´, <u>Large Scale Systems</u>, 4, 1983, pp 27–49.
18. L.T.Duo, W.B.Huang, Q.R.Ping, X.M.Qing and C.Z.Rong, ´A large scale mathmodel for 200 MW thermal power unit´, <u>IFAC Proceedings Power Systems and Power Plant Control</u> No.8, 1987, pp 343–347.
19. M.H.Butterfield and P.J.Thomas, ´Methods of Quantitative Validation for Dynamic Simulation Models´ Parts I & II, <u>Trans Inst MC</u>, Vol. 18, No.4, Oct 1986, pp 182–219.
20. W.J.Turner and M.J.Simonson, ´A compressor model for gas pipeline network simulation´, <u>Proceedings of Transportation Conference</u>, The Institution of Engineers, Australia, 1984.
21. G.K. Lausterer, J.Franke and E.Eitelberg, ´Mathematical modelling of a steam generator´, <u>Proceedings IFAC Digital Computer Applications to Process Control</u>, Dusseldorf, 1980.
22. P.Holl, W.Marquardt and E.D.Gilles, ´DIVA – a powerful tool for dynamic process simulation´, <u>Computer and Chemical Engg</u>, vol. 12, No. 5, 1988 pp 421–426.
23. ´Catalog of Simulation Software´, <u>Simulation</u>, Vol.43, 1984,pp 180–192 and Vol. 44, 1985, pp 106–108.

232

24. Continuous Systems modelling program III, (CSMP III). Program Reference Manual. IBM Corporation, Data Processing Division, New York.
25. M.R.Skrokov (Ed). Mini and Microcomputer Control in Industrial Processes, Van Nostrand Reinhold, 1980.
26. N.R.Prasad and R.D.Dunlop. ´Three phase simulation of the dynamic interaction between synchronous generators and power systems using continuous systems modelling program´, Proceedings Power Industry Computer Application Conference, 1979, pp 29~36.
27. L.P.Smith et al. ´Effective utilisation of two commercially available simulation languages in the modular modelling system development´. Proceedings Summer Computer Simulation Conference, 1983, pp 449~455.
28. R.K.Wood, R.K.M.Thambynayagam, R.G.Noble and D.J.Sebastian, ´DPS: a digital simulation language for the process industries´, Simulation, 42(5), 1984, pp 221~231.
29. A.M.Carlson, ´DAP ~ A Bechtel approach to dynamic simulation´, Proceedings Summer Computer Simulation Conference. July 1982, pp 520~525.
30. V.B.Mantri, R.A.Strain and D.R.Austin, ´Dynamic simulation of Ethylene vaporization system´, Proceedings Summer Simulation Conference, 1983, pp 862~867.
31. R.K.Miyasaki, N.E.Pobanz, D.B.Warren, R.S.Eby and J.R. Jamison, ´Dynamic simulation and verification of a compression liquefaction system for material withdrawal from a Uranium enrichment plant´, Proceedings Summer Computer Simulation Conf. 1983, pp 335~345.
32. G.Kalani, ´Application of microprocessor based control systems for simulation´, IEE Conference Publication No. 226, 1983. pp 222~227
33. C.N.Wei, ´Applications of a dynamic simulation of distillation columns for a new manufacturing plant´, Proceedings IFAC Control of Distillation Columns and Chemical Reactors, U.K. 1986, pp 117~122.
34. T.Shinoda, E.Tose and M.Takahashi, ´Enhancing DCS for simulation Training´, Intech, Sept. 1988, pp 103~105.
35. H.Kreitner and H.Schuler, ´Simulation in process control´ (in German), Chem~Ing Tech, 60, No.8, 1988, pp 613~627.
36. S.Koyama, K.Makino, N.Miki, Y.Iino and Y. Iseki, ´On the parallel processor array of Hokkaido University High~speed system Simulator "Hoss"´, Preprints 8th IFAC World Congress, Vol.11, 1981, pp 137~142
37. H.Kasahara, H.Honda, M.Kai, T.Seki and S.Narita, ´Parallel processing for simulation of dynamical systems´. Proceedings IFAC Digital Computer Application to Process Control, No.6, 1986, pp 527~537.
38. R.M.C.De Keyser, De Coen and Vardiere, ´Multiprocessor simulation of cutter suction dredging ship´, Proceedings IFAC Digital Computer Applications to Process Control, No.6, 1986, pp 307~312.
39. H.Kasahara and S.Narita, ´Integrated Simulation system for design and evaluation of distributed computer control systems´, Proceedings IFAC World Congress, Hungary 1984, vol.5, pp 2669~2674.
40. J.Davidson and J.L.Houle, ´Mathematical model for Simulation of hierarchically distributed process control computer systems´,

Automatica, 24(5), 1988 pp 677–686.

41. J.D.Schoeffler, ´Response of Industrial Distributed Data Acquisition and Control Systems under Upset Conditions´, ISA Transactions, vol 27, No 2, 1988, pp 13–23.

42. J.M.Barbosa et.al., ´Requirements and performance data of control centre EMS configurations´, Proceedings IFAC Planning and Control of Electric Energy Systems, No 3, 1986, pp 275–284.

43. G.Karsai et al. ´Intelligent Supervisory Controller for Gas Distribution System´. Proceedings American Control Conference, 1987, pp 1353– 1358.

44. R. Herrard, J. W. Rickel and T. Garland, ´Glass Annealing Process Simulation Using Expert System : A Glass Industry Application of Artificial Intelligence´. IEEE Trans on Industry Application, 24 (1), Jan/Feb 1988 pp 43–48.

45. Chiyoda Corporation, Japan, ´Process Transient Analysis´, Engineering Document No 2277–72–P1–CL–004 (CPP for Maharashtra MGCC, IPCL/EIL) Dec 1987.

46. J.K.Pal, ´Advanced Control and Supervison of Utility System in Petroleum Industries´, Microprocessor Applications for Productivity Improvement, (ed. V. P. Bhatkar and K. Kant), Tata–McGraw–Hill, 1988.

47. M.Nomura and Y.Sato, ´Adaptive Optimal Control of Steam Temperature for Thermal Power Plants´, IEEE Trans on Energy Conversion, Vol 4, No 1, 1989, pp 25–33.

APPENDIX

PROCESS TRANSIENT ANALYSIS – AN INDUSTRIAL CASE STUDY

Contributed by :
Chiyoda Corporation,
Tsurumi–Ku,
Yokohama 230,
Japan

Prepared by   :   T.Aio and M. Tanaka
Authorised by :   H.Shimuzu

In this section we present an appliction of modular modelling system
(MMS) to a combined cycle power plant. This study [45] was conducted by
M/s Chiyoda Corporation, Yokohama, Japan for the captive power plant of
Maharastra Gas Cracker Complex (MGCC) of M/s Indian Petrochemicals
Corporation Ltd., India. The greater part of this study is reproduced
here with kind permission of Chiyoda Corporation, and Indian
Petrochemicals Corporation Ltd., India.

A.1   System Configuration.

The plant configuration for the dynamic simulation study is shown in
Fig.A.1.

A.1.1 Modelling of equipment: Captive Power Plant consists of major
equipment as shown in Table A.1.   In the dynamic simulation model, main
components such as HRSG, STG, BFW Pump and Piping are rigorously
modelled by programming the differential heat and mass balance equations
and their performance characteristics. But the performance
characteristics of GTG, deaerator, condenser, fuel and PRD system are
given with a correlation curve or constant value because they do not
have so serious effect on the calculation results of simulation. The
performance characteristics of the following valves and dampers are also
considered in dynamic study:

        – Boiler Feed Water Control Valve
        – Attemperator
        – Fuel Gas Control Valve
        – GTG Exhaust Gas Bypass Damper
        – STG Governor Valve
        – Steam Venting Control Valve

A.1.2 Control Systems and Set Points : The main control loops considered
in this model are the same as in actual plant control system (see
Fig.A.1). The information for each loop, e.g.,function, and set points,
is described in Table A.2.

A.1.3 Operation Conditions of Another GTG/HRSG Train : It is considered
that two trains in service are in even load and each train has the same

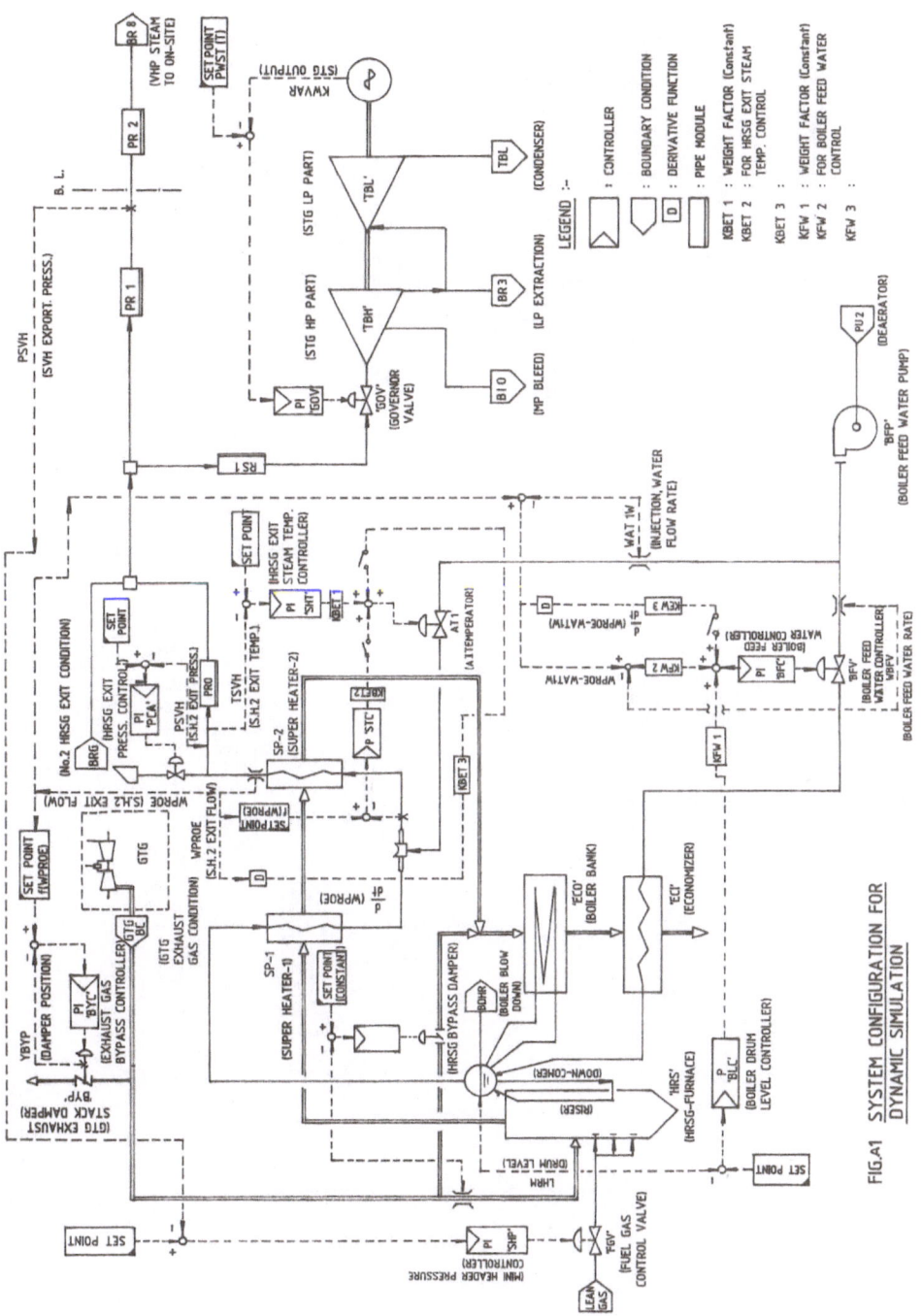

FIG.A1 SYSTEM CONFIGURATION FOR DYNAMIC SIMULATION

236

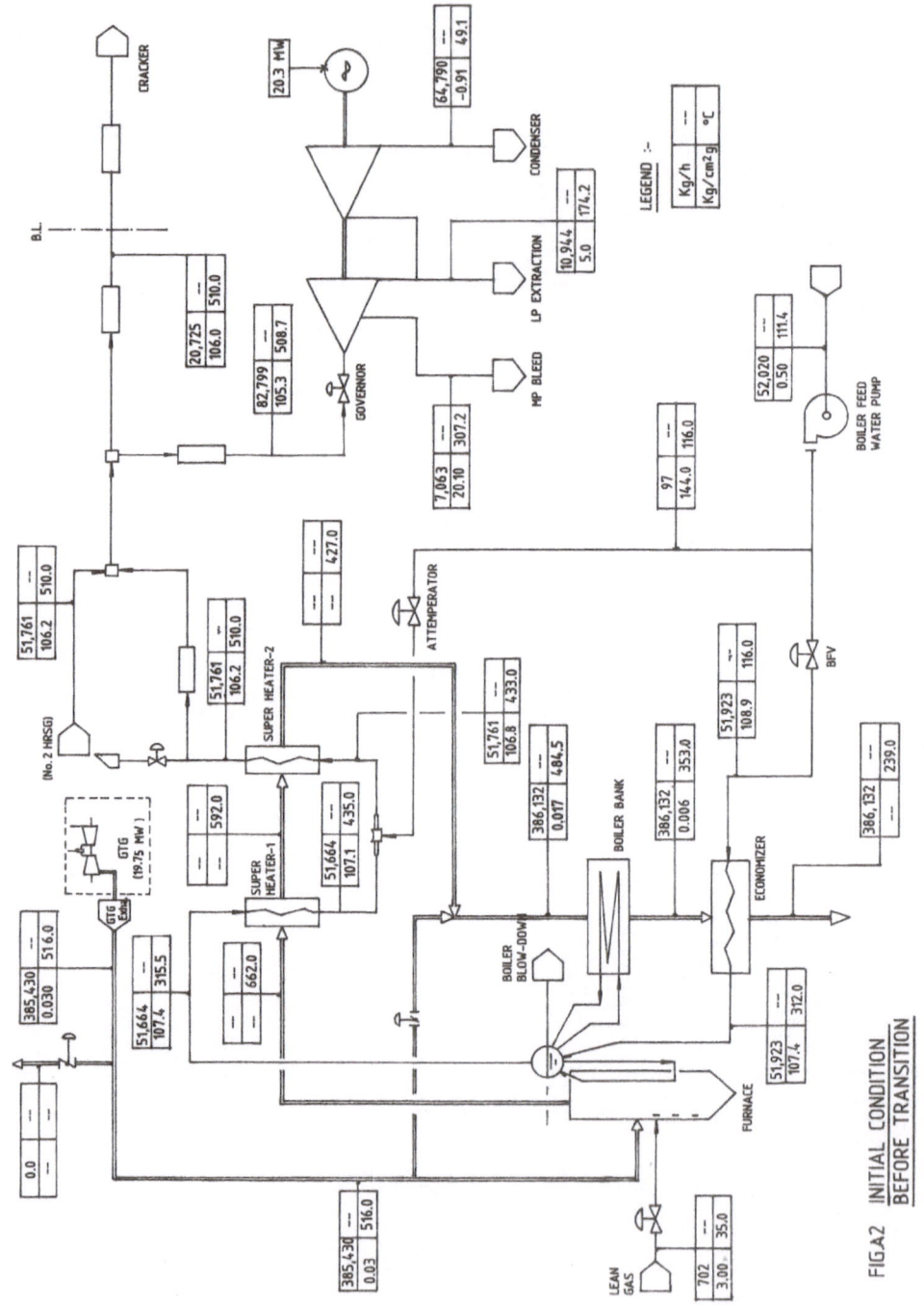

FIG.A2 INITIAL CONDITION BEFORE TRANSITION

Table A.1 : Modelling of Major Equipment

Equipment		Model
GTG	:	Exhaust gas conditions are given in the form of the function of GTG power output
HRSG	:	Differential heat/mass balance equations and its performace characteristics are programmed. Physical data of HRSG are listed in Table A.3.
STG	:	The steady state heat/mass balance equations and performance characteristics curve are programmed because of its quick response comparing with HRSG. (Bleed/Extraction flow rate are taken to be constant)
BFW PUMP	:	Ditto
Deaerator	:	The boundary conditions to be given (Press./Temp./Enth.)
Condenser	:	The boundary conditions to be given (Press.)
PRDS	:	The boundary conditions to be given (Flow)
Fuel Supply:		The boundary conditions to be given (Press.)

Table A.2: Control Loop Description

1. Loop Name	VHP Header Press.Control	HRSG Exit steam temp. control.	HRSG Feed water control.	STG Load control	VHP Steam Venting Control	GTG Exhaust Gas Bypass Control
2. Control Variable	VHP Steam Header Pressure @ B.L. (PSVH)	HRSG Exit Steam (TSVH)	Boiler Feed water flow (WFFV)	STG Power Output (KWAR)	VHP Steam Header Press.@ HRSG Exit (PHRS)	GTG Exhaust Gas Bypass Flow (WFGB)
3. Device	Fuel Gas supply valve	Attemperator	Boiler Feed Water Control Valve	Boiler Feed Governor valve	Vent off valve	Exhaust Stack Damper
4. Set-Point	106.0KG/cmG	510.0 C	(*1)	(20.30MW) (*2)	108.5 Kg/cmG	(*3)
5. Function	A	B	C	D	E	F

table contd...

Table A.2: Control Loop Description (contd....)

6. Controller Action		0.071	0.900	2.000	0.100	0.711	2.000
	P	0.071	0.900	2.000	0.100	0.711	2.000
	I	0.0071	0.0036	0.200	0.005	0.014	0.050
	D	---	---	---	---	---	---

7. Remarks			
	KBET1=1.0	KFW1=0.1	Drum Level Control
	KBET2=0.0	KFW2=1.5	P:0.004
	KBET3=0.0	KFW3=0.0	;0.0
			D; ---

Note: *1: Three Elements Control: Steam Generation Rate/BFW Rate/Steam Drum Level
      *2: Given by Load Sharing Program.
      *3: Depending upon HRSG Steam Generation Rate

A: Main steam header pressure is kept constant by adjusting fuel flow to HRSG with fuel gas control valve.
B: Main steam temperature at HRSG exit is kept constant by injecting boiler feed water into S.H.I. exit steam by attemperator.
C: Boiler feed water is controlled to meet with the steam generation rate required and adjust the steam drum level to the set value.
D: DTG power output is adjusted to the set value by STG governor.
E: When main header pressure at HRSG exit rises above the upper limit, a part of main steam generated is blow off to atmosphere to keep the header pressure below the upper limit.
F: A part of GTG exhaust gas is blown off to atmosphere through bypass damper to avoid the excess heat input by exhaust gas to HRSG when HRS load come down below 38% of MCR.

Table A.3: Physical Data of HRSG

		Superheater-1	Superheater-2
1. Physical Parameter for Superheaters			
Tube outer diameter	(mm)	38	38
Tube inner diameter	(mm)	4.0/4.5/5.0/ 5.6/6.3	4.0/4.5/5.0/ 5.6
Number of rows in direction of gas flow		30	12
Number of tubes over which gas passes per row ..		64	64
Number of tubes the steam flows through per pass		15	12
Tube spacing perpendicular to gas flow direction	(mm)	75	75
Tube spacing parallel to direction of gas flow	(mm)	80	80
Tube length extending into gas path	(m)	4.27	3.17
Width of gas flow passage	(m)	4.5	4.5
Arrangement: 1=Counter flow; 2=Parallel flow		1	1

2. Physical Parameter for Economizer and LP-Bank:

		Economizer	LP-Bank
Tube outer diameter	(mm)	38	38
Tube inner diameter	(mm)	3.2	3.2
Number of rows in direction of gas flow		20	6
Number of tubes over which gas passes per row		60	60
Tube spacing perpendicular to gas flow direction	(mm)	75	75
Tube spacing parallel to direction of gas flow	(mm)	75	75
Tube length extending into gas path	(m)	6.5	6.5
Width of gas flow passage	(m)	4.55	4.55
Arrangement: 1=Counter flow; 2=Parallel flow		1	1

3. Physical Parameter for Boiler

Effective total cross sectional flow area of riser(s)	(m^2)	1.357	
Effective total cross sectional flow area of downcomers	(m^2)	0.719	
Overall height of boiler	(m)	13.6	(Between the Center line of steam drum & lower header)
Inner diameter of drum	(m)	1.408	(Boiler Bank Tube: 95 ton.
Mass of waterwall	(ton)	69	(Membrane wall, piping & headers)

table contd..

Table A.3 contd...

Volume of steam drum	(m^3)	10.38	
Volume of furnace flue gas	(m^3)	276	(Volume of furnace from the bottom of super-heater)
Volume of riser	(m^3)	23	
Volume of downcomers	(m^3)	24	

characteristics.

A.2   The Mathematical Model

A mathematical simulation program has been constructed for the plant model mentioned above. In this dynamic study, the computer software, called "Modular Modelling System" (MMS) is used to make the mathematical equatiions for each equipment and/or the overall plant. MMS was originally developed by the Electric Power Research Institute in USA (EPRI) and is the computer software for use in the analysis of the dynamic performance of fossil and nuclear power plant.

A.3   Simulation Conditions

A.3.1 Study Basis : (1) Plant Conditions before Transition: In all simulation cases mentioned herein, the operation conditions are assumed to be steady at plant normal on lean gas before transition occurs. The process conditions of normal plant operation are shown in Fig.A.2. (2) Relation with Electric Control System: When electrical demand changes occur or one of the generators trips, electrical load command for each generator is also changed and newly set by Electrical Control Systems (ECS), [46]. These electrical transient conditions are considered in the process demand study.
        However, for other transient conditions except electrical power demand or load changes, process dynamic simulation and electrical dynamic simulation can be independently performed, since the order of magnitude of process response time is very slow compared with that of the electrical characteristics change.

A.3.2 The Way of providing disturbance: Conditions for dynamic simulation such as steam flow rate, electric power demand/load changing rates are set as shown in Fig.A.3.

Case 1   Sudden Power Reduction by 11.5 MW

(1) Total electrical power requirement including internal power consumption reduces suddenly from 59.8 MW to 48.3 MW.

(2) Load set point signals to STG and GTG change quickly according to the tie-line control system corresponding to the power requirement after power reduction.
        The changing rate of total power demand for the simulation is 11.5 MW in 0.1 Sec, and power reduction for each generator is as follows:

	From	To
GTG-1	19.75	15.92
GTG-2	19.75	15.92
STG	20.3	16.47

(3) Total steam requirement out of B.L. is constant after transition.

242

CASE 1.   SUDDEN POWER REDUCTION BY 11.5 MW

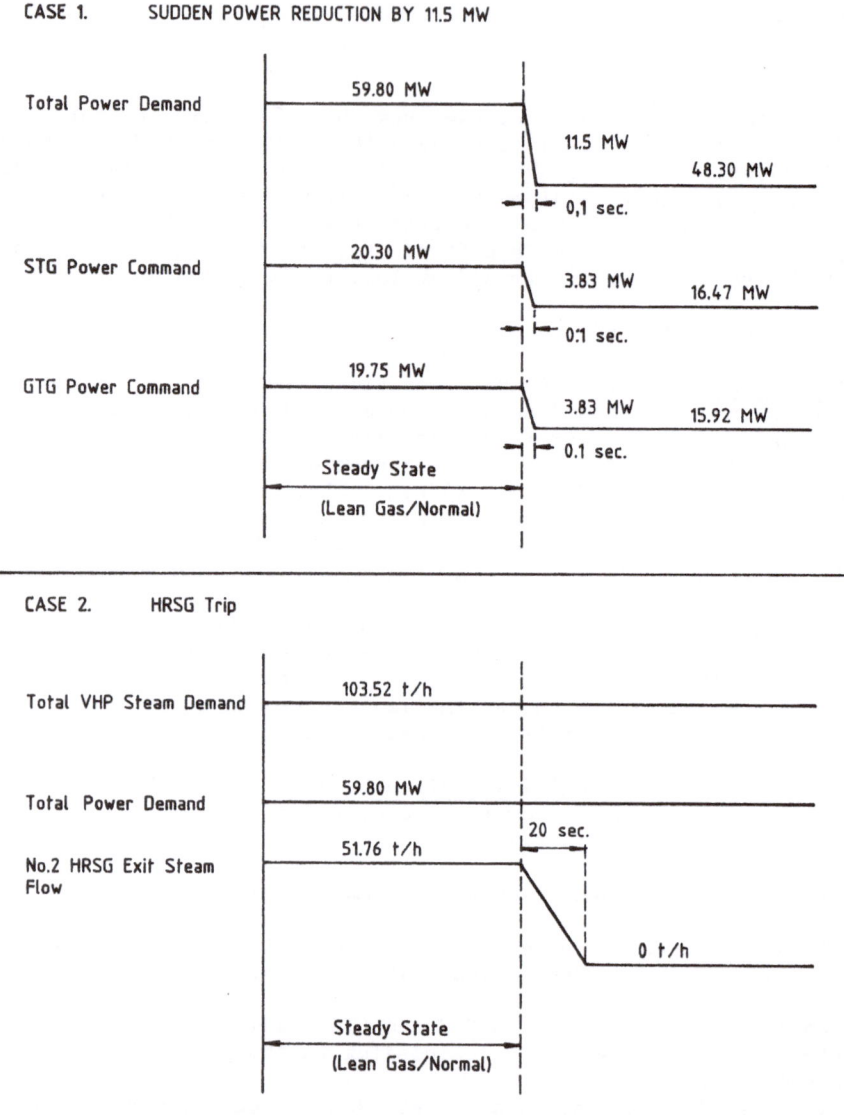

Fig. A.3   SIMULATION CONDITIONS – DISTURBANCE

Case 2 Sudden Steam Reduction by 11.5 T/H

1) Steam requirement out of B.L. reduces suddenly by 11.5 T/H. This change occurs in 0.1 seconds
2) Total power demand is constant during transient.

Case 3 Steam Requirement Increment by 125 T/H in 2 min. (HRSG Load UP)

1) Steam requirement out of B.L. increases at a constant rate by 125 T/H in 2 minutes.

2) Two HRSGs under the normal load operation increase steam generation rate to keep the header pressure constant.

3) Total power reauirement is constant during transient.

Case 4 STG Trip

1) The steam turbine trips in normal operation at a load of 20.30 MW

2) GTG load set-points increase to 20.5 MW quickly to cover the power reduction by STG trip.

3) Steam exporting mode changes automatically from "by STG bleed / extraction" to "by PRDS" after STG trip. Accordingly, VHP steam generating rate increases from 20.725 T/H to 45.48 T/H in 3 seconds after STG trip in this case.

Case 5 HRSG Trip

1) No. 2 HRSG trips.

2) Exit isolation valve of No. 2 HRSG closes in 20 seconds after No. 2 HRSG trips.

3) Total steam demand and electric power demand are constant during transient.

Case 6 GTG Trip

1) One GTG which supplies exhaust gas to No. 1 HRSG trips.

2) STG load set value increases to rated power (25 MW) to cover the power reduction by one GTG trip.

3) Stem position of fuel flow control valve of No. 1 HRSG is fixed for 30 seconds after GTG trip for safe operation to avoid misfiring of the HRSG because of reducing flow rate of air from GTG.

4) FD fan starts and reaches the rated speed within 30 seconds. In 30 seconds after GTG trip, fuel control valve of No. 1 HRSG begins to

control the fuel flow rate again.

Note:- To examine the response No. 1 HRSG, steam flow rate of No. 2 HRSG is kept constant during transient.

A.5   Results of Dynamic Simulation

Some dynamic study results for six cases are summarized in Table A.4. The results show that the Captive Power Plant can be continuously operated after transition with small chnge of steam header pressure and temperature. Detailed descriptions of calculation results are made for each case as below.

Case-1 Sudden power Reduction by 11.5 MW : Fig. A.4 shows the change of STG load command, HRSG inlet exhaust gas flow and temperature after the sudden reduction of electrical power demand in MGCC occurs. In this case, STG load command changes from 20.3 MW to 16.47 MW and GTGs load command from 19.75 MW to 15.92 MW in 0.1 seconds respectively. These command signals are delivered from ECS by the tie-line control to cover the sudden reduction of power consumption in MGCC. And then, the exhaust gas flow and temperature at HRSG inlet will change according to GTG performance as shown on Fig.A.4 after transition.
     Fig. A.5 shows the main header pressure and main steam temperature change during transition. The header pressure at B.L. reaches the maximum of 106.8 kg/cm(2)G due to the reduction of main steam consumption by STG quick load-down in 10 seconds after transition. The main steam temperature decreases to a minimum of 503.5 Deg. C in 60 sec, and the maximum overshoot is 1.0 Deg.C. Also, GTG exhaust stack damper will partially open and blow-off the exhaust gas to atmosphere not to introduce the excess heat input to HRSG because the steam generation requirement for each HRSG decreases to below 38 % of MCR due to the main steam demand reduction by STG load-down. (See Fig. A.6).
     The reason why a part of GT exhaust has to be blown-off is that steam temperature must be always kept constant 510 Deg. C by supplemental firing.

Case-5: HRSG Trip : Fig. A.7 shows the change of steam generation rate from each HRSG. The decreasing curve shows the generation from the HRSG which tripped and the rising one from another HRSG which backs up to keep the header pressure constant during the transition. In this case, the header pressure comes down to it's minimum of 103 kg/cm(2) G in 40 seconds, after one HRSG trip, while the main steam temperature only change from minimum 509.8 Deg.C to maximum 510.7 Deg. C during the transition as shown in Fig. A.8.

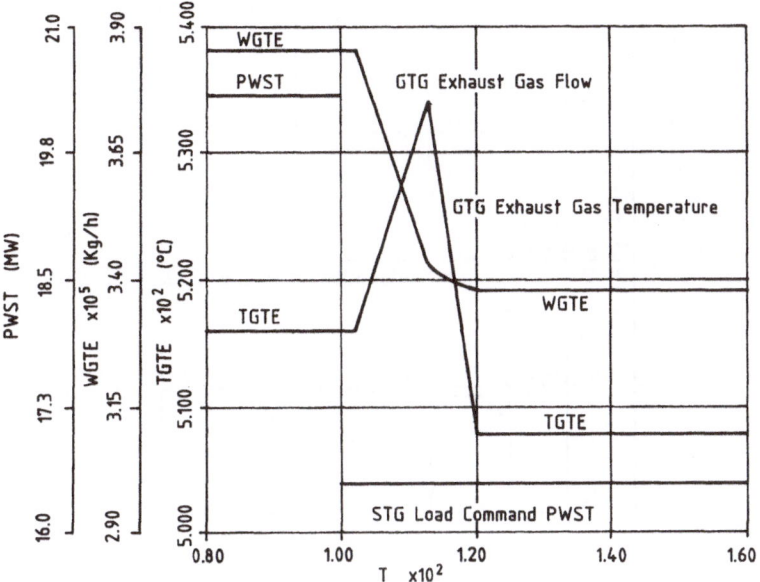

Fig. A.4

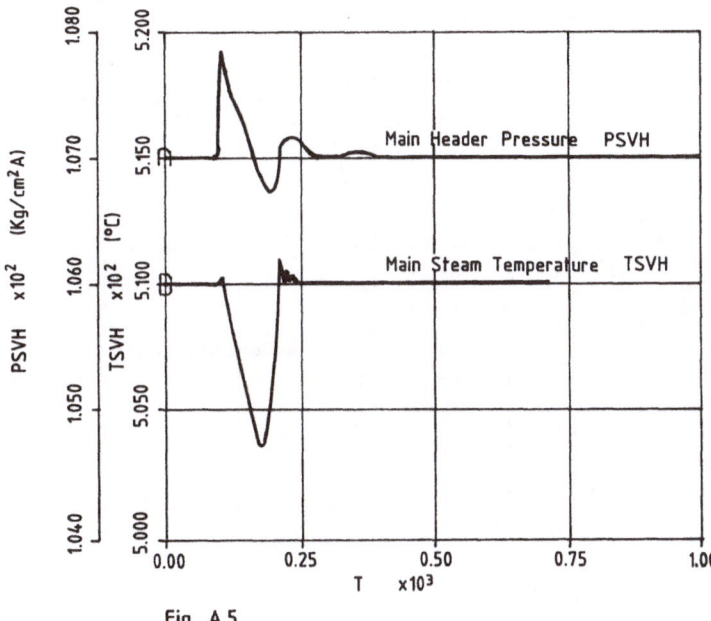

Fig. A.5

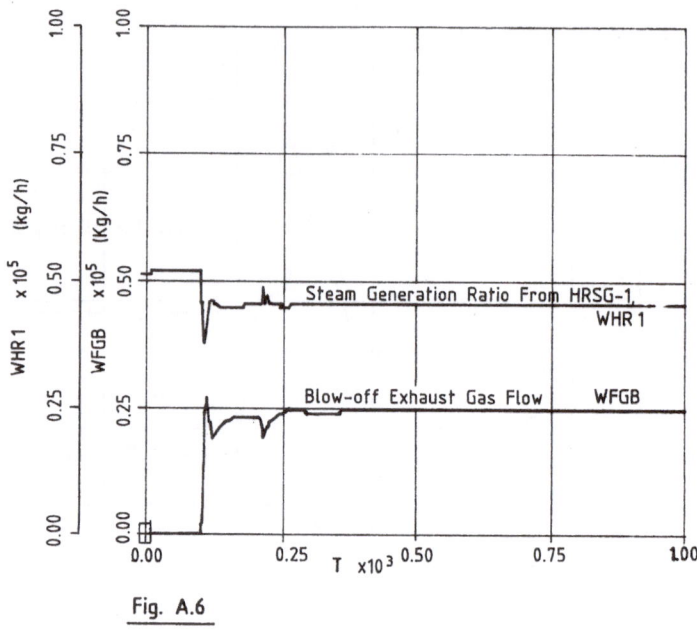

Fig. A.6

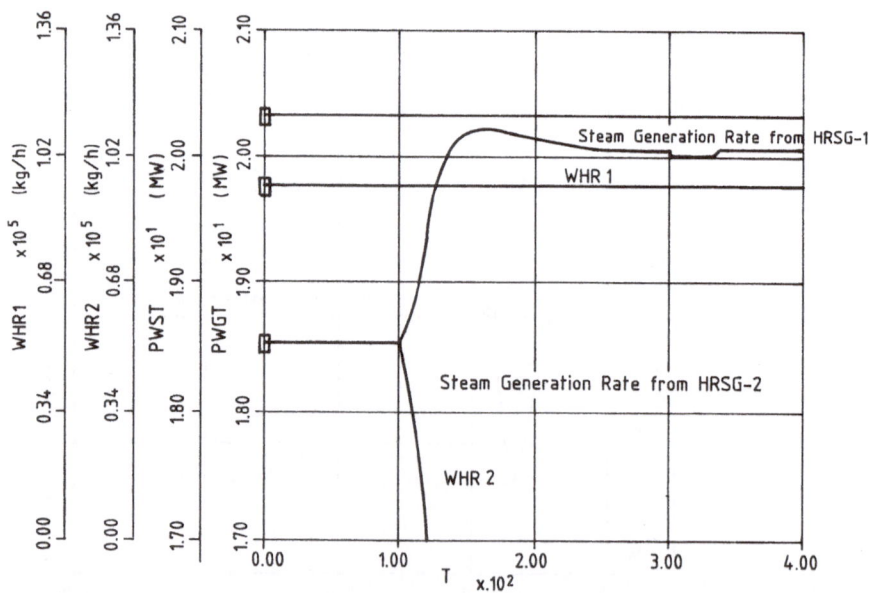

Fig. A.7

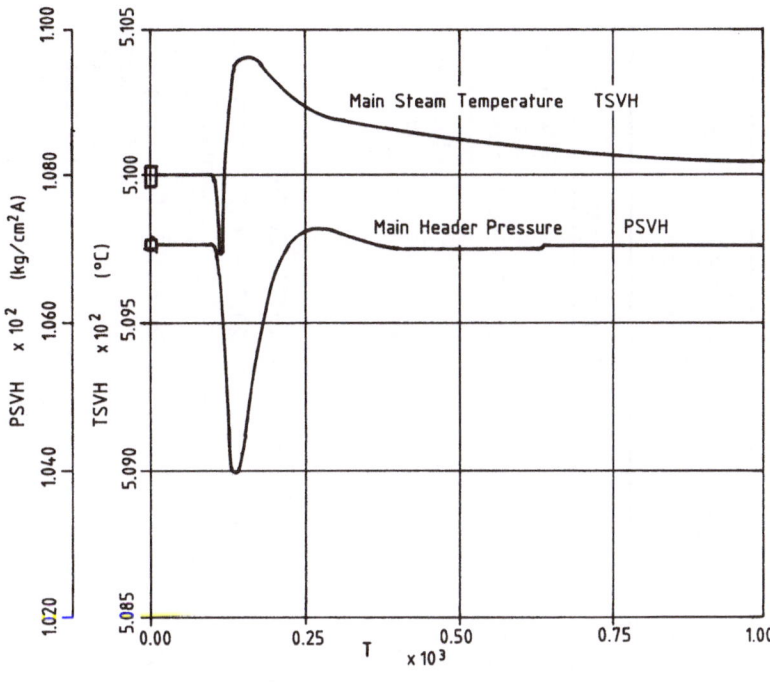

Fig. A.8

Table A.4: Dynamic Study Results Summary

Case Description No.	Deviation from Normal @ B.L.				Steady Status after Transition					Remarks
	Steam Press @ B.L.		Steam Temp. @ HRSG Exit		Steam Press @ B.L. HRSG Exit	Steam Temp. @ HRSG Exit	GTG-1 Output	GTG-2 Output	STG Output	
	Max. K/g	Min. k/g	Max. °C	Min. °C	k/g	°C	MW	MW	MW	
1. Power Demand Reduction by 11.5 MW *	106.8	105,7	510.9	503.5	106.0	510.0	15.92	15.92	16.47	* GTG exhaust is blown off continuously after transition.
2. Steam Demand Reduction by 11.5 T/H	106.4	105.8	510.3	497.0	106.0	510.0	19.75	19.75	20.30	
3. Steam Requirement Increment by 125 T/H in 2 min.	106.3	103.9	510.4	509.9	106.0	510.0	19.75	19.75	20.30	
4. STG Trip **	108.8	106.0	510.5	509.6	108.3	510,0	20.5	20.5	0.0	**GTG exhaust is blown off & excess
5. HRSG Trip	106.2	102.9	510.4	509.7	106.0	510.0	19.75	19.75	20.30	
6. GTG Trip	108.3	104.1	511.1	504.3	106.0	510.0	0.0	20.50	25.0	steam is vented off continuously after transition.

Normal Operating point before transition:

Steam pressure @ BL : 106 kg/cm² G
Steam temperature @ HRSG out: 510°C
GTG-1, GTG-2 output : 19.75 MW
STG output : 20.3 MW

SECTION II

REAL TIME APPLICATIONS

Chapter 7

A LOW COST SELF-ADAPTIVE MICROPROCESSOR CONTROLLER: APPLICATION TO HEATING AND VENTILATION CONTROL

R.M.C. De Keyser
Automatic Control Laboratory
University of Ghent
Grotesteenweg Noord 2
B-9710 Ghent
BELGIUM

ABSTRACT. The availability of cheap microprocessor chips and peripheral interface chips creates new openings for building a low cost but intelligent controller. It is then possible to implement smart advanced control strategies, e.g. self-adaptive control.

A similar system was developed, initially as controller for residence building heating. Thanks to its versatility it is now also used for greenhouse climate control, incubator control and ventilation control in animal houses. Other applications, also in the non-heating sector, will follow.

This paper gives a full description of the system, i.e. the hardware, the system software, the user interface, the adaptive control algorithms, the application to heating and ventilation control and some typical results from real-life experiments.

1. INTRODUCTION

This paper describes the results of a demonstration project. It was the objective to illustrate by means of real-life experiments the feasibility of some concepts of modern control theory and digital instrumentation to low cost applications.

The capabilities of advanced control strategies (i.e. self-tuning and adaptive methods) and computer-based control instrumentation to industrial process control systems have been well established now during the last decade.

However this paper focuses on another potential field of automation and control applications, i.e. in the non-industrial sector such as building automation and agriculture as typical examples. In order to end up with a product which is economically feasible, the designer has to keep in mind straight from the beginning and continuously during the development the important aspect of low cost. Although introducing severe limitations this also stimulates the designer in thinking about smart hardware configurations as well as in writing efficient software without the assistance offered by a powerful operating system.

S. G. Tzafestas and J. K. Pal (eds.), Real Time Microcomputer Control of Industrial Processes, 251–274.
© 1990 *Kluwer Academic Publishers.*

This demonstration project illustrates two important facts : first microprocessor-based digital hardware is today a valuable alternative besides classical analog electronics for low cost automation equipment; and second, as a result of this, the availability of advanced digital control algorithms is no longer a privilege of the larger process control projects.

The paper is organized as follows. In section 2 the system hardware is described with emphasis on the specific measures that were taken in order to end up with a low cost product. The next section deals with the system software showing how the total hardware system is controlled by means of high level language programs (Pascal) without the help of a real-time operating system (which could not be afforded without violating the low cost specification). A selection of several applications which have been realized with the adaptive microcomputer system is described in section 4. For one of them, a residence floor heating control system, detailed information on the rule based expert control strategy with self-learning features is presented. Some real-life experimental data are summarized in the last section.

2. SYSTEM HARDWARE

The hardware is built around the concept of a system bus. The general structure of the Adaptive Microcomputer Controller (AMC) is shown in figure 1. The advantage of using a bus is that almost all connections can be made by means of on-board connectors and that all bus signals are available to each board. This eliminates the need of complex wiring (only two flat-cables are used in the system, respectively for the Keyboard/Display board and the Indicator Leds which are both mounted on the front panel).

The system bus is implemented on the main board, which also contains the power supply, 2 plug-in connectors to the external input/output screw terminals and 4 slots to connect the individual boards to the system bus (of which only 3 are used in the current configuration). This configuration results in a flexible and modular structure in which all boards can be easily and quickly removed without the need of disconnecting the external wiring to sensors or actuators. This is important from the maintenance viewpoint.

Apart from the main board with system bus, there are 4 smaller boards which implement specific functions : 3 identically sized (12 cm x 7 cm) boards which can be plugged into any of the 4 bus slots (further called the processor, input and output board) and a Keyboard/Display board which is mounted on the frontpanel for interaction with the operator. The whole system is very compact and is contained in a (11 cm x 13 cm x 26 cm) wall-mounted case.

2.1. Processor Board

A block scheme of this board is given in figure 2. The well-known MC6809 8/16 bit microprocessor chip is surrounded by 4 other popular VLSI-chips. The standard technique of memory-mapped input-output was

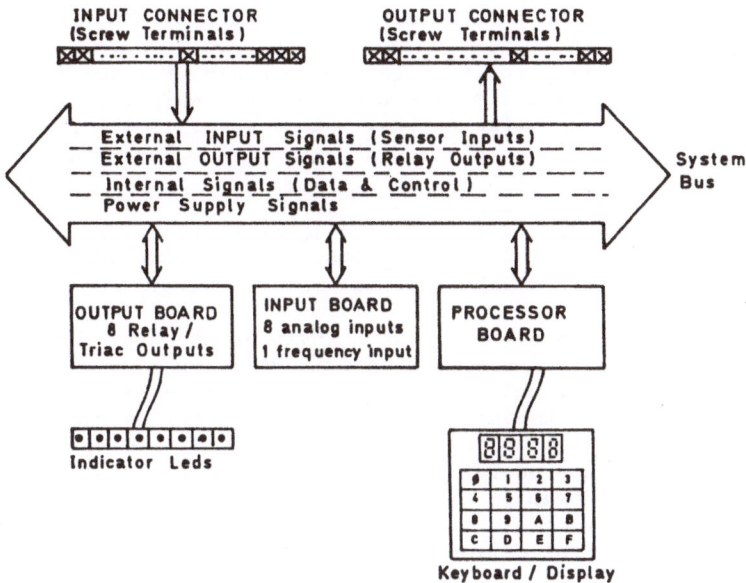

Fig.1 Hardware structure of the AMC

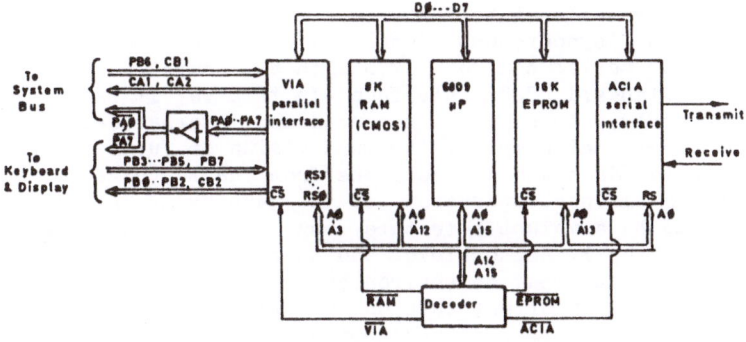

Fig.2 Block scheme of Processor Board

(DØ....D7 : data signals ; AØ....A15 : address signals ;
CS : chip select ; RS : register select)

applied for accessing the peripheral interface chips. The highest
address signals (A14 and A15) are decoded and used to select one out
of the four chips, while the lower address signals are used to select
a specific memory location (for the EPROM and RAM chips) or a specific
register (for the VIA and ACIA chips). This results in the following
layout of the address space (where $ denotes hexadecimal notation) :

```
$C000 ---> $FFFF : EPROM
$8000 ---> $BFFF : VIA (only the lower 16 bytes are used)
$4000 ---> $7FFF : ACIA (only the lower 2 bytes are used)
$0000 ---> $3FFF : RAM (only the lower 8 Kbytes are used)
```

The program code is stored in a 16 Kbyte eprom chip, the program
data in a 8 Kbyte ram chip. The CMOS static RAM chip is plugged into a
28 pin DIP socket with a built-in CMOS watch function, a nonvolatile
RAM controller circuit and an embedded lithium energy source. This
provides a complete solution to problems associated with memory vola-
tility and uses a common energy source to maintain time and date.
There is no battery backup on the system level, so when there is a
power shut-down the processor stops running. This is no severe drawback for the kind of applications that were envisaged (there is indeed
a good chance that the power for the actuators will be off as well).
However it is important that after power start-up the control system
restarts autonomously, having the exact time-of-day available (because
the operator setpoint, e.g. a heating schedule, may be a function of
time) and with the same data as just before shut-down. This is especially true when using self-adaptive control algorithms, because the
controller actions are based on a real-time estimated numerical process model, which should not be lost.

The ACIA chip (Asynchronous Communications Interface Adapter,
MC6850) allows for serial communication with any computer (usually a
PC) which has a standard RS232 interface. In this way it is possible
(albeit not necessary) to send all measured and internal data of the
AMC system to a Personal Computer for supervision or further processing, or to receive data and instructions from the PC (e.g. for remote
operation instead of local operation via the keyboard). This option
has turned out to be a valuable feature, especially to coordinate from
a PC at a hierarchically higher level the operation of several AMC
systems in a larger application environment.

Perhaps the most important chip on the Processor Board is the VIA
(Versatile Interface Adapter, SY6522) because it is the interface
between the processor and all other system functions such as sensor
inputs, relay outputs, keyboard input and display output. The control
and data transfers for all these functions have been realized through
a relatively low number of binary input-output lines (PA0...PA7;
PB0...PB7; CA1, CA2, CB1, CB2). This was only possible when using the
same signals for different functions and multiplexing them in time,
resulting in a less straightforward but cheaper hardware design and an
essentially more complex software. It was however decided from the
beginning that the hardware cost should have a higher weight than the
software cost in the total cost optimisation exercise, to end up with

a low cost product. The reason is mainly because low cost automation is only economically feasible when it leads to a mass product, in which case the software cost is spread over a large number of units. This is perhaps another difference with the design of industrial process control systems where the cost of software becomes more and more important.

Notice that most of the VIA binary signals can be programmed either as input or as output signals. The arrows in the scheme of figure 2 indicate the way they are used in our AMC configuration. The VIA also contains two timers which are essential for the operation of the AMC system. Their role will be explained later.

2.2. Input Board

This board contains three main parts : - the electronic circuits to translate the input signals coming from the external sensors into a suitable 0-5 Volt DC signal; - an electronic multiplexer to select one channel out of the eight input channels; - a voltage-to-frequency converter to convert the selected 0-5 VDC signal into a 0-5 kHz block wave. A block scheme of the board is given in figure 3.

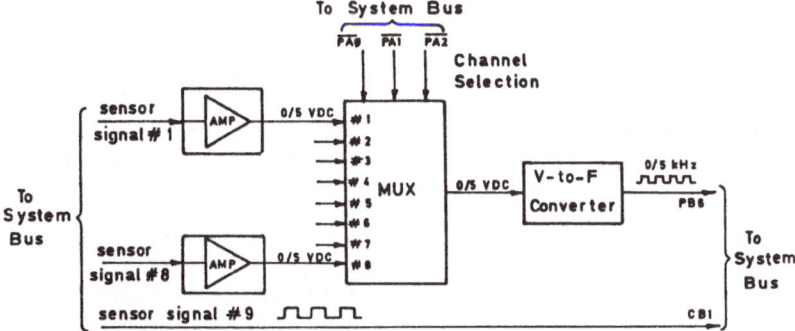

Fig. 3 Block scheme of Input board

As the system was initially built for climate control applications, the currently available input sensors are temperature and relative humidity. The input amplifier was designed such that identical circuits could be used for both type of sensors and all measurement ranges (the only difference being the specific value of some resistors). The eight input amplifiers on the board are thus similar, which is again in favour of the low cost objective. By means of a computer design program it is straight forward to compute for each input amplifier the exact values for the resistors in order to transform the relevant measurement range of each input sensor to the 0/5 VDC output range (e.g. for the outside air temperature : -10/+30 °C ⟶ 0/5 VDC; for the supply water temperature : 0/100 °C ⟶ 0/5 VDC; for the relative humidity sensor : 0/100 %RH ⟶ 0/5 VDC; etc.).

The multiplexer is used to select one out of the eight input channels for further processing. The selection is done by means of the three binary lines ($\overline{PA0}$, $\overline{PA1}$, $\overline{PA2}$) from the VIA chip.

A voltage-to-frequency converter chip is then used to realize a block wave with a frequency proportional to the input voltage. This frequency is directly measured by means of a counter in the VIA chip. In this way the analog sensor signals are converted into a digital value.

An alternative and obviously more direct way to realize this is to apply an analog-to-digital converter chip. This would lead to a higher conversion rate and simpler low-level software. On the other hand the hardware implementation cost would be higher.

Fast conversion is of less importance in climate control applications because the input signals (temperature and RH) are varying very slowly, so the sampling period can be quite long. Moreover as already explained the lower software cost does not counterbalance the higher hardware cost. These arguments motivate the choice of the V-to-F conversion alternative rather than using an A-to-D converter chip. Notice that the accuracy of the conversion is related to the number of block wave periods that is counted, thus depending on the total measurement interval (which is a software parameter rather than a hardware characteristic of the ADC).

There is a special input channel (#9) which accepts a periodic binary signal rather than an analog signal. It is directly connected to input pin CB1 of the VIA, which drives the second timer of the chip. In this way it is possible to measure the period of the pulse sequence. In the climate control application it can optionally be used to measure the flow rate of the supply water by means of a rotating flow meter that gives a pulse for each rotation.

2.3. Output Board

The principle of the output board is shown in figure 4. Eight output switches (relay or triac) are controlled from the VIA binary output signals $\overline{PA0}...\overline{PA7}$ by means of the CA2 signal which acts as 'enable' input to the octal D-type latch. When the 'enable' is high the latch outputs will follow the data inputs $\overline{PA0}...\overline{PA7}$. When the 'enable' is taken low, the outputs will be latched at the levels that were set up at the inputs. In order to control the whole AMC system by means of a single VIA chip, half of the VIA lines have a function that is multiplexed in time, e.g. the $\overline{PA}$ signals are used for the input board circuit (fig. 3), the output board circuit (fig. 4) and for the keyboard/display circuits (fig. 5). Obviously the output switches are not allowed to track the $\overline{PA}$-signal variations continuously which motivates the latch in order to block those unwanted variations.

The remaining part of the circuit is self-explanatory. Notice that each of the eight output switches has a corresponding LED indicating its state on the system front panel.

Although the basic circuit of figure 4 seems quite simple, the actual realization of the output board is quite general. It is possible to select between electromechanical switches (relay) or purely

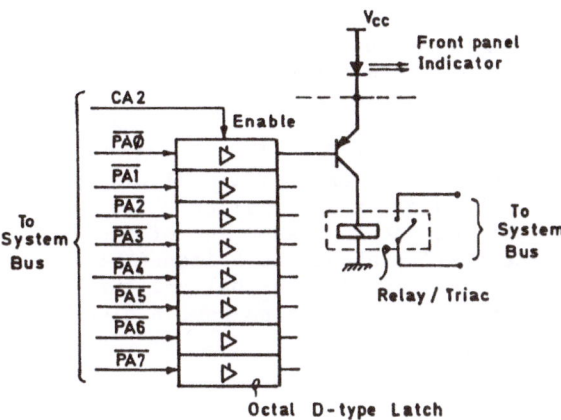

Fig. 4   Block scheme of Output board

electronic switches (triac). The latter leads to lower implementation
costs and does not suffer from mechanical wear when the load has to be
switched on or off frequently. On the other hand the relay seems to be
more robust when switching inductive loads. With these pros and cons
in mind, the actual choice is made according to the specific requirements of the application. Another alternative is to switch both 220
VAC or 24 VAC loads. Notice that for safety reasons the high voltage
(220 VAC) power outputs are isolated from the rest of the AMC system
by means of optocouplers.

   All these alternative configurations can be realized by means of
a single printed circuit board just by using the appropriate
electronic components.

2.4. Keyboard/Display

For the principle of the circuit we refer to figure 5. Again we make
double use of several lines of the VIA to control different things,
obviously for reasons of hardware simplicity. The lines (PB0, PB1,
PB2, CB2) are used for the keyboard decoding but also for enabling the
digits of the display. The lines ($\overline{PA0}...\overline{PA6}$) contain the information

for the 7 segments of each digit while they were also used for
controlling both the MUX on the input board and the states of the
power switches on the output board. The consequence of the low cost
hardware design is the need for a more careful software design,
especially w.r.t. the real-time aspects and the scheduling of the
several functions with time. It is clear from the hardware design that
it is not possible to read the keyboard and set the display simultaneously, nor is it meaningful to display information while setting the
output latch. By the same token during the measuring of the input sensors via the multiplexer, the display should be switched off. Despite
its simplicity, the hardware was designed such that all this can be
realized without conflicts. As an example let us have a closer look at
the operation of the display.

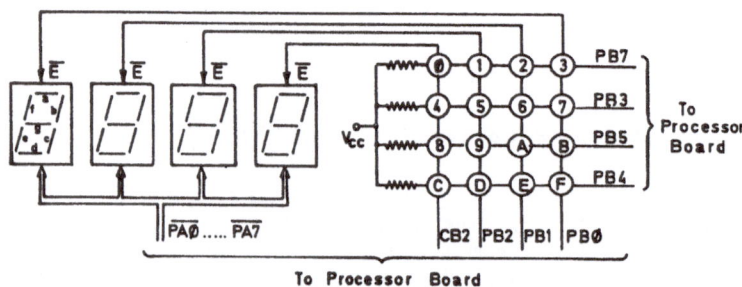

Fig. 5 Keyboard / Display scheme

A particular segment of a particular digit is on if both the $\overline{PA}$-
bit corresponding to that segment and the $\overline{E}$-bit corresponding to the
digit are low. So when using the $\overline{PA}$-lines for controlling the MUX, all
digits are switched off by setting the bits (PB0...PB2, CB2) in the
high state. During keyboard decoding by means of the PB-lines,
interaction with the display is prevented by setting the VIA port PA
in a low state.

The operation of the keyboard decoding is rather classical. To
find out whether a specific key is pressed (resulting in the contact
between a row line and a column line in the matrix) it is sufficient
to set the corresponding column line (one out of PB0...PB2, CB2) in
the low state and read the corresponding row line (PB3...PB5, PB7). A
low state indicates that the key at the crossing of both lines was
pressed.

3. SYSTEM SOFTWARE

All software is developed on a 6809-microcomputer development system
running under a simple but popular operating system. The executable

code is then stored in the 16 Kbyte EPROM chip of the target system. As the target system itself does not run under the supervision of an operating system, the software for the self-adaptive microprocessor controller has to include the low level procedures for the real-time control of all system hardware. Normally this is done in a low level computer language such as assembler.

In this project however all software, including the low level procedures such as timer control, interrupt processing and input/output port manipulation, is successfully programmed in Pascal. The available Pascal compiler has some extensions compared to the standard language which allowed this low level programming.

The heart of the AMC system is a timer in the VIA chip (timer T1) which is interrupting the processor every 20 milliseconds. The structure of the interrupt service software and the different tasks that are initiated by the timer tick is illustrated in figure 6b. The remaining part of the processor time - between the finishing of the interrupt service routine and the arrival of the next timer interrupt - is for the dialog with the user (figure 6a). We refer to this part as the 'background program', while the interrupt handling part of the software is called the 'foreground program'.

3.1. Main Program

The main program (fig. 6a) consists of 2 parts : the initialisation procedures and a loop executing the user interface (dialog) software which is running as a background (low priority) task and which is interrupted every 20 ms by the foreground (high priority) tasks.

At power on, the necessary RAM locations are initialized with default values, the VIA and ACIA chips are properly initialized including the timers T1 and T2 in the VIA and the time-of-day is read from the battery protected CMOS watch (section 2.1). From now on timer T1 is allowed to interrupt the program every 20 ms, while it is waiting for keyboard input from the user.

Reading the keyboard is in fact a function which is done in the foreground program at a rate corresponding to an integer multiple of the basic timer tick. A code corresponding to the key that was pressed is stored in the variable 'Keycode', which is then further processed in the background program. The keys of the hexadecimal keyboard are used to enter numerical values as well as to initiate preprogrammed functions. The exact interpretation of the keyboard input depends on the context in the program. At this stage in the program it corresponds to a function of which some are shown in figure 6a. There are two sets of functions : those which are application dependent and those which are not.

The functions which do not depend on the specific application of the AMC are :
* 'Time' : The user can adjust the time-of-day (eventually day-of-year)
* 'Sensor' : The tuning of the sensor circuits to eliminate component errors is done by software (instead of hardware tuning with trim potentiometers). This leads to both simpler hardware and a less

260

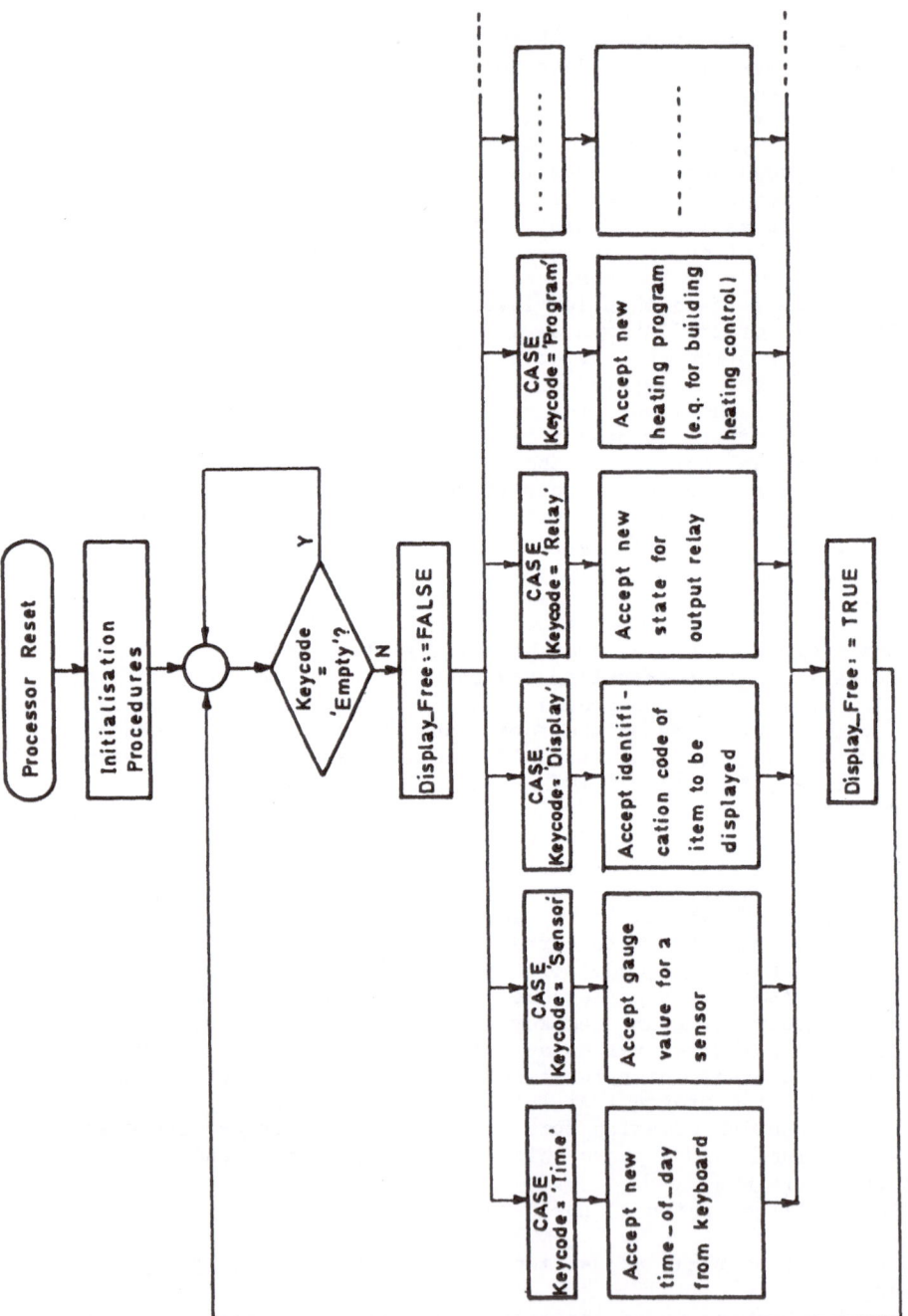

Fig. 6a: Structure of the Background Program

elaborate tuning procedure. The operator just enters the real value of a physical variable (e.g. temperature), the computer measures the sensor value. The difference between both is the calibration error and all further measurements are compensated for this error.
* 'Display' : In the dialog mode, when the user is locally executing some functions at the keyboard, the display is occupied for the interactive dialog. This status is indicated by the logical variable 'Display-Free' being false. However during normal operation the display is used to represent continuously the content of one internal variable (or to represent periodically the contents of several internal variables). These variables can include direct sensor measurements but also indirectly computed information (e.g. boiler efficiency). The 'Display' function then enables the operator to select the variable(s) he wants to be displayed.
* 'Relay' : This function enables the operator to set each of the 8 digital output switches in 1 of 3 modes : automatic, continuously on or continuously off.

The remaining functions that can be initiated by a key stroke are application dependent (e.g. greenhouse climate control requires some other functions than residence heating control). One example for the latter application is the 'Program' key, which allows the user to enter the heating program (desired temperature as a function of time for each independently heated zone in the building).

Notice that remote control of all functions is possible by means of the PC-communication facility.

3.2. Interrupt Service Program

The timer T1 interrupts the main program every 20 ms. The interrupt service procedure consists essentially of 6 parts as illustrated in figure 6b.

Part I is for reading the input sensors and will be explained in the next section. When there is no measurement going on the left part of the flow chart is taken (the variable 'PB6-timer' equals -1).

Part II is for updating the time-of-day information which is stored as the number of basic ticks-since-midnight. Two further variables, one for the background and one for the foreground program, are used to introduce delay periods in the program when required. As an example suppose we want to display a message during 1 second, then remove the message during the next second, then display it again, etc. (blinking display). The 1-second delay periods can be easily implemented with the statements : 'wait timer back:=50; REPEAT UNTIL wait timer back<0'.

Part IV is used to refresh the display. This is done when the counter 'Display Timer' runs through 0. The counter is then reinitialized at a positive value which corresponds to the refreshing period (as a multiple of the basic 20 ms tick). The display segments are thus not continuously illuminated. This is not possible because the display hardware is directly controlled from VIA port PA (fig. 5) without latches. These would be necessary for buffering the information while PA is used for controlling other parts of the hardware but they are

262

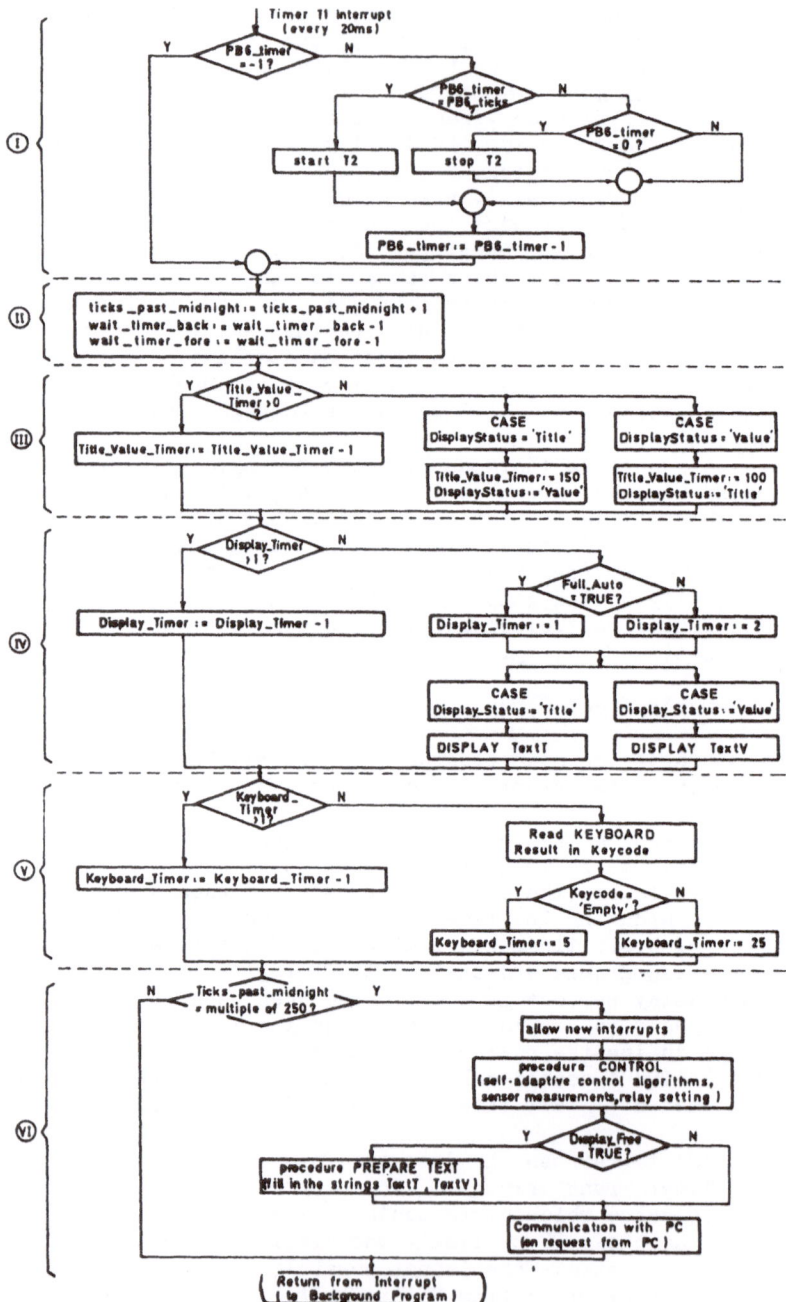

Fig. 6 b: Structure of the Foreground Program

omitted for low cost hardware design reasons. When the refreshing rate is sufficiently high this discontinuous mode of operation is not visible by the user. In the flowchart in fig. 6b it was chosen as 1 or 2 basic timer ticks depending on the state of the logical variable 'FullAuto'. When all relay outputs are in automatic mode, its value is 'true' and the refreshing period is then 20 ms. Otherwise the refreshing period is about 40 ms, resulting in a display which just starts flickering slightly, thus warning the user he left one of the outputs in manual mode. When operating in normal mode (i.e. no dialog with the operator) the system displays the values of internal variables preceeded in time by a 4-character title. Switching between displaying the title and the value is under control of the variable 'DisplayStatus', which can take the values 'Title' or 'Value' alternatively for a certain period of time.

Part III is for managing the Title/Value display period by introducing a counter 'TitleValueTimer' which is initialized at 150 basic tick periods when displaying a value, and at 100 tick periods when displaying a title. As a result each title is shown during 2 seconds, while the corresponding value remains during 3 seconds on the display.

Part V is similar to part IV and is used to read periodically the keyboard. In the example of figure 6b the scanning period is 100 ms during normal operation but 500 ms after a key was pressed (to prevent reading twice the same key).

Part VI is the main part of the foreground program. It contains the control algorithms, including the sensor measurements and the relay setting and also the communication procedure. These parts are described in the next sections. When there is no dialog with the operator, the variable 'DisplayFree' is true and the procedure 'PrepareText' is called. It fills in the 4-character-string variables 'TextT' and 'TextV' with an appropriate title and corresponding value to be displayed. When the AMC system is linked to a supervisory personal computer, receiving or transmitting data via the serial port is the final task for the foreground program. Part VI is repeated every 5 seconds (when the variable 'ticks_past_midnight' is a multiple of 250). It can take one or more seconds to execute depending on the number of measurements, the complexity of the control algorithms, whether PC communication is requested, etc... . Obviously parts I to V of the interrupt service procedure must be executed strictly every 20 ms (and take only some milliseconds to execute). Therefore part VI should start by allowing new interrupts thus resulting in a program structure with nested interrupts and several priority levels. Figure 7 illustrates how processor time is shared by these priority levels.

3.3. Low level procedures

This term is used to indicate the software for measuring the input sensors, setting the output relays, reading the keyboard, writing information on the display and communicating with a PC.

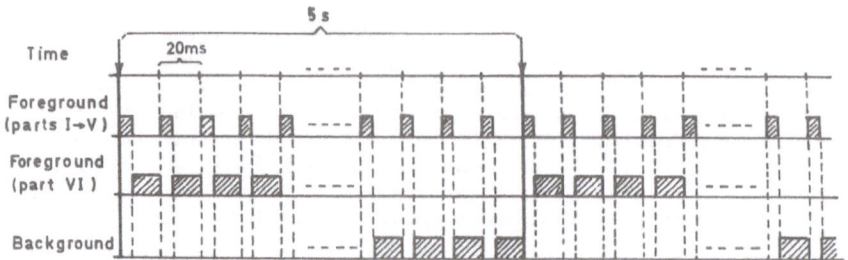

Fig. 7: Processor time and task priority levels

3.3.1. Sensor Measurement. The frequency of the block wave at pin PB6 of the VIA-chip (fig. 3) is measured and converted to a corresponding sensor value. Port PA of the VIA is used to select a MUX-channel and should not change during the counting phase. This port is also used for the display (fig. 5). The sensor measurement is done in part VI of the foreground procedure (fig. 6b) which can be interrupted by part IV where the display is refreshed. This could lead to conflicts. The measurement procedure therefore starts by setting the counter 'DisplayTimer' to a high value so that although it decrements every 20 ms, it will never go through 0 during the frequency counting (fig. 6b, part IV).

The frequency counting itself is done by means of counter T2 in the VIA chip, which counts the negative-going pulses on PB6. It is then sufficient to count the number of pulses during a fixed time interval. This interval is indicated by the constant named 'PB6Ticks' in the flowchart of fig. 8 and a reasonable value is 5 (in timer ticks, resulting in a 100 ms measurement interval per input channel). The flowchart in fig. 8 should be combined with that of fig. 6b, part I.

3.3.2. Relay Setting. The procedure to set the relays is obvious from fig. 4 and needs no further comments.

3.3.3. Keyboard Scanning. In order to detect whether a key is pushed, it is clear from fig. 5 that it is sufficient to make PB0, PB1, PB2, CB2 low and to read PB3, PB4, PB5, PB7. When all 4 bits are high there is no keyboard input. If 1 of the bits is low, say PB3, this indicates that a key from the set ['4', '5', '6', '7'] is pushed. The second part of the procedure has to find out the specific key. This is done by sequentially making the bits PB0, PB1, PB2, CB2 high again and meanwhile watching bit PB3. When PB3 goes high again, say after PB1 went high this indicates that the key at the crossing of the two lines (thus '6' in this example) must have been pushed.

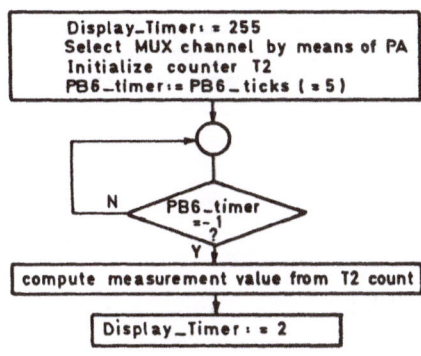

Fig. 8 Procedure for measuring a sensor input

3.3.4. Display Control. The procedure to operate the display is easy
to understand (fig. 5). The data for a specific digit is set on port
PA according to the correspondence : segment [a, b, c, d, e, f, g] ⟵
⟶ [PAO, PA1, PA2, PA3, PA4, PA5, PA6]. The digit is then selected by
making one of the lines PBO, PB1, PB2, CB2 low while keeping the
remaining three lines high. After about a 1 ms delay (realized by
means of an idle loop in the program) in order to make the information
visible, these steps are repeated for the next digit. Executing this
procedure every timer tick (20 ms) leads to a perfectly stable
display.

3.3.5. Communication. According to the philosophy of the whole system
the communication is kept as simple as possible. There is no complex
protocol nor error correcting code necessary taking into account the
kind of applications envisaged. Indeed there is more than enough time
available so that it is perfectly possible (when desirable) to re-
transmit a message or to echo the received data back to the sender in
order to compare it with the source data. Notice that the communication facility is optional and is by no means necessary for the normal
operation of the AMC system.
 The structure of the network is shown in fig. 9. It is a
master/slave configuration : each AMC system can only receive data
from or transmit data to the PC and only on request of the PC. The
primary reason for developing the communication facility was for
monitoring and off-line processing of the AMC measurement data on a
PC. Off course it is perfectly possible to use the facility for remote

control of the AMC system (instead of local control via the keyboard) or for hierarchical control of a number of AMC systems by means of higher level control algorithms in the Personal Computer.

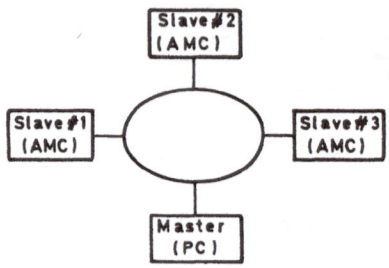

Fig. 9. Structure of the communication network

Each AMC system has in its RAM memory a receive buffer of fixed length and a transmit buffer of fixed length. The structure of a receive or transmit message is thus fixed and predefined. When the master PC wants to receive the data from a slave AMC it sends a 'Transmit' order to the slave. This is just 1 byte, e.g. $12 ($ denotes hexadecimal notation) where the '1' indicates the slave has to transmit data and the '2' is the number (address) of the specific AMC. The PC then waits for the AMC to transmit the data (a fixed number of bytes). When the master wants to transmit data to a slave it first sends a 'Receive' order to the slave. The 'Receive' order consists of 1 byte, e.g. $01, the '0' indicating a receive operation for the slave and the '1' indicating the slave address. The master then waits for an acknowledge from the slave (a copy of the 'Receive' order) and subsequently sends the data. The last byte must be $00 (the reason for this becomes clear when analyzing the flowchart that follows; it is important to notice that $00 cannot be a 'Transmit' or 'Receive' order). After the slave received the data it echoes them to the master and waits for a 'Data-OK' message. This terminates the receive procedure.

Notice that due to the ring structure of the network (occupying only 1 serial communication port in the PC) each AMC system receives all data at its serial port. The communication software in the AMC system has to decide how to process it. The flowchart in fig. 10 indicates how this is done. In order to analyze how this procedure excludes every conflict situation it is important to keep in mind that it is called in part VI of the foreground program (fig. 6b) thus executing every 5 seconds, that the several AMC systems in the network run independently (they are not synchronized) and that the communication itself only takes some tens of milliseconds (at 9600 Baud).

267

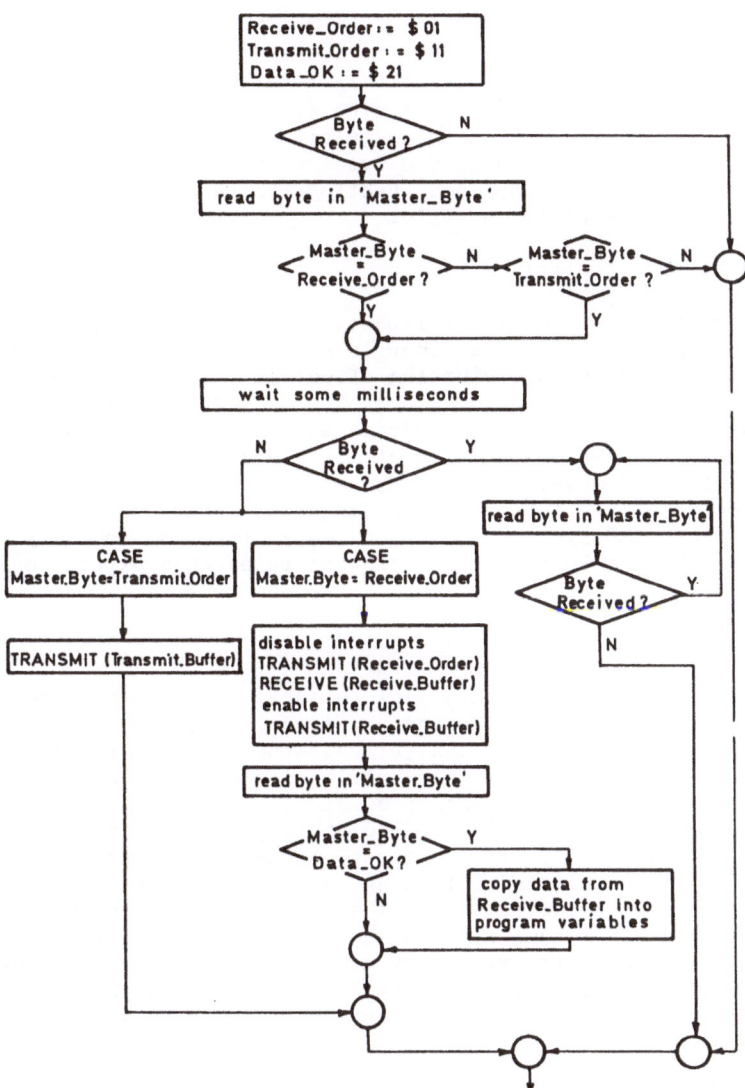

Fig. 10 : Communication Procedure

4. APPLICATIONS

The system described until now is essentially a small general-purpose microcomputer controller. It offers some nice application possibilities for low cost automatic control and regulation in the field of climate control. This section contains a brief summary of some recent applications. The control of a building heating system will be described more in detail as a specific example.

4.1. Greenhouse Climate Control

Temperature and relative humidity are important parameters that should be kept close to a desired value in order to stimulate the growth and quality of the craps. The main disturbing effect is outside weather (solar radiation, temperature, rain, wind). The actuators that can be used for control are essentially heat supply and ventilation windows (eventually sprinkling). The popularity of mini- and microcomputers for the control of greenhouses is steadily growing. Typically in practice control procedures are improved by adding decisions. The combination of one or more AMC systems for basic temperature and humidity control with optionally a PC for monitoring and rule based expert control has obviously a promising future.

4.2. Incubator Control

Temperature is here an extremely important parameter which should be kept within a range of +/- 0.1 ^{o}C around a reference value that slowly varies with time during the incubation period. Also humidity has to be controlled within some %RH. The air is ventilated around the eggs and is heated by means of electrical resistors. The temperature is kept within the specified range by switching the heating power on and off frequently by means of the AMC triac outputs (no mechanical wear). The air is moistened by means of blades turning in and out a water reservoir. The driving motor is controlled by the AMC system on the basis of the measured and desired humidity. The setpoint for temperature and relative humidity can be preprogrammed by the user for the whole incubation period and for several types of eggs (pheasant, duck, hen ...). The system also takes care of some secondary functions such as turning the eggs regularly.

4.3. Sty Ventilation Control

A stable sty temperature is important for low fodder consumption and fast fattening. The dense sty occupation which characterizes modern pig breeding leads to considerable internal heat production coming from the animals. Efficient ventilation is thus of major importance. Applying natural ventilation (which is usually the case) requires that window aperture is adapted in order to keep internal temperature at a specified level. It is well known that in natural ventilation systems the air freshening effect strongly depends on wind parameters such as speed and direction. They represent the main disturbing variables,

while variations in outside air temperature (mainly day/night fluctuations) obviously introduce supplementary perturbations. The task of
the AMC system is to control continuously the position of the
ventilation windows in order to keep the sty temperature between the
specified limits in spite of all disturbances.

4.4. Building Heating Control

Most central heating plants in small buildings and residences are
either controlled by a central room thermostat as the most simple
system or by a mixture valve controlled by a weather compensation
circuit. Large public buildings are more and more controlled by
computer based energy management systems (EMS) which are rather
expensive and cannot be justified on smaller installations(Birtles et
al, 1984). Thanks to its low cost aspects it is not unreasonable to
apply the AMC system for more intelligent control in medium-sized
installations. Compared to thermostatic or weather compensation
control this leads to both a more comfortable and less energy
consuming heating policy.

A typcial configuration that can be controlled by a single AMC
system is given in fig. 11. The residence is divided in 4 heating
zones (each consisting of one or more rooms) which are controlled
independently. Three of them (1-2-3) are floor-heated, the fourth
being heated by a large-surface (low-temperature) radiator. The
following instrumentation is available for the control of the heating
system : temperature sensors for the outside air (T_o), the boiler
water (T_b) and the supply water (T_s); a valve motor (VM) for the 4-way
mixture valve; in each zone a temperature sensor for the room air (T_r)
and an electrothermic valve (ETV) which can shut-off the circuit. The
7 temperature sensors are connected to the input board of the AMC
system, while the 8 relays (triacs) of the output board are used to
switch on/off the circulation pump, the boiler, the mixture valve
(open/close) and the 4 shut-off valves.

The desired room temperature is preprogrammed as a function of
time for each zone independently, which assures that a zone is only
heated when occupied. This is the only information the user has to
enter in order for the system to operate. Building-specific characteristics (such as heating curves that have to be tuned by the user in
current weather dependent control systems; or gain constants for a
modulating mixture valve regulator) are totally absent. All parameters
are identified and optimized by the control software itself from
information in the measured data, resulting in a self-learning or
self-adaptive control system.

As an example let us focus on the calculation of the minimum
supply water temperature T_s necessary to keep a zone temperature T_r at
its desired value T_r^w. The heat balance equation when the zone temperature is in equilibrium at T_r with the supply water at T_s gives :

$$k_1(T_s - T_r) = k_2(T_r - T_o)$$

where k_1 and k_2 are unknown heat conduction coefficients and T_o is the outside air temperature. The ratio $k = k_2/k_1$ can be estimated in real-time by means of a moving average filter:

$$\hat{k}(t) = \hat{k}(t-1) + (1-\alpha) \frac{T_s(t) - T_r(t)}{T_r(t) - T_0(t)}$$

where α is the filter constant ($\alpha < 1$) and t denotes discrete time. The minimum required supply water temperature is then computed from

$$T_s^m(t) = T_r^W(t) + \hat{k}(t) \cdot [T_r^W(t) - T_0(t)]$$

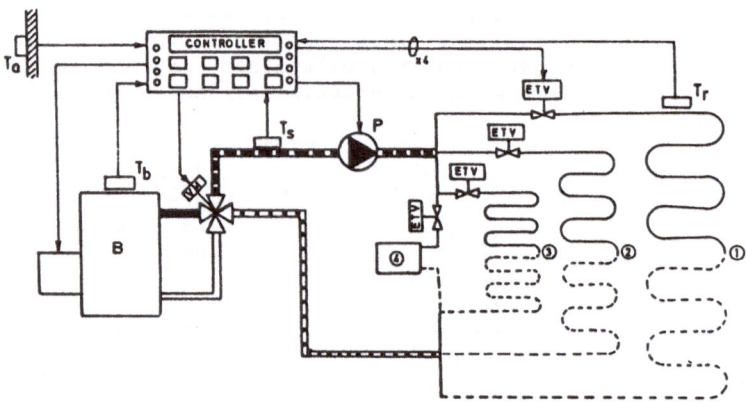

Fig. 11 Configuration of Building Heating Control System

This value can just keep the zone temperature at its specified value and should be increased when the actual (measured) zone temperature has not yet reached its setpoint. The supplement is chosen proportional to the error, resulting in a desired supply water temperature :

$$T_s^W(t) = T_s^m(t) + K \cdot [T_r^W(t) - T_r(t)]$$

This is the setpoint for the mixture valve regulator loop which controls the position of the mixture valve in such a way that the measured supply water temperature $T_s(t)$ follows the desired value $T_s^W(t)$. The boiler is then controlled such that its water temperature $T_b(t)$ is just a few degrees centigrades higher. Although this strategy does not expect a single tuning parameter from the user, it assures that the whole system is always operating at its lowest possible heat output in order to realize the specified zone temperature, thus saving energy. Notice that there is only one mixture valve for the heat supply to four independently controlled zones. Possible conflicts can however be eliminated by shutting off one or more zones by means of the electrothermic valves.

The above strategy is able to keep the temperature in each zone close to its setpoint. As an option it can be overruled by a self-optimizing strategy for setpoint variation. It computes how many hours in advance the heating of a zone should be started in order to attain the new setpoint at the specified time. The underlying ideas were borrowed from the EPSAC method (Extended Prediction Self-Adaptive Control; De Keyser & Van Cauwenberghe, 1981) which is illustrated in fig. 12. The technique is based on : 1. on-line estimation of an input-output model of the process (heating plant) dynamics; 2. use of the model for prediction of the future process output (zone temperature) as a function of the postulated control action (heat supply to the zone); 3. use of the prediction facility for decision-making, i.e. to select the best control action that should be applied to the real process at the current time instant. These steps are repeated the next sampling instant such that new measurement information can be used to improve previous decisions.

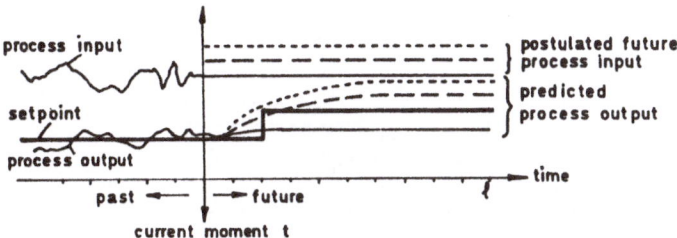

Fig. 12 : Extended Prediction Self-Adaptive Control

The zone dynamics are described by the incremental model

$$\Delta T_r(t) + a_1 \Delta T_r(t-1) + a_2 \Delta T_r(t-2) = b_1 \Delta T_s(t-1) + b_2 \Delta T_s(t-2) + e(t)$$

or in vector notation

$$\Delta T_r(t) = \underline{\phi}^T(t) \cdot \underline{\theta} + e(t)$$

where the data vector $\underline{\phi}^T(t) \equiv [\Delta T_r(t-1) \quad \Delta T_r(t-2) \mid \Delta T_s(t-1) \quad \Delta T_s(t-2)]$

and the parameter vector $\underline{\theta}^T \equiv [-a_1 \quad -a_2 \mid b_1 \quad b_2]$

The notation $\Delta T_r(t)$ denotes the increment $[T_r(t) - T_r(t-1)]$ and the signal $e(.)$ represents zero-mean modelling errors.

The parameter vector $\underline{\theta}$ is estimated by means of a parameter identification method (Isermann, 1981). At the current time instant t, given the estimated $\underline{\hat{\theta}}(t)$ and the postulated control policy $\{T_s(t+k/t); k=0 \ldots 1\text{-}1\}$, it is possible to predict the behaviour of the future zone temperature $\{T_r(t+k/t); k=1 \ldots 1\}$:

$$T_r(t+k/t) = T_r(t+k\text{-}1/t) + \underline{\phi}^T(t+k/t) \cdot \underline{\hat{\theta}}(t); \quad k = 1 \ldots 1$$

where $\underline{\phi}^T(t+k/t) \equiv [\Delta T_r(t+k\text{-}1/t) \ \Delta T_r(t+k\text{-}2/t) \ | \Delta T_s(t+k\text{-}1/t) \Delta T_s(t+k\text{-}2/t)]$

When using this strategy for deciding when to start preheating a zone in order to realize a desired zone temperature increment until a specified time, the postulated control policy $\{T_s(t+k/t) ; k = 0 \ldots 1\text{-}1\}$ is normally the maximum possible supply water temperature. This corresponds to the strategy : start heating as late as possible and with maximum heat input into the zone. This rule is by no means limiting and could be replaced by others. The decision problem is twofold : the selection of the most appropriate time instant t_1 to switch-on heat supply to the zone; and at the end of the heating-up period the selection of the most appropriate time instant t_2 to switch from maximum supply water temperature to the temperature required to maintain the zone temperature in equilibrium at its desired value (fig. 13).

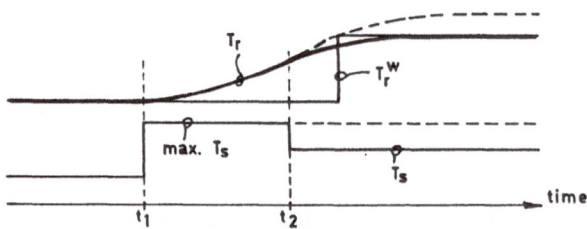

Fig. 13 : Rule based control strategy

Summarizing this section, the control software for the building heating controller can be described as being a simple rule based expert control system(Sutton, 1984; Åström et al., 1986) in which an essential part of the knowledge is obtained by means of self-learning algorithms.

5. REAL-LIVE EXPERIMENTS

The operation of the hardware and software of the AMC system has been thoroughly investigated by means of simulation studies and by real-life demonstration projects. Some results are described in (De Keyser, 1985). Some typical real-life results of the building heating control application are summarized in this section.

The upper part of fig. 14 shows the variation of the outside air temperature T_a during a 1-day period. The day/night difference is over 10°C. The middle part shows the measured temperature in zone 2 during that day as well as the position (open/closed) of the zone shut-off valve. The desired temperature was 19°C all day long. Obviously the control performance can hardly be improved, the natural variations during the night (no disturbances of sun or occupants) being +/- 0.1°C. The lower part of fig. 14 illustrates how a setpoint increment $(19^{\circ}C \longrightarrow 20\ ^{\circ}C)$ is realized in zone 4.

6. CONCLUSIONS

A low cost self-adaptive microprocessor controller was presented. A detailed description was given of the system hardware and real-time software. Special attention was paid to the low cost aspect, how this was a stringent rule in the hardware design and how this introduced interactions that had to be solved by the software. The system being rather general-purpose, it could be applied to several real-life control problems in the field of heating and ventilation. These applications were briefly described. One application, the control of a multizone building heating system was presented with some details. Attention was paid to the control algorithms which are a combination of simple rule based expert control and a self-learning feature to improve the controller's knowledge base. Further investigations will focus on the development of this control strategy for other applications.

REFERENCES

ÅströmK.J., J.A. Anton and K.E. Arzèn (1986), Expert Control, Automatica **22** (3), 277-286.
Birtles A.B., R.W. John and J.J. Smith (1984), 'Before and after study of the performance of an energy management system', CIB-W'79 "Performance of HVAC Systems and Controls in Buildings", BRE Garston U.K.
De Keyser R.M.C., A.R. Van Cauwenberghe (1981), A self-Tuning Multistep Predictor Application, Automatica 17 (1), 167-174.
De Keyser R.M.C. (1985), 'Adaptive microcomputer control of residence heating', CIB-W'79 "Recent Advances in the Control and Operation of Building HVAC Systems", SINTEF Trondheim, 154-164.
Isermann R. (1981), Identification and System Parameter Estimation, Special Issue Automatica **17** (1).
Sutton R.W. (1984), 'Expert Systems for Process Control', in : S. Bennett & A. Linkens (Eds), "Real-time Computer control", Peter Peregrinus Ltd, London, 247-251.
Tokheim R.L. (1983), Microprocessor Fundamentals, McGraw-Hill, New York.

274

Wakerly J. (1981), <u>Microcomputer : Architecture and Programming</u>, Wiley & Sons, New York.

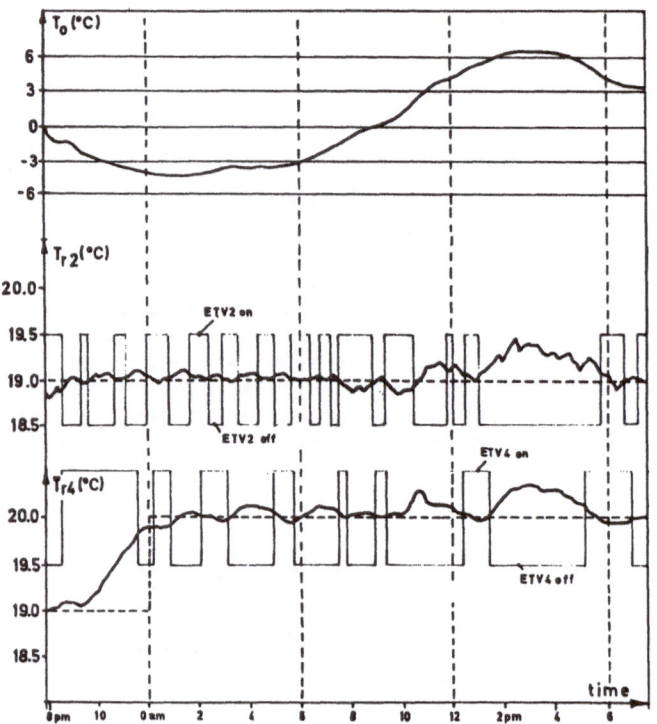

Fig. 14 Experimental results-building heating control

T_0 = outside air temperature

T_{r2} = temperature zone 2 (setpoint 19°C)

T_{r4} = temperature zone 4 (setpoint 19°C ⟶ 20°C)

Chapter 8

SELF-ADAPTIVE MULTI-MICROPROCESSOR CONTROL OF CUTTER SUCTION DREDGING SHIPS

R.M.C. De Keyser
Automatic Control Laboratory
University of Ghent
Grotesteenweg-Noord 2
B-9710 Ghent
BELGIUM

J.L.S. Van Ostaeyen
I.M.D.C. n.v.
Wilrijkstraat 37-45
B-2200 Antwerp
BELGIUM

ABSTRACT. The development of a self-learning computer-based automatic control system for a cutter suction dredger is an important step in the evolution of dredging. The system is based on the latest concepts of control theory, hardware and software.

Commercially available controllers of type PI, whether or not implemented in digital hardware, require careful tuning in order to operate well on this kind of dredger. Therefore in the new automatic controller the classical control algorithms were replaced by more powerful adaptive predictive control methods. The software has been implemented in an on-board multi-microcomputer system.

A spin-off product of the automation project was the development of a comprehensive simulator of the dredging process. It is based on the availability of unique numerical dynamic ship models. These were obtained by active testing aboard of real ships. Afterwards the measured data were processed with identification programs in order to estimate the unknown ship parameters.

1. INTRODUCTION

One of the most frequently used types of large dredgers is the cutter suction dredger, particularly because it can cope with hard materials combined with the advantage of hydraulic transport. Moreover it is the most appropriate type, besides the bucket dredger, for executing the required profiles or cross-sections in canals and docks.

Newly built cutter suction dredgers are becoming more and more powerful and sophisticated in order to increase their efficiency. One of the latest developments is the automatic dredger controller (Van Zutphen, 1983; De Keizer, 1983; Brouwer, 1985; Van Ostaeyen & De Keyser, 1985).

The development of a self-learning computer-based automatic control system realised by means of parallel processing techniques is an important step in this evolution. This high-tech product is the outcome of an intensive cooperation during several years between one of the larger dredging companies in the world, "Dredging International N.V.",

275

S. G. Tzafestas and J. K. Pal (eds.), Real Time Microcomputer Control of Industrial Processes, 275–309.
© 1990 *Kluwer Academic Publishers.*

the engineering and consulting firm "IMDC N.V." and the Automatic Control Lab of the University of Ghent in Belgium.

Extensive use was made of recent developments in control theory, hardware and real-time software. Because of the complexity of the process dynamics the core of the control algorithms was made self-adaptive by combining predictive control strategies with on-line identification (Clarke and Gawthrop, 1979; Åström, 1983, 1986; De Keyser & Van Cauwenberghe, 1985). The hardware and software implementation of these powerful strategies for the control of a dredging ship requires the use of a fast minicomputer or a microcomputer system based on a form of parallel processing (Hwang and Briggs, 1984).

In the field of simulation and real-time control, where it is quite common to find an implicit parallelism in the application, particularly parallel processor configurations have received much interest (Conte and Del Corte, 1985; Dekker, 1983). These configurations offer a high computational performance or, by redundancy at processor level, a more reliable (fault tolerant) operation (Paker, 1983).

This paper gives a survey of the several steps in the development and realisation of the advanced control system. In section 2 the cutter suction dredging process is briefly described. The next section deals with the modeling and identification phase which was necessary in order to build a comprehensive simulator of the process. The simulator realisation by means of a parallel microcomputer system is the subject of section 4. The actual control system with the self-adaptive predictive strategies and their multi-microcomputer implementation on board of the real dredgers is described in the last section.

2. CUTTER SUCTION DREDGERS

2.1. Dredging Equipment

The cutter suction dredger is at the moment the most frequently used dredger : 41% of the world fleet of dredgers are cutter suction dredgers. They have the advantage that they can be used to work on the most varied soil types : from silt to hard rock (with a tensile strength of concrete).

Figure 1 illustrates the main features of a cutter suction dredger (Claeys, 1978) :
- the cutter (1) to dislodge the soil
- the ladder (2) to fix the dredging depth. It holds the shaft for the cutter and also the suction pipe (and pump).
- the spud (3) to fix the length of the cut or step. The dredger swings around the spud and steps forward relatively to the spud at the end of each swing.
- the swing-winches (4), wires (5) and anchors (6) which, by moving the dredger, determine together with spud and ladder position the quantity of soil to be cut per unit of time.
- the pipeline (7) and the sandpumps (8) to transport the dislodged soil over the required distance and height.

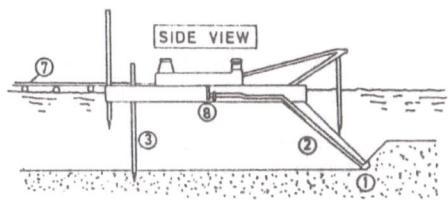
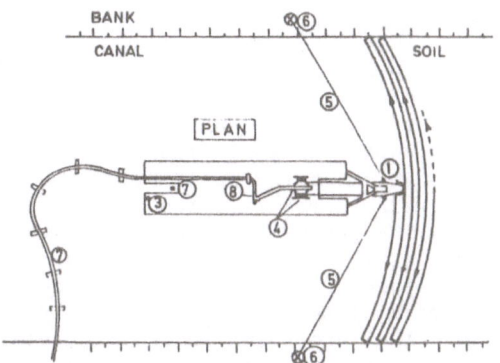

Fig.1 Cutter Suction Dredger

A cutter suction dredger is a stationary dredger, i.e. it remains on the spot while operating. The dredging process can be divided into two parts :
- The mechanical part in which the soil is loosened with a rotating cutter. The movement of the pontoon is brought about by pulling one of the side winches. This causes the ship to rotate around a spud that is fixed into the ground.
- The hydraulic process in which the loosened material is diluted with water and suctioned at the cutter. Via a floating pipeline the species is pumped ashore and furtheron transported by a pipeline to the reclamation area.
The dredging process of a cutter suction dredger proceeds as follows :
- the working spud is placed in the ground, as soon as the ship is situated in the centerline of the cut.
- the dredger rotates around the working spud by pulling either the starboard or the portside sidewinch.
- at the end of the cut the ship steps forward by moving the spudcarriage to the rear.
- the ship swings to the other side of the cut.
- this sequence goes on until the working spud is in its most backward position.
- the ship stops in the centerline and the auxiliary spud is lowered. By lifting the working spud and moving the spudcarriage to the front a new cycle can begin.
Several process variables have to be controlled continuously either manually by the dredgemaster or by an automatic control system :
- the load of the cutter motor
- the concentration of the soil/water mixture in the pipeline
- the load of the swing-winch motors
- the velocity of the mixture in the pipeline
- the load of the pumps and driving diesel engines
- the vacuum before the suction pump
- the tension in the side wires
- the total pressure after the pumps, etc.

2.2. Control Problems

For the control of these parameters the dredgemaster (or the computer) can intervene in several ways : by changing the step length, the ladder depth, the number of revolutions of the cutter, the speed of the pumps. The most important factor for a continuous control is however the swing speed of the ship around the spud. This swing speed determines primarily the quantity of material dredged per time unit, i.e. the production. The swing speed is controlled by means of the current to the starboard and portside winch motors. The input-output structure of the dredging process is illustrated in fig. 2. The on-board computer scans continuously all process outputs (only the most important are shown in fig. 2), computes the required adaptive control actions based on the measured information and sends the necessary control signals to the swing motors.

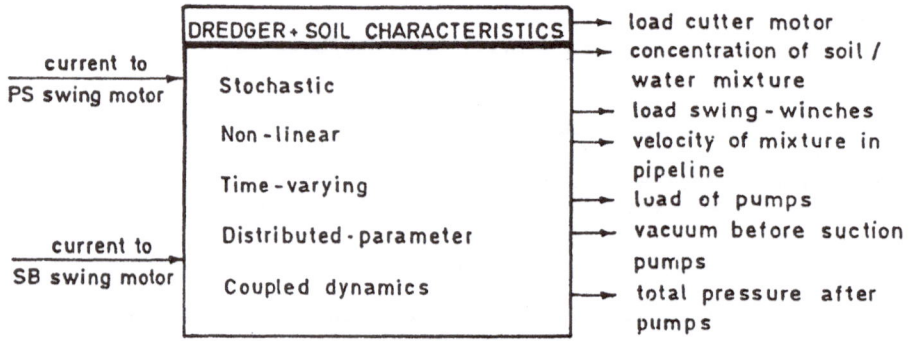

Fig. 2 Input-Output Structure of Dredging Process

The aim of the control policy is to keep the swing speed and thus the production as high as possible. However, the electromotors of the side winches and the cutter may not be overloaded, the velocity of the mixture may not become too low in order to avoid deposition of the material resulting in the blocking of the pipeline and the pumps should not start cavitating.

From the above it is clear that the main task of an efficient control system is to adjust continuously the swing speed by controlling the side winch motors, so that at least one of the critical factors of the dredging process is at its nominal value. In this way the capacity of the ship is fully used; in other words, the production is at its maximum taking into account the available power of the different drives.

The realisation of this objective is not obvious and asks for a continuous adaptation of the control mechanism. Commercially available cutter controllers (type PI) fail to realise this task efficiently due to the extremely random and time-varying nature of the process characteristics and loads. This type of vessel is used to dredge clay, sand as well as rock. The relationship between process input (swing veloci-

ty) and process outputs (cutter load, mixture concentration ...) is
changing almost continuously. The soil characteristics do not remain
constant along the swing and drastical variations in a stochastic
sense are quite normal. This constitutes the whole problem in controlling this kind of process.

To give an idea of the irregularity of the process we refer to
fig. 3. The load of the cutter motor and the mixture concentration at
the measuring-spot in the pipe (near the sandpumps 8) is illustrated
under manual control by the dredgemaster.

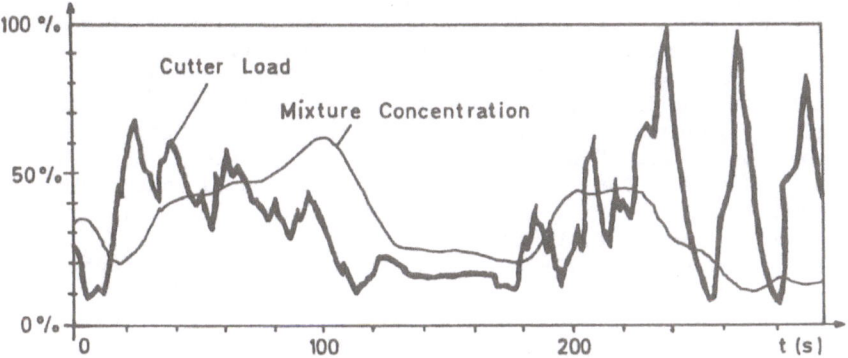

Fig. 3   Irregularity of the Dredger Process

Breaches are another illustration of the variation of process
gains and time constants. At some spots along the cut it is possible
that there is less material so that the production is very low even
when hauling at the maximum swing speed. According to the control
theory this means that the gain between the process input (= swing
speed) and the output (e.g. concentration) is very low. At other
places it is possible that due to breaches the production is very
high, even when the ship is almost stopped. This means an extremely
high process gain.

These working conditions make it necessary for the skipper in
manual control to be extremely attentive, if not the mean production
will decrease. Likewise a good automatic controller has to adapt
itself to the continuously changing situations. Because the control
algorithms are based on the feedback principle, the control parameters, among which the gain, have to be adapted continuously to the
stochastically fluctuating process parameters. This explains why commercially available control systems with pretuned parameters are not
very successful : with increasing process gain the behaviour of the
control loop will be oscillating (instability) and when the process
gain is diminishing the control performance tends to become very slow
(production loss).

To conclude one can state that the aim of the automatic control
of cutter suction dredgers is twofold. First there is the lightening

of the nerveracking task of the dredgemaster : the process takes place under water and has a random character so that it can only be followed up by extensive instrumentation. Secondly, there is the improvement of the mean production by making the process more regular and the continuous adaptation of the available power of the installation to the changing soil characteristics.

3. MODELING and IDENTIFICATION

In order to gain some experience with the system before it was installed aboard of the ship, it was decided to build a computer simulation of the dredging process. A fairly comprehensive dynamic model was developed. The model comprises both the mechanical part (vessel, winches and electric drives, side wires, cutter) and the hydraulic part (pipeline, pumps and diesel drives) of the dredging process. The simulator was built from "white" models based on physical (mechanical, electrical, hydraulic) laws. However in order to obtain some lacking numerical parameters in these models the identification of simple "black-box" models was very helpful.

For modeling purposes the cutter suction dredger was structured into four submodels (fig. 4) :
- the mechanical process : as input we have the currents of both electromotors of the sidewinches, resulting in a swing speed (angular speed around working spud). If we presume that the ladder remains at constant depth and the angular speed of the cutter does not vary, only the hauling speed will influence the dredging process (next to the soil characteristics).
- the cutting process : the swing speed determines the cutterload. When the swing speed increases, there is more soil to be cut so that it is clear that the torque on the cutter is higher. As a matter of fact the influence of the soil characteristics is very important : the behaviour in clay is very different from that in rock, even though the motor in both cases is running on full load.
- the suction process : the material that is cut is sucked. Depending on the soil characteristics, the shape of the cutter, the angular speed of the cutter and the swing speed, more or less material is sucked. This results in a specific concentration (density) of the mixture that is pumped.
- the hydraulic process : the density of the mixture determines all parameters of the hydraulic process, consisting of the pumps and their drives and the pipelines. An important parameter is the velocity of the mixture, which may not be too low in order to avoid deposition in the pipeline. If no action is taken, the pipeline will be blocked so that intervention from outside is necessary.

For all these subprocesses a numerical model was made. From the on-board measurements the unknown parameters were estimated with the maximum likelihood identification method. The goal of this analysis was not only to learn about the dynamic process but also to lay the foundation for the computer simulation of the dredging process, which will be described in section 4.

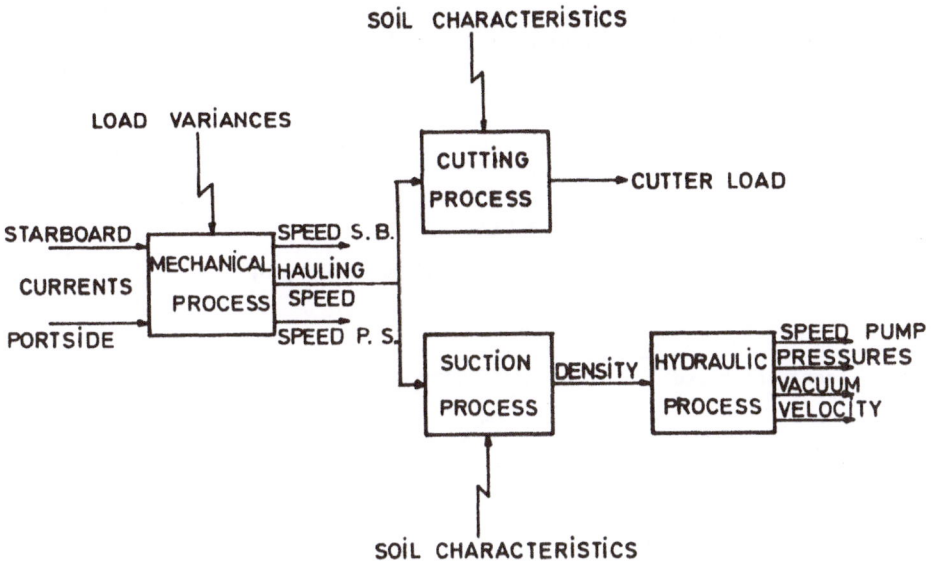

SOIL CHARACTERISTICS

LOAD VARIANCES

CUTTING PROCESS → CUTTER LOAD

STARBOARD ─── MECHANICAL PROCESS
CURRENTS ───
PORTSIDE ───

SPEED S.B.
HAULING SPEED
SPEED P.S.

SUCTION PROCESS │DENSITY│ HYDRAULIC PROCESS

SPEED PUMP PRESSURES
VACUUM
VELOCITY

SOIL CHARACTERISTICS

Fig.4 Structuring the Dredging Process into Submodels

As an example of the use of "white" modeling techniques the hydraulic submodel is briefly described, while "black-box" identification is illustrated by means of a part of the mechanical subprocess.

3.1. Modeling

The hydraulic system consists mainly of the pumps, the pumpdrives and the pipeline. The static behaviour of the pumps and pipeline can be described by the pump and pipeline characteristics, whilst for the dynamics of the mixture and the pumpdrive the differential equations can be derived from Newton's second law.

3.1.1. <u>Pump Characteristics</u>. The pressure H_p of a centrifugal pump can be described as a function of the flowrate Q and the speed of the impeller (ν=speedratio=actual speed/nominal speed) :

$$H_p = H_p(Q,\nu).$$

In this way the well known H-Q graphics can be drawn (fig. 5, constant speed portion).

When pumping a mixture of water and sand, the pump characteristics change by concentration so that

$$H_p = H_p(Q,\nu,\rho)$$

where ρ = density of the mixture (kg/m^3).

Furthermore the mean grain size d_m plays an important role as can be seen by comparing figs 5a and 5b. So that finally the pressure head of a centrifugal dredge pump can be described as :

$$H_p = H_p (Q, \nu, \rho, d_m)$$ [1]

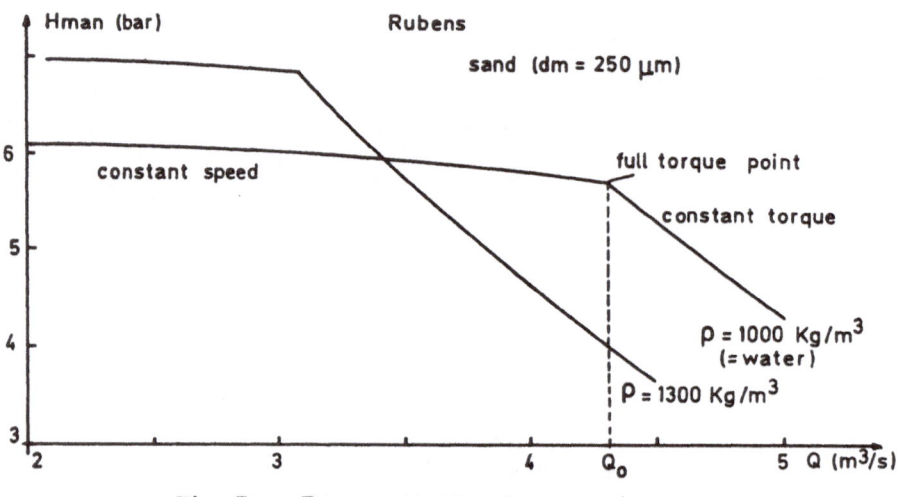

Fig. 5a Pumpgraphic for sand

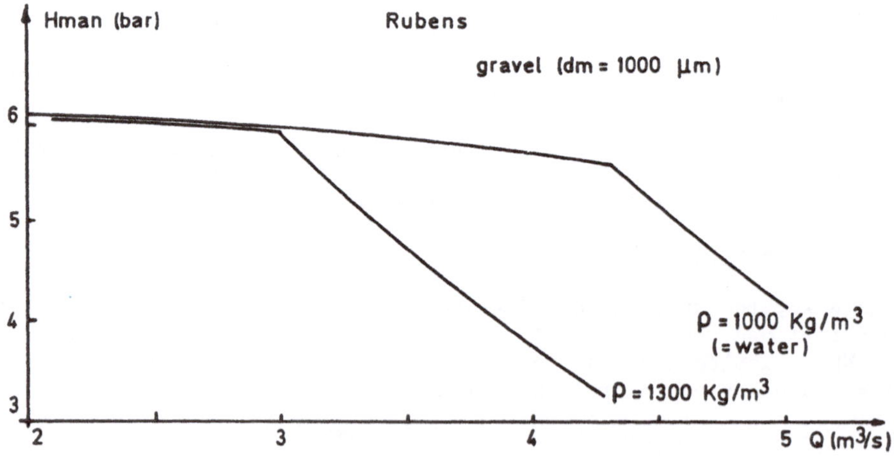

Fig. 5b Pumpgraphic for gravel

Another characteristic of a dredge pump is its efficiency :

$$\eta = \eta(Q, \nu, \rho, d_m) \qquad [2]$$

Since $P = H_p Q/\eta$, where P is the power of the pump drive, it follows that

$$P = P(Q, \nu, \rho, d_m)$$

When a dredgepump is driven by a dieselengine with a torque limiting device, the characteristics consist of two portions : a constant speed portion and a constant torque portion.

The intersection of the two lines is the so called "full-torque point", this is the point where the diesel runs at maximum power.

The constant-speed line for specific ρ and d_m follows directly when applying equation [1] with $\nu=1$.

The constant-torque line can now be calculated as follows :

$$\frac{P}{\omega} = M_o = \frac{P(Q, \nu, \rho, d_m)}{\omega} \qquad [3]$$

where ω is the angular speed and M_o is the nominal torque of the dieselengine :

$$M_o = \frac{P_o(1-\varepsilon)}{\omega_o} \qquad [4]$$

where P_o : nominal power
ε : power reduction due to friction losses in the transmission
ω_o : nominal angular speed of the dieselengine

Substituting [4] in [3] yields :

$$P(Q, \nu, \rho, d_m) = \nu P_o(1-\varepsilon) \qquad [5]$$

With this formula and $H_p(Q, \nu, \rho, d_m)$ the constant-torque line is defined. When several pumps are in series the total pressure head at a certain flowrate is the sum of the heads of the different pumps.

These relations concern the static behaviour of the pump only; for the dynamics a differential equation has to be constructed.

3.1.2. Differential Equation of the Dieselengine. Considering the diesel engine, the gearbox and the pump as a system, Newton's second law yields that the sum of torques equals the moments of inertia times the angular acceleration.

In this case there are 3 torques acting on the system : the external torques (diesel drive and pump action), the damping torque and the spring torque of the shafts.

Since their values are very small compared with the others, the last two torques are neglected. To obtain the differential equation of the system all torques are related to the pump. This is done by multiplying the moments of inertia by the square of the transmission ratio n, leading to :

$$(I_d \, n^2 + I_g + I_p) \, \frac{d\omega}{dt} = M_d n - M_p \qquad [6]$$

where $\quad I_d$: moment of inertia of the diesel engine
$\quad\quad\quad I_g$: equivalent moment of inertia of the gearbox
$\quad\quad\quad I_p$: moment of inertia of the pump impeller
$\quad\quad\quad M_d$: torque of the diesel engine
$\quad\quad\quad M_p$: torque of the pump

When I_{eq} is the equivalent moment of inertia we get :

$$I_{eq} \, \frac{d\omega}{dt} = M_d n - M_p$$

In the case of constant dieselengine torque we have $M_d = M_o$.

The pump torque is given by equation [3]

$$M_p = \frac{P(Q, \nu, \rho, d_m)}{\omega}$$

Hence the differential equation [6] becomes :

$$I_{eq} \, \frac{d\omega}{dt} = M_o n - \frac{P(Q, \nu, \rho, d_m)}{\omega}$$

or

$$\frac{d\nu}{dt} = [P_o - \frac{P(Q, \nu, \rho, d_m)}{\nu}] \, \frac{1}{I_{eq} \, \omega_o^2} \qquad [7]$$

From this equation the speed of the pumps can be derived. We must remark however that this is only true when the operation point lies to the right of the full torque point, i.e. in the constant torque range $(Q > Q_o)$.

When the flowrate is less than Q_o, the speed governor of the dieselengine keeps the number of revolutions constant : $d\nu/dt = 0$.

3.1.3. Pipeline Characteristics. The pipe is a distributed parameter system. It is modeled as a series of N elementary pieces with length L/N, where L is the total pipeline length (discretisation along the pipe length coordinate). For the sake of simplicity the detailed description is not given in this paper. Rather a simplified lumped parameter model is briefly described just to indicate some ideas. All parameters (which normaly vary along the pipeline length coordinate)

should be interpreted as being averaged over the pipeline length.
The pipeline resistance can then be given as follows :

$$H_1 = \Sigma \xi_i \frac{\rho v_m^2}{2} + \Delta p_1 \, L + \rho g H + Z g (\rho - \rho_w) \qquad [8]$$

where H_1 : the total head losses
ξ_i : hydraulic loss factor for pipe obstructions (bends, etc...)
H : static head
Z : suction depth
L : pipe length
ρ_w : density of the water
Δp_1 : pipe resistance per meter. This factor depends on the local pipe cross section A, the mixture velocity v_m, the density and the grainsize d_m of the mixture.

The (static) operation point of the hydraulic system is the intersection of the pump characteristic and the pipeline resistance for a given density.
With equation [8] and the corresponding pump graphs, the vacuum and all intermediate pressures (in between the pumps) can be calculated.
The pressure difference $(H_p - H_1)$ between the pressure head of the pumps and the pipeline resistance results in a force $(H_p - H_1)*A$ which is used to accelerate or decelerate the liquid mass in the pipeline.
Applying Newton's law to the total mass in the pipeline gives :

$$(H_p - H_1) \, A = \frac{d}{dt} (M v_m) \qquad [9]$$

The mass of the mixture in the pipeline is $M = \rho A L$.
With $v_m = Q/A$ the pipeline differential equation results in

$$\frac{d}{dt} (\rho Q) = \frac{A}{L} (H_p - H_1) \qquad [10]$$

More details on the hydraulic model can be found in (Van Ostaeyen and De Keyser, 1983).

3.2. Identification

In order to estimate unknown numerical coefficients, identification methods from stochastic control theory were applied. The experiments were done aboard of the cutter suction dredger "Rubens" belonging to the fleet of the Belgian company "Dredging International N.V.". It is one of the biggest dredging companies operating world-wide and the tests were done while the ship was engaged in dredging operations in

the Suez Canal (Ismailia, Egypt, 1980) and in the harbour of Lazaro
Cardenas (Mexico, 1981).

Some major specifications of the vessel are :
- built in 1977
- overall length 101 m
- installed power 11436 KW
- maximum dredging depth 25 m

The identification of simple dynamical models describing the
mechanical part of the dredging process was done step by step :
- the model of the unloaded (autonomously running) winches
- the model of the vessel in an unloaded swing (i.e. without cutting)
- the model of the vessel in a loaded swing
- the model of the cutter load
- the model of the suction concentration

During the experiments with the unloaded winches and vessel,
active testing with PRBS-type input sequences was allowed. The models
under load conditions had to be identified from normal operating
records.

The parameter estimation was done off-line with a computer
package that contains several well-known identification methods (De
Keyser et al., 1979). The technique used here is essentially the
maximum likelihood/prediction error identification method described in
(Åström and Bohlin, 1965). It estimates the parameters in the input-
output model

$$A(z^{-1})y(t)=B(z^{-1})u(t-d)+C(z^{-1})e(t)+m \qquad [11]$$

where t denotes discrete-time index (t = 0, 1, 2 ...)
  d denotes time-delay index (d $\geqslant$ 0)
  y, u and e : output, input and uncorrelated noise sequences
  m : constant offset in the signals
  $A(z^{-1})$, $B(z^{-1})$, $C(z^{-1})$: polynomials in the backward shift
  operator with the zero order parameters $a_o = c_o = 1$ and $b_o = 0$
The unknown parameters are collected in the parameter vector

$$\underline{\theta}^T=[a_1 \ a_2 \cdots \ a_{na} \vdots b_1 \ b_2 \cdots \ b_{nb} \vdots c_1 \ c_2 \cdots c_{nc}]$$

and estimated by minimizing the prediction error criterion :

$$V(\underline{\theta}) = \frac{1}{2} \sum_{t=1}^{N} \varepsilon^2(t) \qquad [12]$$

with $\varepsilon(t)=y(t)-y^*(t/t-1)$ (1-step-ahead prediction error)

$$\varepsilon(t) = \frac{A(z^{-1})}{C(z^{-1})} \ y(t) - \frac{B(z^{-1})}{C(z^{-1})} \ u(t-d)$$

and N the total number of collected input-output samples. An estimation of the noise variance σ_e^2 is then given by

$$\sigma_e^2 = \frac{2}{N} \min_{\theta} V(\underline{\theta}) \qquad [13]$$

The minimization is done efficiently by a Newton-Raphson search procedure.

Several criteria were used to validate the results and to look for the most reasonable model structure (model order n_a, n_b, n_c, dead-time index d and signal offset m). Elementary tests such as interpretation of the resulting parameters (gain, time constants) and visual inspection of the model output compared to the real output were combined with statistical criteria such as Akaike's Criterium (Åström, 1980) :

$$AIC = \frac{N}{2} (2.8379 + \log \sigma_e^2) + 2(n_a + n_b + n_c) \qquad [14]$$

or the average loss value

$$ALV = \frac{V(\underline{\theta})}{N} = \frac{\sigma_e^2}{2} \qquad [15]$$

As an example let us describe briefly the results for the model of the vessel in an unloaded swing. The parameters were estimated in the model [11] with :
- input signal $u(t) = I_h(t) - I_v(t)$, i.e. the difference between the armature currents of hauling and veering swing winch motors
- output signal $y(t)$ = swing velocity $v_h(t)$ of the dredger (in m per minute)

The choice of the input signal needs some further explanation. The armature current of a winch motor is proportional to the driving torque and in the first instance thus also to the tractive power of the side wire. The last assumption is only valid to a certain extend, i.e. when the winch dynamical forces are small, requiring reasonable accelerations of the winch and neglectable inertial forces compared to the forces acting on the vessel. The difference between the hauling and veering side wire forces then results in a signal representing the swing torque of the vessel around the spud.

The swing velocity was computed by differentiating the swing position available from the gyro-compass. Figure 6 shows an example illustrating that the velocity information is totally hidden in the noise. By means of digital filtering it was possible to produce an acceptable signal, see fig. 7. It is interesting to illustrate the effect of the filter on the estimated model.

Weak Filter .If the raw v_h-signal is only filtered weakly resulting in a noisy process output for the identification, typical results are as in table I.

288

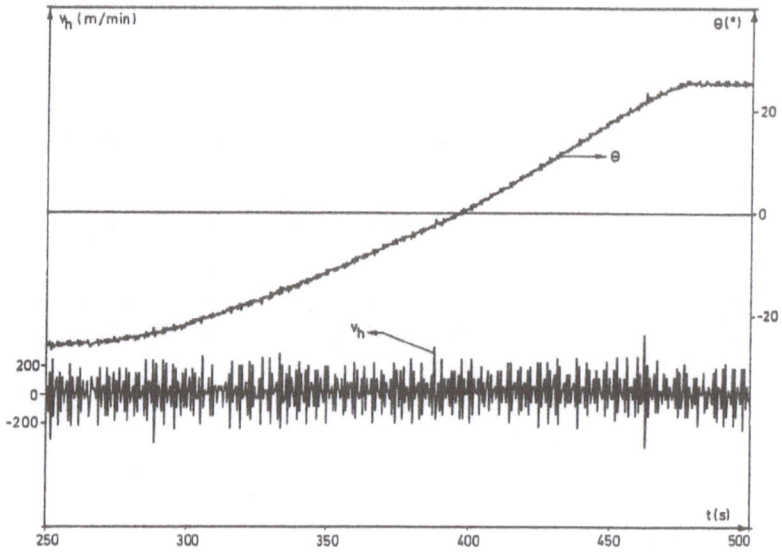

Fig. 6 Swing position (θ) and differentiated swing position (v_h)

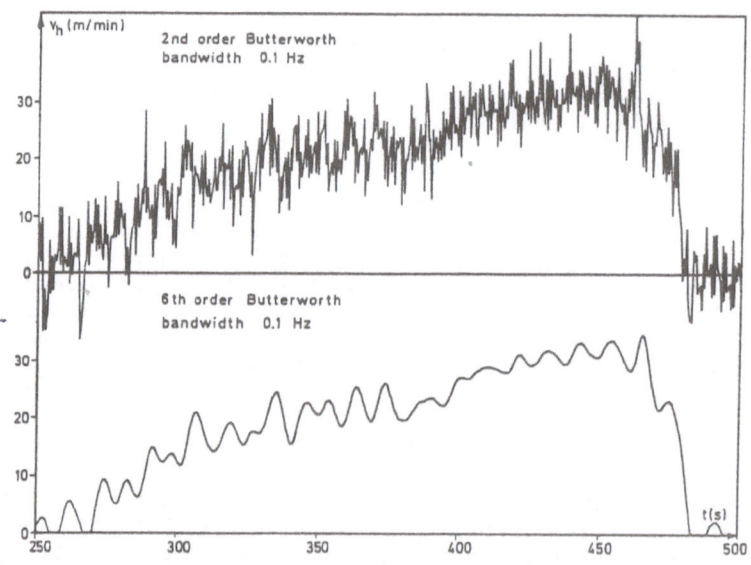

Fig. 7 Velocity signal after digital filtering

The results are obtained from the identification of a time series of 280 samples (sampling period T_s=0.5s). The dead-time index was found to be d=6 (3 seconds). It is striking to observe when searching for the best model orders, how the lowest AIC is obtained for a high degree of the noise polynomial n_c, indicating the importance of the stochastic disturbances in this experiment.

Table I : Typical results for weakly filtered v_h.

n_a	n_b	n_c	AIC
1	1	1	509.8
1	1	0	540.0
1	2	1	511.5
2	1	1	554.0
1	1	2	500.5
1	1	3	499.5
1	1	2	501.0
1	1	1	510.0

Strong Filter. If the raw v_h-signal is filtered strongly, typical results are as in table II for d=6 and in table III for d=10 (respectively a dead time of 3 and 5 seconds).

Table II : Typical results for strongly filtered v_h (d=6).

n_a	n_b	n_c	AIC
1	1	1	502.2
1	2	1	503.8
1	1	1	504.1
1	1	0	501.8
1	1	0	503.8
2	2	0	503.4
2	1	0	501.8

Table III : Typical results for strongly filtered v_h (d=10).

n_a	n_b	n_c	AIC
1	1	1	500.4
1	1	0	499.3

Two effects are important :
- AIC now prefers a low-order noise model (n_c=0)
- the dead-time index seems to be rather 10 (as compared to 6 for the previous experiment). This should be explained as the effect of the important phase lag introduced by the low-pass filter.

As a conclusion we may state that the interpretation of the identification results can give us important feedback on the suitability of the filter. As a result a moderate filter was designed, reducing the noise to a fair level but without introducing too much supplementary dead time. Five time series (3 starboard swings and 2 portside swings) were then processed and used for identification, giving the result :

$$A(z^{-1}) = 1 - 0.904 \ z^{-1}$$

$$B(z^{-1}) = 0.017 \ z^{-1} + 0.007 \ z^{-2}$$

$$C(z^{-1}) = 1 + 0.36 \ z^{-1}$$

This corresponds to the equivalent continuous-time model

$$\frac{v_h(s)}{I_h(s)-I_v(s)} = \frac{0.25 \ e^{-3s}}{1+29.7 \ s} \ (\frac{m/min}{A})$$

Figure 8 gives the output of this model for a portside swing (showing excellent results taking into account the simplicity of the model).

The identification of all other parts of the dredger is described in detail in (De Keyser et al, 1986).

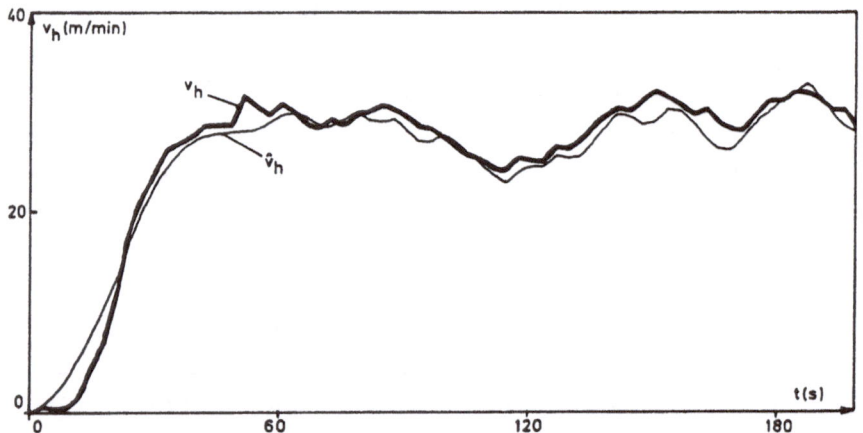

Fig. 8 Model output compared to real output (unloaded portside swing)

4. MULTIPROCESSOR SIMULATOR

The simulator runs in a multi-microcomputer system which is linked to a dredging desk with instrumentation similar to that on board of a real cutter suction dredger. It allows for the study and visualisation of dynamical effects that occur during the dredging operation.

The simulation consists essentially of three parts : a model for the dredger mechanics (side winches and motors, cables, the vessel inertia forces, the cutting process, ...); a model for the hydraulics (suction pump, sand pumps + diesel engine drives); a model for the dredge delivery pipeline.

As the dynamics of the dredging process are quite complex, a fairly performing simulation system is needed. Traditionally for such a simulation an analog or hybrid computer was used. The actual prices for digital equipment (particularly for microprocessors) however make it tempting to implement the simulation on a digital computer, thus avoiding the known disadvantages of an analog/hybrid computer configuration (maintenance, scaling, aging of amplifiers, ...). To approximate the computing speed of analog computers for simulation, very fast sequential computers or computers based on a form of parallel processing (Hwang and Briggs, 1984) are needed.

At the Laboratory of Automatic Control a parallel processor system, consisting of a number of 8 bit microprocessors (MUMPSY : Multi Microprocessor System) was developed (De Keyser and Van de Velde, 1987). As in parallel processing and multitasking different tasks need to synchronize and communicate (Hansen, 1973; 1977), PASCAL primitives to ensure proper process interaction were implemented. Provided with this software, the configuration is used as a simulation workstation.

4.1. Structure of the Simulation Model

The three submodels of the simulation are strongly coupled with each other as is illustrated in fig. 9.

Moreover each submodel has a number of inputs (indicated by Im, Ip, Ih) and outputs (Om, Op, Oh) which are links to the outside world (a real-life dredging desk with the most important instrumentation as seen by the dredgemaster aboard of a cutter). A functional description of each submodel is now given with reference to fig. 9.

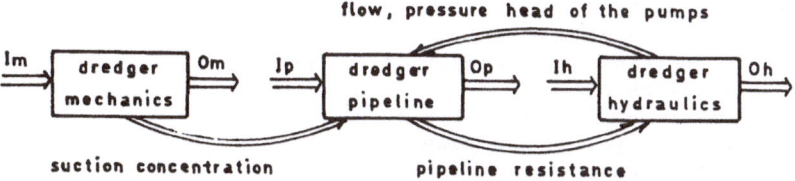

Fig. 9   Block scheme of the simulation model

Dredger Mechanics. The external inputs to this block are :
- the armature currents of both the portside and starboardside winch motors which alternatively act as hauling and veering winch.
- the characteristics of the soil.
    The block contains the necessary differential equations and logic functions for modeling the dynamics and logics of :
- the side winches and their electrical drives
- the swinging of the vessel around the spud
- the cables which act as the link between the winch rotation and the ship rotation

- the load of the cutter motor as a result of the swing speed and the soil characteristics
- the effect of the cutter load on the movement of the dredger
- the resulting suction concentration of the soil/water mixture.
    The external outputs of the block are :
- the angular speed of the side winch motors
- the swing speed and position of the ship
- the load current of the electrical cutter drive.

Dredger Pipeline. The external inputs to the block are :
- the characteristics of the soil.
    The block contains a lumped parameter description of the pipeline which is actually a distributed parameter system. The pipeline with total length of 770 m is divided in cells of length 10 m. Each cell has its own set of parameters such as local mixture concentration and velocity, local effective pipe section (there may be material sediment in the pipe) and local pipe resistance. The mathematical models in this block calculate essentially the evolution with time of these parameters.
    The external outputs of the block are :
- vacuum before the (submerged) suction pump
- intermediate pressure between the delivery pumps aboard of the ship
- total delivery pressure after the pumps
- mixture concentration, mixture velocity and material sediment along the pipeline. These can be visualized in order to observe the state of the pipeline. It is a unique realization in the dredging world, offering quite interesting supplementary information which is of course not available aboard of a real dredger.

Dredger Hydraulics. The external inputs to this block are :
- the characteristics of the soil.
    The block contains all necessary differential equations for modeling the dynamics of :
- the suction pump and the two delivery pumps
- the diesel engine driving the pumps.
    The external outputs of the block are :
- the angular speed of the pumps and diesel drives
- the flow of the soil/water mixture.

    There are two additional tasks in the simulator software : the interface software to the dredging desk and the dialog software with the operator. The interface software takes care of all analog and digital input/output. Each sampling instant the computed simulator outputs Om, Op and Oh are sent to the analog measurement instrumenta-tion (similar to the instrumentation available aboard of the real ships). The steering actions of the user at the dredging desk are then read and used as input to the simulator for the next integration step. The dialog software is for numerical and graphical output on a compu-ter display and for input of characteristic parameters of the dredging process, such as the grain size of the soil. These parameters can be changed by the training master in order to simulate different operating conditions. All input and output is done in real-time during the normal course of the simulation.

4.2. Computer Implementation

The differential equations in the submodels are solved by means of a 4th order Runge-Kutta method with fixed integration step. The simulation has to run in real-time in order for the user at the dredging desk to experience the same effects as on board of the real dredger. For this reason the routines of the dredger mechanics and dredger hydraulics must be scheduled to run each 0.5 seconds real-time, while the pipeline routines have to run each 1 second real-time. The process interface task reschedules every 0.5 seconds while the dialog programs run in the background at a lower priority level.

In Table IV some run-time requirements on a MC6809 microprocessor with 1MHz system clock are given for the 4 tasks. The total run time requirements are thus 7280 ms equivalent 1MHz-6809 processor time per 1000 ms real time. It should be noticed that an earlier computer configuration, a PDP11/34 minicomputer linked to a PDP11/23 minicomputer, was not able to run the simulation in real time.

Table IV : Run times on a 1MHz-6809 microprocessor

Task	run time/ cycle	cycles/ second	run time/ second
mechanics	720 ms	2	1440 ms
pipeline	3440 ms	1	3440 ms
hydraulics	800 ms	2	1600 ms
interface	400 ms	2	800 ms

A multi-microprocessor system was developed with hardware configuration as shown in Table V. Some details on the hardware and software of the multi-microprocessor system are given in the next sections.

Table V : Run times on the multi-microprocessor system

Task	Hardware	1MHz 6809 equivalent	run time/ second
mechanics	2MHz-6809	2	720 ms
pipeline	2MHz-6809 +AM9511-FPP	4	860 ms
hydraulics	2MHz-6809	2	800 ms
interface	1MHz-6809	1	800 ms

4.3. Hardware Structure of MUMPSY

Several microprocessors (MC6809), each of them with a local bus, local memory and eventually peripherals (e.g. floating point processor), are connected to a fast common bus with a 4 Kbyte common memory area. All local busses are modular, so extra peripherals can easily be provided for. One of the microprocessors is fully equiped (mass storage interface, serial communication lines, ...) and acts as the "master". Normally this processor will also provide the real-world interfaces (digital and analog I/O), that are necessary in real-time simulations and control applications. The program code for the "slave" processors is downloaded through common RAM, under control of the "master" processor. Once the different tasks are downloaded and started, the common RAM area is used as a means of communication between the different microprocessors. A bus arbiter and a number of bus adapters (one for each local bus) allow the microprocessors to access the global bus, on a fixed priority base. Notice that, as far as the multiprocessor hardware is concerned, the notion of "master" and "slave" does not exist. Once the application is started, all processors have equal possibilities.

The hardware configuration of the multi-microprocessor system is explained in detail in (De Keyser & Van De Velde; 1987). The structure of the system is depicted in fig. 10.

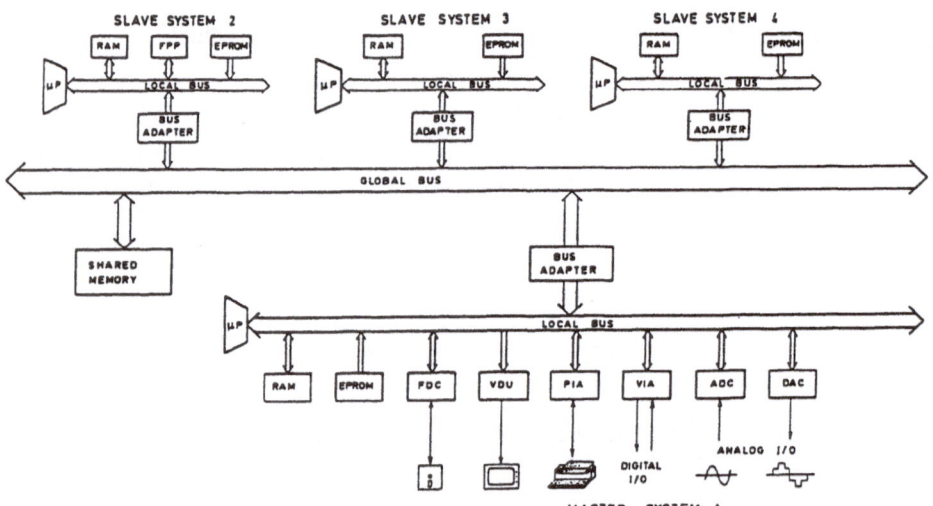

Fig. 10 MUMPSY multiprocessor system

4.4. Software Aspects

This section deals with some specific aspects of software development for multiprocessor applications. Most of the software is written in a high level language (PASCAL). The programs that run in the slave systems are developed on the master system. Afterwards they are downloaded, through common memory, into the slave's local memory. This goes under control of the master processor.

To perform this downloading a special utility was written. This utility runs on the master processor. It will read a program from diskette and present it to the slave. On the slave side, a small monitor program initializes the slave environment, puts the downloaded program at the proper memory location and starts execution.

In multitasking and multiprocessor applications some specific problems that are not present in sequential processing have to be solved. Most of these problems have to do with task synchronization and communication. It is e.g. unacceptable that two tasks change the same data at the same time. To solve these problems a number of high-level multitasking primitives, based on the semaphore principle (Hansen, 1973) are implemented. They are written in PASCAL as functions and procedures that can be called from a PASCAL application program. The primitives allow for :
- task synchronization ("Rendez-Vous") : two or more tasks wait for each other before continuing execution.
- communication with synchronization ("Mailbox") : two or more tasks wait for each other and exchange data before continuing execution.
- communication without explicit synchronization ("FIFO buffer") : data are sent from one process to another through a buffer. The process sending the message does not wait until the message is received.

The basic utilities for downloading and the primitives to synchronize tasks and to enable communication between tasks, make it possible to use the multiprocessor configuration efficiently.

4.5. Simulation Results

A typical result of the simulator is illustrated in fig. 11. Some simulator outputs (fig. 11c) are compared to the corresponding real-life dredging parameters (fig. 11b) that were measured aboard of the cutter suction dredger "Rubens". The simulator was fed with the same input signal (i.e. mixture concentration , fig. 11a) as measured aboard. The outputs considered are the vacuum before the suction pump (VAC), the mixture velocity in the pipeline (v_m) and the angular speed (n_d) of the diesel drive of the pumps.

The simulator has been extensively used for studying the dredging process and for developing and testing the new automatic cutter control system before installing it on board of real ships. The simulator could also be useful for training the operators and dredger personnel (computer aided instruction and training).

5. ADAPTIVE CONTROL

Taking into account the complexity of the process dynamics the closed loop control of cutter suction dredgers is a real challenge for applying adaptive and self-learning strategies. Because of the multiple control loops and because of the numerical complexity of adaptive regulators, the realisation was done by means of parallel operating microcomputers.

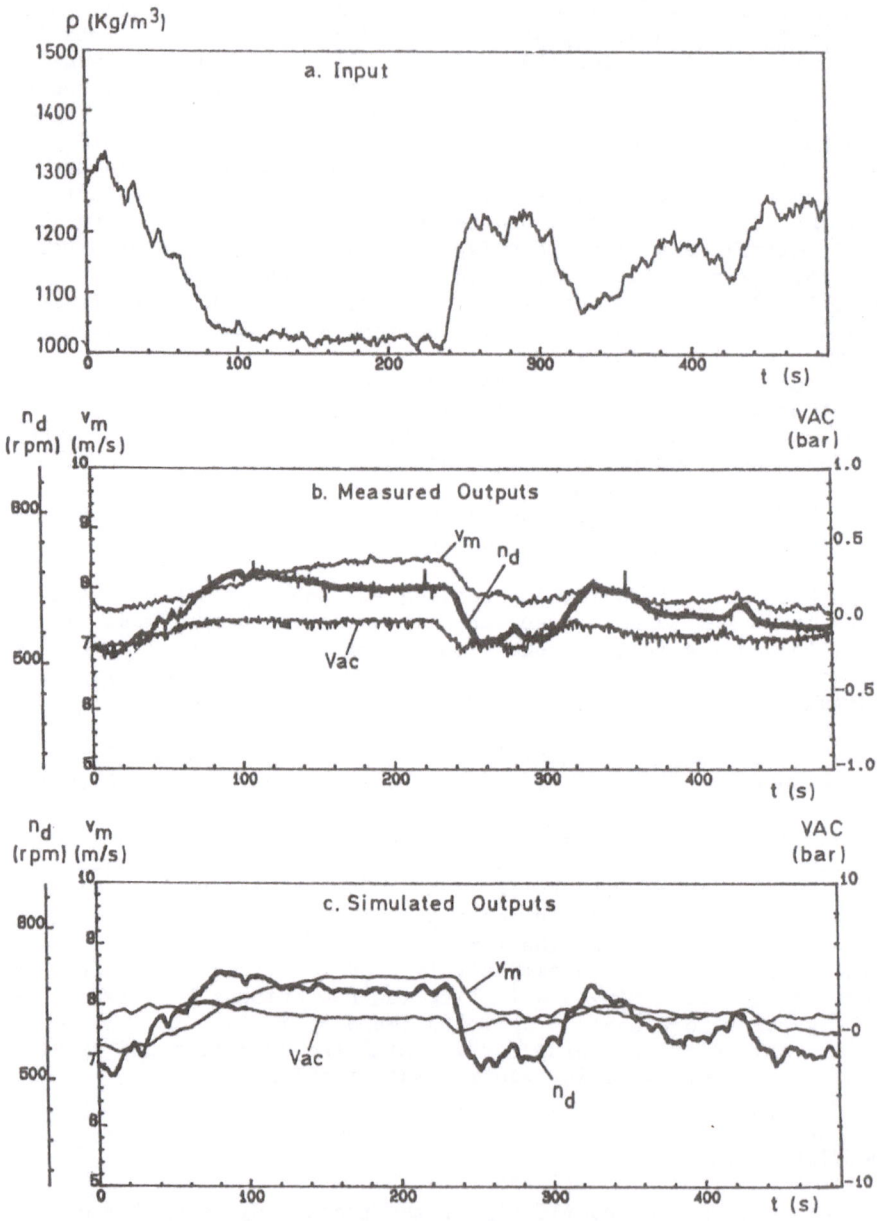

Fig. 11 Evaluation of the simulator

5.1. Structure of the Control Loops

Although computer control of a dredging ship has several additional
aspects and secondary control functions, the most important loops are
depicted in fig. 12.
        This is essentially a master/slave structure with a slave loop
controller, that aims to haul the dredger with a certain swing speed
and several master controllers, which compute the desired swing speed
from the load of the different parts of the dredger.
        The slave controller has as setpoint the desired swing speed.
Since the accurate measurement of this parameter is rather difficult,
this signal was replaced by the angular speed of the hauling winch. As
output of this slave controller, there is the current of both side
winch motors. The reaction to special situations, as running on the
cutter, fast stopping and starting, is included in this controller.
Full attention was paid to the safety of the installation, to the cut
out or bad operation of the instrumentation. When the computer itself
fails, the shut-down control of the dredger is taken over by hardware
that stops the dredger in the shortest possible time.
        The master controllers compute the optimum swing velocity in
order to keep the dredger parameters at their optimal value. The
controlled parameters of the actual Adaptive Automatic Cutter Controller are the cutterload, the concentration (density), the velocity
of the mixture, the intermediate pressure (between suction pump and
the first main dredger pump) and the vacuum of the suction pump.
        All controllers are based on the EPSAC principle (Extended
Prediction Self-Adaptive Control).

5.2. Adaptive Predictive Control

The general principle of self-tuning control is shown in fig. 13
(Åström and Wittenmark, 1984). A suitable process model structure is
postulated. The best model parameters are estimated in real-time from
the measured process input-output data. This is done by means of one
of the well-known recursive identification methods. The estimated
parameters are then used to calculate the control law, which is in the
dredger application a predictive control strategy.
        The regulator obtained is called a self-tuning regulator because
it has facilities for tuning its own parameters. The regulator can be
thought of as being composed of two loops. The inner loop consists of
the process and an ordinary linear feedback regulator. The parameters
of the regulator are adjusted by the outer loop, which is composed of
a recursive parameter estimator and a design calculation.
        Long-range predictive control methods are based on the principle
of fig. 14 which illustrates the EPSAC method described further on.
They are characterized by the following strategy :
- At each sampling period t, a forecast of the process output over a
  long-range time horizon (l sampling periods) is made in the control
  algorithm, based on an estimated mathematical model of the process
  dynamics. Moreover it is a function of the control scenario proposed
  to apply in the future.

298

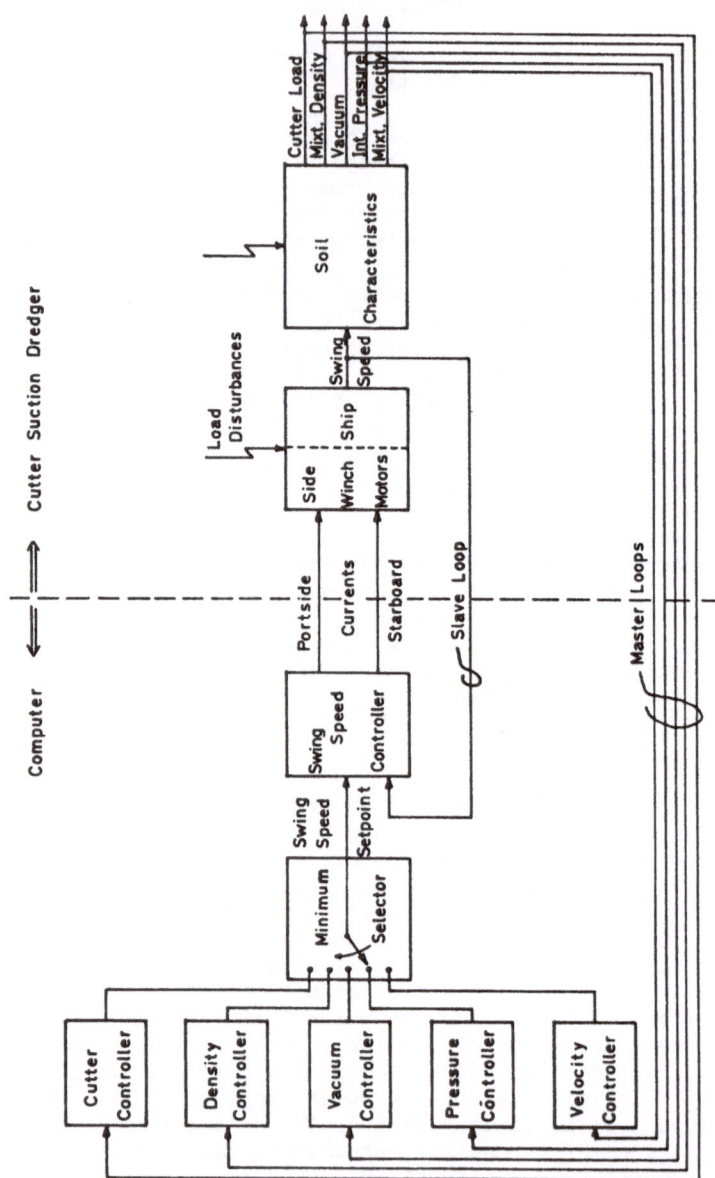

Fig. 12 Structure of the Control Loops

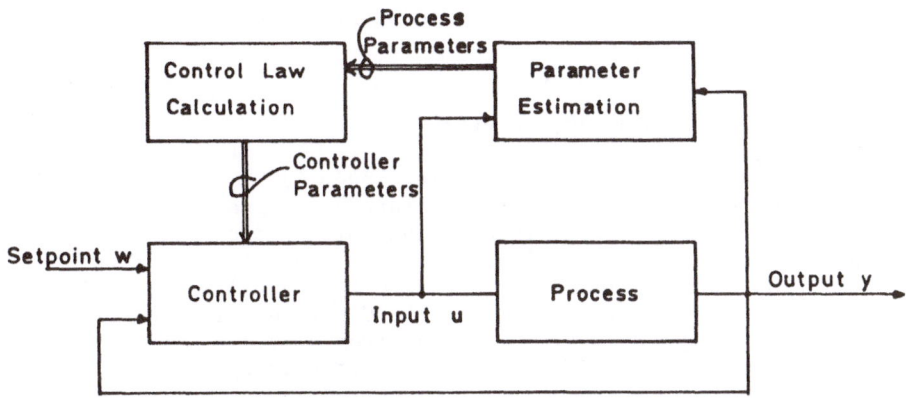

Fig. 13 Block diagram of a self-tuning regulator (STR)

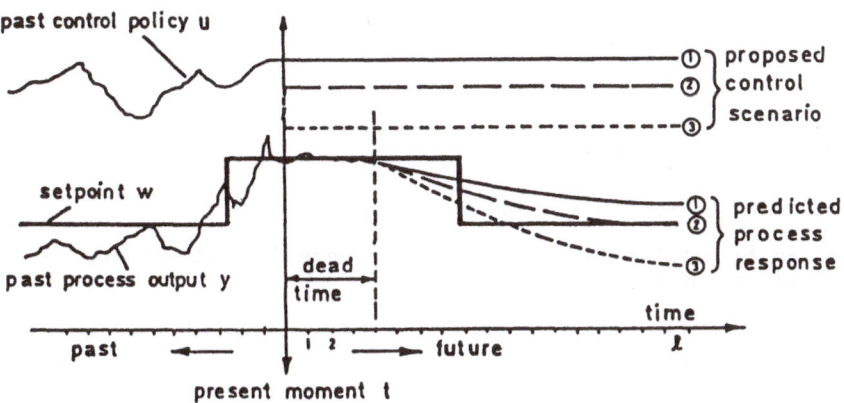

Fig. 14 Long-range predictive control

- From several proposed control scenario's, the strategy is selected
  which brings the predicted process output back to the setpoint in
  the "best" way according to a control objective of form

$$\min_{\Delta u(t)} \sum_{k=d}^{1} \left[w(t+k) - y^*(t+k/t) \right]^2 + \mu \Delta u^2(t)$$

where $\Delta u(t) = u(t) - u(t-1)$ is the current control input variation and
where $y^*(t+k/t)$ denotes the predicted process output (the output
$y(t+k)$ predicted at time t).

- The control action for the first sampling period of the best
  strategy is then applied as a control action to the real process
  input at the present moment. At the next sampling instant the whole
  procedure is repeated, leading to an updated control action with

corrections based on the latest measurements (receding horizon strategy).

The process is modeled by :

$$A(z^{-1})y(t) = B(z^{-1})u(t-d)+v(t) \qquad [16]$$

with t discrete-time index, d time-delay index (d>0), y(.) and u(.) process output and input and v(.) a disturbance signal. It includes the effect of all unmeasured disturbances (stochastic noise as well as stepwise load disturbances) and of linearisation offsets. If necessary the model [16] can be extended with a term $D(z^{-1})d(t)$ where d(.) denotes a measurable disturbance. The EPSAC method will then automatically include feedforward action from d(.).

The 1-step-ahead predicted output is computed by means of the prediction model :

$$C(z^{-1})y^*(t+1/t)=C(z^{-1})y(t)+z[C(z^{-1})-A(z^{-1})]\Delta y(t)+B(z^{-1})\Delta u(t+1-d) \qquad [17]$$

with A(.), B(.), and C(.) polynomials in the shift operator z^{-1} of order n_a, n_b, n_c and :

$$\Delta y(t) = y(t)-y(t-1); \quad \Delta u(t) = u(t)-u(t-1).$$

The parametervector :

$$\underline{\theta}^T = [c_1 \ldots c_{nc} | -a_1 \ldots -a_{na} | b_0 \ldots b_{nb}]$$

is estimated by means of the recursive (extended) least-squares method with the estimation model :

$$\Delta y(t) = \underline{\phi}^T(t) \cdot \hat{\underline{\theta}}(t) + \eta(t) \qquad [18]$$

The measurement vector $\underline{\phi}(t)$ is given by :

$$\underline{\phi}^T(t) = [\varepsilon(t-1) \ldots \varepsilon(t-n_c) | \Delta y(t-1) \ldots \Delta y(t-n_a) | \Delta u(t-d) \ldots \Delta u(t-d-n_b)]$$

where $\varepsilon(t) = y(t)-y^*(t/t-1)$ is the 1-step-ahead prediction error. Notice that the predictor equation [17] can be replaced by :

$$\Delta y^*(t+1/t) = \underline{\phi}^T(t+1) \cdot \hat{\underline{\theta}}(t)$$

with $\Delta y^*(t+1/t)=y^*(t+1/t)-y^*(t/t)=y^*(t+1/t)-y(t)$.

The polynomial $C(z^{-1})$ can be estimated or can be fixed a priori as a design polynomial (a special case being $C(z^{-1})=1$) (De Keyser and Van Cauwenberghe, 1985).

For k=2,3,...l the process output is predicted by means of the multistep predictor :

$$A(\tilde{z}^{-1})[y^*(t+k/t)-y^*(t+k-1/t)]=B(z^{-1})\Delta u(t+k-d) \qquad [19]$$

where the special operator z^{-1} only operates on the first of both time arguments in $y^*(t+k/t)$, i.e. :

$$\tilde{z}^{-1}y^*(t+k/t) = y^*(t+k-1/t).$$

Details on how to use this alternative predictor are given elsewhere (De Keyser and Van Cauwenberghe, 1982). Notice that l is the length of the prediction horizon (fig. 14).

We are thus able to predict the process output over the range $\{y^*(t+k/t), k=1,\ldots d,\ldots l\}$. The next step is to calculate the control action u(t). At this stage it is important to draw the attention to the fact that the prediction values $\{y^*(t+k/t), k=d,\ldots l\}$ depend on the postulated future control policy $\{u(t/t)\ldots u(t+l-d/t)\}$, suggested at the present moment t. An extremely simple strategy for this (ficticious, hypothesized) future control scenario is :

$$\Delta u(t+k/t) = 0 \quad \text{for} \quad k>0$$

or u(./t) remains constant from now on (as suggested in fig. 14). Of course this strategy only imposes a constraint on the imaginary future control actions we suggest at the present moment. The actually realized control policy u(t), u(t+1), ... will certainly not remain constant in the future (receding horizon effect!).

Thanks to the superposition principle, the predicted process output $y^*(t+k/t)$ can be thought of as consisting of 2 parts :

$y_t^*(t+k/t)$, the "transient" part of the predicted output, which is the effect of the "no-further-control" policy, i.e.
$\Delta u(t/t)=\Delta u(t+1/t)=\ldots \Delta u(t+l-d/t)=0.$

$y_c^*(t+k/t).\Delta u(t/t)$, the "controlled" part of the predicted output, which is the effect of the stepwise variation $\Delta u(t/t)$ of the control input.

As a result of this simple strategy the long-range output prediction $\{y^*(t+k/t), t=1,\ldots l\}$ only depends on $\Delta u(t/t)$ which will now be computed so as to minimize a specified control cost criterion. One possible candidate cost function which has a strong intuitive appeal to the designer is :

$$V[u(t/t)] = \sum_{k=d}^{l} [w(t+k)-y^*(t+k/t)]^2 + \mu\Delta u^2(t/t) \qquad [20]$$

where μ is the only tuning parameter left. It has a direct and simple physical interpretation to the user : it controls the speed by which the controller reacts to setpoint variations and load disturbances.

The control law minimizing [20] can be written in closed form as :

$$\Delta u(t/t) = \frac{\sum\limits_{k=d}^{1} y_c^*(t+k/t)\left[w(t+k)-y_t^*(t+k/t)\right]}{\sum\limits_{k=d}^{1} y_c^{*2}(t+k/t)+\mu} \qquad [21]$$

where $y_c^*(t+k/t)$ (the "controlled" part) corresponds to the discrete-time step response coefficients of the system :

$$z^{-d}B(z^{-1})/A(z^{-1}) \qquad [22]$$

and $y_t^*(t+k/t)$ (the "transient" part) is the k-step-ahead predicted process output, computed with [19], assuming $\Delta u(t/t)=\Delta u(t+1/t)=\ldots=\Delta u(t+1-d)=0$.

Summarizing, the EPSAC algorithm consists of :

- Estimation of $\hat{\underline{\theta}}(t)$ in [18] with RLS.

- Computation of $y_t^*(t+k/t)$ with the multistep predictor [19].

- Computation of $y_c^*(t+k/t)$ from [22] for k=d,...1.

- Computation of the control law [21].

A more general treatment of the strategy is given in (De Keyser and Van Cauwenberghe, 1985; Van Cauwenberghe and De Keyser, 1985; De Keyser and Van Cauwenberghe, 1986).

5.3. Computer Implementation

All software runs on a parallel computer system. The principle is shown in fig. 15.

It consists of a 68000 processor on a 16/32-bit bus (VMEbus) and one or more 6809 processors, each active on an own local 8/16-bit bus (EURObus). Each EURObus system is connected to the VMEbus through an interface board (VIOC = VME bus/IO channel convertor).These boards map the 64 Kbyte address spaces available on each of the EURObusses into the VMEbus memory. In this way the VME processor can access RAM and memory-mapped I/O peripherals on the different EURObusses as if they were part of its own memory map. To make the access completely transparant for the programmer the interface boards essentially perform two functions :
- The asynchronous word access (16 bit) on the VMEbus side is transformed into two synchronous byte accesses on the EURObus side.
- By means of a DMA (Direct Memory Access) technique, bus conflicts are avoided that might arise when both the local processor and the VMEbus processor are accessing the EURObus.

Because the VMEbus processor can communicate with the different EURObus processors through their RAM area, multiprocessor applications are possible. As can be seen from fig. 15, an EURObus processor cannot directly communicate with the other EURObus processors. All messages must pass through the VMEbus processor. From a hardware point of view the VMEbus processor could be seen as the "master" processor and the EURObus processors are considered to be "slave" processors.

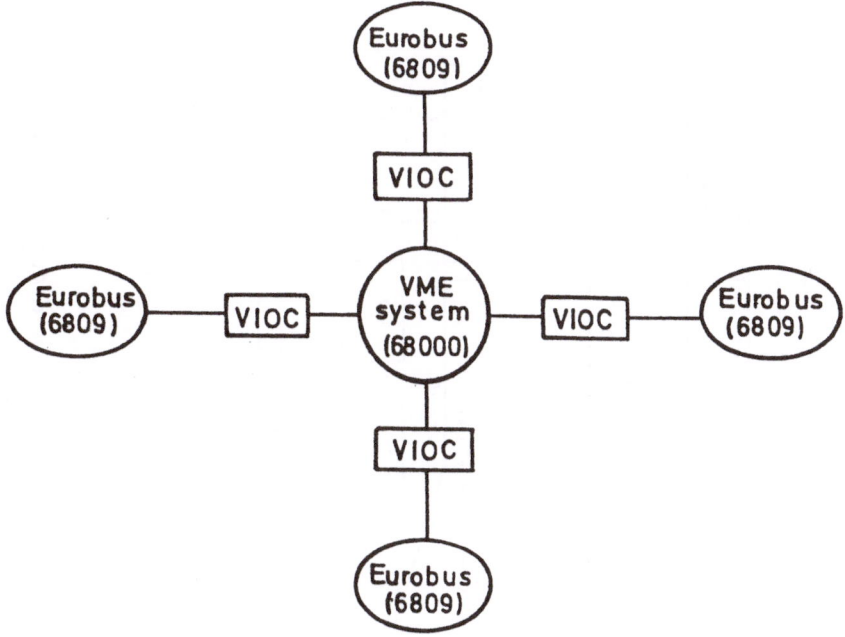

Fig. 15 General Structure of VME-based multi-processor system

The configuration is very tempting for advanced real-time automatic control applications, where the fast "master" processor can perform the more complex tasks (e.g. optimisation, complex calculations, identification and prediction algorithms, ...) whereas the "slave" processors do the more simple jobs (I/O, filtering and scaling, slave control loops). An additional advantage of splitting up the tasks in this way, is that all the hardware for real-world interfacing (analog and digital I/O) resides on the 8-bit side, resulting in less expensive equipment.

The configuration for the implementation of the adaptive control system of the dredging ship is shown in fig. 16. Only one EURObus is linked to the VMEbus. The development of the software can be done either on the VMEbus processor or on the EURObus processor. In the final application both processors work together.

5.4. Selected Results

For the evaluation of the controller extensive measurements on board of the cutter suction dredger RUBENS were performed.

304

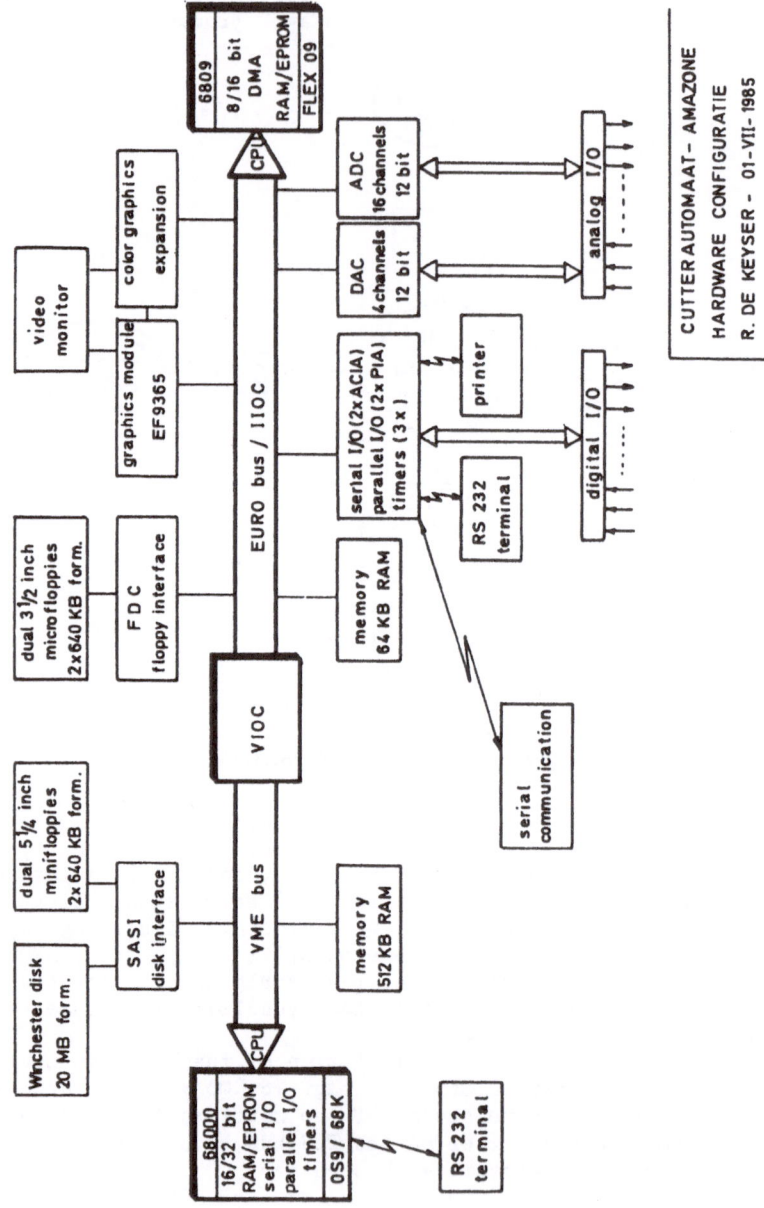

Fig. 16 VME-based multi-processor system for dredger simulator

At first the slave loop controller was examined. Figure 17 shows
the behaviour of the dredger in steady state conditions but with
strongly stochastic load.

From this it is obvious that the adaptive automatic controller
keeps the swing speed well around its setpoint. Furtheron the reaction
speed to set point changes of the hauling speed is improved compared
to manual control. Because of the fact that the current of the side
winches is controlled directly, the hauling forces in the side wires
are better kept below the limits so that breaking of a cable occurs
less frequently.

As for the master controllers, figs 18 and 19 show respectively
the control of the cutter load and of the concentration. The outcome
could be compared to the corresponding signals for manual control presented earlier in fig. 3 (notice the scale difference). Especially the
performance of the density regulation is unachievable by manual control because of the important deadtime. Indeed this control loop is
hindered by a severe transportation lag because the measuring instrumentation is located near the pumps (indicated by 8 in fig. 1), which
is about 50 m away from the suction inlet (indicated by 1 in fig. 1).
Moreover the deadtime changes due to the varying mixture velocity in
the pipe (between 4m/s and 9m/s). The controller has to be robust
w.r.t. the unknown, varying deadtime.

From the experiments it was evident also that the control system
is reacting well to varying soil characteristics. It is here that
classical PI controllers failed.

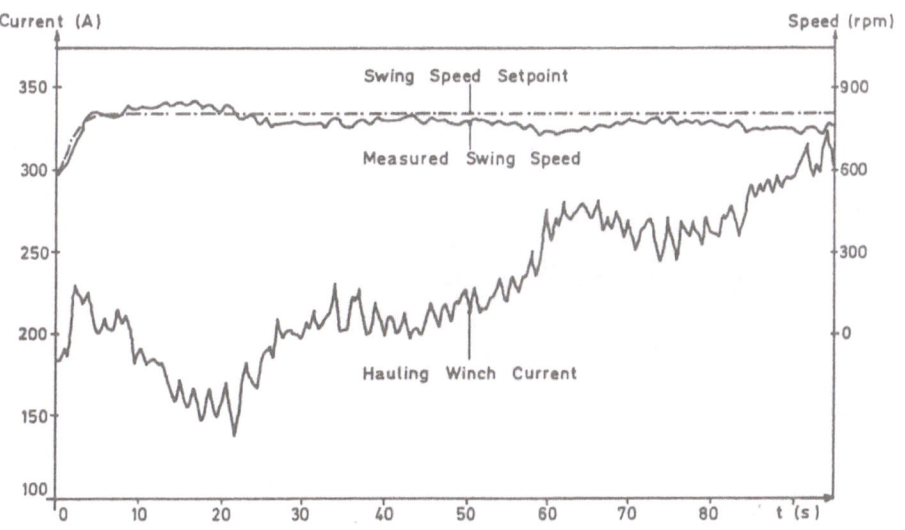

Fig. 17 Evaluation of Slave Loop Controller

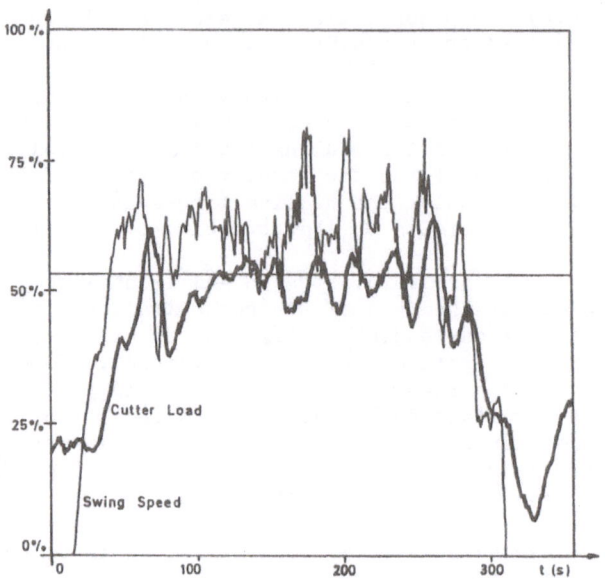

Fig. 18 Master Controller of the Cutter Load

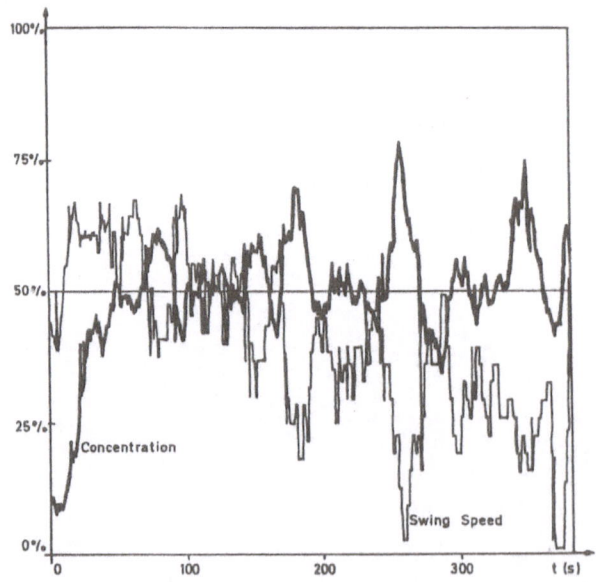

Fig. 19 Master Controller of the Concentration

6. CONCLUSIONS

The realisation of a computer control system for large cutter suction dredgers was a project with several important aspects of modern control theory and application.

A sensible combination of a priori process knowledge, white modeling and black box identification techniques led to suitable simulation models of the ship and its environment.

A low cost though powerful simulator was built by using a multi-microcomputer configuration. The simulation has been extensively used for studying the dredging process and for developing and testing the advanced control algorithms.

These control strategies are based on the combination of real-time parameter estimation with predictive control techniques, thus resulting in a computer control system featuring self-learning and adaptive properties. Because of the highly stochastic and time-varying nature of the dredging process it could be expected that adaptive control might lead to a better overall performance when compared to classical PID control. This was confirmed by several on-board experiments.

When compared to manual operation with skilled skippers the conclusions depend very much on the work situation. Sometimes the computer beats the human operator, especially in the long run and when the disturbances are more or less stationary. There are other situations such as dredging soil with a tendency of breaching, which is an extremely nonstationary phenomenon, where the heuristics and expertise of an experienced human operator is preferable compared to a purely mathematical control procedure.

This takes us to the main topic for further improvement : the development of a higher level expert control system which could be realised quite independently of the existing system thanks to the multiprocessor structure. A promising implementation is the use of a 5th generation language (e.g. Prolog) in a separate computer linked to the actual on-board computer system. Rule based expert control, mainly for use during exceptional work conditions, could be placed on top of the adaptive controllers in order to combine efficiently the flexibility of logics and heuristics with the power of adaptive predictive control.

ACKNOWLEDGEMENTS

The development project of the adaptive automatic cutter controller is an initiative of the Belgian contractor "Dredging International n.v.". It was realized in collaboration between "IMDC n.v." and the Automatic Control Lab of the University of Gent.

REFERENCES

Åström K. and T. Bohlin (1965). 'Numerical identification of linear
    systems from normal operating records'. In : P.H. Hammond (Ed.),
    Theory of Self-Adaptive Control Systems, Plenum Press, New York,
    96-111.
Åström K.J. (1983). Theory and Applications of Adaptive Control - A
    Survey. Automatica 19 (5), 471-486.
Åström K. and B. Wittenmark (1984). Computer Controlled Systems :
    Theory and Design, Prentice-Hall, Englewood Cliffs.
Åström K.J. (1986). 'Adaptation, Auto-tuning and Smart Controls'. CPC
    III Asilomar, Elsevier, 427-466.
Brouwer K. (1985). 'The Selection of Dredging Equipment'. Marintec
    China '85 Conference, Vol. 5 : Ports, Dredging and Cargo
    Handling, Lloyd's of London Press, London.
Clarke D.W. and P.J. Gawthrop (1979). Self-tuning Control. Proc. IEE
    126 (6), 633-640.
Conte, G. and Del Corte D., Multi-processor Systems for Real-time
    Applications (D. Reidel Publishing Company, Dordrecht, 1985).
De Keyser, R.M.C., F. D'Hulster, J. Heyse and A. Van Cauwenberghe
    (1979). 'Experiments with a Software Package for Process
    Identification and Control'. In : M. Cuenod (Ed.), Computer Aided
    Design of Control Systems, Pergamon Press, Oxford, 473-478.
De Keyser, R.M.C. and A.R. Van Cauwenberghe (1982). 'Simple Self-
    tuning MultiStep Predictors'. In : G. Bekey, G. Saridis (Eds),
    6th IFAC Symposium on Identification and System Parameter Estima-
    tion, Washington DC, pp. 1558-1563.
De Keyser, R.M.C. and A.R. Van Cauwenberghe (1985). 'Extended
    Prediction Self-Adaptive Control'. In : H.A. Barker, P.C. Young
    (Eds), Identification and System Parameter Estimation, Invited
    session on Applications of Adaptive and Self-tuning Control, Per-
    gamon Press Oxford, 1255-1260.
De Keyser, R.M.C. and A.R. Van Cauwenberghe (1986). 'Towards Robust
    Adaptive Control with Extended Predictive Control'. In :
    Proceedings of the 25th IEEE Conference on Decision and Control,
    Athens, Greece.
De Keyzer C. (1983). 'A User Friendly Automatic Controller for Cutter
    Dredgers'. Europort Congress Dredging Days Amsterdam, paper D4,
    18 pp.
Dekker L., 'Parallel Processors in Engineering Sciences', Proc. of the
    IMACS Symposium on Simulation in Engineering Science, 21-31,
    1983.
Hansen, P., Operating System Principles (Prentice-Hall, Englewood
    Cliffs, 1973).
Hansen, P., The Architecture of Concurrent Programs (Prentice-Hall,
    Englewood Cliffs, 1977).
Hwang, K., Briggs, F., Computer Architecture and Parallel Processing
    (McGraw-Hill, New York, 1984).
Paker, Y., Multi-microprocessor Systems (Academic Press, New York,
    1983).

Van Cauwenberghe, A.R., R.M.C. De Keyser (1985). 'Self-Adaptive Long-
Range Predictive Control'. In : Proceedings of the 1985 American
Control Conference, Boston, MA.; IEEE Service Center, Piscataway,
Vol. II, pp. 1150-1160.
Van Ostaeyen J.L.S., R.M.C. De Keyser (1983). 'A Simulation Model for
the Hydraulic Process of Cutter Suction Dredgers'. In : Europort
Congress Dredging Days, Amsterdam, paper D1, pp. 1-22.
Van Ostaeyen J.L.S., R.M.C. De Keyser (1985). 'The Design of an
Adaptive Cutter Controller'. Marintec China '85 Conference, Vol.
5 : Ports, Dredging and Cargo Handling, Lloyd's of London Press,
London, 25 pp.
Van Zutphen A.C. (1983). The Influence of Comprehensive Automation
Systems on Cutter Suction Dredging Operations", Europort Congress
Dredging Days Amsterdam, paper D6, 20 pp.

Chapter 9

ADVANCED CONTROL IN THERMAL POWER PLANTS

Motomiki Uchida
Kyushu Electric Power Co., Inc.
Thermal Power Department
1-82, 2-chome, Watanabe-dori
Chuo-ku, Fukuoka 810 Japan

ABSTRACT. This chapter describes present status of automation and its background as well as future view of advanced techniques at thermal power plants. As for the implementation of modern control theory at actual power plants for the improvement of dynamic performance, methods of steam temperature control using linear quadratic regulator or feedforward controller based on linear programming principle are respectively introduced with their control performance. In the present state of the arts, actual application of the advanced techniques, such as diagnosis and expert system to be described in the text, will largely depend on the development of technology in future.

1. INTRODUCTION

This chapter describes a perspective on the present status of advanced control systems applied in thermal power plants. As an introduction, overview of the development of process control from its early stage to today's computer controlled system is provided. In the early stage of automation, on-off control was widely used in process industry. Regulators with proportional, integral, and derivative actions became common in the 1930s.

Second World War brought about a significant impact on process control. For instance electronic instruments were applied for process control: controllers were tuned based on Ziegler-Nichols type rules. Automation system was widely accepted by the engineers, because it could be installed quite easily.

The first computer controlled system was installed in 1959 for oil refinery. However, computers at this period were so unreliable that they had to be used only in supervisory mode or set point control mode.

In 1962, direct digital control system appeared in chemical industry. A complete analog instrumentation for process control was replaced by computer in this case. Advances in integrated circuit realized more reliable minicomputers and finally microprocessors. At the same time, performance-price ratio has improved remarkably. Consequently, this will be a strong driving force for development of advanced process control

S. G. Tzafestas and J. K. Pal (eds.), Real Time Microcomputer Control of Industrial Processes, 311–335.
© 1990 *Kluwer Academic Publishers.*

system and other advanced maintenance technology such as artificial
intelligence and expert system.

As to main theme of this chapter, i.e., an advanced control system,
there remains still a lot of points to be solved in the future. At present, time consuming engineering efforts are required for applying an
adaptive control in industry. An adaptive control is still in its in-
fancy. It is expected that a new concept of an adaptive control theory
suited to application for large scale industry such as thermal power
plants will be available in the near future.

2. PRESENT STATUS OF AUTOMATION IN THERMAL POWER PLANTS

2.1. Essential Conditions for Automation

It is rather difficult to describe the history and present status of
automation in Thermal Power Station in the world. Important things which
we have to consider for successful automation are (1) incentive for
automation, (2) well-suited piping arrangement for automation, (3) reliability of a computer, (4) its cost performance, (5) computer software
for large scale system, (6) a control system with hierarchical structure
for automation, (7) cooperation of boiler and turbine designer, instruc-
tors for test run, and system designers of utility company. First, let's
go into some details about these matters.

2.1.1. Incentive for Automation. After World War II, scaling up of unit
sizes for economical merits and higher steam conditions for improving
plant efficiency have been developed. This causes the appearance of
supercritical boilers. The development of supercritical boilers requires
reducing starting loss by the establishment of proper start-up systems
and shorter start-up time. Furthermore, it requires that complicated
plant operations such as switching from start-up system to normal system, turbine warming-up, start and synchronizing, and increasing turbine
inlet pressure in accordance with loading up are progressed correctly.
If a plant fails to supply electric power to a power system in a scheduled time, the power system must be operated less economically. Improvement of reliability for start-up and shutdown of a high efficiency unit
is the chief aim not only for automation but also for localizing the
extent of a failure detected by diagnostic system.

2.1.2. Improvement of reliability of equipments for automation. It is
easy for a plant to run in normal operation. For example at constant
load, plant can continue to run even if a control valve sticks.

But there is high possibility to cause troubles during procedure of
plant start-up and shutdown because all equipments run over the wide
range. In order to achieve a successful automation, all equipments must
have an ability to perform normal action correctly: reliability is
needed. Also all equipments are to be checked exactly whether its action
is normal or not.

In this sense, the reliability of limit switches mounted on each
local equipment is important. Since slight movement of equipment

position is detected by using on-off action of a limit switch, bad installation of limit switch, dust, and extremely high or low ambient temperature lead poor reliability and may cause troubles with automation.

It was not until torque switch had become to be used to detect
motor operated valves' closing position judging from its closing torque
that the foundation of automation was established.

2.1.3. Plant software (Engineering).

Fossil-Fired Plant construction and
operation consists of many phases such as planning, designing, manufacturing, construction, test run, commissioning, and scheduled outage. It
was often the case that problems raised at each phase weren't reflected
so much on the next phase or planning of a new plant.

Some troubles especially raised at a test-run often require the
improvement of a plant design. The key of successful automation for
thermal power plant is to improve the plant design in order to reduce
distorted operations caused by the lack of consideration at designing
for safer and simpler operations.

The essence of computer software for automation program is established by utilizing the logical complex judgement operator have had to
make to coordinate each equipment. Thinking about the more advanced
automation by computer, it is important for a successful automation to
construct piping system so as to be easily automated and supervised. And
it is also necessary for it to develop the detector and its software for
air venting process of boiler and pumps. Also from the aspect of life
management and life extension, it is necessary to mount a pilot valve
with register for warming-up avoiding the delicate intuitional operation
by operator. Fully reflecting the analytic result of control theory, it
is very important to select a size of boiler tubes and to design the
proper arrangement of heating surface in order to balance heat absorption of each heating surface during plant start-up in consideration of
shortening of plant start-up time, reducing starting loss, and controlling consumed life by thermal stress. As mentioned above, if it becomes
clear that a boiler has such problems, countermeasures against them
should be taken and these experiences should be reflected on the design
of a new plant in order to make effort to build an ideal plant for the
economical operation of power system. From this point, it is most important for manufacturer and utility personnel associated with planning,
designing, manufacturing, construction, test run and planned outage to
study ideal plant hardware and software as the premise of complete automation, making arrangements and discussing in detail together.

2.1.4. Computer software for large scale system.

a. Event oriented program. At the beginning of the development of DDC, a
computer program for control checked repeatedly a flow chart from beginning to end. That was a time oriented program. So it was applied only
to a small scope of automation and a countermeasure against a serious
event occurred in a process was delayed. As the solution of these problems, event oriented program that checked whether the control mode
should be switched or not by interrupting CPU when an event occurred was

developed. At that time it was a key point for computer to process effectively on real time after interruption by an occurrence of an event.

b. Software system structure. For controlling a large scale system, computer system should be based on the function of priority interrupt. Furthermore software system structure should be a hierarchical structure as shown in Fig. 1 and consists of supervisory program, executive control program and actual action control processor.

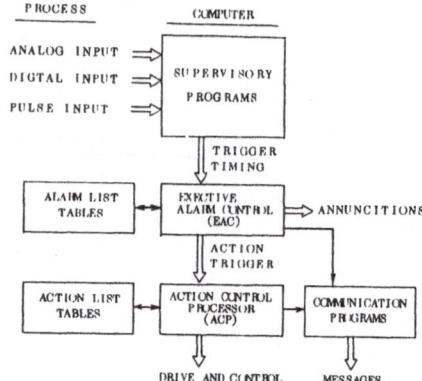

c. Fill in the blank. Usually startup engineer who has know-how about start-up and shutdown of an actual plant doesn't know well mechanism of computer software. Therefore, it is necessary that computer software have

Fig. 1   Software system structure

an ability to be added easily some program to present one by start-up engineer for successful automation. Since automation program became able to be expressed simply by plant table on which conditions of various sequences were filled in the blank, programming has become easy.

d. Back ground processor. It is necessary that this kind of table can be changed easily by the experience of actually controlled start-up and shutdown during the test run. The development of a Back Ground Processor in order to modify the table during on-line control played a great role on the successful automation.

The development of Off-line Back Ground Processor that can evaluate control programs before installing computer in site has also had an impact on the successful automation. By using pseudo signals, the off-line simulation test of the start-up procedure is able to be carried out as if fluid temperature, flow rate, pressure and so on were changing as the result of the opening or closing valves and start or stop of motors.

e. Reliability and cost performance. It goes without saying that reliability and cost performance are the vital issue when it comes to the stage whether full automation by control computer should be adopted or not. Recently advanced technology and massproductivity by electronics industries has remarkably improved the reliability and cost performance of control computers. So the system of executing a thousand of jobs using many computers has been applied widely.

2.1.5. Hierarchical system. Successful automation requires various premises. For this purpose, it is finally important to establish reasonable and more reliable Hierarchical Control System Structure shown in Fig. 2 installing appropriate computers in each structure. And it is also important to be easily supervised and controlled by operators and to establish an operator friendly system. Therefore, almost of all manipulations can be done in the central control room. The photograph (Fig. 4) shows one of the most advanced central control room and it adopts an advanced plant supervisory board to supervise and control collectively installing some CRTs that provide required information during start-up. This shows the evidence that during start-up automation has progressed to be executed smoothly without manual operation.

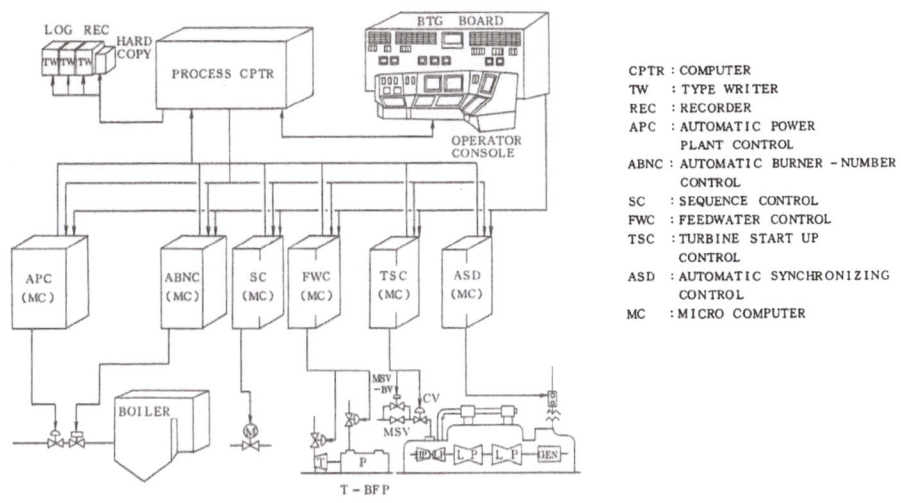

CPTR : COMPUTER
TW : TYPE WRITER
REC : RECORDER
APC : AUTOMATIC POWER
        PLANT CONTROL
ABNC : AUTOMATIC BURNER - NUMBER
        CONTROL
SC : SEQUENCE CONTROL
FWC : FEEDWATER CONTROL
TSC : TURBINE START UP
        CONTROL
ASD : AUTOMATIC SYNCHRONIZING
        CONTROL
MC : MICRO COMPUTER

Fig. 2   Configuration of automatic control system

2.2. Progress of Automatic Control at Thermal Power Plants

2.2.1. Progress of automatic control in scope.

a. Enlargement of application. Automation has been progressed step by step according to the progress of automation related technology mentioned in section 2.1. At first, reliability of computer was not so good as to apply it as plant control use, and it was applied as a data logger. After that, it made us more familiar with digital information than analog one. Secondary, as a first step for plant automation, computer became to be used to supervise whether boiler and turbine start-up process was progressing correctly, instead of depending only upon operator. At last, after reliability of supervision by means of computer was confirmed, computer became to be used to control the actual process. The process that could easily be automated and might be better controlled than operators was chosen for computer automation. As these application's results were confirmed, another application has been enlarged. Table 1 shows its history of progress.

b. Substantial subsystem. Besides enlargement of application, substantial subsystem to support computer control is essential. It is important to make up hierarchical system by adopting computerized subsystem and

Table 1 Progress of automatic control at thermal plants of Kyushu Electric Power Company

PROGRESS	INTRODUCTION OF COMPUTER	AUTOMATIC TURBINE START CONTROL	AUTOMATIC BURNER CONTROL	AUTOMATIC CONTROL OF SUBCRITICAL THERMAL UNIT	AUTOMATIC CONTROL OF SUPERCRITICAL THERMAL UNIT	TOTALLY COMPUTER AUTOMATED SYSTEM OF SUPERCRITICAL THERMAL UNIT
APPLIED UNIT	SHINKOKURA 1, NOV, 62	KARATSU 1. SEPT, 67	OITA I, JUL, 69 OITA II, JUL, 70	KARATSU II, JUL, 71	KARATSU III, JUL, 73	SENDAI I,JUL, 74 SHINKOKURA II, OCT, 76 BUZEN I, 77 BUZEN II, JON, 80
COMPUTER CONTROL	• INTRODUCTION OF COMPUTER (DATA LOGGER)	• PEFORMANCE MONITOR • TSM * • ADOPTION OF AUTOMATIC TURBINE START ANALOG CONTROLLER		• TSMC ** • BSMC ***	• TSMC • BSMC **** COMPUTER CONTROL IN UNIT STARTING • TURBINE START BY DDC • SETTER SET CONTROL OF BOILER	• MANAGEMENT OF TRURINE LIFE • TURBINE STAR TBY SCHEDULE CALCULATION • TOTAL COMPUTER CONTROL AT UNIT START AND STOP • OPTIMAL CONTROL
SUBSYSTEM	TSM * TURBINE START-UP MONITOR TSMC ** TURBINE START-UP MONITOR AND CONTROL BSM *** BOILER START UP MONITOR BSMC **** BOILER START UP MONITOR AND CONTROL		• BURNER REMOTE MANUAL CONTROL	• INTENSIVE ADOPTION OF SEQUENTIAL CONTROLLERS		• SEQUENTIAL CONTROL BY COMPUTER KICK CONTROL
					• AUTOMATIC BURNER-NUMBER CONTROL	
MAN-MACHINE COMMUNICATION				• ADOPTION OF AUTOMATIC SYNCHRONIZER (MANUAL OPERATION)	• AUTOMATIC SYNCHRONIZATION • ADOPTION OF MICROCOMPUTER IN NUMBER LOGIC	• EXCITATION, SYNCHRONIZATION AND BUS TRANSFER BY COMPUTER CONTROL
				• GRAPHIC AUXILIARY BOARD OF BTG	• ADOPTION OF CRT	• IMPROVEMENT OF MAN-MACHINE COMMUNICATION
PATROL-SAVING					• OBSERVATION OF VIBRATION OF IMPORTANT MACHINERY	• ADOPTION OF MAIN TURBINE LUB CHECK MONITOR • OBSERVATION OF MAJOR MACHINES' VIBRATIONS AND ABNORMAL SOUNDS
ROTATION OF ROUTINE OPERATION					• ABNORMAL AND ROUTINE OPERATION BY SEQUENTIAL CONTROLLER	
PREVENTION OF TROUBLE EXTENSION	PLANT INTERLOCK AUTOMATIC START OF AUXILIARY MACHINERY				• AUTOMATIC RUNBACK • FCB (FAST CUT BACK) • TURBINE SPEED CONTROL IN THE ABNORMAL CONDITION	• AUTOMATIC CONTROL AT RESTARTING AFTER UNIT TRIP OR FCB

programmable controller in order to make information easily transmitted and reduce the main computer's job, and as a result the system can deal with more complicated job. For a practical use, substantial subsystems for water treatment equipment, burner ignition control, automatic burner number control and so on are necessary.

c. Adoption of friendly man-machine interface. The purpose of plant automation is to perform a reliable start-up and shutdown operation. Another purpose is to have supervisory function on a plant condition without any difficult operation, that is, operator can easily understand what computer is aiming at and whether plant is progressing correctly for it or not. In order to do that, it is very important to adopt a supervisory system with friendly man-machine interface. For example, adoption of advanced plant Supervisory Board makes easier to understand present plant's situation, which is mainly composed of colour CRT display and shows progressing sequence and related system charts, also shows machine to start in flickering, and draws related analog quantities.

d. Labor saving of patrol on start-up. To adopt full automation in power plant, it is necessary to make it patrol free, while it used to be indispensable for operator to check the starting up machinery. It is effective to install an accelerometer based on magneto striction phenomena to watch vibration of machinery at bearings. And also to check the noise of starting up machinery, it is necessary to install equipment which we can listen to the noise of machinery at central control room. Regarding vibration during normal operation, it is a normal practice to pick up and check the machine automatically when it exceeds the alarm point. It is also important that rub-check of steam turbine at 400 rpm could be checked at central control room.

e. Automation of routine and abnormal operation. It is desirable to automate routine operation such as condenser back washing. Abnormal operation such as high pressure water heater one half capacity operation usually occurred in the case of heater tube leak, and air heater water washing which needs careful attention to temperature, moisture, valve position and so on might be better to be automated.

f. Prevention against extension of accident at thermal power plants. Classical countermeasures for it include plant-interlock and automatic start-up of spare machine in the case of machine's trip. Especially automatic start-up system of spare lubrication pumps for steam turbine is popular. If all power plants on electric power system tripped, service interruption would occur in wide range. On the other hand, starting up of power plant needs power supply to start up big machinery. These relationship would make up a vicious circle, and would make service interruption longer. To prevent this, Fast Cut Back and auxiliary load operation have been adopted, and these are included into countermeasures for the extension of accident. Fast Cut Back and auxiliary load operation mean that when something happen on electric power system, power plants parallel off and keep operation supplying own auxiliary load. The system using Artificial Intelligence or Expert System will be able to control the extension of accident much better in future.

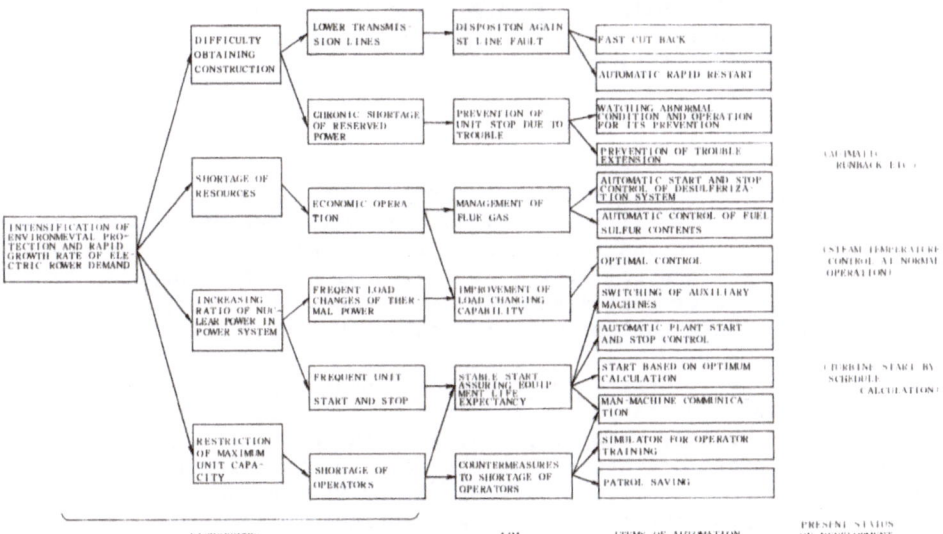

Fig.3 Trend and items of automation for thermal power units

318

2.3. Present Status of Automation in Thermal Power Plants

As mentioned above, the full scale automation system using control computer may be the only measure that achieves high reliability in the case of labor shortage and performs correct manipulations when a supercritical boiler starts up. Nowadays, control computer is capable to handle a lot of information. However, most important job of the control computer is to supervise the status of whole control system and decide what should be done at every instant.

Configuration of automatic control system should be a hierarchy system as shown in Fig. 2, in which a computer plays a role of operators' brain and issues orders to several digital or analog controllers to take adequate manipulations. In a hierarchy system, the control computer orders commands instead of plant operators to APC (Automatic Plant Controller), ABNC (Automatic Burner Number Controller), subcomputer and sequence controllers. In this system, operators can take over start-up of the boiler by ordering commands to APC, ABNC and so on in the event of computer failure.

2.3.1. Control boards.

In a totally computer automated control system, control boards should be arranged so that fewer operators can supervise the plant status. The main control board for normal operation is called BTG (Boiler Turbine Generator) board. Fig. 4 shows the BTG board at the Shinkokura #5 thermal power plant of Kyushu Electric Power

Fig. 4 BTG board at the Shinkokura #5

Company. During the unit start-up, several sequences run simultaneously, while controllers work all the time. Consequently, operators must supervise all of the valves and instruments. It is ideal that the operators can watch the whole control status at a glance. For this purpose, several CRT display the present status and in the near future operators will be able to manipulate the control station on the CRT display.

2.3.2. Man-machine interface by chime and announcing system.

In a totally computer automated control system, the most important thing is that the operators correctly recognize what is going on at every instant. Computer must inform operators what the computer is going to perform. A chime and announcement system was adopted for this purpose. When the start-up system progresses to the stage mentioned below, the computer rings a chime to arouse operators' attention and make announcement.
a. When the computer kicks the sequence to start,

b. When the sequence controller completes the configuration of an important system,
c. When important pumps or fans are about to start,
d. When main valves open or close,
When the progress lamp of automatic start-and-stop control (ASC) console flashes at the break point; i.e., the computer informs operators that the next stage is ready to start and is waiting for operator's approval, an operator is supposed to press the flashing lamp button if he permits to start the next sequence.

On these occasions, the computer rings a chime and flashes the corresponding lamp to notify an operator the location. At the same time the computer displays messages about it on the CRT. Fig. 5 shows a flow chart of the man-machine interface using a chime and a flashing lamp. When timing conditions to start a sequence controller are satisfied, a computer rings a chime, flashes the progress lamp, and announce if it is a break point, and displays unsatisfied conditions in violet, if there were any.

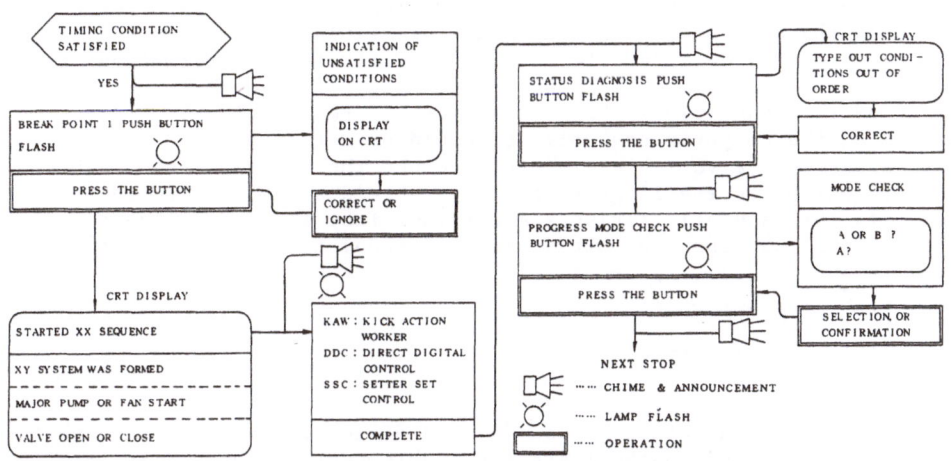

Fig. 5 Man-machine interface

2.3.3. Progress mode selecting system. There are several cases when the computer must consult with the operators which one of the two possible ways should be selected. For instance, after furnace purge the computer can not find out whether hot clean up should be performed or not. The computer flashes the progress mode select PB (Push Button) and displays message "Mode check; Hot clean up complete?" on the CRT in yellow, and rings a chime at the same time. If hot clean up is unnecessary, the operator is supposed to set the hot clean up complete PB. Then, the computer omits to perform hot clean up. The progress mode selecting system furnishes flexibility of operations of the unit start-up and shutdown.

3. ADVANCED TECHNIQUE FOR THE IMPROVEMENT OF DYNAMIC PERFORMANCE

3.1. Steam Temperature Control

3.1.1. Linear quadratic regulator. In a large capacity high-pressure high-temperature boiler for electric power generation, deviations of steam temperatures at the boiler outlet must be kept within one or two percent of their rated values to maintain the nominal operating efficiency and insure the safety and the maximum equipment life of the plant. The main purpose of the boiler control is to allow the increase or decrease of steam generation as fast as possible in response to the load command from the power system's dispatch center, while satisfying the above-mentioned operating conditions. However, since a modern thermal power plant usually includes many control loops with significant mutual interactions within the boiler process, it is not easy under the conventional PID controller to fully compensate for these interactions to satisfy the required steam conditions for a large and fast changes in plant load. To solve this problem, Linear Quadratic Regulator has been applied to the supercritical thermal power plants. In the system, system identification or state space representation of the plant is performed based on the AR (Autoregressive) model describing the system dynamics. The order of AR model is determined by AIC (Akaike's information criterion). In this system, the plant, consisting of the boiler-turbine process and the conventional PID controller, is regarded as the objective system of the computer control. It should be noted that MWD is included as a pseudo-state variable, in order to use the information of MWD changes which are the largest and measurable disturbances to the plant. The data for AR model fitting are obtained from the system identification experiment. Then, a multivariate AR model to the vector data series X(n) (n = 1, 2, ..., N) are fitted.

$$X(n) = \sum_{m=1}^{M} A(m)X(n-m) + U(n) \tag{1},$$

Neglecting the innovation in Eq. (1) and dividing X(n) into two subvectors, i.e., an r-dimensional state-variable vector x(n) and an l-dimensional manipulated-variable vector y(n), we obtain

$$\begin{bmatrix} x(n) \\ y(n) \end{bmatrix} = \sum_{m=1}^{M} \begin{bmatrix} a_m & b_m \\ * & * \end{bmatrix} \begin{bmatrix} x(n-m) \\ y(n-m) \end{bmatrix} \tag{2},$$

where the symbol * expresses submatrix of A(m) irrelevant to the state-variable vector x(n). From Eq.(2) we obtain

$$x(n) = a_1 x(n-1) + \ldots + a_M x(n-M) + b_1 y(n-1) +$$

$$+\ldots + b_M y(n-M) \qquad (3),$$

The optimal state-feedback gain matrix is determined by Dynamic Programming procedure using the state equation and the quadratic criterion function defined below,

$$J = E \sum_{i=1}^{K} (X'(i)\ Q\ X(i) + Y'(i-1)\ R\ Y(i-1)) \qquad (4),$$

where E denotes the statistical expectation, Q and R are non-negative definite weighting matrices.

3.1.2. <u>Application of linear programming to temperature control</u>. Boiler process forms a typical mutually interacting multivariable system, and appropriate adjustment of PID controllers is difficult. Temperature deviations are caused by transient thermal-hydraulic imbalance of working fluid initiated by MWD change signal from the power system's dispatch center. The appropriate feedforward control signals making use of the MWD change can be determined by means of Linear Programming as follows.

Fig. 6 shows dynamic behaviors of SHT and RHT. In each of the frames in the center of Fig. 6, that relate MWD, FR-D, SP-D, GD-D to SHT and RHT, are shown the dynamic responses of the temperatures to a stepwise increase of each of the corresponding manipulated variable.

Fig. 6 suggests possibility of suppressing fluctuations of the SHT and RHT by providing appropriate manipulation signals to FR-D, SP-D, and GD-D, which act to cancel the temperature deviations due to MWD changes.

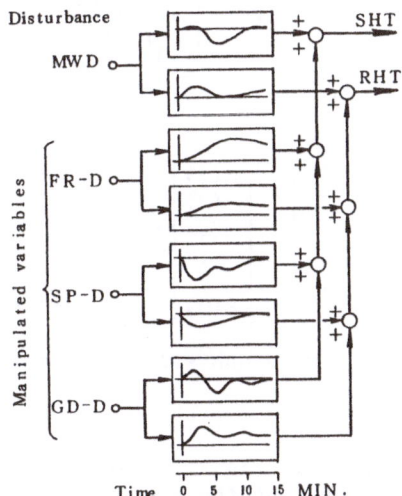

Fig. 6 Response of steam temperatures to step change of process inputs

Suppose that a stepwise MWD change v takes place at time t=0 and we manipulate the FR-D, SP-D, and GD-D by $u_1(k)$, $u_2(k)$, and $u_3(k)$ at t=k. Then, the following equation, which relates steam temperatures to the manipulated variables, is obtained:

$$\begin{bmatrix} x_1(s) \\ X_2(s) \end{bmatrix} = \begin{bmatrix} m_1(s) \\ m_2(s) \end{bmatrix} \cdot v + \sum_{k=0}^{s-1} \begin{bmatrix} h_{11}(s-k) & h_{12}(s-k) & h_{13}(s-k) \\ h_{21}(s-k) & h_{22}(s-k) & h_{23}(s-k) \end{bmatrix} \begin{bmatrix} u_1(k) \\ u_2(k) \\ u_3(k) \end{bmatrix} \qquad (5),$$

where $x_1(s)$ and $x_2(s)$ are the deviations of SHT and RHT from their set points at t=s, $m_1(s)$ and $m_2(s)$ are the response of SHT and RHT at t=s to a unit step change of MWD at t=0, and $h_{ij}(\ )$'s denote the indicial response function of the i-th temperature to the j-th manipulated variable. From Eq.(5) we get

$$X(s) = M(s) \cdot v + \sum_{k=0}^{s-1} H(s-k)\, U(k) \qquad (6),$$

where

$$X(s) = \begin{bmatrix} x_1(s) \\ x_2(s) \end{bmatrix}, \qquad M(s) = \begin{bmatrix} m_1(s) \\ m_2(s) \end{bmatrix}$$

$$H(s-k) = \begin{bmatrix} h_{11}(s-k) & h_{12}(s-k) & h_{13}(s-k) \\ h_{21}(s-k) & h_{22}(s-k) & h_{23}(s-k) \end{bmatrix}$$

$$U(k) = \begin{bmatrix} u_1(k) \\ u_2(k) \\ u_3(k) \end{bmatrix}.$$

The equations from the time point s=o to s=m of Eq.(6) are collectively expressed as Eq.(7).

$$\begin{bmatrix} x(1) \\ x(2) \\ \cdot \\ \cdot \\ \cdot \\ x(m) \end{bmatrix} = \begin{bmatrix} M(1) \\ M(2) \\ \cdot \\ \cdot \\ \cdot \\ M(m) \end{bmatrix} \cdot v + \begin{bmatrix} H(1) & 0 & \cdots & 0 \\ H(2) & H(1) & 0.. & 0 \\ \cdot & \cdot & \cdots & \cdot \\ \cdot & \cdot & \cdots & \cdot \\ \cdot & \cdot & \cdots & 0 \\ H(m) & H(m-1) & \cdots & H(1) \end{bmatrix} \begin{bmatrix} U(0) \\ U(1) \\ \cdot \\ \cdot \\ \cdot \\ U(m-1) \end{bmatrix} \qquad (7)$$

or in a more abbreviated matrix expression as Eq.(7)'.

$$X = M \cdot v + H \cdot U \qquad (7)'$$

As shown in Fig.2, we have three manipulated variables, FR-D, SP-D, GR-D, whereas the steam temperatures to be controlled are two, namely SHT and RHT. Therefore, each of the matrices H(i) (i=1,2,.....,m) in Eq.(7) has the size of 2×3 and the resultant matrix H in Eq.(7)' has the size of 2m×3m, i.e., the number of the manipulated variables (3m) is greater than the number of equations (2m). This means that Eq.(7)' has no unique solution. In other words, the designer of the control system can choose a solution that shows desirable control performance.

The problem of finding optimal manipulation can be formulated by Linear Programming (LP) in the following way. If we want to keep the temperature deviations at t=s within ±d(s) (2×1 vector), then the next equation must be satisfied.

$$-d(s) \leq M(s) \cdot v + \sum_{k=0}^{s-1} H(s-k) \cdot U(k) \leq d(s) \qquad (8),$$

The object function of LP problem is defined so as to satisfy the condition that the manipulation which provides prompter damping is the better solution. It is easily understood that the amplitudes of permissible steam temperature deviations with respect to time lapse, can be completely determined by the control system designer's proposal by specifying d(s) at his will. This is an outstanding feature of the LP method in the design of a steam temperature control system.

3.2. High-Response Excitation System

As a power plant has become to be built in remote site and capacity of a power station has increased, increase of capacity and length of transmission lines is highly required. A high response excitation is one of economical means and practically used. When accidents of transmission lines happen, voltage of power system is reduced and impedance of transmission lines is increased because of the removal of the section where accidents had happened. As the result of that, an electrical output of a generator is reduced and the generator is accelerated, and cause easily steps out. As a countermeasure for this, strengthening excitation of a generator as soon as possible and raising a terminal voltage is effective when power system becomes unstable. As the result of this, the generator output is increased and its acceleration is controlled and power system stability can be improved. High-Response Excitation system is a system which detects the terminal voltage and the phase angel of a generator as soon as possible when unstable states such as power system accidents happen and controls properly and promptly the excitation of the generator, therefore it can improve the dynamic and transient stability of power system. In the case of AC exciter it can be realized by using an exciter field current feedback control system and a power system stabilizer.

4. ADVANCED TECHNIQUE FOR MAINTENANCE AND OPERATION

4.1. Auto Tuning Controller

There are some examples of Auto Turning Controller that has been studied in field tests or developed as products. There is no suitable theory how to tune the parameters of controllers for multi-variable control system. Therefore, it should be carried out to tune the controllers by trial and error at the site. Consequently, it should require a long period until tuning is completed. So, some systematic and efficient tuning procedure will be available in the actual thermal plant. Many tuning procedure for linear system have been presented. However, they can not be applied directly to the actual thermal power plants, because it is no linear, multi-variable and strongly coupled system. One of the practical tuning system, so to speak, computer aided tuning system has been developed and has carried out the field tests at thermal power plants in Japan.

This tuning system aims at shortening tuning time and grasping quantitative controllability. This system adjusts parameters setting so-called control areas that are obtained by time integration of control deviation (difference between objective and control output) as evaluation function. In the case of changed parameters, behavior of evaluation function is confirmed by simulation tests or field tests in order to decide control areas for next step corresponding to that parameters. However, characteristics of a plant is non-linear, so the relation of changing trend between parameters and evaluation function doesn't change uniformly under the influence of the change of other parameters by plant non-linearity.

A. Halme reports that self-tuning algorithm based on modified minimum variance criteria have been implemented and tested in a commercial microcomputer based process automation system in Finland. The controlled system is generally characterized by the linear SISO difference algorithm. Optimal control signals are obtained by the so-called minimum variance control. When implementing the control law, the system order n and delay k have to be chosen by the operator. The commercial system has been developed by VALMET Instrument Ltd., Finland. The prototype of this tuning station has been tested in industries in a paper factory and mining plant. However, it will be useful for the single input single output system of auxiliary equipments in thermal power plants.

An ordinary process controller is a linear device. When the range of control is so wide that the nonlinear nature of the process is strongly manifested, the ordinary controller is not very effective. The new microprocessor-based controller from Taylor Instruments have been developed to accommodate the nonlinearity of process.

Basically, the action of Taylor's new adaptive gain controller is similar to that of a PID controller in the region of setpoint. The gain is characteristically low and constant, and the loop is tuned for fastest response. But when error on deviation exceeds a certain magnitude, the controller gain will start increasingly dynamically with the error. Programming function is also introduced to eliminate the nonlinearity of the process.

L&N has developed a self-tuning controller that continuously optimizes PID constants. The controller learns in a manner similar to that used by a skilled operator. During the tuning procedure, a controlled upset is introduced into the system to determine system time constants, and it forms an internal model of the process. The controller then watches the system response and necessary corrections are made to the original set of constants.

Self-tuning is a learning technique that enables the controller to identify the dynamics of the process and automatically compute the PID tuning constants required to achieve the process time response selected. Self-tuning minimizes, but does not eliminate, the skills and interaction required by the operator during initial setup, but assures optimum control under a wide range of conditions during operation. Of special interest is the self-diagnostic capability that is used to verify the controller operation and aid in troubleshooting. For example, the controller will recognize a power loss during the self-adaptive mode of operation and will automatically switch into a safe mode of operation to protect the process. When power is restored, the controller will return the process, again automatically, to the previous setpoint before switching into the self-adaptive mode.

4.2. Intelligent Operation System

The problems to be solved on the practical use of a real diagnosis system still remain at present. However, various means for lightening operators' burdens such as recommended information system for optimal and safety operation, a practical annunciator suppression system when many annunciators ring simultaneously at an accident are practically used. A few examples are introduced below.

4.2.1. <u>Diagnostic function</u>. Formerly analog APC system for boiler control at thermal power plants was widely adopted. Recently digital APC system tends to have been adopted because of reduced cost of IC and advanced reliability of a digital control computer. Characteristics of the digital system are described below; (1) Analog system has few flexibility in the case of highly developing and remodeling control functions because of one-to-one correspondence between control function and hardware, (2) Maintenance work of analog system is troublesome because of its drift, (3) Control device in digital system is compact and economical, (4) Digital system can easily have redundant system and separate CPU in each function group, so can have advanced reliability, (5) Self diagnosis function can be easily set in the digital system. And typical self diagnosis function of digital APC is described below.

a. Diagnosis of control signals. Output from a controller is transferred to an actuator through E/I converter. This diagnosis function checks whether analog output from the controller is the same signal as a computer provide or not.

b. Diagnosis of input signals. Reliability of process input signals is guaranteed by checking whether these signals are in the range between upper limit and lower limit or their changing rate are normal or not.

c. Diagnosis of processing system for analog input and output. This diagnosis is executed by adding a standard voltage as a test signal in every control intervals. The diagnosis of its input part is executed by checking whether the standard voltage can be read correctly as an analog input. And the diagnosis of its output part is executed by checking whether the output signal is confirmed correctly by A/D converter in the case of standard voltage output.

d. Diagnosis of control processing program. Control program is called at constant intervals and executes control processing. In order to check whether normal processing is executed cyclically or not, its diagnosis is executed by so-called Watch Dog Timer system which judges that the control processing is suspended when RC timer composed of hardware reaches its limit in each processing cycle.

It is also common to have a fail safe function such as control state at present is maintained by automatically changing its mode to manual mode during the exchange of module because of diagnosis results and an ability to change its control program on line.

4.2.2. <u>A multi-level ANN</u>. A new artifical alarm system has been developed and put into practical use, which can deal with alarm messages as an excellent and experienced operator would propose to achieve safe and reliable power plant monitoring functions. It is especially effective during unstable conditions such as start-up and shutdown, boiler trip, turbine trip and so on as well as during normal operating conditions.

During normal operating conditions, the conventional alarm system has not any special problem. On unstable plant conditions, however, so many process quantities become abnormal simultaneously and there is a possibility that this flood of information disturbs correct interpretation and correction of plant conditions by the operator.

A Multi-Level Alarm Information Processing System is a computer aided monitoring system which can lighten operator's burden during the non-steady state as well as during

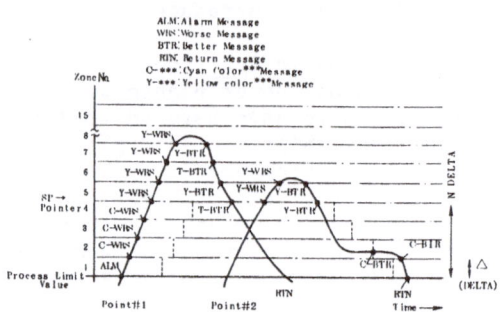

Fig. 7 Alarm message output control

the steady-state of the plant. This new alarm system is currently applied and implemented to Eraring Power Station, Australia by Toshiba processing computer (TOSBAC-7/40). In the conventional alarm system, the process variables are judged whether they are in alarm or in normal stage. Once a process variables enter into the alarm region, the point(s) value and its alarm limits are trending on the alarm CRT. In this new alarm system, however, the alarm region is decided by the unit called DELTA or significant change value which is assignable for each energy process variable as shown in Fig. 7. By using this configuration, this system is considered that the alarm tracking information in the alarm region can be output to the operator. Therefore this new alarm system can call the operator's attention to the alarm tracking information by such output messages as WRS (Worse), BTR (Better), and so on. ALM (Alarm) and RTN (Return) messages have the same meaning as those in the conventional alarm system.

Zone No. is defined to be equal to the degree of penetration of a process variable in the alarm region, where energy point is normalized by this unit DELTA. This Zone No. can be regarded as the degree of urgency among the process variables in the non-dimensional unit.

SP (Suppression) pointer is a variable which change its value from O to N. In emergencies, SP pointer can control the excess alarm messages by suppressing the alarm messages under the value of SP pointer.

4.2.3. <u>Combustion management system</u>. The system which gives information about combustion management to an operator is practical for the optimal combustion, especially for a mixed coal fired boiler. Systems mentioned below are practically used.

a. Automation of intervals for soot blow operation. The degree of purity of heat transfer surfaces is constantly calculated from heat absorption rate. And judging from the result of this, soot blowing is done at proper intervals. This causes the reduction of steam consumption for soot blowing, fuel cost, and the boiler tube erosion.

b. Indication of flue gas temperature at each part of a boiler. By indicating decrease of flue gas temperature at each part of boiler, especially boiler outlet flue gas temperature, the margin against initial deformation point of ash is evaluated. Consequently, the creation of a bird nest can be prevented.

By indicating the exhaust gas temperature from economizer, efficiency management of electric precipitator can be performed and the operation that prevents the air heater erosion is possible.

c. Indication of predicted exhaust gas characteristics at a coal blending ratio. In the case of blending different kinds of coals, it is difficult to keep fuel ratio (C/H), melting point of ash, ash ratio, slagging, fouling within the limits that a boiler permits. There is also restriction in electric precipitator. Since the characteristics in the case of blending coals, however, can not be obtained by linear interpolation method, the prediction of exhaust gas characteristics is effective on the operation management of a boiler and electric precipitator.

d. The guidance for a high efficiency operation. An idealistic method of the guidance for high efficiency operation has not been established yet. There are some guidances that consider the relationship among fineness of coal, heat loss in exhaust gas, heat loss due to combustible in refuse, supply of power to a mill, and wear of mill roll.

e. Supervision of the state of combustion. For example, computer graphics gained by processing a wide view picture taken by a video camera and optical fiber are used for the purposes of flame detection.

In the case of coal fired boiler, it is difficult to catch exactly its combustion state. Therefore, if a higher reliability of flame detection can be established, the rate of stabilizing oil will be reduced at a lower load and that will be more economical.

f. Indication of operational state of a mill. The wear degree of rolls and segments of a mill is predicted by supervising its operational state. Besides that, the malfunction of a mill can be predicted, therefore the effect of prevention from an accident can be expected and rational management of maintenance schedule is possible.

g. The operational management of boiler auxiliaries. The characteristics of aged deterioration on the performance of larger auxiliaries are graphed for the prediction of their malfunctions. And the margin against surging is confirmed by using a pressure-flow rate characteristic curve for fan such as forced draft fan.

4.2.4. <u>Start-up schedule management by Fuzzy theory.</u> The start-up schedule for a steam turbine is decided under the management of its life consumption by arranging the difference between boiler outlet steam temperature and turbine metal temperature.

Considering the life consumption by thermal stresses of a turbine, the profile of metal temperature of the turbine when it starts up depends on the state of cooling process of a turbine at its shutdown and the periods of out-of-service. Furthermore, thermal stresses at its start-up are changed by the boiler outlet temperature and the process of the turbine speed up because heat transfer coefficient depends on a turbine speed. These thermal stresses can be exactly calculated. This procedure, however, is troublesome. When turbine shutdown schedule is known, the tendency of thermal stresses can be predicted by boiler steam temperature and speed up curve of a turbine. Since stress margin in that case can be predicted by selecting properly a membership function of Fuzzy theory, a conventional start-up schedule of a turbine is modified and a proper one can be developed.

The system with learning system becomes such a most desirable system as thermal stresses at start-up can be calculated in the case of the modified start-up schedule and the membership function can be modified automatically.

4.3. The Life Management of a Boiler

The life management of a turbine is generalized. Recently thermal power plants have been operated for intermediate load duty and the life management of a boiler has become required to be considered. Objects of its management can be a water separator and final superheater outlet header. Water separator is a thick-walled part exposed in a lower temperature range, so that it is the most fatigue damaged part by thermal stresses. On the other hand the final superheater outlet header is heavy thick-walled components and exposed in high temperature atmosphere, so that creep damage of these parts is larger.

Internal fluid temperature, pressure and the inside-outside temperature difference of components are measured. Using these data, life consumption by thermal stress fatigue and by creep damage are calculated.

4.4. Management of Vibration of Rotating Machinery

VIBRAROMA was established in 1977 for research of the measurement of vibration of rotating machines and its diagnosis by some companies participated in UNIPEDE, and the exchange of its information has been done in that organization since then. Since present units have become to be operated with great care because of postponement of new plants by slower increase of electric demand, request for the diagnosis of vibration of rotating machines has been increased. Also some accidents by the concentration of stresses such as cracking of a turbine rotor shaft, fretting fatigue of a generator have strengthened it. Therefore there is a growing tendency to develop a technology of a early fault detection system by diagnosis of vibration. If a technology of predicting the occurrence of shaft cracking can be established, life of rotor becomes longer and reasonable maintenance can be done. Finally it is considered that the development of the program will pay.

Technologies such as measurement of high frequency component of vibration, regression analysis, pattern recognition and so on have been tried in the analysis of abnormal vibration. It is desirable to measure vibration both vertically and horizontally and reduce the number of datum for theoretical model. In the development of supervisory system, there are three steps such as presentation of information to operators (1st step), presentation of analytic datum to plant engineers (2nd step), automatic diagnosis system with the function to transfer datum to specialists in a central research center (3rd step).

Since vibration analyzer is carried to a site in Japan and locus of vibration at turbine speed up of a test run can be presented automatically in the polar coordination, patterns of vibration can be understood just after the test run. Theory of adjustment of vibration is practically used and vibration is adjusted easily by influence coefficient method from the locus of a turbine vibration.

A micro computer-based vibration diagnosis system of rotating machinery has succeeded in field test in Japan. Abnormal vibrational behavior can be detected at early stage by continuously monitoring frequency spectra using a modified Walsh to Fourier transform. It is also possible to discriminate automatically types of abnormal vibrations by comparing measured vibration characteristics with predetermined malfunction patterns memorized in the microcomputer. Furthermore, suitable correction weight to be attached to each rotor for trimming of unbalanced vibration can be calculated by the influence coefficient method after just one trial start-up of the rotating machinery.

Table 2 Areas for on-line monitoring of incipient failures

On-line monitors now in place and successfully operating	Problems current the subject of ongoing R&D for on-line monitoring	Areas not yet the target of significant monitoring activity
Shaft vibration	Bearing failure	Boiler slagging and fouling
High temperatures (stationary component)	Shaft fatigue cracking	Tube leaks (low pressure)
Generator core overheating	Turbine water	High temperature (rotating components)
Low oil or water flows	Turbine corrosion	Air/gas system instabilities
Gross over-stressing of turbines, boilers	Tube leaks (high pressure)	Pump cavitation
Shorted generator field turns	Turbine blade vibration	Armature support system fatigue

4.5. On Line Diagnostic Monitoring in US. Fossil Plants

Diagnostic monitoring is advancing for all areas of the power plant.
Electric Power Research Institute gives an overview of on-going work in
the area of fossil plant on line diagnostics in the United States utility industry as shown in table 2.

Although many of the technique are in early stage of development,
several others have already demonstrated their usefulness in detecting
failure and helping utilities schedule their maintenance more effectively.

5. SOME EXPERIENCES OF ACTUAL PLANT CONTROL

5.1. Steam Temperature Control

The application of Advanced Digital Control mentioned above to the steam
temperature control of an actual plant is described below.

5.1.1. Linear quadratic control. Above mentioned advanced digital control system named ADC-AR system by linear quadratic regulator has been
successfully applied to five supercritical power plants of Kyushu Electric Power Co. in Japan (Total output 2,700 MW) since 1978. In Fig. 8
shows the control performance of the ADC system compared with that of
the conventional PID controller at the field test of Shinkokura #3. Considerable improvement of SHT and RHT control can be seen in the figure.

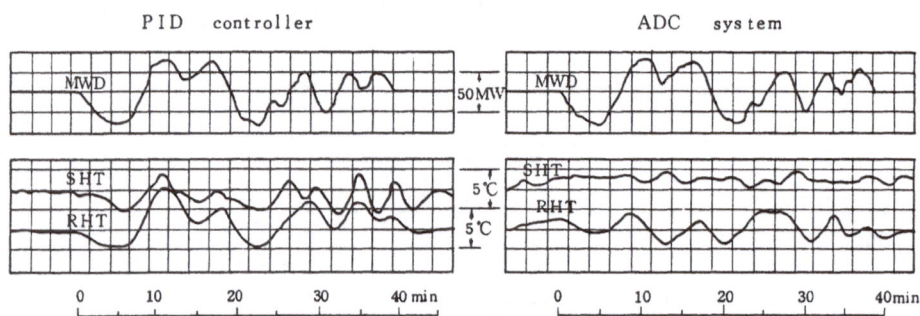

Fig. 8 Comparison of control performance of the ADC system
and the conventional PID controller

5.1.2. Application of linear programming. Feedforward control signals
determined by means of linear programming was applied to the temperature
control of Buzen #2 500 MW supercritical constant pressure boiler to
suppress the initial temperature deviation due to MWD change.

In addition to it, linear quadratic regulator based on AR model was
applied to the whole system, consisting of the boiler-turbine process,
conventional PID controllers, and feedforward control by linear programming.

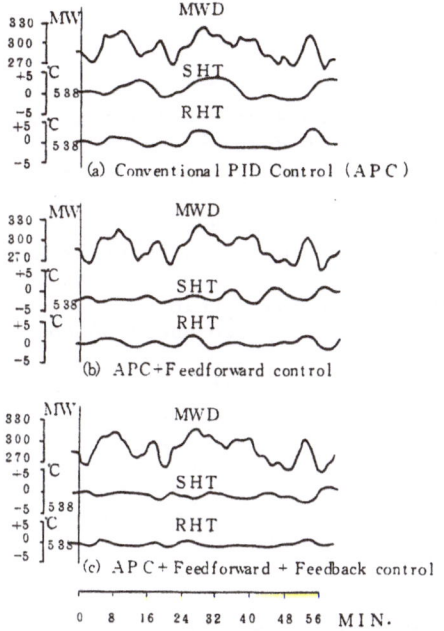

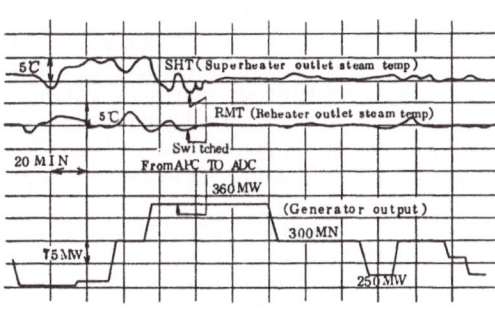

Fig. 10 Comparison of PID control (APC) and optimal control (ADC) in commercial operation

Fig. 9 Comparison of control performance of random MWD change

This control system is called ADC-ARLP system. Fig. 9 shows the control performance of these three control types, that is: (a) PID control only, (b) PID + linear programming, (c) PID + linear programming + AR model (ADC-ARLP). Remarkable improvement of SHT and RHT control can be seen by adoption of advanced control with linear programming and quadratic regulator. Fig. 10 shows a record of the plant in routine operation under the load command from the power system's dispatch center. In the figure the control system is switched from the conventional PID controller to the ADC at the time point shown with arrows. Considerable improvement of the control performance by means of the optimal control is clearly observed in the figure.

5.1.3. <u>500 MW supercritical variable-pressure plant</u>. After a series of elaborate preliminary studies, the ADC system was applied to a 500 MW supercritical variable-pressure plant.

In the case of variable pressure boilers, the energy level stored in boiler varies largely according to generator output level compared with constant pressure boilers. Consequently, temperature control of variable pressure boiler becomes difficult. It turned out not to be sufficient only to apply linear interpolation to cope with time varying characteristics of the dynamics of the boiler. The optimal measures against the difficulty of the steam temperature control of the variable pressure boiler with time varying parameters will be practical Model Reference Control System mentioned in section 6.1.

6. FUTURE OF ADVANCED CONTROL SYSTEM

6.1. MRACS for Temperature Control

Almost all of papers only guarantee the convergence of the solution in such way that the deviation will converge as time goes by, provided the system parameters are constant. To the worse, tremendous labours are required in applying the theory to actual plants. In the case of steam temperature control of thermal power plants, system parameters vary according to output of the generator. Temperature deviation must be kept within allowable limit all the time even when the system parameters are varying. To cope with these difficulties, author has developed a new practical way of Model Reference Adaptive Control System based on computing network.

In this system, adaptive feedforward control signals are synthesized by closed circuit of computing network expressed in the observable canonical form whose time varying parameters are estimated adaptively by least square estimation as quadratic curve. As this system can be realized with less labour than the conventional ones, it is expected to promote the application of the modern control theory to actual plants.

6.2. Learning System

A thermal power station can be kept at any load for a long period to identify parameters under normal operation. The state of start-up, however, can not be maintained for a long period and its dynamics cannot be identified. And it depends on not only the start-up process but also the shutdown process. It is necessary to control life management during start-up considering mainly a turbine metal matching. Besides, reducing start-up time, starting loss and so on are also important during start-up. Usually thermal stress of a turbine can be calculated mathematically but it is troublesome. To make it easier, using the membership function of Fuzzy Theory is effective on controlling start-up schedule. But it is necessary to change automatically membership function by experiences of start-up and shutdown in the past by means of learning system. Proper supply of fuel to a boiler is a key point to shorten start-up time. Time delay of a system response during start-up is much larger than during normal operation and operations in the past have an effect for a long time. Therefore, it is desirable that fuel is supplied properly in advance and boiler is started up as quickly as possible within the range of thermal stresses permitted, learning a boiler characteristics with reference to its dynamics during start-up in the past. In order to reduce the starting loss, overall control of a start-up process must be rationalized. As described above, the development of application of learning system to the start-up process is expected to be realized.

6.3. High Efficiency Operation

One of themes during normal operation of a unit is high efficiency operation in company with an advanced ability of response to load change.

Ways for high efficiency operation are mentioned below:(1) predicting state of a unit by KALMAN filter; i.e., its efficiency can be predicted properly and the most suitable operational state can be searched from the correlation of controllable variables, (2) the most suitable operational state can be searched under the condition that the sum of output of some turbines is constant and also as many turbine as possible operate with fully opened main steam control valves, (3) in the case of sliding pressure boilers, steam pressure during load chang is permitted to be manipulated to some extent so that steam temperature and proper pressure control for better efficiency can be realized utilizing its flexibility, (4) an operation to regulate turbine output in accordance with steam temperature is worth being considered. The development of other methods for high efficiency operation is expected at present because energy cost is increasing.

6.4. Diagnosis

As a design method for failure detection in dynamical systems, A. S. Willsky proposed a generalized likelihood ratio (GLR) approach to detection and estimation of jumps in linear system. Some failure modes may be mathematically modelled as abrupt changes of the system configuration parameters and/or the inputs. In this case the failure characteristics can be well diagnosed by using GLR approach. To overcome the complexity of the GLR algorithm, a method using a hierarchical application of GLR for on-line diagnosis was proposed by Kumamaru.

For the failure modes which cannot be explicitly described by mathematical representations, Kullback Discrimination Information (DKI) can be adopted as the model distortion measure to detect dynamic failures based on input-output modelling.

There are a lot of methods for failure detection as described above and theoretical considerations have been advanced. Problems in practical use of them are considered to be as follows.

a.Time variable system. When deviation from normal states occurs, its influence appears on a lot of detections, so that causes or positions of failure can be detected by catching their changes.

A thermal power unit, however, is considered as a time varying system whose parameters varies as load changes. Since a failure can not be confirmed only by the change of its state, a failure is not to be detected until deviation grows so large that an operator comes to believe that some failures must have happen. Therefore identification of present states is required to be done as precisely as possible.

b. State prediction. State variables in each parts of a plant are mutually connected with equations of mass balance and heat balance. However, these measured variables don't satisfy these equations practically because information gained through detectors has errors. And also the deviation taken by detectors changes with time. Therefore it is necessary to predict present states.

c. Noises. Although a thermal power plant is considered as a closed loop, various kinds of colored noise are added from outside of the plant. These noises circulate in this closed loop. Whitening filter

makes it possible to find real parameters of the plant.
d. Automatic control. Many parts of a plant is considered as various
kinds of closed loop and they are controlled locally. Sometimes it is
difficult to detect the failure of these parts because its failure is
compensated by automatic control. Therefore it is necessary to supervise
not only the change of states but also the action of automatic control.
e. Large scale system. A thermal power station is a large scale system
and sometimes all parameters change at the same time and as the results
the detectable limit of failure becomes bigger. This means difficulty of
failure detection.

It is important for the practical use of a thermal power station
diagnosis to predict the degree of detection errors made from those problems and to set reasonable threshold level that decide whether the
failure is occurring or not. It is also desirable for practical diagnosis on failure detection to have the function that predicts how the
state of failure will develop in the future and that indicates proper
measures against it.

6.5. Expert System

In the case of restoration of an accident at a substation, the damaged
section is isolated from power system and an electricity is supplied
from the other sections that are not damaged through one of any routes
to reach these sections without overload operation of transformers and
transmission lines. Consequently,the failure can be restored without
perfect shutdown of a substation. Since in the case of a thermal power
station, however, it is necessary to consider whether the damaged section can be isolated from the system or not and how the power plant can
maintain its operation, a system should have an ability to do complicated judgement. But it is difficult to formulate it. Since the type of a
failure depends on its position where failure occurred, there are a lot
of measures of the restoration for it. Consequently, it is difficult and
impractical in a large process such as a thermal power plant to input
measures of the restoration for all failures to a computer. Therefore, a
simulation test has been carried out as a demonstration of applying
Expert System to a relatively small integrated system so far.

It is desirable that a new language of computer will be developed
to express easily the logic to decide a measure of the restoration for a
failure. If the computer software algorithm for a new expert system can
not be developed, which can make a production rule by itself when the
computer is given a system diagram, the position and the type of a possible failure, and the way of thinking for the detection of a failure
and the restoration, it is difficult to realize the expert system for a
thermal power plant.

REFERENCE

(1) K. J. Åström, 'Process Control - Past, Present, and Future', 1985, IEEE, August 3/9
(2) J. Rocca, 'Steps to Automation', the ASME Winter Annual Meeting, November 16-20, 1969, Los Angeles, U.S.A.
(3) M. Uchida, N. Kato and others, 'Totally computer Automated Control System in a Thermal Power Unit', IFAC 8th Triennial World Congress, Aug., 1981
(4) N. Karashima, Y. Kogure and others, 'Newly Developed Comprehensive Automation Technique Applied to Hirono Thermal Power Station', American Power Conference 43rd Annual Meeting, Chicago, U.S.A.
(5) K. Kawai and others, 'Operator Friendly Man-Machine System for Computerized Power Plant Automation', IFAC '84 World Congress, Budapest, Hungary, Jul., 1984
(6) M. Uchida, H. Nakamura and K. Kawai, 'Application of Linear Programming to Thermal Power Plant Control', Proceedings of IFAC 8th Triennial World Congress, Kyoto, 1981
(7) H. Nakamura, M. Uchida and others, 'Optimal Control of Thermal Power Plants', '86 ASME Winter Annual Meeting, Dec., 1986
(8) T. Tani, K. Kawai and others, 'Model Reference Adaptive System for Collector Temperature Control of Solar Thermal Electric Power Plant', 5th SICE Symposium on Adaptive Control, January 25-26, 1985, Tokyo
(9) T. Fujiwara and others, 'Development of Controller Tuning System for Thermal Power Plant', Mitsubishi Juko Giho 18, No.2, March 1981
(10) N. Andreiev, 'A New Dimension: A Self-Tuning Controller That Continually Optimizes PID Constants', Control Engineering, August 1981
(11) K. Oxby, K. Kawai and others, 'A Multi-Level Alarm Information Processing System Applied to Thermal Power Plant', IFAC/IFIP/IFORS/IEA Conference on Analysis, Design and Evaluation of Man-Machine Systems, Baden-Baden, Sept., 1982
(12) 'Report of IERE Workshop on Vibration in Rotating Machines', IERE Workshop on Vibration in Rotating Machinery, Bristol, 14-16 June, 1983
(13) N. Kurihara and others, 'A Microcomputer Based Vibration Diagnostic System for Steam Turbines and Turbo-Generators in Power Plants', IEEE Trans. Pwr. App. Syst., Vol. PAS-103, No.6, June, 1984
(14) A. F. Armor, 'On-Line Diagnostic Monitoring in U.S. Fossils Plants', IERE/CEGB Workshop on Vibration in Rotating Machinery, Bristol, England, June 14-16, 1983
(15) M. Uchida, T. Hirosaki, Y. Toyoda and H. Nakamura, 'Practical Model Reference Feedforward Control System of a Boiler by the Computing Network', Transaction of The Institute of Electrical Engineers of Japan, Vol. 108-D, No. 1, 1988
(16) A. S. wilsky and H. L. Jones, 'A Generalized Likelihood Ratio Approach to Detection and Estimation of Jump in Linear System', IEEE Trans. Autom. Control, Vol. AC-21, pp108-12, 1976
(17) K. Kumamaru, T. Soderstrom, S. Sagara and H. Yanagida, 'Fault Detection of Dynamical System Based on Model Discrimination Approach', Report UPTEC, 84-123R, 1/29, 1984

Chapter 10

STATUS REPORT ON REAL-TIME CONTROL IN INDUSTRIES: ELECTRIC POWER SYSTEMS

H. Nakamura
Kyushu Denki-Seizo (Electric Manufacturing) Company
19-18 Shimizu 4-chome
Minami-ku, Fukuoka
815   Japan

ABSTRACT. An overview on the present status of electric power system control is provided from the view point of computer application. The text describes in the first place the features peculiar to the power system on-line control, as well as the control concept. Then the hierarchical control system configuration to meet such requirement, which consists of central, regional, and local control centers, is explained together with the functions of recent control computers. The theory and practice of on-line automatic control in the normal operating condition, such as load frequency control, economic load dispatch control and voltage and reactive power control, are stated followed by the on-line system security assessment and control problems. The text mentions some theoretical background for security control, such as the state estimation and load flow techniques. Future development is briefly referred to in the closing remarks.

1. INTRODUCTION

1.1. Electric Power System

A system consisting of electric power generation plants, transmission lines, substations, distribution lines, and consumers is collectively called an electric power system.

In a large electric power system, the electric power is usually produced by various kinds of power plants, such as hydro, thermal, nuclear, geothermal, solar, etc., which convert raw energy into electric energy – a high quality energy easy to use – by means of generators connected with hydro, steam, or gas turbines. Because of various restrictions, such as the location of the energy resources, the right of way, environmental constraints, etc., power stations are often located far away from load centers.

In order to transmit the electric power effectively from the power stations to the loads and supply it to the consumers which are located in a vast area, high-voltage primary transmission lines, intermediate-voltage secondary transmission lines, and distribution lines are built

S. G. Tzafestas and J. K. Pal (eds.), Real Time Microcomputer Control of Industrial Processes, 337–362.
© 1990 *Kluwer Academic Publishers.*

which also cover a vast-spread areas. In addition, a large number of
substations, such as primary substations, secondary substations, and
distribution substations are installed at suitable points in the power
system to minimize transmission losses and keep the specified voltage at
the customers as well as to ensure enough reliability of operation.

Today, electric power systems are usually interconnected aiming at
the effective use of power resources, the easier operation based on the
power exchange, and the larger power reserves for nominal and emergency
operating conditions. Power pools in America, interconnected grids in
European Continent, England, Japan, etc., are the typical examples of
these interconnected power systems. In this respect, an electric power
system may be said a largest man-made system on earth.

1.2. Features of Power System Control

The main purpose of the electric power system operation and control is
to deliver the consumers high quality electricity at low prices. The
quality of electricity is evaluated by the stability of supply. In
other words, electric utilities are required to supply the consumers
stable, constant-voltage, constant-frequency power at the price as low
as possible.

Since production and consumption of the electric power takes place
simultaneously, it is indispensable for the stable and reliable power
supply to keep at every instant the balance of demand and generation of
the whole power system; the balance in the active power (P) is required
in order to keep the power system frequency within allowable ranges,
while the balance in the reactive power (Q) is necessary to keep the
voltages at the main supply buses within their target ranges.

To meet the above requirement, a large number of components of the
power system, such as the generators, transformers, transmission and
distribution lines, regulators and controllers involved in the system,
are designed to be operated in a consistent manner, so that each of
these components plays its proper role, to realize the expected functions of the integrated power system. This is an outstanding feature
which distinguishes control problems of the electric power systems from
those of other systems.

1.3. The Role of Computers in Power Systems

As mentioned above the target of the power system operation is to maintain reliable power supply to the consumers, while achieving high operating efficiency.

To cope with such requirement in a huge and complex power system, a
supervisory control system is widely adopted in which control computers
play important roles in supervising system conditions, making decisions,
issuing instructions and control signals to the system components, and
in logging the operating data at the electric stations, for a reliable
and effective operation.

Recent rapid advance in data processing and increase of data storage capacity of the computers, together with the improvement of the data
communication speed, have brought forth an innovation in the power

system control. In other words, the hierarchical configuration, which enables us to overcome the problems inherent to a large-scale wide-spread system, has become possible based on such development in computer and communication techniques.

As for the use of the computers, a large number of them are in operation ranging from the super-mini control computers at the central control room to the medium-scale control computers or multi-functional microprocessor systems at the local and district control centers including the remote terminal units in the electric stations.

Although many multi-purpose computers are being used in electric utilities for system planning and operation scheduling, we restrict our discussion in this article to the computers or computer systems concerning the real-time operation and control of the power system.

2. POWER SYSTEM CONTROL CONCEPT

As described in the previous section, the roles or functions of the computers in the power system are roughly divided into two categories: (1) the control in normal operating conditions, such as automatic load-frequency control (LFC), economic load dispatch control (ELDC), voltage and reactive power control (VQC), etc., and (2) the control in abnormal conditions such as the security assessment and control of the system. In the latter are included security monitoring based on the on-line state estimation and load-flow checking, preventive control which prevents the power system from falling into emergency states, and also the system stabilizing control in the emergency state, including switching operation of the transmission lines, load shedding, urgent commissioning of the standby generators, etc.

From the aspect of automatic control the relationship between the above-mentioned control functions can be expressed as shown in Fig. 1. In the figure, the blocks LFC and VQC situated at the lowest control level are operated in completely automatic modes, while the ELDC block in the intermediate level does not directly act to the power system itself, but provides instructions to LFC and VQC blocks.

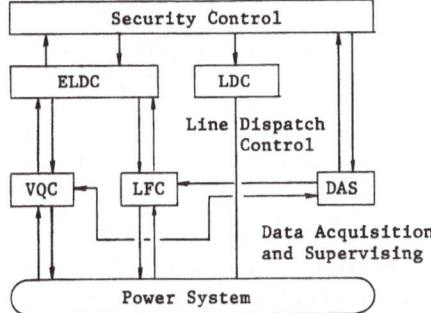

Fig. 1. Conceptual block diagram of
power system control functions

Generally speaking, the LFC block or the VQC block is respectively responsible for keeping the instantaneous balance of demand and supply in active or reactive power, while the ELDC block issues to the generators control signals of their active power outputs so as to minimize the final operating cost of the integrated power system by taking into consideration the power losses in the transmission lines. Thus, the control interval of the ELD is longer, ranging from a few minutes to ten or fifteen minutes, than that of the LFC, which is chosen to be several seconds.

The Security control block situated at the highest level in Fig. 1 is given the highest priority in power system control: it may sometimes require to mend the cost minimum allocation of generator outputs for security purpose, or for the purpose of the system security enhancement against the possible contingencies. The functions of the automatic control under security constraints are mostly performed directly by the large-capacity computer systems in the central dispatch room with the help of the information from regional supervising control centers which operate the system by the instructions from the central dispatch room. As to the configuration of the computer system for power system control, a detailed description will be provided in the next section.

2.1. Automatic Control in Normal Operating Conditions

Load-frequency control (LFC). In order to keep the system frequency and the tie-line load flow within the specified ranges, the control computer always computes the amount of the power generation requirement and issue instructions of generation adjustment to the generators for LFC use in the system.

Economic load dispatch control (ELDC). Considering the power generation cost function of each generating unit in the system, the computer computes most economical allocation of the system load and issues instructions to the generating units so as to achieve the minimum cost operation of the whole system. The constraints on the generator output power, its load following capability, transmission line load flow, and bus voltages are taken into consideration in the computation. Computation is usually performed on the basis of the load prediction of two or three hours ahead of time.

Bus voltage and reactive power control (VQC). Although the frequency of the power system is almost the same at any point of the system, the voltages at the power supply buses vary according to the local conditions of the balance of the demand and supply of the reactive power. Reactive power that influences the bus voltages varies depending upon the properties of the loads connected to the bus, the power flow in the transmission lines to the bus, and other factors. Thus, the bus voltage adjustment at the local bus is performed in the first place with the combination of the transformer's on-load tap changing and the switching of the shunt capacitors or shunt reactors connected to the bus. Such voltage control is carried out by the automatic voltage regulator installed at the substation bus: at the principal electric stations in the power system the control from the central dispatch room is also employed in combination with the local voltage regulators.

2.2. Security Control

A power system remains in its normal state for more than 99 percent of the time. In this state the balance between the real and reactive power demand and supply is carefully kept: as a result, the frequency and bus voltages are kept within their prescribed values.

However, if the margin of power generation falls below some thresh-hold, certain amount of disturbances caused by, for instance, the loss of generation by accidental generator tripping or the loss of transmission lines by lightning or other unpredictable causes, would bring about significant imbalance between the power demand and supply that exceeds the threshhold which sometimes leads the overall system to an abnormal condition.

One of the most important functions required for the computer system is to prevent the system from falling into such abnormal conditions. Should the system fall into abnormal conditions, the control system is required to take actions to restore the normal operating conditions as fast as possible by means of the most appropriate operations performed automatically (if possible), or manually by providing proper guidance to the operators. Such system operation is collectively called "security control". The concept of system security control usually consists of the functions of emergency evaluation, contingency evaluation, emergency control, preventive control, and restorative control. The more detailed description of these functions will be given in the later section.

As mentioned above the computer system installed in the central dispatch room performs most of the control in the normal state and the security evaluation on the basis of the information from the computer systems at regional and local control centers. Although automatic control in the emergency and restorative states has been undertaken with the cooperation of the computers in all hierarchical levels, the present state of the art has not still reached to the level that allows realization of such system.

3. CONTROL SYSTEM CONFIGURATION

3.1. Hierarchical Control System

The modern supervisory control system for a large electric power system usually has hierarchical configuration. It consists of two or three layers with the system's dispatch center at the top, several intermediate dispatch centers or regional dispatch centers in the lower levels. In addition, supervisory control centers are placed for the operation of unmanned substations and hydro-power stations. Most of these dispatch and control centers are equipped with multi-computer systems which are linked with digital communication systems to form an integrated computer network. Such system configuration is indispensable to perform sophisticated functions required for the operation and control of the power system which spreads over a vast service area.

Fig. 2 and Fig. 3 show typical configurations of hierarchical control systems[1]. In Fig. 2 is shown a decentralized three-layer system with one central dispatch room, three regional dispatch centers,

342

and eight local dispatch centers. In this example system, fifty supervisory control centers for unmanned local substations and six control centers for hydro-power stations are also involved for the remote control of these stations.

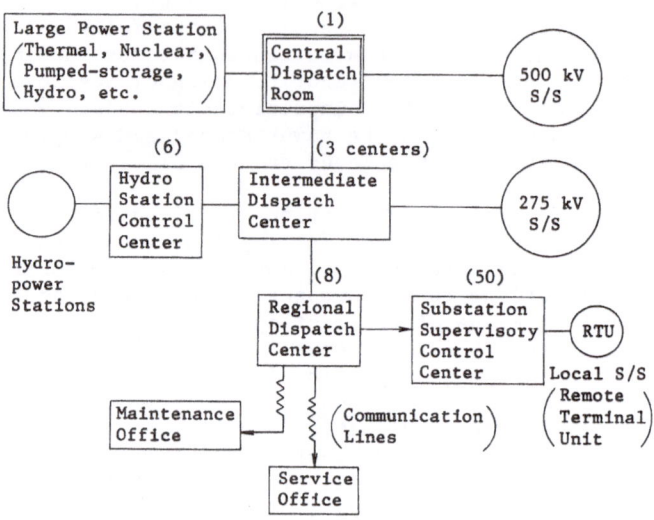

Fig. 2. Decentralized three-layer hierarchical system

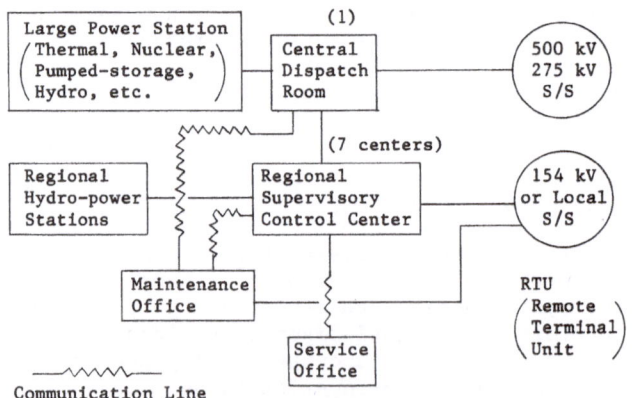

Fig. 3. Centralized two-layer hierarchical system

Fig. 3 shows an example of the two-layer hierarchical control system. In this system the regional dispatch centers and the supervisory control centers in Fig. 2 are put together to form larger-scale regional supervisory control centers which remotely control unmanned local substations and hydraulic power stations.

3.2. Central Dispatch Room

Fig. 4 shows an example of the computer system installed in the central dispatch room which takes charge of supervisory control of the overall power system.

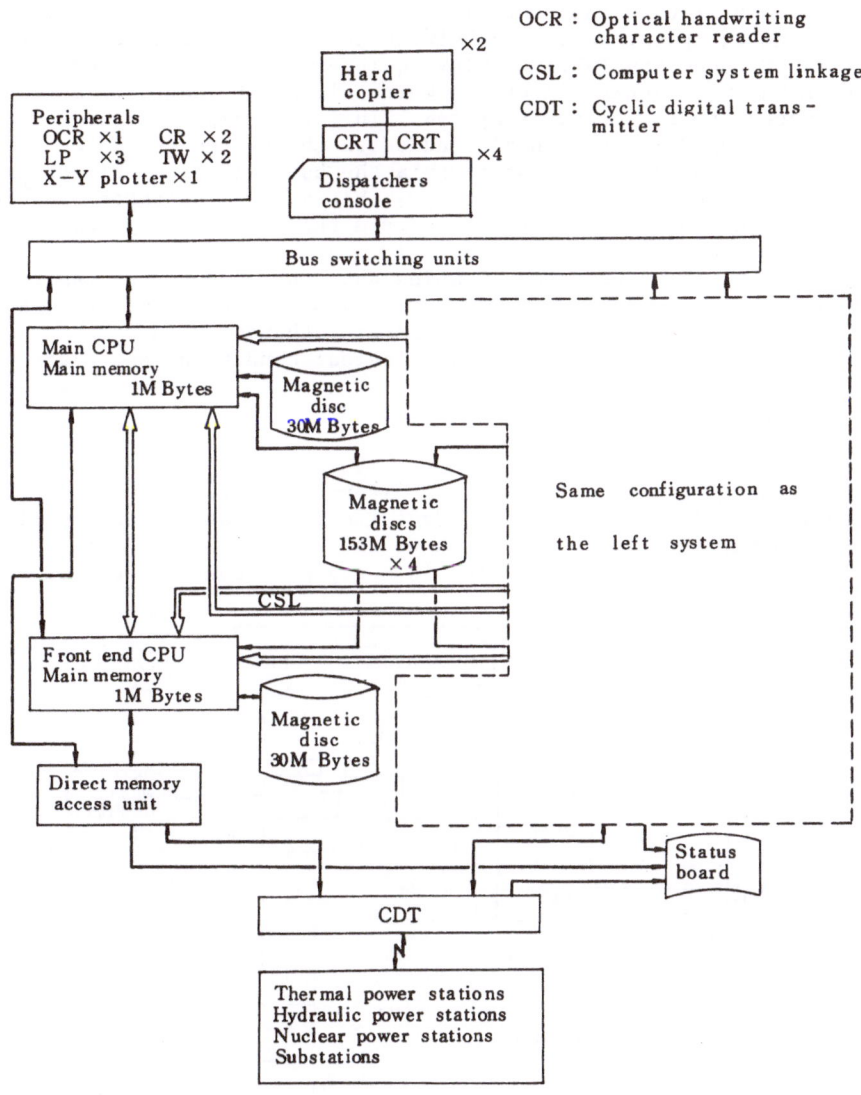

OCR : Optical handwriting character reader

CSL : Computer system linkage

CDT : Cyclic digital transmitter

Fig. 4. Computer system in the central dispatch room

344

As shown in the figure, a duplex system is usually adopted to ensure high reliability and easy maintenance work: sometimes a triplex system is employed in the advanced control room for a large-scale power system. The duplex system shown in Fig. 4 is composed of two main computers and two front-end computers. In normal operation these computers share the task, while in the case when one of the computers happens to fail by accident, its task is automatically taken over by the other without any loss of data and control functions by the magnetic discs and system linkage equipment installed between the central processing units (CPU). CRT (Cathode Ray Tube) display equipment with hard copiers is employed as the man-machine interface which allows the operator an easy and quick access to the computer system. In this example system, Dispatcher's Training Simulator is also added to the system which receives the data from the actual power system via the bus switching unit to make the training a more vivid one. Nowadays, this kind of operator training simulator is becoming a common practice from the aspect of system security improvement[2].

In addition, there are some examples that the computer system in the dispatch center working with weather radars, processes weather data such as the rainfall, atmospheric temperature, thunder clouds and thunderbolts, etc., to grasp the meteorological conditions necessary for the power system operation.

3.3. Regional Control Center

A regional control center covers the territory of several lower level supervisory control centers which control local substations and small-scale hydro-power stations. Fig. 5 shows the block diagram of an example of the hardware configuration in a regional control center.

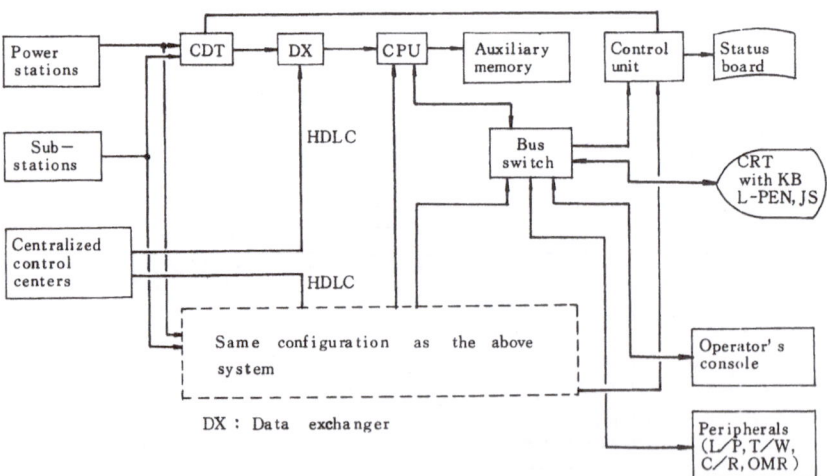

Fig. 5. Block diagram of the hardware configuration in a Regional Control Center

The features of the system shown in the figure are:
(1) a fully duplicated structure to ensure high reliability and also to minimize the influence of the software maintenance on the system functions; (2) a man-machine interface that allows the operator easy access to the computer by means of the operator's consoles which are equipped with high-fidelity color CRT display; (3) High-speed data communication between the central or regional control centers using HDLC (High level Data Link Control).

3.4. Substation Supervisory Control Center

In the case of a three-layer hierarchical system for a large power system, supervisory control centers are placed as previously shown in Fig. 2.
The main tasks of these control centers are the operation of the local unmanned distribution substations and/or hydro-power stations, etc., as well as the monitoring of the status and conditions of local electric stations.
Fig. 6 shows an example of the system configuration of a supervisory control center. The computer system in the supervising control center is similar to that in the upper-layer control center except that in this system local stations to be controlled are connected to the computer system in the control center via TC (Tele-Control) equipment and communication channels.

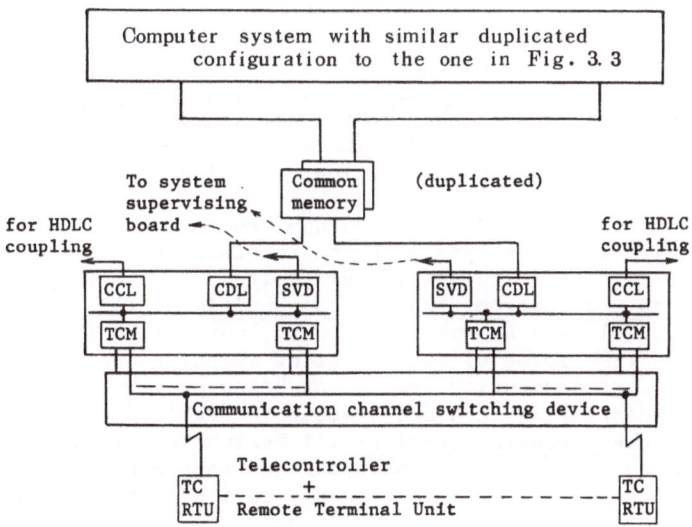

CDL : Computer Interface Unit
TCM : Tele-Controller Master Unit
SUD : Supervisory Display Unit
CCL : Communication Control Unit,   TC: Tele Controller
HDLC: High level Data Link Control Unit

Fig. 6. Configuration of local supervising control center

The system operation is performed in accordance with the instructions from the control center: the monitored information on the local stations is transferred to the upper-layer control center after being processed.

In the computer system of the control center and in each of the RTU (Remote Terminal Unit), microprocessors are employed for pre-processing of the data in order to reduce the load of the main computer and communication channels. Presently, investigations are going on, as the pre-processing means at the site, to develop intelligent functions of the RTU, such as data suppressing or alarm suppressing in the emergency conditions of the power system by providing the monitored information with priority to match the situation. Application of the concept of "Expert System" seems to be quite promising for this purpose.

3.5. Tasks of the Hierarchical Computer System

Main tasks of the hierarchical system described above can be summarized as follows:

Off-line computation: (1) Short term (daily and weekly) scheduling for Economic Load Dispatch (ELD) based on the short-term load prediction, estimation of available hydro power, power exchange schedule mutually agreed upon by the power pool members; (2) Planning of generator commitment considering spinning and standby power reserve, bus voltages and power flow in the transmission lines, etc.

On-line power system control: (1) Frequency and tie-line power control (LFC), Economic load dispatch control (ELDC) based on the short-term load prediction and consideration of the various constraints in the plant operation; (2) Bus voltage and reactive power flow control.

On-line power-line dispatch control: (1) Manipulation of switches (circuit breakers and line switches) in the normal operating condition for the planned maintenance works or temporary switch operation to meet the unexpected change in the load condition; (2) Line dispatch and/or operation guidance to the operator in the abnormal state such as the switch operation to restore the normal frequency, to remove the overloading of the transmission lines or apparatus, or to ensure continuous power supply to the customers against system faults, etc.

Supervising and recording of the system conditions: (1) Supervising of the frequency, voltages at specified buses, power flow in the transmission lines and transformers, the amount of spinning and standby power reserves; (2) System security assessment based on these measurement; (3) Recording and displaying such measurements on CRT's (Cathode Ray Tube).

Assessment of the operation results: (1) Evaluation of the operation results based on the statistics on power generation and load, power exchange calculation and rate adjustment, review of the operation results.

3.6. Modern Control Computer

Today, 32-bit super-minicomputers are widely adopted as the control computer in the on-line real-time SCADA (Supervisory Control And Data Acquisition) system for power system control and operation scheduling.

The functions required for these computers are high reliability, high
speed in data processing, fast response in data access and output,
flexibility in expanding the system, the ease and simplicity in programming, maintenance, and operation.

From the standpoint of computer facility, such requirements may be
summarized as follows: (1) to realize high capability and high processing speed which make possible large-scale numerical computations or
simulations, while ensuring quick real-time response characteristics;
(2) to offer the user an architecture and the operating and support
systems which allow easy programming; (3) to hold high reliability,
compactness, and low installation cost which are the primary requirement
to a control computer.

In order to meet such requirement, control computers for SCADA
system are furnished with the following features:
(1) Architecture: use of farmwares, adoption of decentralized configuration of processor groups for Input/Output control and communication
control, employment of multi-processor system and virtual storage system
for the parallel processing of large-scale computation;
(2) Processor and main memory: realization of high-speed large memory-
size hardware by the adoption of CMOS (Complementary Metal Oxicide Semiconductor) type elements and high-speed, high density, low-power VLSI;
reduction of processing time by means of the techniques such as cache
memory (buffer memory), memory interleave for multiprogramming, pipeline type parallel computation control which are common practice in
general purpose computers.

For instance, a recent CPU (Central Processing Unit) for the SCADA
system consisting of four ACP (Arithmetic Control Processor), with the
processing speed of 5.6 MIPS (Million Instructions Per Second) per each
ACP, has the overall processing speed of approximately 18 MIPS, which is
a figure competent to a large capacity general purpose computer.

4. THEORY AND PRACTICE OF ON-LINE AUTOMATIC CONTROL

4.1. Load Frequency Control (LFC) of Interconnected System

Most of the power systems today are interconnected with neighboring
areas via tie lines. The aim of interconnecting several power systems
to form a large power pool is to take full advantage of mutual assistance.

The operation of the power pool from the standpoint of LFC is performed in the following way:
(1) Under steady operating conditions, each member of the power pool is
responsible to balance the active power demand and supply of its own
area, except the scheduled portion of the power exchange.
(2) For a large load change, all the members of the power pool cooperate
during the load change so as to keep the frequency change minimum with
the stored kinetic energy and the generators' governor action involved
in the whole power pool: when the transient period has died out, the
LFC set of the power pool member in which the load change took place
detects the amount of load change, i.e., the amount of the power generation adjustment necessary to restore the frequency and the scheduled

tie-line power, and issues control signals to the generators of its own area.

The most common control strategy in use today for the LFC of an interconnected power system is the one called TBC[3] (Tie-Line Bias Control). Now, let us consider the concept of the TBC system. Suppose a load increase ΔL[MW], in a power system caused the frequency drop ΔF[Hz]. Then the relationship between ΔL and ΔF is expressed as

$$\Delta F = -\Delta L / K, \tag{4.1}$$

where K is a constant called "load-frequency characteristic constant" or "load-frequency constant" in short. K is the constant peculiar to the system which is dependent on the kinetic energy of the rotating machines in the system, increased generation by the governor action of the generators, and the decrease of system load due to the frequency drop.

In Fig. 7 is shown a two-area interconnected power system consisting of system A and system B: each of them has the load-frequency constant K_A and K_B, respectively.

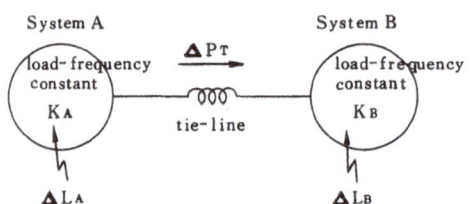

· frequency deviation

$$\Delta F = -\frac{\Delta L_A + \Delta P_T}{K_A} \qquad \Delta F = -\frac{\Delta L_B - \Delta P_T}{K_B}$$

Fig. 7. Frequency deviation in the two-area interconnected system

Suppose load increases of ΔL_A and ΔL_B took place in each of system A and system B simultaneously and after the transient state they finally caused the frequency drop ΔF and the deviation of the tie-line power ΔP_T from the scheduled value.

Then, the relationship between the load increase and the frequency drop is expressed as

$$\Delta F = -\frac{\Delta L_A + \Delta P_T}{K_A} = -\frac{\Delta L_B - \Delta P_T}{K_B} = -\frac{\Delta L_A + \Delta L_B}{K} \tag{4.2}$$

Accordingly,

$$\Delta L_A = -K_A \, \Delta F - \Delta P_T$$

$$\Delta L_B = -K_B \, \Delta F + \Delta P_T \tag{4.3}$$

Equ.(4.3) shows that, if K_A and K_B are known and ΔP_T and ΔF are measured, the amount of the load increase ΔL_A and ΔL_B can be calculated. ΔL_A (or ΔL_B) is called area control error in system A (or system B).

By eliminating ΔF, ΔP_T, ΔL_A, ΔL_B from Equ.(4.2) and Equ.(4.3), we obtain the following relationship

$$K = K_A + K_B, \qquad (4.4)$$

which means that the load-frequency constant of the inter-connected system is equal to the sum of the load-frequency constants of the member systems. The above discussion can be easily extended to the N-area system. If we denote the area control error in the i-th system by $(ACE)_i$, i=1,2,...,N, then the equations of the N-area system corresponding to Equ.(4.3) and Equ.(4.4) are written as

$$(ACE)_i = -K_i \, \Delta F + \sum_{\substack{j=1 \\ j \neq i}}^{N} \Delta P_{Tj} \qquad (4.5)$$

$$K = \sum_{j=1}^{N} K_j, \qquad (4.6)$$

where ΔP_{Tj} denotes the deviation of the power from the scheduled value on the tie-line connecting system i and system j. Since $(ACE)_i$, i=1,2,...,N in Equ.(4.5) means the amount of the load change in the i-th member of the interconnected power system, if each pool member adjusts its own power generation to cancel its ACE, then the power pool will finally recover the original operating condition.

In the TBC (Tie-Line Bias Control) system, one of the pool members, usually the largest one situated nearly in the center of the pool, is in charge of maintaining the specified frequency (Flat Frequency Control), while the rest of the pool members control their generation so as to cancel their own ACE's (TBC). Actually, the computer in the central dispatch room of each pool member calculates the ACE on the basis of ΔF, Ki, and the sum of ΔP_{Tj} and issues control signals to the generators of its own.

4.2. Economic Load Dispatch Control (ELDC)

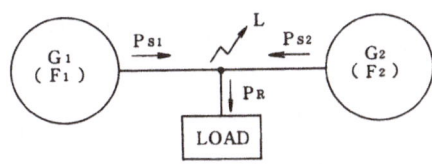

Psi, Ps2 : generator outputs
PR : load
L : transmission loss
F1, F2 : fuel cost of the generators

To get the general idea of the problem, let us consider, as the simplest case, the ELDC of a two-generator (two-thermal power plant) system shown in Fig. 8.

If we denote the generator output power by P_{S1}, P_{S2}, the load by P_R, the power loss in the transmission line by L, and the fuel cost of the generators by F_1, F_2, then the following relationships apply between them:

Fig. 8. Two-generator system model

$$P_R + L - (P_{S1} + P_{S2}) = 0 \qquad (4.7)$$

$$F = F_1 + F_2 = f_1(P_{S1}) + f_2(P_{S2}). \qquad (4.8)$$

The problem is how to obtain P_{S1} and P_{S2} which minimize the total fuel cost F in Equ.(4.8) under the constraint of Equ.(4.7), the balance of power demand and supply.

Here, we define a function C described below:

$$C = F_1 + F_2 + \quad P_R + L - (P_{S1} + P_{S2}) \qquad (4.9)$$

By differenciating Equ.(4.9) partially in terms of P_{S1} and P_{S2} and putting the results zero, we obtain the following equations:

$$\frac{\partial C}{\partial P_{S1}} = \frac{dF_1}{dP_{S1}} - \lambda(1 - \frac{\partial L}{\partial P_{S1}}) = 0$$

$$\frac{\partial C}{\partial P_{S2}} = \frac{dF_2}{dP_{S2}} - \lambda(1 - \frac{\partial L}{\partial P_{S2}}) = 0 \qquad (4.10)$$

From Equ.(4.10), the optimal load allocation that makes the total fuel cost F minimum is given as follows:

$$= \frac{dF_1}{dP_{S1}} \frac{1}{1 - \partial L/\partial P_{S1}} = \frac{dF_2}{dP_{S2}} \frac{1}{1 - \partial L/\partial P_{S2}} \qquad (4.11)$$

In Equ.(4.11), dF_i/dP_{S_i} (i=1,2) means the rate of the fuel cost increase of the i-th generator against the increase of its power output: in this sense, dF_i/dP_{S_i} is called the "incremental fuel cost" of the i-th generator. The coefficient $1/(1-\partial L/\partial P_{S_i})$ in Equ.(4.11), which is always larger than 1.0, is the factor that modifies the incremental cost dF_i/dP_{S_i} by taking into account the transmission line loss L: it is called "loss penalty factor" or simply "penalty factor". By extending the above two-machine case to the general N-bus system, we obtain the following equations which correspond to Equ.(4.7) and Equ.(4.8).

$$\sum_{k=1}^{N} P_{Rk} + L - \sum_{k=1}^{N} P_{Sk} = 0 \qquad (4.12)$$

$$F = \sum_{k=1}^{N} F_k = \sum_{k=1}^{N} f_k(P_{Sk}), \qquad (4.13)$$

where P_{Rk} denotes the load at the k-th bus, P_{Sk} the power generation at the k-th bus, L the sum of the transmission loss in the whole system, and F_k, which is the function of P_{Sk} denotes the generation cost at the k-th bus. In the above discussion, we assume that $\sum P_{Rk}$ is the sum of the loads in the system to be allocated by ELDC to the generators for LFC: here, $\sum P_{Rk}$ is obtained by subtracting, from the total system load, the base load component which is usually produced by nuclear plants and other uncontrollable hydro, geothermal power plants. In the similar way to that in the two-machine system, the optimal dispatch formula for the N-bus system is obtained as follows:

$$\lambda = \frac{dF_1}{dP_{S1}} \frac{1}{1 - \partial L/\partial P_{S1}} = \cdots = \frac{dF_n}{dP_{Sn}} \frac{1}{1 - \partial L/\partial P_{Sn}} \qquad (4.14)$$

The value of λ is determined so as to satisfy Equ.(4.12). In the actual system, it is the problem how to determine the penalty factors in Equ.(4.14) since they vary depending upon the operating conditions of the power system. One systematic approach for determining the penalty factors is given in the reference [4] at the end of this article.

Basically, ELDC is performed by the large-capacity computer installed in the central dispatch room: the computer computes the optimal allocation of the load to the generators and issues them control signals for generation adjustment via communication networks.

λ in Equ.(4.14) is determined to satisfy the balance of power demand and generation by considering the transmission loss; the incremental fuel cost and the penalty factor at each generation bus are the function of the power generation. Therefore, the computation of the optimal load allocation satisfying Equ.(4.14) and Equ.(4.14) requires an iterative procedure which starts under certain assumed initial values. In the actual ELDC, the upper and lower limit of each generator output power and its load-following capability must be taken into account: in addition, in thermal power plants there sometimes exist discontinuity in the generation increase or decrease due to burner switching, feedwater pump commissioning, etc. The ELDC in the actual power system performs a sophisticated tasks by considering such various factors. This is the reason that the ELDC requires a large-capacity high-speed computer.

4.3. Voltage and Reactive Power Control

For the stable power system operation, control of the bus voltage is important from the following aspects:
(1) Appropriate and effective operation of the apparatus involved in the power system, (2) Reduction of power loss in the transmission line by means of reasonable reactive power distribution, (3) Enhancement of the power system stability by maintaining the bus voltages within specified ranges: this is especially important when the system is subjected to large disturbances like accidental tripping of generators or loss of transmission lines.

Just like the system frequency fluctuations depend mostly on the balance of the active power, the bus voltage is dependent on the balance of the reactive power. However, different from the case of the system frequency, which can be assumed almost the same over the whole power system, the bus voltage presents quite strong locality: for example, the imbalance of the reactive power at a certain bus directly influences its voltage. Thus, the problem of the bus voltage control can be reduced to the problem of reactive power adjustment at each bus. In this sense, the bus voltage control is often referred to as "VQC", which is the abbreviation of "Voltage and Q (reactive power) Control".

The principal sources of the reactive power are the generators with excitation control, shunt capacitors or reactors installed at the substations. On-load tap changing of the power transformers at substations is also effectively employed for the purpose of voltage regulation. Recently, the static voltage regulator, consisting of the combination of a shunt capacitor and a non-linear shunt reactor utilizing ferroresonance effect, has been attracting attention because of its continuous reactive power regulating property.

As a rule there are two types of VQC: one is the local VQC and the other is the VQC run by the instruction from the central dispatch room.

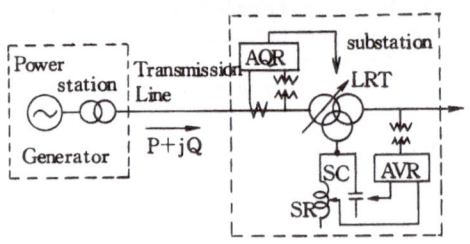

Fig.9 shows an example of the local VQC, in which the bus voltage is regulated with AVR (Automatic Voltage Regulator) by manipulating shunt capacitors or shunt reactors, while the reactive power flow is adjusted with AQR (Automatic Q Regulator) by changing the transformer tap.

SC : Shunt capacitor
SR : Shunt reactor
AQR : Automatic Q Regulator
AVR : Automatic Voltage Regulator
LRT : On—load Tapchanging Transformer

Fig. 9.
Local voltage regulator

In a large power system having a great number of generation and load buses, it is sometimes a problem to hold proper coordination between the actions of the local controllers: thus, the control computer in the central dispatch center supervises the bus voltages and the reactive power flow in the whole system and issues control signals to the local VQC equipment, so as to avoid their competing or mutual interactions. Fig. 10 shows the block diagram presenting the flow of the supervising and controlling signals between the dispatch room computer and the local VQC equipment.

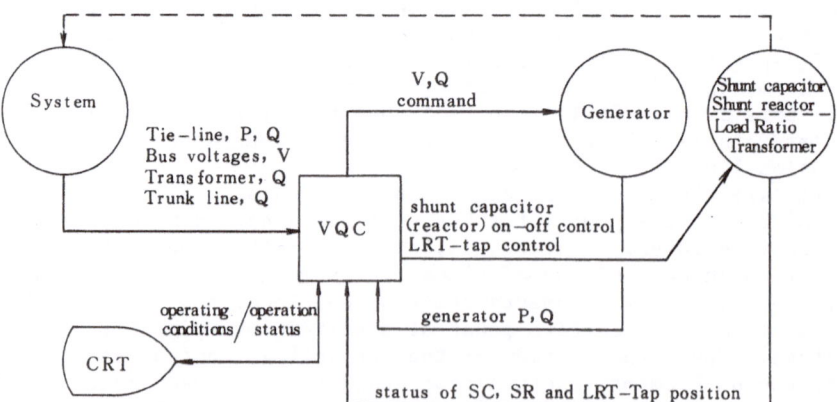

Fig. 10.   Integrated voltage and reactive power control system

As for the centralized VQC, there are several control strategies: one of them is shown in Fig. 11. (5)

In this control algorithm, the control computer in the central dispatch room computes a value called "Decision Functions", which is employed as a measure to assess the extent of the deviations of the bus voltages from their specified values: Decision Function (DF) is defined as

$$DF = \sum_{j=1}^{N} (V_j - V_{oj})^2 / B_{vj}^2,$$

$$(4.15)$$

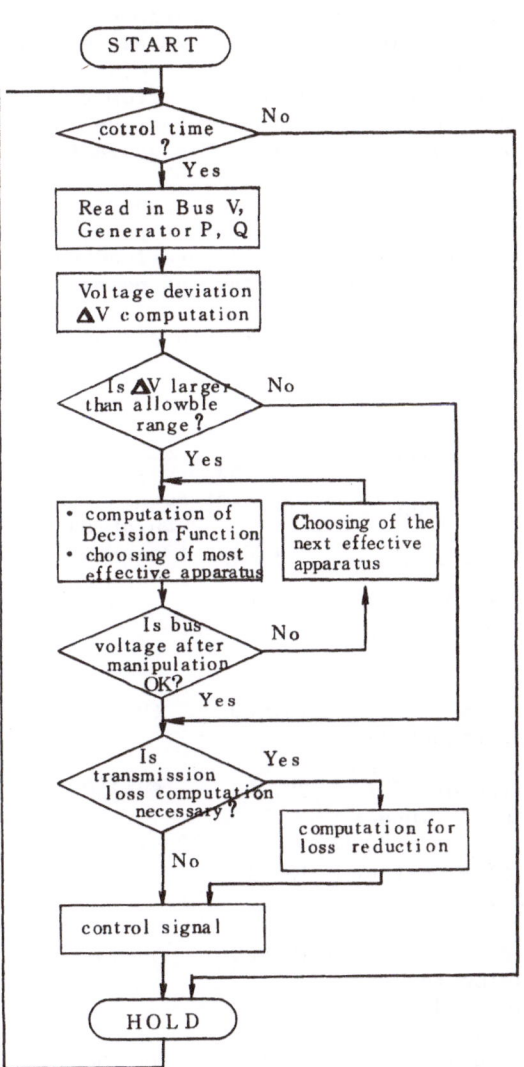

where the suffix j denotes the bus number in question and N is the number of the controlled buses, and V_j = measured bus voltage, V_{oj} = nominal value of the bus voltage, B_{vj} = allowable range of the bus voltage.

The apparatus whose manipulation reduces DF the most is selected as the first candidate to be manipulated. In order to calculate DF, we perform load flow computation by providing the node injection of the reactive power Q, which corresponds to the unit manipulation of the Q adjusting equipment: computation is carried out for every node one after another, for instance, for the k-th node by providing Q_k at node k: based on the result, the estimated DF's are compared to determine the candidate node whose manipulation gives the maximum DF. The above procedure continues repeatedly until the DF and the voltages at the specified buses become less than the pre-determined threshold values. The actual control is performed step by step by repeating the computation, manipulation, and checking of the result at each step.

Fig. 11. VQC based on Decision Function

5. SECURITY CONTROL IN ABNORMAL OPERATING CONDITIONS

5.1. Large-Scale Outage of Electric Power System

A great number of components in the electric power system, i.e., insulators on the transmission lines, transformers, circuit breakers, line switches, lightning arresters, etc., are exposed to the severe natural environment, such as thunder storms, heavy rainfalls or snow-falls, strong winds, scorching or freezing atmospheric conditions, which often cause component failures directly or indirectly as a result of the deterioration of the apparatus or equipment.

Usually, the faulted components are immediately cut off from the system with circuit breakers which are energized by the protective relaying system: thus the normal operation of the system is guaranteed except for the localized faulted portion.

However, it may happen that some sorts of faults cause abnormal voltage dips or frequency variations of overall system, or stepping out of the synchronous generators, which result in successive overloading of the apparatus or transmission lines, leading to the outage of power supply over a vast service area.

The table below shows such kinds of large-scale blackouts experienced in the world in these ten years or so.

Examples of large-scale black-out

Place	Date	Primary Cause	Suspension of Power (MW)	Duration (Hour–Min.)
New York	July 13 1977	Succession of lightnings in multi-areas	5,800	25H
Quevec, Canada	Sept. 20 1977	Failure of a potential transformer	10,000	6H 30M
France	Dec. 19 1978	Overloading of transmission lines	29,000	8H 30M
Quevec, Canada	Dec. 14 1982	Failure of a current transformer	15,240	7H
Western U.S.A.	Dec. 22 1982	Falling of a transmission tower by strong wind	12,400	3H 30M
Sweden	Dec. 27 1983	Line switch arm falling off due to superheating	11,400	5H 20M

It is observed from this table that, under certain unfavorable
system conditions, even a simple failure of an apparatus or equipment
can be a trigger to cause the overloading of transmission lines or
transformers, loss of system stability, voltage and frequency disturbances, which might extend to a large-scale blackout.

Therefore, in order to avoid such disasterous state of the power
system operation, it is important to analyze the sequence, the cause and
consequence, of the failure, from its beginning to the suspension of the
power supply over a large areas, and establish proper countermeasures
which should be flexible enough to meet the ever-changing conditions of
the objective system.

According to the analysis of the large blackouts so far experienced, the principal cause and consequence of the failures are summarized as shown in Fig. 12.

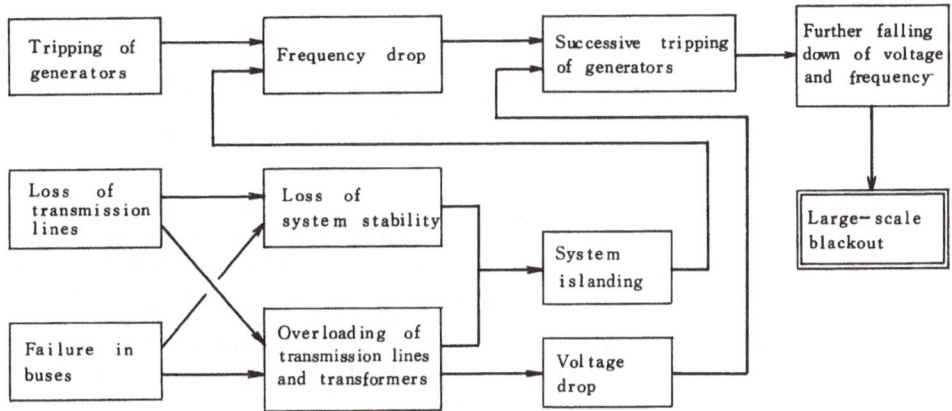

Fig. 12. Cause and consequence of large-scale blackout

5.2. Concept of Power System Security Control

The fundamentals for maintaining system security is to construct, in the
system planning stage, a system which is strong enough against possible
faults. However, too much investment should be avoided from the economical standpoint. Thus, the enhancement of the system security by the
help of computer control combined with the establishment of the proper
protective relaying system becomes indispensable. As a matter of fact,
the function of security monitoring and control is one of the most
important roles of the computer system in most modern power system
control centers.

The concept of "security control of the power system" are originally advocated and discussed systematically by T.E. Dy Liacco in 1967,
more than twenty years ago, right after the large-scale blackout which
hit the north-eastern America in 1965. Today's power system security
control systems are, more or less, planned, designed, and implemented
based on this concept.

According to Dy Liacco[6] power system may be operated in three different modes, or states, depending on its condition; i.e., the preventive state, emergency state, and restorative state, as defined below (See Fig. 13):

(1) Preventive state: This state is also referred to as "normal state", where the system holds enough margin of active and reactive power sources with no overloaded apparatus or transmission lines in the overall system. In the normal state, the control computers in the central control room periodically checks the security level: in other words, it checks whether the intact system holds enough operating margin based upon the data transmitted from the regional or local control centers. This function is called "security evaluation". State estimation technique to be described in the later section is a powerful tool for this purpose.

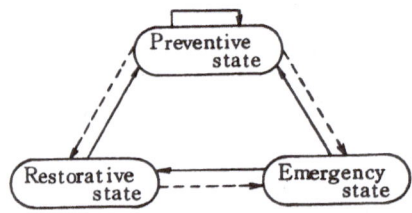

Fig. 13
Operating modes or states of power system

If it is anticipated that the system will possibly run into abnormal conditions due to some kinds of contingencies, the computer issues an alarm signal and indicates the operator appropriate preventive measures to be taken to restore the operating margin by enhancing the security of the system. Such functions are referred to as "emergency evaluation", "security monitoring", and "preventive control". Fast load-flow computation technique is effectively employed for this purpose.

(2) Emergency state: If unexpected heavy load or successive contingencies take place in such a case when, for some reasons, the preventive actions fail to ensure the system enough margin of active and reactive power sources, the power system on occasion enters "emergency state". In this state, abnormal frequency deviation and bus-voltage drops are experienced over the whole system because of the imbalance of demand and supply in active and reactive power: overloading of some system components are also seen. As shown in the previous Fig. 12, a diagram illustrating the cause and consequence of a large-scale blackout, unless some appropriate countermeasures, such as urgent commissioning of the standby generators, or load shedding in the worst case, are initiated to bring the system back to the normal state, the system may go into extremis state, i.e., the total blackout.

Presently, power system operation in the emergency state are performed not automatically by the instructions from the computers, but manually by the human operator who makes decisions and operations on the basis of the information or the guidance delivered from the computer system.

(3) Restorative state: If large-scale blackout of the power system takes place by misfortune as a result of a chain of unfavorable events,

the actions to restore the system must be taken in the most effective
way. The state in which the blacked-out system is under restoration
process is called "restorative state". In this state, the tripped generators are resynchronized to the local system, and cut-off loads are
picked up, then the isolated local systems are tied one by one to form
the original power system configuration. As in the case of emergency
state, the system operations in the restorative state are still carried
out mostly by human operators supported by the information or guidance
from the computer system.

5.3. Real-Time Network Security Analysis[7]

As described above, the operations to maintain the power system security
in the abnormal state are still performed by human operators. However,
the decision making of the operator largely depends upon the information
or guidance provided by the computer. Besides such roles in the abnormal conditions, the computer system in the normal operating condition,
regularly monitors the conditions of the system on the basis of the
information from the local control centers and assesses its security
margin preparing for the possible preventive control. In this sense,
the functions of on-line real-time network security analysis are the key
factor that determines the reliability, availability, and serviceability
of the computer control system from the security point of view.

Fig. 14 is the flow diagram expressing the procedure of the security analysis. The function of each steps in Fig. 14 is as follows:

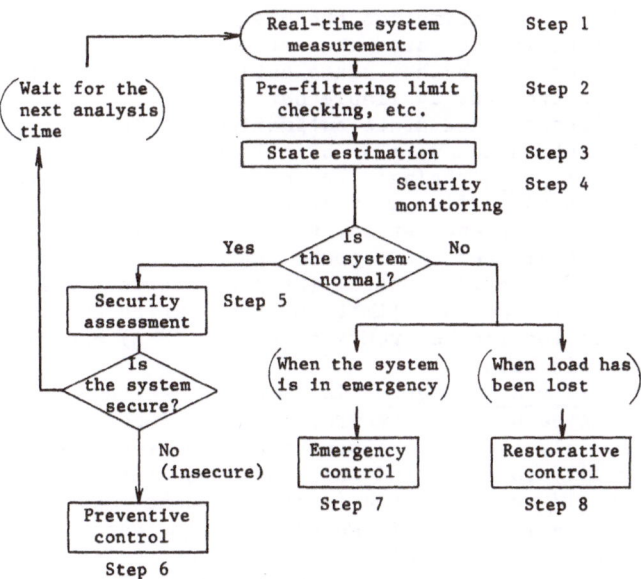

Fig. 14. Flow diagram of security analysis

Step 1. Real-time system measurement: Measured data form location are telemetered to the control center computer: the measurement consists of line power flows, line current flows, bus power injections, bus voltage magnitudes, the on-off status of circuit breakers and line switches, etc.

Step 2. Filtering: In the control center computer, the telemetered data are checked on the basis of their reasonability and consistency; obviously abnormal data due to meter failure or other reasons are eliminated in this stage.

Step 3. State estimation: State estimation consists of two stages, namely, determination of system topology, and the generalized state estimation. State estimation is defined to be a mathematical procedure for computing the best estimate of the state variable data which are corrupted with errors.

The noise or error sources which contaminate the telemetered data are failures in metering or telemetry, errors in the measuring instrumentation, noises in the communication system, delays in the transmission of data, non-simultaneous measuring, etc. It sometimes occurs that the Kirchhoff's First Law does not hold at a certain bus; in other words, the sum of the incoming and outgoing active or reactive power measured at the bus is not equal to zero. Such fact necessitates the estimation of system variables with highest likelihood, based on which the computation for system analysis to follow is carried out.

Bus voltage magnitudes and phase angles measured from the reference bus at normal steady state are chosen as the state variables to be estimated, since the necessary active and reactive power flows in the transmission lines can be computed using these variables.

Prior to the state estimation the observability analysis, bad data detection and identification are necessary: the observability analysis is necessary to find out whether state estimation of the objective system is possible from the available data. The above two functions are included in the "generalized state estimation":. The concept and the method for state estimation will be briefly discussed later.

Step 4. Security monitoring: Based upon the real-time system measurement, this step identifies whether the system is normal or not. If the system is in an emergency state, the analysis for Emergency Control (Step 7) is initiated; when it is found that some loads have been lost, the analysis for Restorative Control (Step 8) starts. The normality checking is also performed by "limit checking" which checks whether the specified transmission line currents or powers (MVA) are well under the prescribed threshold values.

Step 5. Security assessment: In order to assess whether a normal operating state is secure or insecure, a set of important and possible disturbances are selected. The system security is defined in terms of the system's ability to withstand such contingencies: in other words, whether danger would appear in the event of a next contingency is identified in this analysis. In this sense, Security assessment is also referred to as "Contingency evaluation". For the selection of the contingencies, dispatcher's experience and judgement are employed: application of expert system seems to be effective for this purpose. The system's capability to withstand the proposed set of contingencies is assessed from the result of Load Flow analysis by "limit checking" described before: fast load flow computation technique is employed in the analysis. In the present practice, only the static effects of contingency are taken into account, only considering on the steady state after the transient dies out.

Step 6. Preventive control: When, as a result of security assessment, the system is judged to be insequre, i.e., there is at least one contingency which can cause an emergency, preventive control analysis determines the corrective actions to be taken in order to cancel a danger which would appear in the system in the event of next contingency. As mentioned above, the principal objective of the preventive control is to avoid a cascade trip i.e., the subsequent cascade trip to one or a few lines, which would happen if a contingency security is not met. The application of "On-lien Optimal Power Flow" technique to be described later seems promissing for this analysis.

Step 7. Emergency control: The proper corrective actions to make the system normal is obtained as the output of this analysis. Theoretically the static problem might be solved in a centralized manner by "Optimal Power Flow Techniques" and by "changes in network topology": the former is ready for use, while the later, which aims, as one of the corrective actions, to find the proper change in network configuration by switch operation, seems to be still far from implementation. As to the transient emergency control, it is still at a research stage.

Step 8. Restorative control: Presently the restorative control is completely carried out by human operator, and will remain in the same situation for a long time, because a rather quick response is required to meet properly to a lot of possible various system conditions. A knowledge-based expert system architecture is expected to make a breakthrough in this field.

5.4. State Estimation

From the above discussion it is apparent, the state estimation technique provides the basis to most of the security analyses. Accordingly, most of the modern power system control centers are furnished with this function. The principle and application of the static state estimation will be briefly given below.

In the static state estimation, the variation of the state variables along the time passage is not taken into account. Thus, we assume the following relationship:

$$\underline{Z} = \underline{h}(\underline{x}) + \underline{V}, \tag{5.1}$$

where, $\underline{Z}$ denotes the set of measurement, $\underline{x}$ the vector of state variable, $\underline{h}$ the mathematical relation of $\underline{x}$ to the measured variables, and $\underline{V}$ the vector of measurement errors. The estimate of the state vector $\underline{x}$ is obtained by minimizing the so-called weighted least square function as

$$\underline{J}(x) = (\underline{Z} - \underline{h}(\underline{x}))'\underline{W}(\underline{Z} - \underline{h}(\underline{x})), \tag{5.2}$$

where, ' denotes matrix transpose and $\underline{W}$ is (m x m) positive definite diagonal matrix called weighting matrix: each element of $\underline{W}$ is chosen to be the reciprocal of the error variance, expressing the uncertainty of the measurement. The estimate of the state $\hat{\underline{x}}$ which minimize $J(\underline{x})$ is obtained from the following equation, which is obtained by differentiating Equ.(5.2) in terms of $\underline{x}$, and putting the results zero:

$$\underline{H}'(\hat{\underline{x}})\underline{W}(\underline{Z} - \underline{h}(\hat{\underline{x}})) = 0, \tag{5.3}$$

where, $\underline{H}(\hat{\underline{x}}) = \underline{h}(\underline{x})/\underline{x}\big|_{\underline{x} = \hat{\underline{x}}}$.

$\underline{H}(\cdot)$ is called Jacobian matrix.

Iterative computation method called Newton Gauss method is often used to solve Equ.(5.3), in which the following relation is used:

$$\hat{\underline{x}}_{i+1} = \hat{\underline{x}}_i + (\underline{H}'(\hat{\underline{x}}_i)\underline{W}\ \underline{H}(\hat{\underline{x}}_i))^{-1}\ \underline{H}'(\hat{\underline{x}}_i)\underline{W}(\underline{Z} - \underline{h}(\hat{\underline{x}}_i)) \tag{5.4}$$

Other methods such as the one using non-linear programming, or the simplified one employing linearized approximation

$$\underline{h}(\hat{\underline{x}}) = \underline{H}\ \hat{\underline{x}} \quad \text{in} \quad \hat{\underline{x}} = (\underline{H}'\ \underline{W}\ \underline{H})^{-1}\ \underline{H}\ \underline{W}\ \underline{Z} . \tag{5.5}$$

In the process for solving Equations (5.4) and (5.5), the computation of inverse matrix $(\underline{H}'\ \underline{W}\ \underline{H})^{-1}$ is the problem. Since the dimension of $\underline{H}$ is quite large, say as large as several hundreds or several thousands, various methods have been considered to save computation time: they are, (1) the method utilizing the sparsity of matrix $\underline{H}$, (2) the one to solve the problem after reducting the matrix size by system decoupling, (3) the one using an algorithm that does not necessitate inverse matrix computation, etc. For the detection of system topology change or bad data included in the measurement, various approaches have been suggested. Among them, the methods that utilize the residual defined by $\underline{r} = \underline{Z} - \underline{h}(\hat{\underline{x}})$ are most common: $\underline{r}$ is obtained as a result of the static state estimation explained above. In the method the transmitted data are checked one by one, and the measurement which causes dominant decrease of $\underline{r}$, if exists, is picked up as a bad one.

5.5. Load Flow Computation[8][9]

The steady-state power flows in the transmission lines or transformers connecting the systems with various classes of voltages are mathematically obtained by solving a set of nonlinear equations called power flow (or load flow) equations, which describe the power (active and reactive) balance of the system.

In the power flow equations power injections or voltage magnitudes at the buses are specified: then, the magnitudes and phase angles of buses, and consequently the line power flows, that satisfy the specified conditions are obtained as a solution. Since the power flow equations are nonlinear, iterative procedure is usually employed to get the solution: for instance, in Newton-Raphson method, an equation with the form of $\underline{A}\ \underline{x} = \underline{y}$ is derived for iterative computation, where A is a $(2n \times 2n)$ matrix called Jacobian, n the number of nodes, $\underline{x}$ a 2n vector consisting of voltage-magnitude and phase-angle errors, y a vector consisting of active and reactive power mismatches.

In the above equation, Jacobian matrix, which is obtained based on the bus-frame admittance matrix, is a very large sparse matrix with large number of zero elements. Accordingly, efficient time-saving algorithm for solving large sparse matrix, such as optimal ordering of row and columns of matrix A, triangular factorization, etc., have been developed.

Optimal Power Flow (OPF),[10] one of the extension of general Power Flow, seems to be a powerful tool for security control.

OPF minimizes an objective function under the constraints which are set up from security consideration. Operation cost, or the sum of load sheds if feasible solution does not exist, is taken as the objective function. Constraints include security in the intact system and under contingency. The results of OPF are generators' active power, voltage magnitudes at voltage magnitudes at reactive generation buses, tap position of LRT, the amount of load shedding, which minimize the objective function and meet the constraints. In particular, these results define the actions to be applied to meet security.

The concept of on-line OPF is stated below with an example of its application to static preventive control problem: the principle is to express the change of constrained variable after trip versus the changes of control variables before trip. For instance, we compare I_L^K, the current in line K after the trip of line K, with I_L and I_K before trip: then, we compute the corrections on the control variables to bring I_L^K within its allowable limits. Similarly, we define V_i^K, the voltage magnitude at the generator bus i, when a generator unit tripped at bus K: then, we compute the corrective actions on the control variables to bring V_i^K within its limit. The principle is general and valuable for any type of constraint. The practical implementation is similar to the emergency control.

6. CLOSING REMARKS

In this chapter, an overview of the electric power system control was provided. As described in the text, recent 32-bit super minicomputers are installed in most power system control centers as the kernel of the integrated control system. However, in the peripheral devices composing the system, a large number of microprocessors are in use. Multi-microprocessor based tele-control and tele-metering systems or communication controllers in the control centers, remote terminal units in the substation, supervisory control system for distribution line switch control are some examples of these kinds. Another important field of the microprocessor application in power system is the protective relaying system. With its various advantageous functions over the conventional hardware-based equipment, the microprocessor-based equipment has been replacing the conventional one. The appearance and recent progress of one-tip digital signal processors will promote this tendency. From these respects it can be said that power system control and protection will not function without the application of microprocessors.

As to the state of the art in the power system control engineering, the application of Expert System seems quite promising among many techniques. In CIGRE, an international organization on power system problems, established a study committee on expert system for power system analysis and techniques in 1986, which consists of the experts from all over the world. According to the present authors opinion, Expert system will be quite useful in the field of on-line power system security assessment and control. In other words, the application of Expert system is expected to offer a breakthrough in this field.

REFERENCES

(1) T. Sugiyama and others: 'Hierarchical Computer Control for Power System', IFAC 8TH TRIENNIAL WORLD CONGRESS, August 24--28, 1981 Kyoto, CS-2.1 CS-34--CS-44, (1981).
(2) T.E. Dy Liacco: 'A Dual-Purpose Simulator for Power System Operator Training and for Performance Testing of Control Computer Systems', ibid, 95.3 p.xx-49--xx-53, (1981).
(3) N. Cohn: Control of Generation and Power Flow on Interconnected Systems, John Wiley & Sons, New York, (1971).
(4) Olle I. Elgerd: Electric Energy Systems Theory, An Introduction (Second Edition), McGraw-Hill Series in Electrical Engineering, (1982).
(5) K. Kumai, and K. Ode: 'Power System voltage control by using a process control computer', IEEE Trans. Pwr. App. Syst., Vol. PAS-87, pp.1985--1990, (Dec. 1968).
(6) T.E. Dy Liacco: 'The Adaptive Reliability Control System', IEEE Trans. Pwr. App. Syst., Vol. PAS-86, No. 5, pp.517--531, (May 1967)
(7) Felix F. Wu, and A. Monticelli: 'Recent Progress in Real-Time Network Security Analysis', IFAC SYMPOSIUM ON POWER SYSTEMS & POWER PLANT CONTROL, PLENARY PAPERS 2, p.10--16, Aug. 12--15, 1986, BEIJING CHINA.
(8) Y. Wallach: Calculations & Programs for Power System Networks, Prentice-Hall, New Jersey, (1986).
(9) Alan George, and Joseph W. Liu: Computer Solution of Large Sparse Positive Definite Systems, Prentice-Hall series in Computational Mathematics, (1981).
(10) J. Carpentier, and G. Cotto: 'Modern Concept for Security Control in Electric Power System', IFAC Symposium on CONTROL APPLICATIONS FOR POWER SYSTEM SECURITY, 102-01 FLORENCE 26--28 Sept., (1983).

Chapter 11

STATUS REPORT ON REAL TIME CONTROL IN STEEL INDUSTRY

Dr. K. SAITO
Omika Works, Hitachi, Ltd.
2-1, Omika-cho 5-chome,
Hitachi-shi, Ibaraki-ken
319-12 Japan

ABSTRACT. The steel industry was the first introducer of a process computer system in which real time control was adopted. With the development of the microcomputer thereafter, it was made possible to operate and control directly, through a digital controller using a microcomputer, motors, electromagnetic valves, etc. which drive the process.

The controller or the control unit which consists of the micro-computer is different from the conventional controller which is hard-wired in it. The control logics of the former are made by software which makes it very flexible and reliable. The control system used in the steel industry is normally featured by its hierarchical configuration which is composed of a few process computers that control the overall process and of many microcomputers.

This article mainly relates to the configuration of this control system, using a real time microcomputer and its control effects.

In particular, the technical development of the microcomputer has made it possible to control, with a very rapid sampling pitch, the speed of DC motors and AC motors through Direct Digital Control using a micro-computer. This technique has been successfully put into practice in Japan since around 1980, and now has been put into practical use also in the U. S. A. and in Europe.

It may safely be said that the steel industry, whose growth has been in parallel with the progress of microcomputer, has a technical potential which allows it to pull ahead of other industries not only from the standpoints of productivity, reliability, maintainability, etc. but also in all other respects.

The modern control theory which had hardly been applied can now form the subject of ordinary applications thanks to the recent rapid development of the microcomputer. This situation will be explained with examples. We have also added an explanation of knowledge engineering which has been partially put into practical use in the steel industry, though it still remains in its early stage.

S. G. Tzafestas and J. K. Pal (eds.), Real Time Microcomputer Control of Industrial Processes, 363–395.
© 1990 *Kluwer Academic Publishers.*

1. Introduction

Computer control systems in the steel industry are now applied to all
control systems which control and manage the entire steel plant. Aided
by the remarkable progress of the microprocessor's basic techniques, the
multi-microprocessor system the distributed system, the centralizer
multiplex system and their complex systems have come to be employed.

On the other hand, with the development of the microcomputer, it
was made possible to operate and control directly, through a digital
controller using a microcomputer, motors and other devices which drive
the process.

2. Control System Configuration

In order to facilitate a full understanding of real time control using a
microcomputer applied in the steel industry, we will first explain its
historical evolution of development and then add required explanations
consecurively along with concrete examples of applications.

2.1. The Historical Evolution of Development of Control System
Configurations

With the control elements changing, the system configuration has also
undergone changes. Specifically transistors came into use as control
elements around 1960, enabling highly responsive and accurate control
systems of the hard wired type to be manufactured. About 1965, transistors were replaced by integrated circuits(IC). Moreover, the advent of
microcomputers using large scale integrated circuits(LSI) has changed
the control systems step by step.

Since 1965, computers have had their reliability improved and
highly reliable systems can now be expected. In place of the conventional analog or digital wired logic control systems, the computer-aided
direct digital control(DDC) system has come to be applied. For process
control, however, the microcomputer has begun to be used instead of the
setup control or DDC process computer.

Given in Figure 1. is the transition of rolling mill control system
designs.[1] The history of computerization of rolling mill control
systems may be classified roughly into three periods:
1st period(1965-1970): Process computer & wired logic age
2nd period(1975-1979): DDC age
3rd period(1975-1979): Microcomputer age-(1)
         (1979-   ): Microcomputer age-(2)
In Figure 1., the control systems can be classified into the setup
computer level, control operation signal level, power control level and
sensor level.

At the operational control level, the Automatic Positioning Control
(APC), the Automatic Gauge Control (AGC), etc., belong to the arithmetic
operation control level.

The sequence control and manual control logics are usually processed by means of a programmable logic controller. The power control
level comprises the control of DC or AC motors with thyristors, control
of induction motors with a motor control center(MCC) and control of

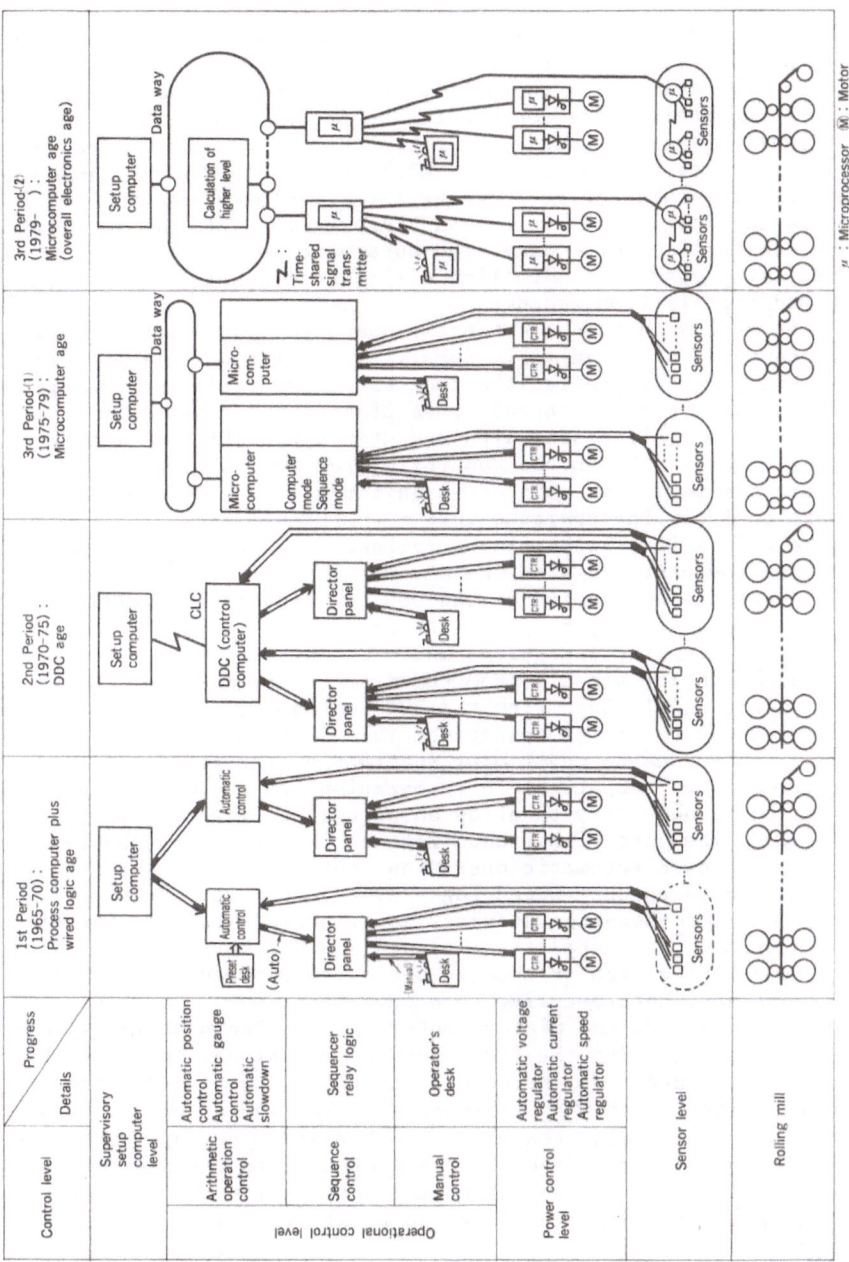

Figure 1. Transition of Mill Control Systems

magnet valves. The sensor level covers detectors such as the rolling mill machine position sensors and steel material sensors, etc.

In the 1st period, the control system is linked with the computer through the automatic operation panel. The process computer at the setup computer level prepares the rolling pass schedule and sets it on the automatic operation panel, based on information relative to steel ingots obtained from the computer at a higher level, information actually measured in the rolling mill line, and information obtained by tracking throughout the rolling mill line. The automatic operation panel is used to automatically operate the rolling mill in accordance with the pass schedule and signals from sensors.

In this period, both the automatic operation panel and control panel are composed of wired logic comprising semiconductors and relays. Therefore the period is called the process computer plus wired logic age. During that period, iron and steel plants were progressively computerized on an overall basis, with computer applications expanded to the rolling process. Though the control system for rolling mills was of wired logic, the period is of great significance in that iron and steel plants were entirely incorporated into a computer system.

In the 2nd period, automatic operations were programmable as software and were computerized with the advent of small size control computers. This period is designated as the DDC age because the rolling mill was directly controlled by computer software. In this period, the DDC mainly covered automatic operations while manual operations were performed through control signals from the desk and by means of the wired logic on control panel. Even with the DDC shut down, rolling could somehow be carried out by manual operation. This is because a single computer was inevitably used to cover many sections of the rolling line due to the expensiveness of computers. In preparation for a computer shut down, the minimum logic necessary to allow an automatic operation is provided as wired logic. The most significant feature in the period, however, is a remarkably improved automatic operation control performance as compared with the conventional wired logic type automatic system.

The arithmetic operation accuracy was improved, as a matter of course. For example, advanced feed-forward control was materialized for the first time by the high arithmetic operation accuracy and highly flexible functions, both peculiar to a computer.

In the 3rd period, the microcomputer age has begun at last. This period can be divided into the first and second halves. In the first half, the microcomputer was introduced and the system was designed on a separate basis and linked with a high-speed time shared signal transmitter. In the second half, the microcomputer had been applied in the form of a microprocessor, rather than a computer, to the desk or power control and even to the concentration and rationality check of sensor signals. In this age, processors are separately located all over the control system.

In the first half of the 3rd period, the microcomputer was introduced. Every section of the rolling equipment is provided with a microcomputer which processes both the manual and automatic operation control signals. The microcomputer is required to effectively process not only the arithmetic operation control but also sequence control which, conventionally, have been processed on the control panel. As a result, a microcomputer with both computer and sequencer modes has been born. Many microcomputers are separately located and the high-speed time-shared signal transmitter (data freeway) is used for intercomputer linkage and for linkage with the setup computer. With the computers decentralized, however the system came to be so designed that control signals could be easily centralized by means of a time-shared signal transmitter.

On the other hand, the intercomputer linkage allows a trouble-shooting computer to monitor the conditions of the microcomputers. In addition, such a complicated control arithmetic operation that cannot be processed by a microcomputer is performed by a larger control computer after collecting data from the microcomputers. The controlled output is then transmitted to the microcomputers.

Thus, a system can be designed so that microprocessors and large to medium and small size computers will mutually allocate the task of processing according to the process level.

In the second half of the 3rd period, microcomputerization is being accelerated further and even minor loop controls on the thyristor Leonard panel are being digitalized in the image of a microprocesser, rather than a microcomputer. In addition, the control desk signals are also becoming highly semiconductor-electronics-oriented so that micro-computers are employed to centralize the signals in the control desk and to make rationality and failsafe checks. Moreover, there is a tendency toward introducing microcomputers into the sensor groups for the ration-ality and failsafe checks. The electronics and microcomputers employed in each control block allows an easy interlock linkage with the time-shared signal transmitter on a 1 : 1 basis.

In other words, the present time may be called an age of overall electronification in the sense that the control systems consist entirely of electronics and microcomputers. The time-shared signal transmitters which have been substantially introduced have effectively greatly reduced the control cable volume and installation costs.

2.2. Control Systems Configuration

The distributed microprocessors are linked through the DFW (Data Free Way) bus linkage and STU (Serial Signal Transmission Unit). The linkage and location of microprocessors are as illustrated in Figure 2.

The minor loop controls can be microcomputerized efficiently by making boundaries with the power control such as the thyristor Leonard unit which is small in size and lower in voltage by using hybrid ICs. This will allow automatic optimum minor loop gain compensation, all of which have been unavailable in conventional analog circuits.

2.2.1. Multi-Micorprocessor Distribution System[2]. The advantages of the multi-microprocessor distribution system can be summarized in five points.
(1) Distributed parallel processing which avoids the drop in response which is otherwise linked in a large size system, thereby securing a quick response.

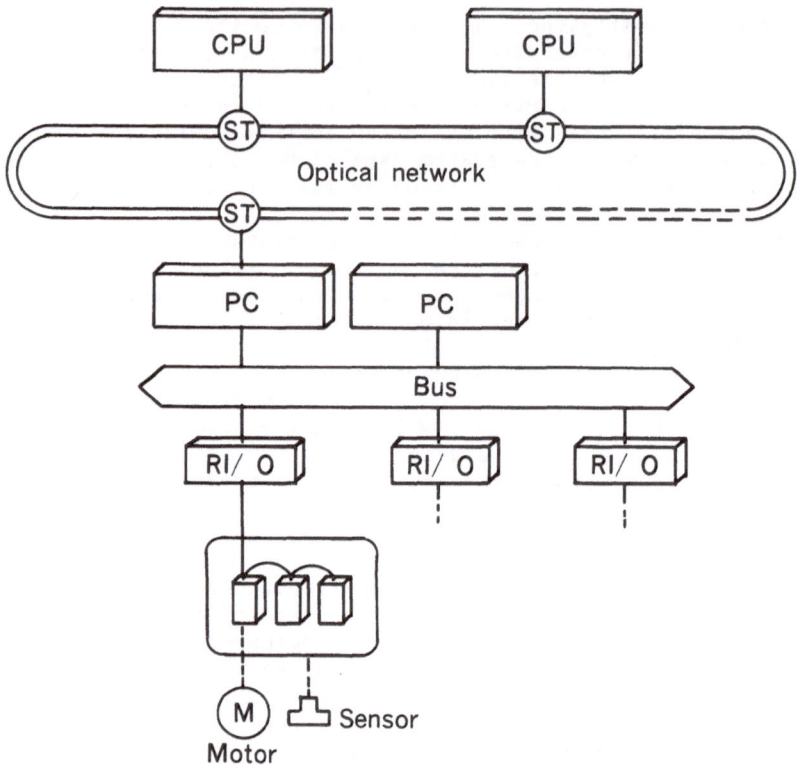

Note
   CPU : Central processing unit
   ST   : Station
   PC   : Programmable controller
   RI/O : Remote process input/output device

Figure 2. Typical Composite System in a Steel Plant

(2) Even when the system is abnormal, the error can be localized, thereby preventing the entire system from stopping, facilitating the build up of a distributed system, and securing dependable quality.
(3) System expansion is facilitated because influence over the existing section is minimized by connecting an additional microprocessor to the network.
(4) Adoption of microprocessors which have an excellent performance/cost ratio permits the build up of a large scale system at a low cost.
(5) Adoption of a universal microprocessor permits the effective utilization of commercially available software, thereby improving software productivity.

2.3. Example of an Actual Application

An actual application of microcomputers in the rolling mill will be
explained by taking up instances of hot rolling mills, cold rolling
mills, and processing lines.

2.3.1. Microcomputer System Configuration of Hot Strip Mill[2]. Figure
3. shows an application of microcomputers to the hot strip mill. As the
setup computer, a large-size computer is provided to make setup calculations and information processing. Signals from this large computer are
given to the microcomputer through the data freeway. These signals are
information relative to the line equipment settings, such as screwdown
opening, sideguide opening and rolling mill stand speed based on the
schedule calculation results.

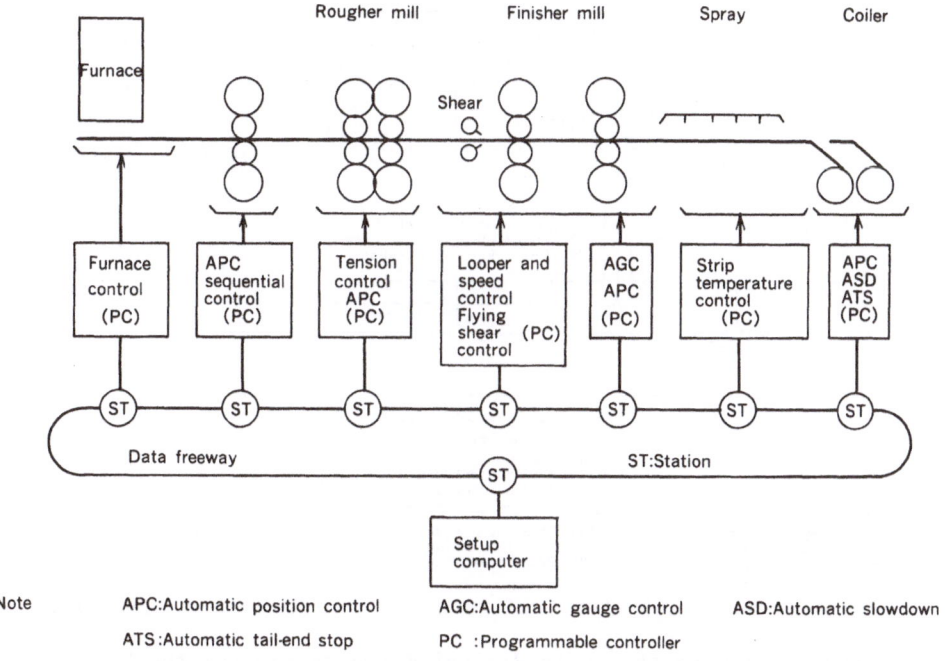

Figure 3. Typical System Configuration for Hot Strip Mill

The microcomputer has the functions of automatic gauge control,
automatic tension control, looper and speed control and flying shear
control as well as sequence control.
This microcomputer is designed to give commands directly to the
power electronics or contactors for the direct drive of motors, magnet
valves and equipment.

2.3.2. Microcomputer System Configuration of Cold Mill[3]. Figure 4.

370

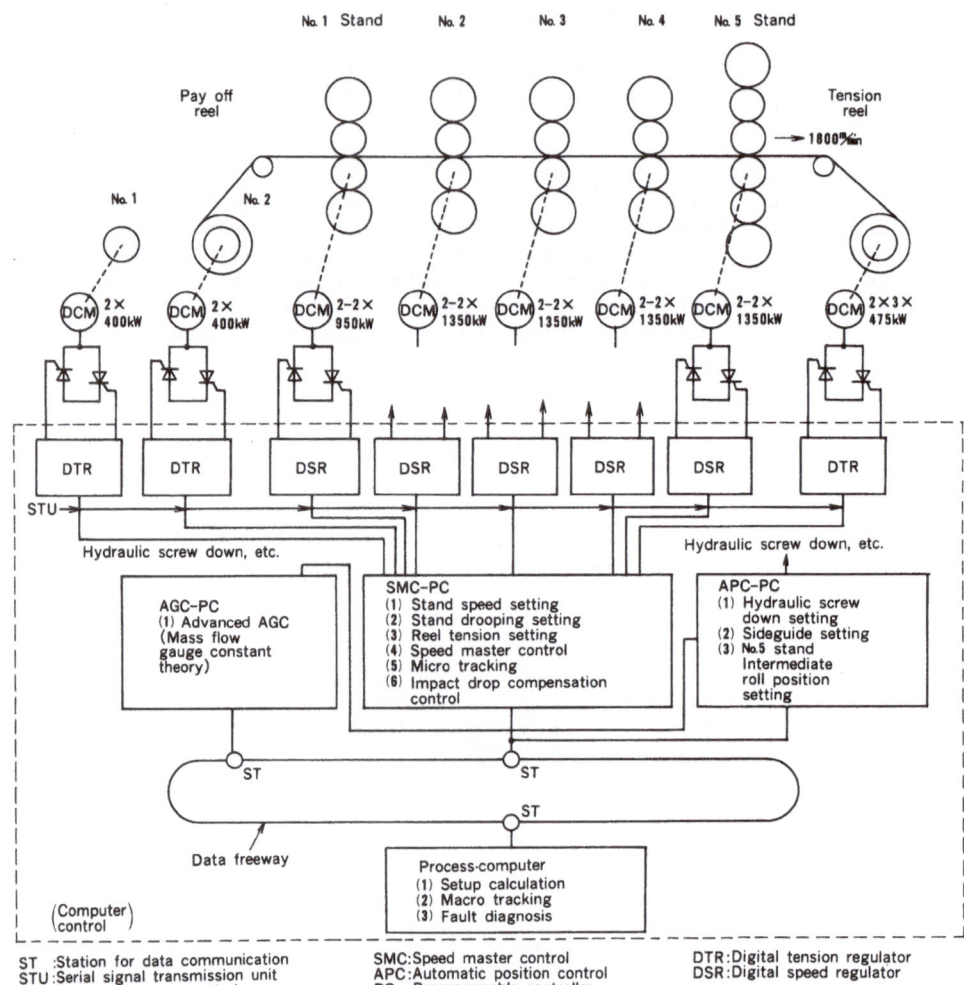

Figure 4. Control System Configuration of Five-Tandem Cold Mill

is a block diagram of the fully digital drive system for the 5-tandem cold mill which employs digital speed regulators. The hardware configuration of the speed and tension control systems is composed of a programmable controller (PC) for speed master control (SMC). A speed command is given to the PC from the process computer through a data freeway. Subject to this command value, the necessary reference speed and tension values are given to the digital speed and current regulator at each stand through the STU according to the rolling schedule. To generate these references, the SMC directs the PC to carry out the programs for stand speed rheostat (SSRH) and master rheostat (MRH) operation, vernier speed and successive speed compensation, etc. The STU has a transmission rate of 128 points/2.4 ms. Between the SMC and digital regulator units, consisting of digital controllers, signal information can be easily transmitted by the STU which can permit 32 to 128 signals to be transmitted on one pair of twisted wires. Thus, the amount of conventional external control cables in use can be decreased by about 30%.

In addition, the speed dropping characteristics can be given to each stand speed regulator as a function of current and speed by the PC. A variety of information (e.g. voltage, current and speed data, etc.) on the drive control system can be transmitted to the host process computer. Based on this information, trouble-shooting is easily performed. All of these data are transmitted by a multiplex signal transmitter through the STU between the PC and digital speed regulator and through the data freeway between a PC and a process computer.

2.3.3. Microcomputer System Configuration of Processing Line[4]. Figure 5. shows the processing line system configuration. This processing line anneals thin steel plate continuously. The DC motor to drive the line is controlled by digital regulators, using a microprocessor. The programmable controllers are employed in key control functions, such as the tension control of strips, speed pattern, tension pattern, acceleration/deceleration, compensation, reel control, etc., to carry out digital computations in our efforts to increase accuracy.

3. Microprocess or Based Fully Digitalized Speed Control for Motor Drives[5]

Practical application of fully-digital speed control equipment to power electronics motor drive systems using microprocessors has been pursued, concentrating on iron and steel making plants with the aims of higher accuracy, higher performance, and improved reliability, among other things. Since fully-digital speed control equipment was put on the market in October, 1981, by Hitachi, Ltd. in Japan it has manufactured and delivered a lot of all-digital speed control equipment to the iron and steel making industry, or cold tandem mills, hot tandem mills, processing lines, etc., and has greatly enhanced plant operation efficiency, product quality and yield rate. This paper deals mainly with the effects obtained when applying a digital drive system to an iron and steel manufacturing plant. It also describes new techniques, such as new control algorithms aimed at higher performance and reliability, a new linkage system by optical cable and digitalization of AC variable speed control systems.

3.1. Background and Application Techniques

3.1.1. Background. The variable speed control system for motors in steel manufacturing plants is an important part of the plant operation and has significant effects on the product. In a plant where products are manufactured using several variable speed control systems, it must be kept in mind that, while respective accuracies and responses are important, good coordination between different machines is more

372

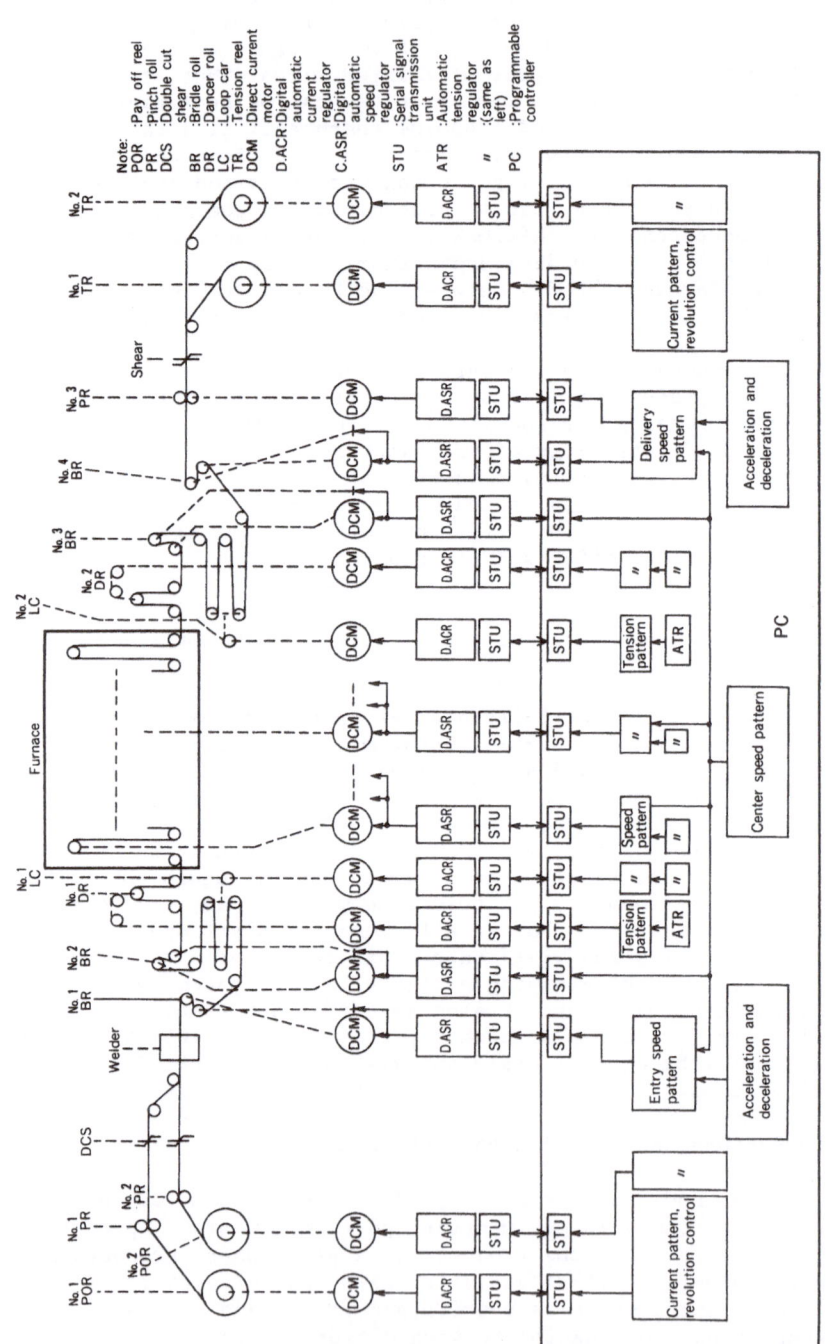

Note:
POR :Pay off reel
PR :Pinch roll
DCS :Double cut shear
BR :Bridle roll
DR :Dancer roll
LC :Loop car
TR :Tension reel
DCM :Direct current motor
D.ACR:Digital automatic current regulator
C.ASR :Digital automatic speed regulator
STU :Serial signal transmission unit
ATR :Automatic tension regulator
″ :(same as left)
PC :Programmable controller

Figure 5. Simplified Driving and Control System

important from the viewpoint of improving operation efficiency, yield rate and product quality. From such a viewpoint, various improvements were carried out.

3.1.2. Application Techniques. A high degree of accuracy, as shown in TABLE I[5], has been attained and great improvements have been made in product quality, economic operation, etc. through industrial applications of highly sophisticated digital speed control facilities. For example, excellent speed accuracy and trouble diagnosis, which were regarded as difficult on analog speed control equipment, can now be available economically without any decrease in response. Digital techniques provide similar effects when renewing facilities in which only the control section needs to be replaced without renewing the power equipment (motors, thyristor convertors and etc.).

TABLE I   Comparison of Performance Between Analog and Digital Control

Main performances (measured values) are compared between analog and digital controls.

Item		Analog control	Digital control
Speed detector error	At 100% speed	±0.3%	±0.014%
	At 1% speed	±0.3%	±0.0088%
Overall accuracy		±0.7%	±0.02%

Note: Percentage is with respect to rated value.

3.1.3. Effects of All-Digital Control System. Figure 6. shows the requirements of the plant operation in tandem mills, and the performance of digital speed control equipment satisfying these requirements. A particularly good example is the all-digital control system of the tandem cold rolling plan, which has led to the following. (1) Improvement of the operation efficiency through reduction in the number of strip breakages during strip threading, threading time and acceleration/deceleration time. (2) Improvement of the yield rate and product quality by the reduction of the offgauge length during threading and checking of the strip surface for scratches at very low speed. (3) Improvement of maintenance and check-ups through decreased down time, by the improved diagnosis and monitoring and no need of drift adjustment. The speed matching characteristics between stands of the Leonard equipment in a tandem rolling mill has been noticeably improved. Therefore, disturbance of the AGC (automatic gauge control) has been reduced.

374

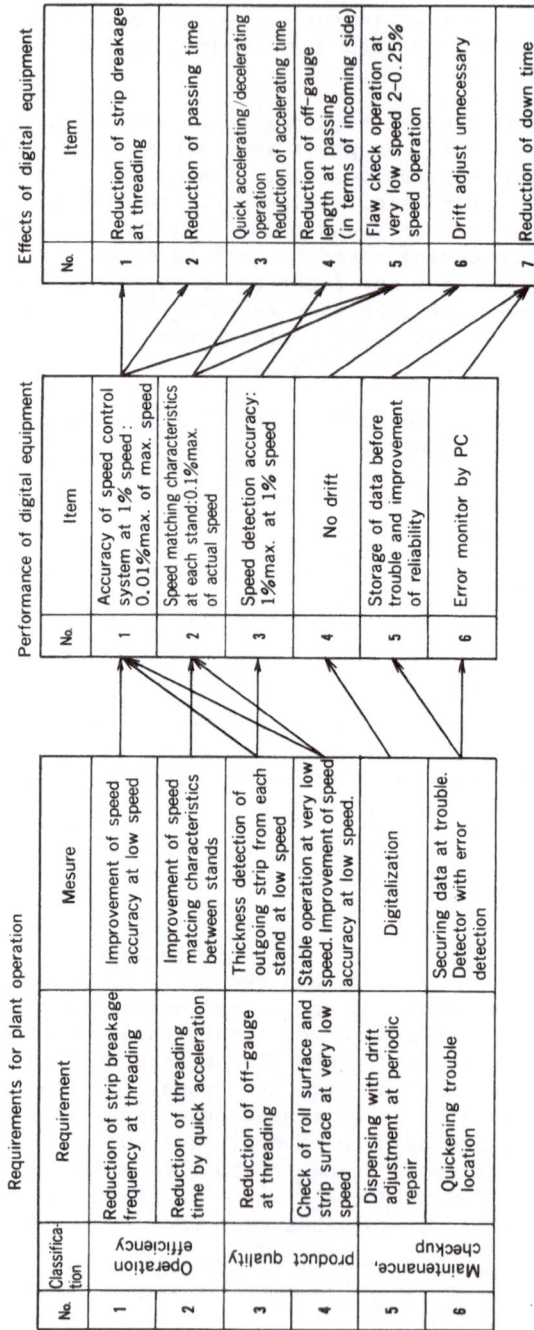

Figure 6. Effects of Digital Speed Control Device

Figure 7. illustrates the low level of disturbance and indicates that tension between stands is very stable at the time of acceleration, even when the AGC or ATC (automatic tension control) are overridden.

Figure 8. is an oscillogram indicating the superior speed matching characteristics of the digital Leonard equipment. Its operation is excellent with a speed deviation between stands of ±0.1% or even less when accelerating.

The digitalization techniques stated above are suitable for not only new facilities but also can be used when replacing facilities.

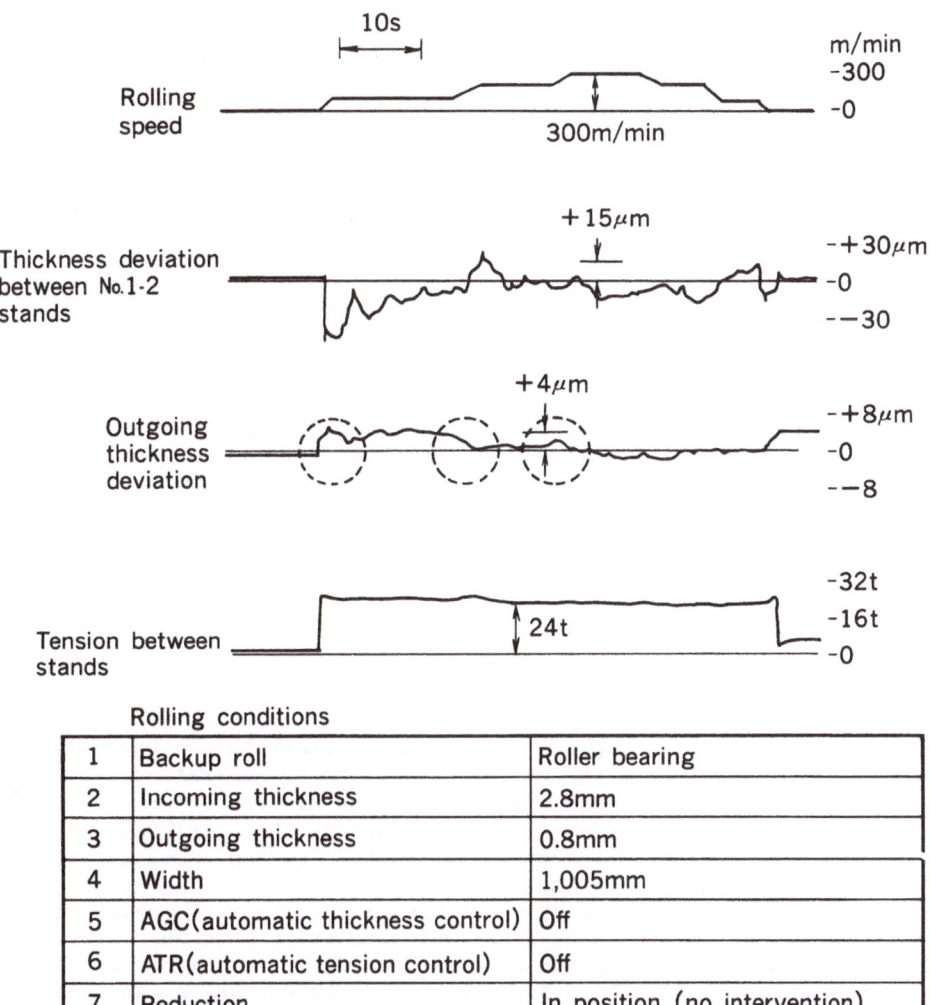

Rolling conditions

1	Backup roll	Roller bearing
2	Incoming thickness	2.8mm
3	Outgoing thickness	0.8mm
4	Width	1,005mm
5	AGC(automatic thickness control)	Off
6	ATR(automatic tension control)	Off
7	Reduction	In position (no intervention)

Figure 7. Speed Matching Characteristics of Digital Speed Control Device (1)

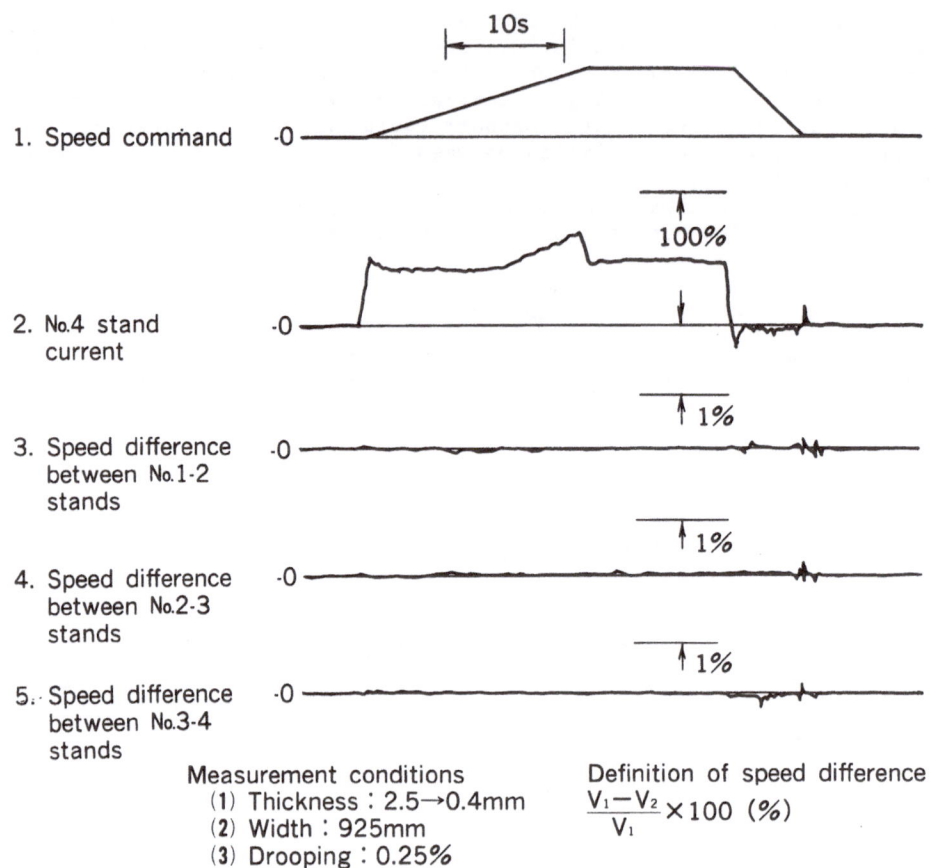

1. Speed command

2. No.4 stand current

3. Speed difference between No.1-2 stands

4. Speed difference between No.2-3 stands

5. Speed difference between No.3-4 stands

Measurement conditions
(1) Thickness : 2.5→0.4mm
(2) Width : 925mm
(3) Drooping : 0.25%

Definition of speed difference
$$\frac{V_1 - V_2}{V_1} \times 100 \ (\%)$$

Fig. 8 Speed Matching Characteristics of Digital Speed Control Device (2)

Figure 9. compares typical configuration drawings and operation oscillograms of equipment for which the M-G (motor generator) type has been totally digitalized. The oscillograms indicate that for digital Leonard equipment, the variation in thickness at the accelerating stage is reduced even when AGC is overridden, thereby giving an excellent result.

Looking at the larger picture of the tandem mill as a whole, this means that, since influence over the outgoing strip thickness is greatly affected by the speed deviation of a certain stand, particularly the first and final stands, the speed deviation rate appears as an outgoing strip thickness deviation rate. Therefore, decreasing the speed deviation rate is directly translated into an improvement in product quality.

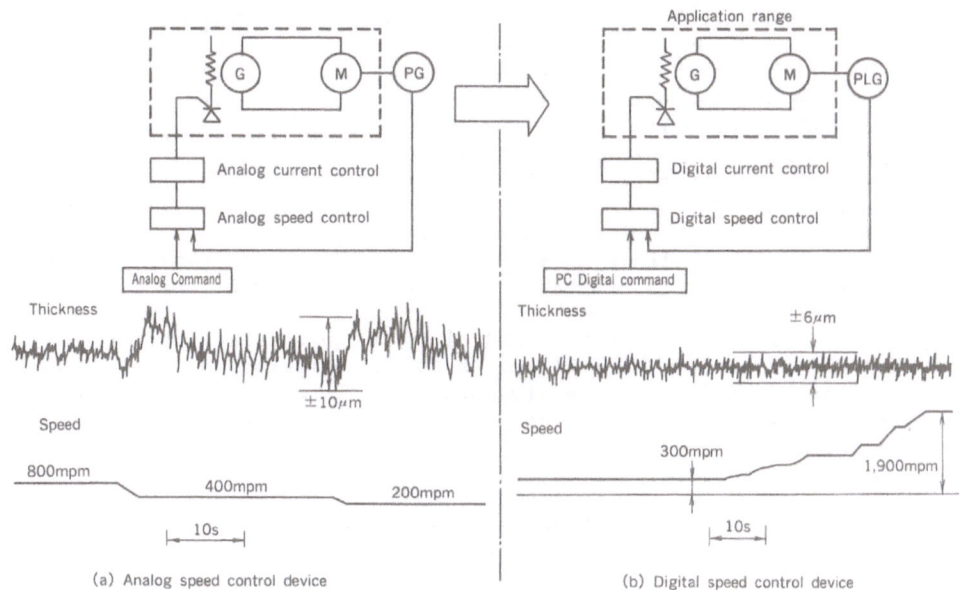

(a) Analog speed control device   (b) Digital speed control device

Figure 9. Typical Application of Digital Speed Control Device to M-G Drives

Thus, because the speed deviation of the said stands at a mill acceleration time is ±0.7% with an analog type or ±0.1% with a digital type as
shown in TABLE I and Figure 6., the mill outgoing thickness deviation is
improved by about 40%, thereby improving the yield considerably[3].

To improve the RAS (Reliability, Availability, Serviceability), the
digital control system allows a failure to be diagnosed earlier and a
failed printed card to be localized, with the trouble-shooting function
which gives useful data for locating the real cause of the system fault.
In addition, the number of cards which are annunciated in failure is
decreased due to the wider use of LSIs. This trouble-shooting function
can not be realized in the analog control system.

Operating conditions for the digital speed regulator are of three
categories: (1) ready to operate before starting; (2) operation after
starting; (3) failed after a fault has taken place. Each operating
status is diagnosed. For (1), a prestartup diagnosis is performed to
ascertain that the hardware is properly functioning. For (2), a normal
service diagnosis is carried out to determine that detected values and
controlled variables are normal in relation to the operating conditions
and control data and detected values are stored. For (3), an abnormality diagnosis is traced back to find the cause of the failure using
stored control data, detected values and failure mode data.

The diagnostic data under each operating condition can be displayed
by the digital speed regulator through the console interface.

3.1.4. Digital Control for AC Machine. Recently, as a high performance AC variable speed control method, a vector control having a quick response equivalent to that on a DC machine has been gradually introduced, replacing the DC machines because of the advantages of induction machines such as excellent serviceability, high environmental resistance, and lighter weight than DC machines.

With this vector control, the motor magnetic flux is estimated based on a motor control model and the torque and magnetic flux are subjected to a decoupling control in accordance with each command. Therefore the control circuit is complicated, requiring much time to adjust. Thus having to mount a speed sensor on the machine, and poor maintainability, constituted problems.

To cope with them, digital control by microprocessor has been studied. One feature of digitalization using microprocessors is that high optional functions are easily mountable. It can be advantageously applied to digital vector control with an automatic tuning function and all-digital cyclo-converter type vector control in a main rolling mill drive motor or other large AC speed change system.

3.1.5. AC Design of Main Rolling Mill Drive Motor. Strict control performance is needed in a main rolling mill drive motor since it requires a large overload capability (a steep impact load is applied to it), and quick reverse rotation and accurate and wide variable speed range are also required. Therefore mostly DC Leonard equipment was employed.

On the other hand, DC machines have commutators and brushes which restrict large capacity design, higher speed and simplification of motor maintenance. Thus a study of applications of AC variable speed drive systems free from these limitations has been under way.

As a drive system satisfying the above mentioned control requirements, a fully digital cyclo-converter type vector control is most appropriate. This AC drive system is at the stage where it can be compete sufficiently with the DC machine system from the viewpoint of response and cost, thanks to progress in the techniques of vector control (used recently as a high performance speed regulation method for induction machines), in digital control techniques involving the application of microprocessors to thyristor Leonard equipment and improved power factor control techniques for cyclo-converters.

Figure 10. shows a system configuration of a 72-arm cyclo-converter of circulating current control type which now has been developed. In the main circuit, two sets of three phase full-wave bridges are connected in series and such units are connected in inverse parallel through DC reactors. The DC reactor is used for limiting the circulating current which is allowed to flow between the positive and negative converters. A vector control method is adopted because it is capable of obtaining a control performance equivalent to that of static Leonard equipment. Also a fully digital type has been used in which decoupling, control and adaptive control are easily introduced, in addition to the improvement of control and accuracy.

The above control configuration is composed in a fully digital form as a multi-processor system, using microprocessors and high speed signal processors.

379

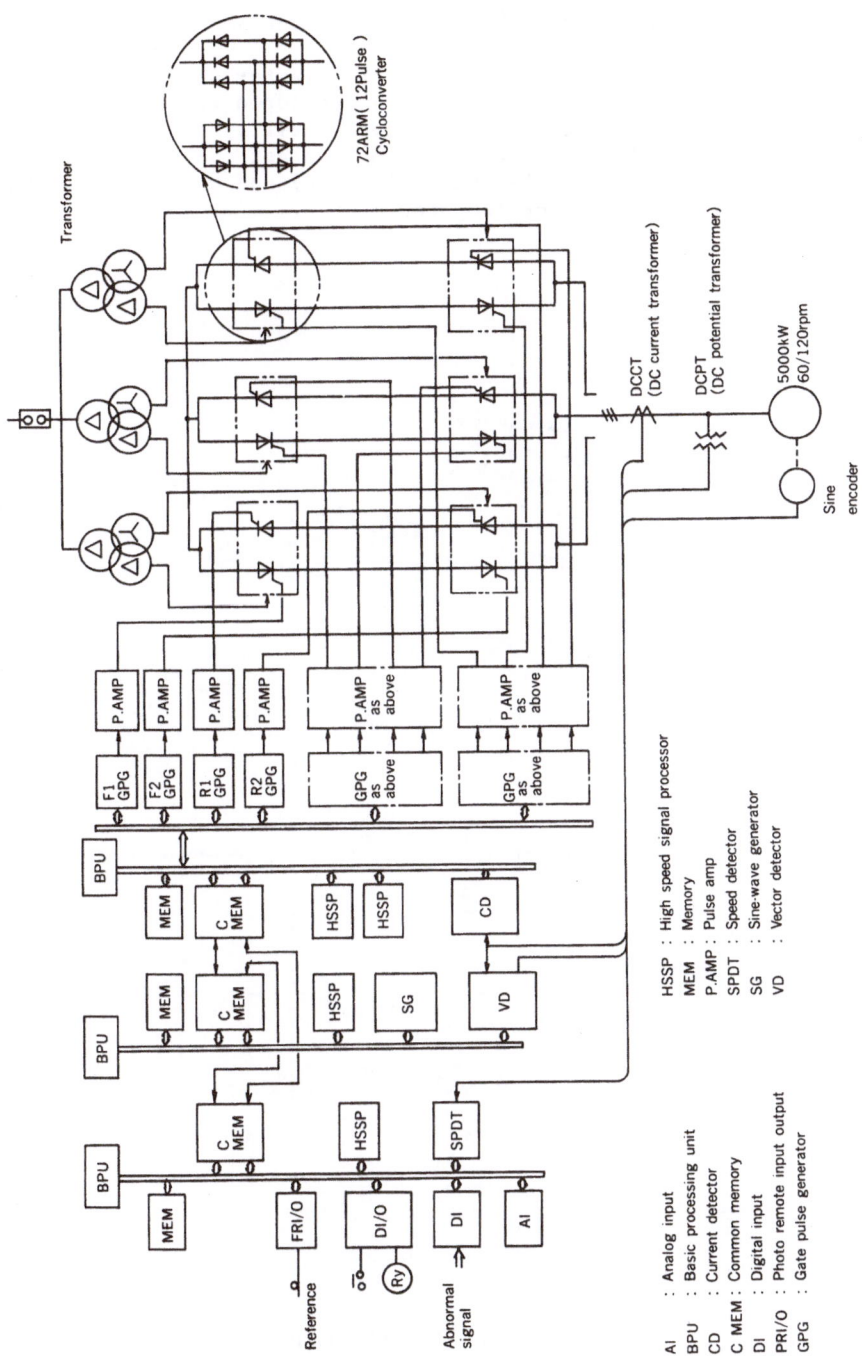

Figure 10. Configuration of Vector Control Cyclo-Converter System

AI   : Analog input
BPU   : Basic processing unit
CD   : Current detector
C MEM : Common memory
DI   : Digital input
PRI/O : Photo remote input output
GPG   : Gate pulse generator

HSSP : High speed signal processor
MEM   : Memory
P.AMP : Pulse amp
SPDT : Speed detector
SG   : Sine-wave generator
VD   : Vector detector

3.2. Programmable Controller Using Micro Computers[6]

The mill plant control system, having a high performance level, reliability and maintainability, has been developed from the viewpoint of manufacture for economic operation, expandability and a short execution period. Figure 11. illustrates the mill plant control system configuration. The programmable controller (PC) and the bus coupling device constitute a multi-microcomputer system.

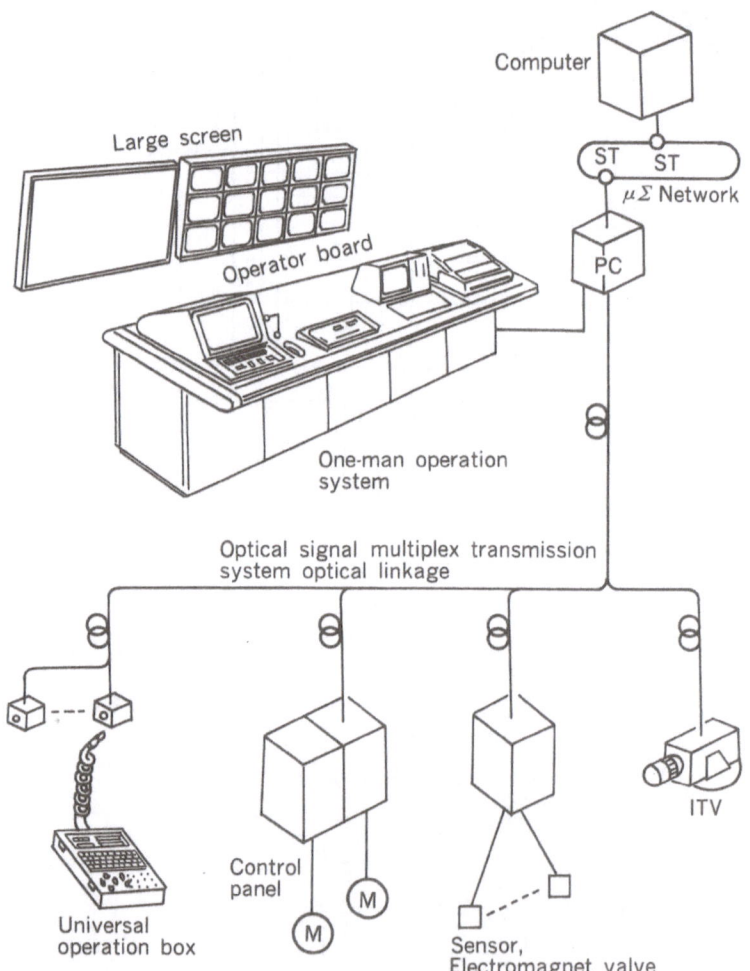

Note PC : Programmable controller   ITV : Industrial television

Figure 11. Configuration of Mill Plant Control System

With the PC as the controlling brain, an operator board and a large size screen are provided to enable man-machine operation. A connection is made with a host computer via a local area network and a small scale optical transmission device connects the sensors, electromagnetic valves, motor control panel and universal operation box.

There are many types of PC in the world. To simplify understanding the HISEC-04M, developed by Hitachi will be described below as a typical example of a PC.

The PC (HISEC-04M) is standardized with a D-type processor and is optionally equipped with a G-type processor. The D-type processor is provided with an interactive programmer PADT (Programming And Debugging Tool) using POL (Problem Oriented Language) which permits a data flow for easy programming.

The G processor is used for linking information devices with data processing as a main objective. The D processor and G processor are installed in a single CPU (Central Processing Unit) and share respective functions but perform their controls independently. TABLE II gives the specifications of the two processors.

TABLE II Specifications for HISEC 04-M

The speed ratio of the sequence instruction with the conventional HISEC 04-E is double.

Item / Type		HISEC 04-M	
		D processor	G processor
Processor	DDC	O	-
	Data processing	-	O
Application		High speed DDC	Data processing (CRT display, typewriter output, high order linkage)
Max. memory capacity		64 kwords (POL)	512 k, 2 Mbytes
Instruction execution (µs)	Sequence (POL)	1.0	
	Add	(POL) 2.0 (Assembler) 1.5	
	Multiply	(POL) 6.0 (Assembler) 9.8	
Language		POL	POL & C
Linkage	Host computer	µΣNetwork/DFW	
	Between controllers	Bus coupling/µΣNetwork/STU	
	Between terminals	STU/PIO (remote)	

Note
DDC: Direct digital control
DFW: Data freeway
POL: Problem oriented language
STU: Serial signal transmission unit

4. Advanced Control System and its Application

4.1. Optimum Regulator

Though the application of the optimum regulator to the rolling control
has formed the subject of various studies from about 15 years ago, they
were always centered on simulation off line. Application fields of the
optimum regulators using off-line simulation have been reported such as
thickness control in cold strip mills[7], looperless control in hot
strip mills, tension control between stands in structure mills, looper
control in hot strip mills and so on.
　　We now introduce an example in which the optimum regulator was
applied to an actual process. This is an example where it was applied to
the AGC, in particular to the Feed Forward (F.F.) control in the finishing train of a hot strip mill[8]. Figure 12. represents the configuration of the AGC system and Figure 13. shows the block diagram.

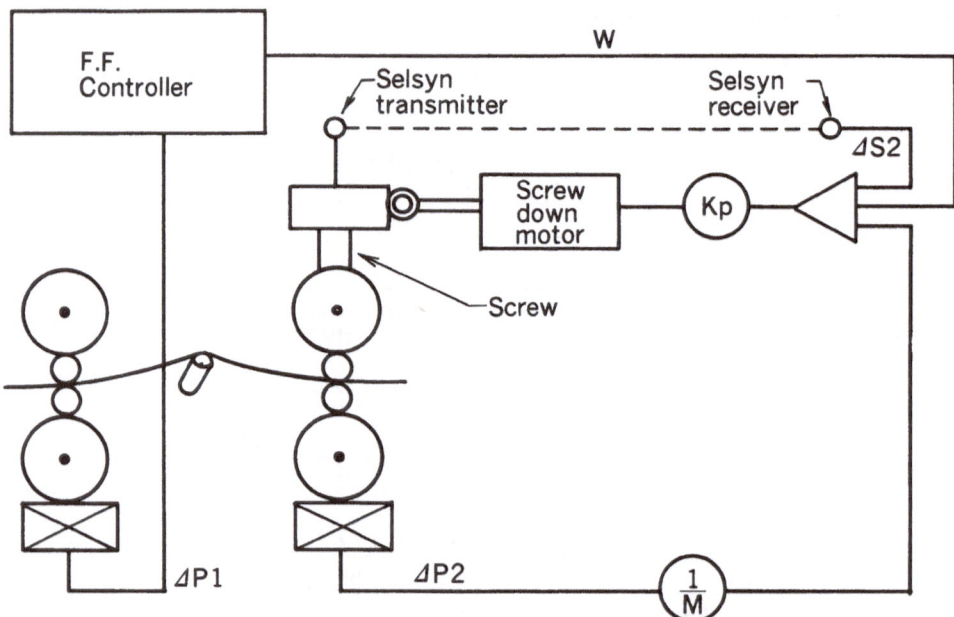

Figure 12. Configuration of Optimum Control System

　　The newly added W indicates the input for optimizing the control
loop. The control loop in Figure 13. can be expressed by the following
equation:

$$\frac{d}{dt}\Delta S(t) = -\frac{M}{M+Q}\cdot K_p \Delta S(t) + K_p W(t) - K_p \cdot h_d(t) \qquad \cdots\cdots(1)$$

$$\Delta h_c(t) = \frac{M}{M+Q}\, \Delta S(t) + \Delta h_d(t) \qquad \qquad \dots\dots(2)$$

In view of simplification, the response Gv(s) of the screw down system is neglected here.
The evaluation function J is represented by the following equation:

$$J = \int_0^\infty [\Delta h_c^2(\tau) + \gamma W^2(\tau)]\ \ (\gamma \rangle 0) \qquad \qquad \dots\dots(3)$$

The input W is to be obtained to minimize J.
In the equation above, γ represents a weighting factor. Also,
h   : Strip thickness on delivery side
S   : Position of screw
P   : Roll separating force
ΔPD : Disturbance of roll separating force
Δhd : Disturbance of strip thickness
M   : Coefficient of mill stiffness
Q   : Plasticity coefficient of the material
Km   : Deformation resistance
KA   : Gain of AGC control loop
Kp   : Gain of screw position control loop
Δhc : Variation of strip thickness on delivery side
    In this case, status variables are $\Delta s(t)$ and $\Delta h'd(t)$. $\Delta hd(t)$ is an estimation of the roll separating force for the front stand.

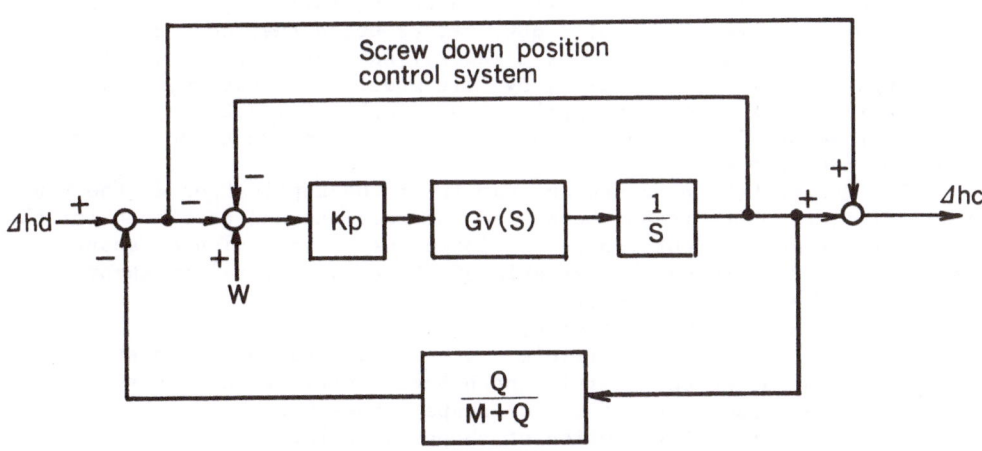

Figure 13. Block Diagram of AGC System

In solving the equation above, the following optimum control law, Wopt, is obtained:

$$W_{opt}(t) = \left[-\frac{M}{M+Q}\Delta S(t) - \right.$$

$$\left. -\frac{M}{M+Q}K_p\delta \cdot \int_0^{t_D} exp\left(-\frac{M}{M+Q}K_p\cdot\delta\tau\right)\Delta h_d(t+\tau)d\tau \right] \times (\delta-1) \qquad \ldots\ldots(4)$$

Here it is assumed that: $\delta^2 = 1+1/\gamma$

While the conventional method (Feedback AGC) outputs the command signal of the screw position $-(M+Q)/M\cdot\Delta h_d(t)$ in considering the real time disturbance $\Delta h_d(t)$, this optimum control law adopts a Feedforward control where the screw position command signal is no other than the mean value weighted by $exp(-K\tau)$ of the disturbance $\Delta h_d(\tau)$ from real time t up to the time $t+t_D$.

It has been reported in actual application results that detection of the roll separating force at the finishing mill stand No.4 and Feedforward control of screw position at the stand No.5 were useful for removal of skid marks.

4.2. Adaptive Control (AC)

The steel making process has promoted, from the earliest stages, the analysis of processing and indentification of parameters by widely executing computer control using mathematical models. Consequently a variety of learning has resulted. On the other hand, adaptive control has recently begun to be applied to the controlling of drive motors in rolling mills. A full-scale application to the rolling process control is a future problem. It should, however, be noted that some documentation in the steel industry gives a general name of adaptive control to learning, adaptive modification, dynamic model modification, etc.

We introduce herein an application example of model preparation, parameter identification and gain regulation which are to be regarded as bases of adaptive control.

The document(9) presents an example of the application of the Von Karman filter, where it has been used and self-learned to search for the shape effect coefficient (degree of influence of variation of shape manipulated variable on the variable of shape parameters) in shape control of a cold rolling mill.

Let us suppose the following;
$\hat{X}_K$: Predictor of effect coefficient after modification of shape
$\overline{X}_K$: Predictor of effect coefficient before modification of shape
Z_K : Measured value of variation of shape parameters
H_K : Measured value of variation of shape manipulated variable
K_K : Filter gain

The effect coefficient X_K is self-learned each time the shape control operation (control period: 2 sec) is affected.

$$\hat{X}_K = \overline{X}_K + K_K(Z_K - H_K \cdot X_K) \qquad \qquad \dots\dots(5)$$

Figure 14. shows a block diagram of the filtering mechanism.

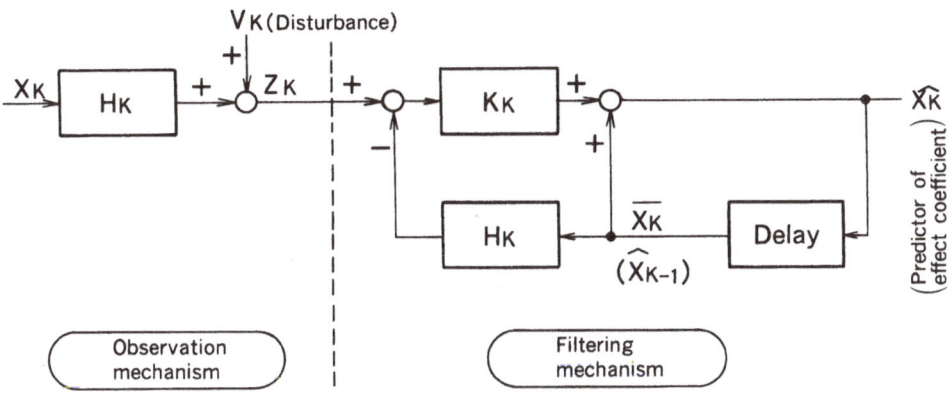

Figure 14. Block Diagram of Filtering Mechanism

Document[10] reports on the tests made on real machines in applying adaptive control intended for cold rolling mills using the 5-tandem mill. Figure 15. shows the control system. This is a feedback control system in which the strip thickness and tension between stands are monitored all through the rolling process from threading up to the passing through of the material, in modifying roll speed and optimizing the roll gap as against the deviation thereof. This is called adaptive control which is similar to MRAC (Model Reference Adaptive Control).

4.3. Automatic Gauge Control for a Cold Rolling Mill System Using Optimum Control Theory

In the gauge-control for a cold rolling mill system, the system provides only feedback control on specific stands. However, efforts to increase productivity and yield have been made by exactly detecting the gauge error of each stand by using sophisticated control techniques based on predictive control theory and optimum control theory.
    The block diagram of the system functions is shown in Figure 16.
    In the following section, details of this system are described[11].

4.3.1. Massflow Gauge. Measuring the gauge of each stand is the basic function of the AGC system. In previous systems, gauge detection using

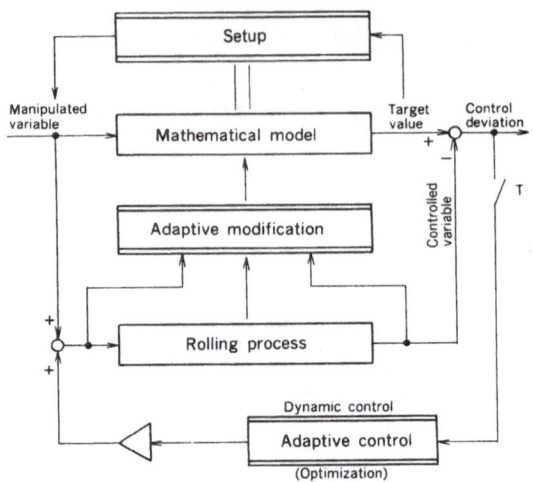

Figure 15. Control System

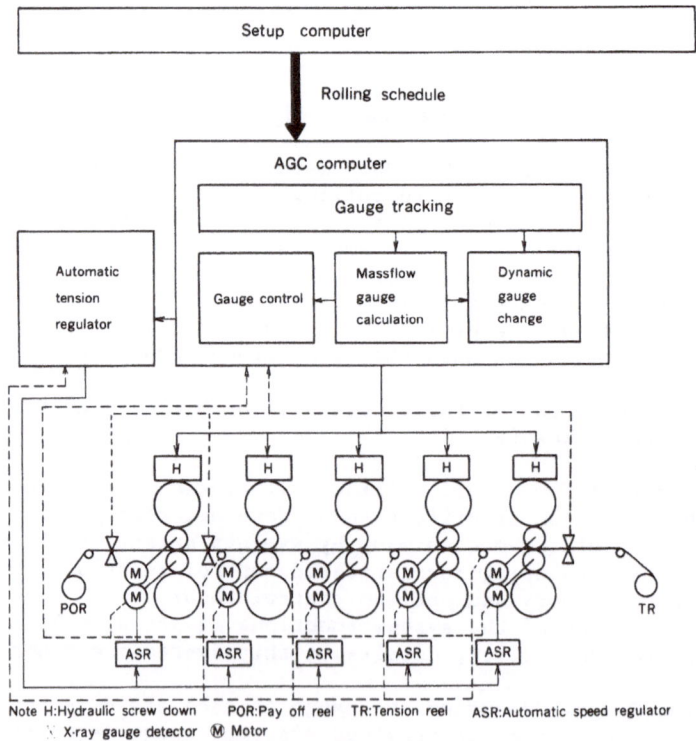

Note H:Hydraulic screw down   POR:Pay off reel   TR:Tension reel   ASR:Automatic speed regulator
X-ray gauge detector   Ⓜ Motor

Figure 16. Diagram of Advanced Gauge Control System

Hooke's law was used. However, this method has a weak point in that the detection gauge error increases for any roll eccentricity, screw position zeroing, and roll heating up. Therefore detection based on the massflow constant law is used for the purpose of minimizing gauge measuring errors. Assuming that the entry massflow is equal to the delivery massflow at each stand, the massflow gauge is calculated as follows:

$$f_{Di} = f_{oi} + (\partial f/\partial H)_i \Delta H_i + (\partial f/\partial t_f)_i \Delta t_{fi} + (\partial f/\partial t_b)_i \Delta t_{bi} \qquad \ldots\ldots(6)$$

$$h_{Di} = H_i v_{Ri-1}(1+f_{i-1})/\{v_{Ri}(1+f_{Di})\} \qquad \ldots\ldots(7)$$

$$h_{mi} = h_{Di}\{1 - (\partial f/\partial H)_i (h_{di} - h_{pi})/(1+f_{Di})\} \qquad \ldots\ldots(8)$$

where, fDi, hDi are the first approximate values of forward slip and delivery gauge at each stand respectively.

The basic value of forward slip foi is estimated from the setup calculations of each coil using a model derived from the Bland and Ford equation expressed in a more convenient form for computing. The basic value foi is modified from Eq.(6) using the gauge and tension error. Also, using the modified slip, the massflow gauge is calculated using Eqs.(7) and (8).

The massflow gauge calculated using Eq.(8) has an error due to the adopted mathematical model which has a difference between the X-ray measured value and the calculated value. Therefore if the coefficient of the equation error is represented by η, then it can be applied to Eq.(7) and its adaptive coefficient can be calculated as follows.

$$\eta = 1 + 0.25\,(h_{xF}/h_{mF} - 1) \qquad \ldots\ldots(9)$$

The effect of an adaptive modification of a massflow gauge with η is shown in Figure 17., close agreement was obtained between the observed and calculated values when the adaptive coefficient was used.

4.3.2. Automatic Gauge Control (AGC). For the purpose of making this system into a synthetic system which is superior in gauge accuracy and running stability, the AGC function has been designed with two mode control systems for changing from the low-speed rolling stage to the high-speed rolling stage.

One mode at low rolling speed continually controls the interstand tensions and quickly minimizes the gauge error of each stand by regulating the roll gaps. If the sampling time of the control is represented by T, the entry gauge by Hi, and the delivery gauge by hi, then the delivery gauge which changes every second during a period $kT \le \tau \le (k+n)T$

388

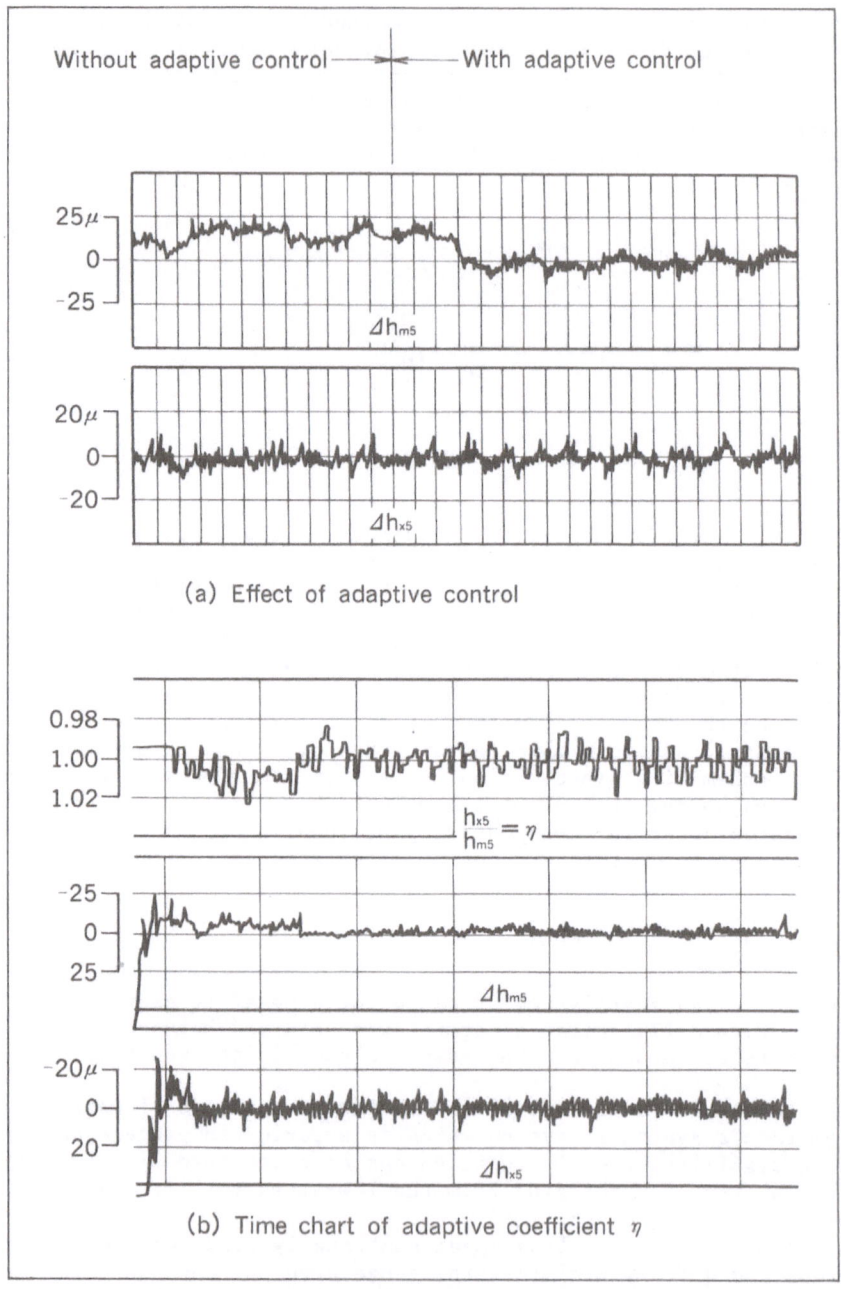

(a) Effect of adaptive control

(b) Time chart of adaptive coefficient η

Figure 17. Effect of Adaptive Modification for Massflow Gauge

can be estimated by introducing a function defined by the following equation:

$$\Delta h_i(\tau) = \Delta h_i(kT) + ((\sigma h/\sigma h)_i(\sigma H/\sigma\tau)_i(\tau - kT) +$$

$$+ (\sigma h/\sigma s)_i \Delta s_{ci}[1 - exp\{-(\tau - kT)/T_{si}\}] +$$

$$+ (\sigma h/\sigma t_b)_i \Delta t_{ci-1}[1 - exp\{-(\tau - kT)/T_{Ni-1}\}] +$$

$$+ (\sigma h/\sigma t_f)_i \Delta t_{ci}[1 - exp\{-(\tau - kT)/T_{Ni}\}] \qquad \dots\dots(10)$$

Where Δs_{ci} = control output of roll gap of No.i stand at $\tau = kT$;
Δt_{ci} = control output of front tension of No.i stand ;
T_{si} = time constant of roll position regulator ;
T_{Ni} = time constant of tension regulator ;
n   = number of sampling periods for which gauge error is estimated by Eq.(10)

Based on the estimated gauge error from Eq.(10), the most suitable conditions for control output, Δs_{ci}, Δt_{ci}, minimize the evaluation function defined by the following equation:

$$J = \int_{kT}^{(k+m)T} \sum_{i=1}^{m} \{\Delta h_i(\tau)\}^2 d_\tau \underset{\Delta s_{ci} \Delta t_{ci}}{\rightarrow Min} \qquad \dots\dots(11)$$

Where m = last stand number

The driving values Δs_{ci}, Δt_{ci} which satisfy the above relation can be obtained by solving (2m-1) equations which are derived from $(\sigma J/\sigma\Delta s_{ci})=0$ and $(\sigma J/\sigma\Delta t_{ci})=0$. On the other hand, the control method during high-speed rolling is a feed-forward control that regulates roll speed of the (i-1)th stand to keep a constant massflow which is disturbed by the entry gauge error of the i-th stand. The control equation is shown as follows:

$$\Delta v_{Ri-1} = h_{pi}(1 + f_{Di})v_{Ri}/\{H_i(1 + f_i)\} - v_{Ri-1} \qquad \dots\dots(12)$$

If a gauge error should occur at the last stand, the roll speed of the No.4 stand will be regulated by the following equation based on the massflow gauge and the X-ray measured gauge at the last stand.

$$\Delta v_{R4} = G_1 v_{R4} \cdot FBX + G_2 \Delta h_{m5} \qquad \dots\dots(13)$$

Where

$$FBX = FBX^{-1} + v_{R5} \cdot \Delta h_{X5} \qquad \dots\dots(14)$$

Several experiments on rolling using this system were made at the five-stand tandem cold mill at the Mizushima Works of Kawasaki Steel Corp. Using the optimum control theory, the off-gauge length (initial length) at the final strip is 3.4 m on average.

Comparing these results with those of the previous conventional system, which was designed with a preset control and conventional AGC, the off-gauge length was improved by 40 % as shown in Figure 18. The gauge error of the new system was 0.42-0.92% of the desired value as shown in TABLE III. It is therefore superior to the conventional system by 0.2-0.7%.

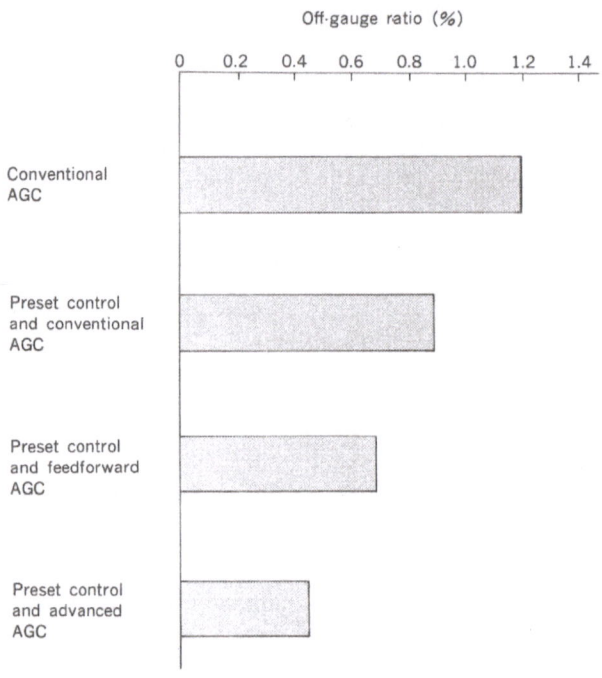

Figure 18. Change in Gauge Accuracy by Improvements in Control System

TABLE III   Comparison of Gauge Accuracy

(gauge tolerance %)

Gauge (mm)	Conventional AGC	Advanced AGC
$h_F \leqq 0.5$	1.58	0.92
$0.5 \leqq h_F \leqq 1.2$	0.89	0.67
$1.2 \leqq h_F$	0.60	0.42

4.4. Knowledge Engineering Applied Techniques

In recent computer control systems, considerable function improvements
and capacity increases have been realized through progress in hardware
techniques. The labour required for development and maintenance of
software is increasing more and more at the present stage.

As regards productivity, serviceability and quality improvement of
software, a new technique applying knowledge engineering has been developed. The knowledge and know-how of specialists concerning the plant
and objects to be controlled were handled as data and treated as a group
of knowledge data (hereinafter referred to as "knowledge data base").
This is independent from the procedure program, thereby coordination
which unifies the operation method and improves productivity, the
serviceability and the quality of software.

This section describes the rule-type control method developed by
using this knowledge engineering approach[2], and the effects of its
application to steel plants. The rules referred to here are those which
describe what control to make in which plant situation and the rule-type
control refers to the method by which rules are accumulated as a knowledge data base. A control is then made to match the particular circumstances while comparing the current plant situation with the knowledge
data base.

4.4.1. Rule Type Control Method. Generally speaking, the knowledge about
a plant which is to be controlled can be classified into the following:
(1) Knowledge which is easy to turn into a numerical formula model such
    as formulae, physical rules and control algorithms.
(2) Knowledge which is hard to turn into a numerical formula model such
    as know-how, common sense and experience.

The know-how in (2) above is knowledge belonging to an individual
specialist concerning the plant to be controlled. It is not optimized
and is often altered. It is also not universal nor systematic, and is
often subject to exceptional processings. Therefore its incorporation
into software is problematic.

The rule type control method has been developed to solve these
problems.
(1) Knowledge is handled as data and is expressed as rules. Concrete
    expression of rules takes the form of an "IF (condition) THEN
    (conclusion)" and, if the plant situation matches the (condition),
    the (conclusion) is deducted.
(2) By turning the rules into data and subjecting them to a centralized
    management, the rules can easily be referenced, added or modified.
(3) With the rule base built up beforehand, the current plant data are
    compared with the rule base at a control in real time and, by the
    deduction of a conclusion from a matching condition (hereinafter
    referred to as a interference), a control output optimum to the
    current plant is obtained.

Figure 19. is a conceptual diagram of the rule type control method.
Plant data are stored in the current data base by the process input/
output function. The current situation data and rules compared by an
inference feature is called a rule base controller and, from a matching

condition, a conclusion is deducted. The conclusion obtained is delivered to the plant by the process control output function.

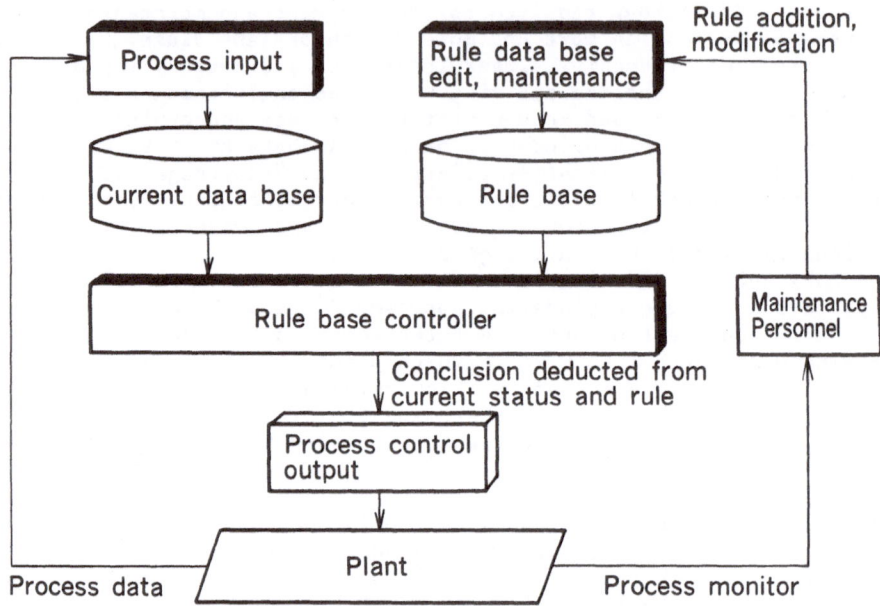

Figure 19. Conceptual Diagram for the Rule-Type

After starting the control, the maintenance personnel proceed to supervision and, by editing and maintenance of the rule base as necessary, can modify or add rules easily.

4.4.2. Typical Application of the Rule Type Control Method. This rule type control method has been applied quite successfully for the first time in the world to real time control of the finishing line for the billet rolling plant of Mizushima Works, Kawasaki steel, Japan.

Figure 20. shows the layout of the finishing line. In this line, a defect in a billet is detected by the inlet defect detector FD1 and, if it is judged to be in need of remedying, it is removed by two grinders GR3 and GR4. A billet judged to be satisfactory by the defect detector FD1 goes straight through the transfer table TB and is unloaded onto the sorting line inlet SL. A billet judged to be in need of remedying by the defect detector FD1 is sent to either grinder GR3 or GR4. The defective billet sent to the transfer table TB is shifted by the transpositioner TM3 to the carriage CA3 or CA4. A defect in a billet is removed by grinder GR3 or GR4 and the billet is then returned to the transfer table by the transpositioner TM4. The billet with the removed defect is re-checked by the outgoing defect detector FD2 and is sent to the sorting line inlet SL when normal. Alternatively it is sent to the finishing line FL when erroneous, to be corrected by the grinder GR4. Between the transfer table TB and the carriage, temporary storage yards CA3, CA4, AT3 and AT4 are provided.

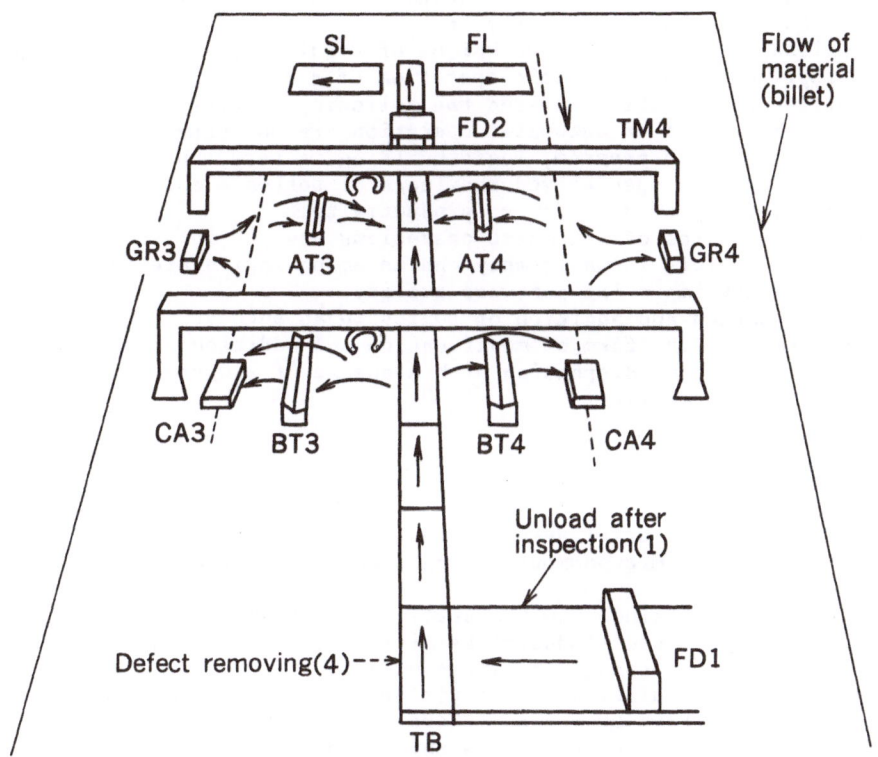

Note
SL : Sorting line inlet
FL : Finishing line
TM_1 : Transfer machine
AT_1 : Storage yard, after removing defects
GR_1 : Grinder
CA_1 : Carriage
BT_1 : Storage yard, before removing defects
FD_1 : Defect detector
TB : Transfer table
$Suffix_1$ denotes machine No.

Figure 20. Layout of Precision Billet Finishing Line

Important points of the automatic operation of the finishing line are; minimizing the wait time for the grinder, efficient transportation of materials for which defect removing is not required and shortening or curtailment of operation of faulty equipment. Such automatic operation processing has been built up as a rule base.

An online process control technique of a rule type control method has been applied to the automatic operation of a finishing line of a billet rolling plant. This provided the following results.

(1) Because the logics of automatic operation are not expressed by rules independent to the program, logic building is easy, thus dispensing with the need for specialized knowledge of software and thereby considerably improving software productivity.

(2) Because the logics of automatic operations are put in order as rules, the logical inconsistency checks among logics are facilitated, thereby improving software quality.

(3) The modification and addition of logics in an automatic operation are possible in the form of modification and addition of rules in the rule base, thus dispensing with a change of program and greatly improving servicability.

5. Conclusions

We have so far given a brief description on concrete examples of real time control in which microcomputers are used in the steel making industry.

We have also explained how the application of the microcomputer has been so rapid in the steel industry by giving its historical background and evolution. A great change was introduced by cheap microcomputers which has enabled calculations to be performed at high speed. That is to say, the driving of both DC and AC motors has been made possible by utilizing full-digital microcomputers and thereby contributing to a large improvement in response time and accuracy. This has allowed a great step to be made in the improvement of accuracy of rolled products.

On the other hand, the modern control system which had not been applied now brings about a variety of effects, being adaptable in many ways.

Further the knowledge engineering put into many applications in the steel industry has resulted in numerous advantages such as the improvement of software productivity.

We can therefore expect many other further merits to be granted by these microcomputers.

6. References

(1) T. SAKURAI et al. : 'Direct Digital Control System in Steel Rolling Mill'
      Hitachi Review Vol. 28 (1979) No.5 pp.240.
(2) S. Nisitomo et al.: 'Recent Computer Control Technology in the Steel Industry'

Hitachi Review Vol.34 (1985) No.4
pp.194~200.

(3) K. SAITO et al.   : 'Application of Fully Digital Speed Regulators to Tandem Cold Mills'
IEEE Transaction on Industry Application
Vol. 1A-20 No.4 July/August 1984 PP.785~794.

(4) T. HONDA et al.   : 'The Most Advanced Control Techniques for Steel Processing Lines'
Hitachi Review Vol.32 (1983) No.2 PP 83~86.

(5) T. SUKEGAWA et al.: 'Recent Microprocessor-Based Control for Motor Drives'
Hitachi Review Vol.34 (1985) No.4 PP.187~193.

(6) Y. ASAKAWA et al. : 'New Development of Rolling Mill Control System'
Hitachi Review Vol.34 (1985) No.4 PP.181~186

(7) M. TAMURA   : 'Optimal Thickness Control in 5-Stand Cold Tandem Mill'
Lecture on the 23rd Conference of Plastic Working Association. 343. (1972).

(8) R. TAKAHASHI and Y. MISAKA :
'Development of Gaugemeter AGC'
Plasticity and Working, Vol.16 No.168 (1975).

(9) Y. NAGANUMA et al.: 'Development of Shape Control Method in Cold Rolling Mill'
Lecture on the 29th Plastic Working Association (1978).

(10) Y. MOROOKA, K. NAKAI, S. KITAO :
'Adaptive Control in Cold Rolling'
Lecture on the 14th Conference of Automatic Control Association 3024, (1971).

(11) N. KITAO   : 'Advanced Gauge Control System in Cold Tandem Mills'
Hitachi Review Vol.28 (1979) No.5 PP.245~249.

Chapter 12

PROCESS MONITORING AND CONTROL OF GAS PIPELINE NETWORKS

G. Lappus and G. Schmidt
Technische Universität München
Lehrstuhl und Laboratorium für Steuerungs- und Regelungstechnik
P.O. Box 202420, 8000 München 2, F.R. Germany

ABSTRACT

Gas transmission by high-pressure pipeline networks is a large-scale, nonlinear, distributed-parameter process. Process monitoring, fault detection/diagnosis, and control of such networks have become complex tasks. Their solution requires sophisticated tools, which are based on the application of advanced theoretical methods as well as modern computer and communication technologies. Microprocessors and microcomputers (including personal computers) have taken over the role of formerly hard-wired front-ends and large process control computers. Though the solution of specific problems, such as optimal, predictive network control, still requires the use of the latter ones, it is worth to review process monitoring and control of gas transmission networks in a chapter of this volume as many essential subtasks, such as dynamic simulation, state estimation, and fault detection/diagnosis, can be solved by means of today's powerful microprocessors and microcomputers.

CONTENTS

S. G. Tzafestas and J. K. Pal (eds.), Real Time Microcomputer Control of Industrial Processes, 397–433.
© 1990 *Kluwer Academic Publishers.*

1. INTRODUCTION

Natural gas is one of the four most important sources of energy all over the world. However, consumption areas are usually situated far away from gas reservoirs and production sites. This fact has created a challenging transportation and distribution task. Gas transmission by high-pressure pipeline systems is a large-scale, nonlinear, distributed-parameter process. An example of its complexity can be given by considering a spatially discretisized process model: it would be composed of a hundred up to a thousand nonlinear differential equations describing a process characterized by transportation and settling times of several minutes up to some hours, along with hundreds of technical and non-technical constraints.

System reliability, fail safe gas supply and economic dispatching are general operational objectives. Due to the complexity of today's gas distribution and transportation pipeline systems (comparable with electric power systems in structure and extension), it is obvious that the operation of these systems must be based on modern computer and communication technology as well as sophisticated methods of process monitoring and control, taking into account the real process dynamics.

First, we describe the considered networks in section 2. Then, in section 3, we briefly discuss SCADA (supervisory control and data acquisition) systems and telemetry systems as commonly used today by industry. We have tried to distinguish between basic supervision and control tasks (section 4) and advanced dispatching tools (section 5). The discussion of these modern techniques, in particular dynamic simulation (section 6), state estimation (section 7), and fault detection/diagnosis (section 8), is our main concern. However, we will also briefly review model parameter estimation (section 9), load prediction (section 10), and optimal network control (section 11). An outlook to future knowledge-based process monitoring and control (section 12) concludes this chapter.

2. HIGH PRESSURE GAS TRANSMISSION NETWORKS (HPGTN)

The general structure of national/international gas supply systems is schematically shown in Fig. 2.1. We can decompose such a large-scale system into several subsystems which are usually owned and operated by individual companies:
- gathering systems,
- transportation systems,
- regional distribution networks,
- municipal distribution networks.

The first three pipeline systems are operated at high gas pressures (10 ... 80 bar), while municipal networks (with the exception of some circular or main pipelines) are operated at low gas pressures (< 1 bar). This chapter merely considers gas transportation systems and regional gas distribution networks. For both types we use the general term "high-pressure gas transmission network", abbreviated HPGTN.

To transport gas over long distances, single or parallel pipelines are used. Gas flow is unidirectional, pipeline loops hardly occur, and there are only a few branches to other pipelines. Transient flow conditions are mainly caused by time-varying offtake flows and the switching of compressor units. For regional gas distribution, complex pipeline networks are employed. There are only a few supply points, but a large number of offtakes. Gas flow is transient; it may even occasionally change its direction in some sections of the pipeline network. In general, a HPGTN is composed of different technical subsystems, such as
- pipelines,
- gas import stations (supplies),
- gas export stations (offtakes),

- gas import/export stations (e.g. underground storages),
- compressor stations,
- regulator stations,
- valve stations,
- gas blending stations.

Most of these stations are very complex dynamic systems within themselves. In particular, this holds true for compressor stations. Here, however, it is sufficient to treat them in a simplified, aggregated manner with respect to the network monitoring and control problem.

Unbranched, homogeneous sections of a pipeline, usually called *pipelegs*, are typically characterized by a length l of 5 to 20 km, a diameter d of 10 to 150 cm, and a roughness factor k of 0.001 to 0.02 mm. Most pipelines are underground. Some networks cover regions with varying geodetical elevations. The total length of all pipelines is between 50 km (very small networks) and some 1000 km (transportation systems or large networks). Gas is fed into a HPGTN at *import stations* which are controlled by setting either pressure or flow. Fig. 2.2 shows a typical station along with its instrumentation. *Gas export stations* are pressure controlled (e.g. gas delivery to another network) or uncontrolled (e.g. supply of industrial plants). A typical station is shown in Fig. 2.3.

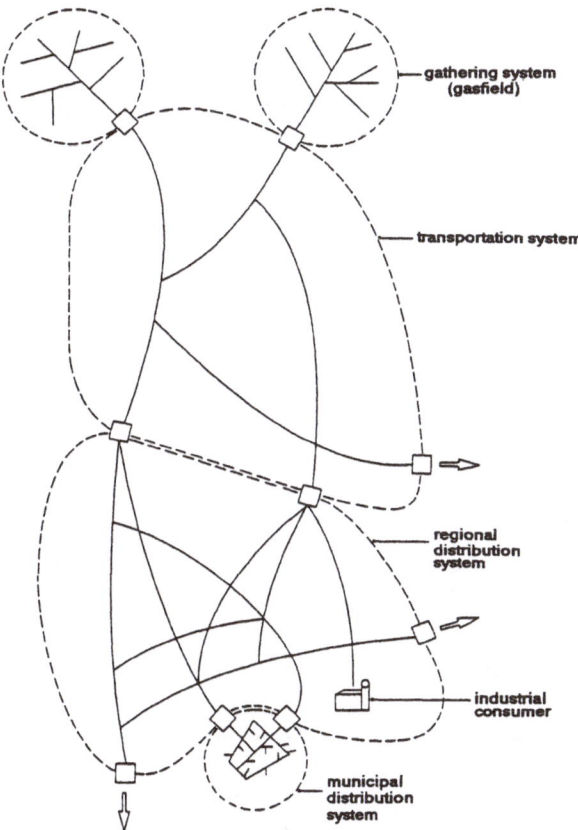

Figure 2.1. Schematical structure of national/international gas transmission systems.

Gas storages are either import or export stations, depending on the actual mode of operation. Descriptive parameters in our context are the maximal input or output flow rate, the actual quantity of stored gas and the maximal capacity being available for short or long term operational purposes. *Compressor stations* provide the pressure required for gas transport. Each station is usually composed of several individual compressor units which can be interconnected to serial and/or parallel compressor systems, forming so-called compressor plants. A typical configuration is shown in Fig. 2.4. Modern, fully automated compressor stations (eventually plants) are controlled by flow or suction or discharge pressure; older stations are sometimes controlled by setting individual machine speeds. *Regulator stations* are incorporated in the network for pressure reduction purposes, e.g. at network entries with lower admissible maximal pressure. These stations are simi-

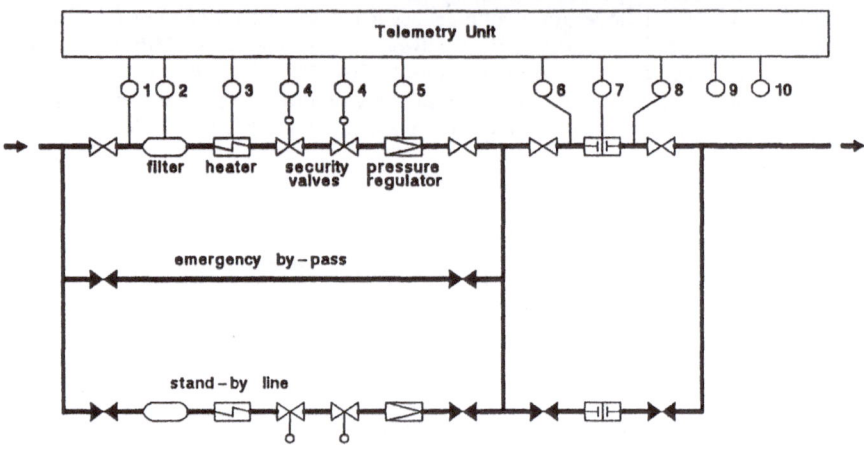

Figure 2.2. Typical gas import station and its instrumentation: (1) suction pressure, (2) pressure difference, (3) temperature, (4) valve position, (5) pressure setpoint, (6) gas temperature, (7) gas flow rate, (8) discharge pressure, (9) air temperature, (10) local power fail indicator.

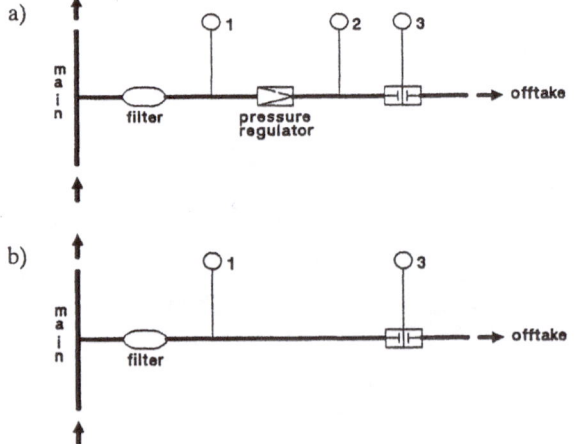

Figure 2.3. Typical pressure controlled (a) and uncontrolled (b) gas export stations and their instrumentation: (1),(2) pressure, (3) flow rate.

lar in structure to gas import stations. *Valve stations* are used to close a pipeline segment or to interconnect several pipeline segments, see Fig. 2.5. In some networks, *gas blending stations* are used to mix two or more gas streams of varying quality to produce a gas of uniform quality.

3. SCADA AND TELEMETRY SYSTEM

This section discusses typical Supervisory Control And Data Acquisition (SCADA) systems, the telemetry equipment of HPGTN, and the *basic functions* required by process monitoring and control. A SCADA system is usually defined as a measurement and control system that covers great distances, as is the case with oil or gas pipelines. This definition was intended to distinguish these systems from ordinary centralized measurement and control systems. With the advent of so-called distributed control systems (DCS), this previous distinction has lost its proper meaning (Laduzinsky, 1986). However, we have chosen to retain this familiar term.

Historically, there are three eras in the development of supervisory control and telemetry equipment for gas transmission systems. In the early days, i.e. before 1960, little equipment was needed. Just a few measurement data were transmitted to a "central control station"; remote control was usually done manually via telephone by local station operators. The period from 1960 to 1975 was the era of the traditional SCADA systems enabling centralized dispatching. These systems incorporated special hard-wired telemetry systems and process control computers, which,

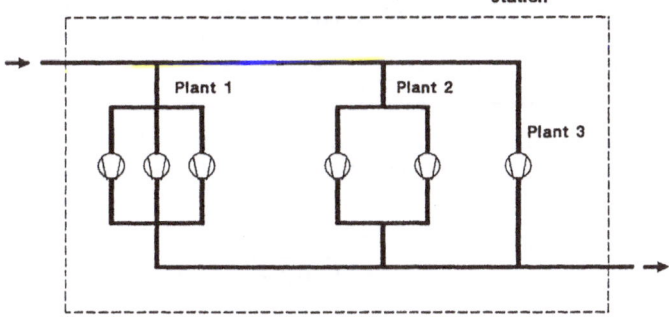

Figure 2.4. Simplified scheme of a typical compressor station.

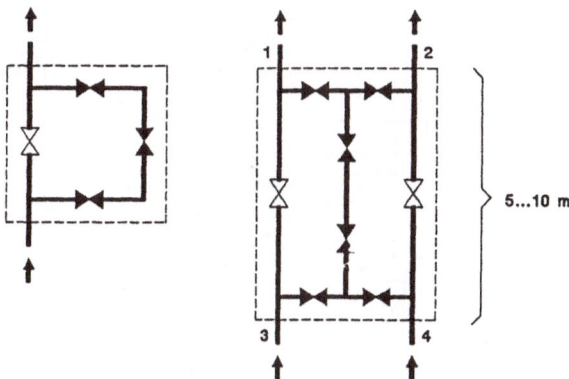

Figure 2.5. Simplified schemes of typical valve stations.

however, were used mostly for data acquisition and data storage purposes. In the third era, which began approximately in 1975, most companies introduced new SCADA and telemetry systems based on distributed microprocessors and minicomputers, see, for instance, Turner (1974), Liaugaudas (1979), Desaultes (1981), Kaier (1987).

Of course, these three periods cannot be strictly separated. There are forerunners and latecomers and, for economical reasons, old equipment (especially expensive data transmission systems) was always integrated into the new system. Today, we can observe the advent of the next era, which is essentially characterized by wide area networks (WANs), local area networks (LANs) and distributed data and information processing.

A generalized block diagram of a typical modern SCADA and telemetry system for gas transmission networks is shown in Fig. 3.1. It is composed of three major subsystems:
- remote control and meter stations,
- the telemetry system,
- the central control station.

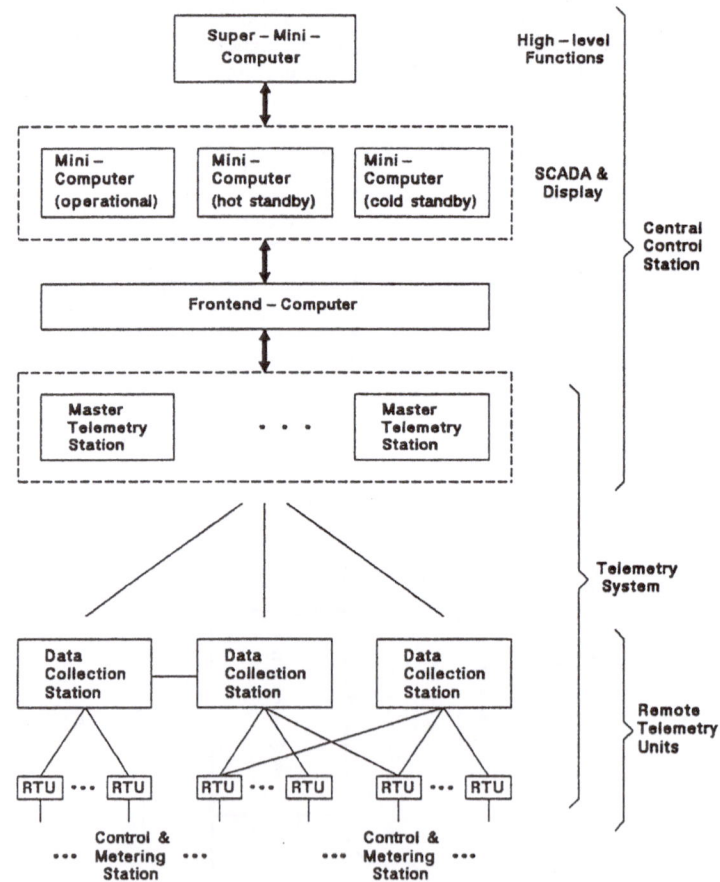

Figure 3.1. Block diagram of a typical SCADA and telemetry system for a HPGTN.

The central control station is the heart of the total system, where a three level hierarchy of computers is generally found:
- front-end computers,
- process control computers,
- high-level control computers.

Front-end computers, e.g. 8 or 16 bit microprocessors, are used for interfacing and controlling the telemetry system. Basic process monitoring and control functions are assigned to process control computers. Usually, we find a hot-standby double computer and double disc system, sometimes even a third back-up computer. Computers are connected via an interprocessor link which has to detect faults in the computer system, switch over in such cases and exchange information. One computer is in master mode operation, the other in hot-standby. Real-time network control and monitoring is performed in the foreground of the master computer. In its background basic support routines are running, e.g. plotting, data manipulation, disc backup, etc., for instance, see Eichner (1985). Another up-to-date method is the connection of all computer systems via a local area network (LAN), where the computer manufacturer provides hot-standby as well as takeover functions. Powerful minicomputers or super-minis are required for modern high-level process monitoring and control aids on the utmost level.

An important part of the central control station is the *man-machine-interface*. The old wall-mounted display and control panels have vanished. Instead, we find colour graphic CRTs and hierarchically ordered graphics of the network and its equipment. Actual corresponding measurement data are included. However, most companies rely not only on CRT displays, but also include additional modern wall or table-mounted displays showing the total network structure with only the most essential switching and alarm conditions. Some companies use *active* table-mounted displays with a digitizer tablet below the surface enabling the direct selection of the CRT graphics (Hoffmann and Ferenz, 1981).

Today, most remote stations, such as compressor, gas import or export stations, are fully automated, permitting unattended operation. Such stations are connected to the telemetry system by so-called RTUs (remote terminal unit or remote telemetry unit). Modern RTUs are modular, microprocessor-based computers with a variety of plug-in boards providing all functions required. Typical functions are:
- encoding and decoding of data,
- transmitting data from the remote station to the central control station, such as
    # pressures,
    # flows, accumulated flows,
    # temperatures,
    # gas quality parameters (BTU values),
    # status signals (on, off, open, closed, local leak alarm, etc.).
- receiving commands from the central station, such as
    # change a pressure or flow setpoint,
    # open or close a valve,
    # initialize a compressor start-up or shut-down sequence, etc.

Data is transmitted every 10 ... 300 seconds using a polling procedure initiated by the master station. During actual remote control inputs, a shorter polling cycle is required. However, there is a trend to instantaneous data transmission from RTU to the central station which was used formerly in alarm cases, only.

Data transmission is done via cables, or, with an increasing trend, wireless (Baur, 1987). FM-modulation is typical for regional data collection, while time-multiplex transmission is used for sending data from data collection stations to the central station. Bit sequential synchronous modes, with a variety of codes and redundancy checks, provide the necessarily high security for data transmission. In addition, nestet or alternating data transmission paths and fixed repetition schemes are employed.

4. BASIC PROCESS MONITORING AND CONTROL FUNCTIONS

Basic process monitoring and control functions are hierarchically distributed within a HPGTN. Such functions are realized on the station level as well as on the network level. *Network control* refers to the control of the entire gas transmission network. Data is transmitted from the remote meter and/or control stations to the central dispatching station. There the information is processed and the results displayed to the dispatcher. The dispatcher addresses the remote stations to transmit commands, e.g. changes in the setpoint values. The *station control level* accepts these commands from the master station, processes them, and passes them on to either a station device, e.g. a regulator, or the *plant or unit control level* in case of a compressor station, see Fig. 3.2 and 3.3. Compressor stations are very complex and so is the related monitoring and control task which is described in detail by Binder (1970), Binder and Brandt (1974), Moellenkamp (1976), as well as Matsumura and co-workers (1979). Here, we can only summarize the main features.

Unit Control. Unit control provides the basic functions necessary for the safe and reliable operation of each compressor unit which consists of a driving engine and a compressor engine. Typically, unit control is composed of three functions: a) monitoring, b) sequencing and c) control. The *monitoring subsystem* is responsible for the protection of the compressor units. It monitors all vital parameters, such as pressure, temperature, speed, vibration, oil levels, etc. A unit alarm or shut-down is initiated if any of these parameters exceed pre-set limits. The *sequencing subsystem* controls the start-up and shut-down procedure of the unit, which is usually rather complex because of the many technical subsystems forming the compressor unit and the high degree of safety required for unattended operation. The *control subsystem* is responsible for the normal mode operation of the unit, i.e., all variables must be kept within the admissible operating regime, see Fig. 3.4. The most common method of load control involves varying the speed of the compressor machines. The speed command is received from the plant level control. Other forms of load control are, for example, pocket control, controlled timing of the suction valves in compressor cylinders or angle control of the inlet guide vanes. Temperature control and surge control are two other essential controls at the unit level.

Plant Control. The plant control coordinates several parallel or serial compressor units forming a so-called plant. As the dispatcher can address a whole plant, he has usually not to deal

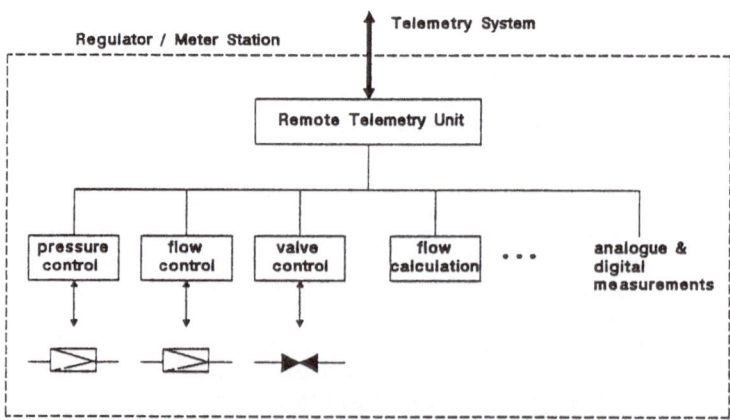

Figure 3.2. Scheme of the local monitoring and control functions of a regulator/meter station.

with the individual units. Plant control has two main tasks. First, it automatically interconnects or disconnects compressor units as required by the instantaneous pressure and flow, taking into account the physical limitations and the varying efficiencies of the different units, see Fig. 3.5. Second, plant control has to assign individual gas flow setpoint values to the parallel units, taking into account the required total gas flow and the usually different efficiencies of the units. This flow splitting task is discussed in detail by Gagne (1975).

Station Control. Station control provides main monitoring functions, e.g. shut-down of the whole station and enable or disable remote restart, main sequencing functions, e.g. start-up of a plant, and main control functions, e.g. feedback control of the total gas flow or of the discharge pressure. Related setpoint values are received from the central control station, processed and passed to the plant control level.

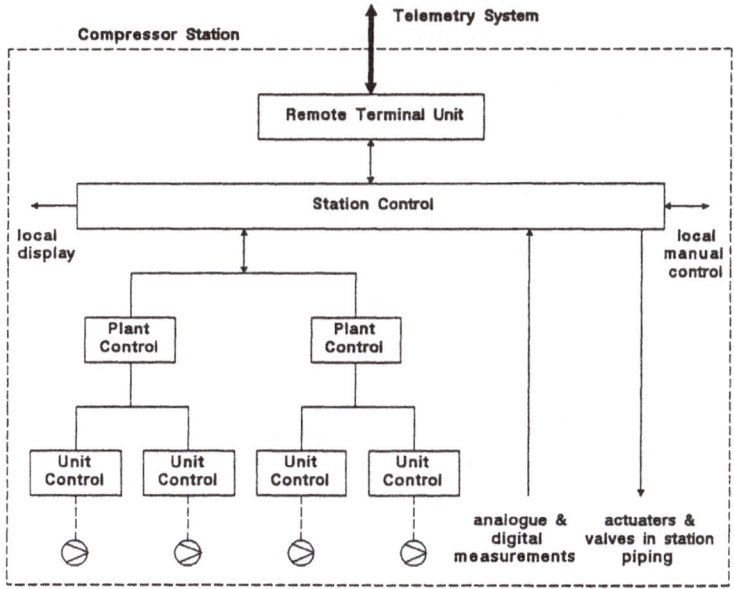

Figure 3.3. Scheme of local monitoring and control functions of a compressor station.

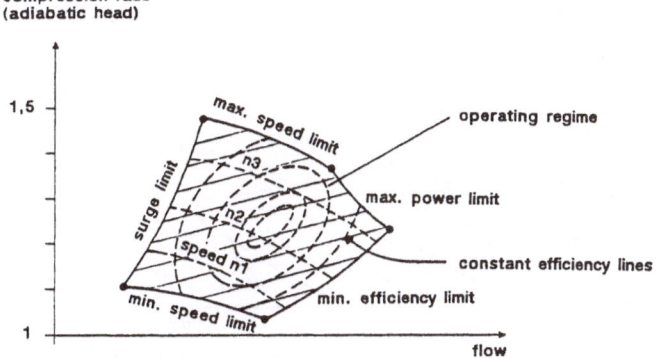

Figure 3.4. Typical envelope of a centrifugal compressor unit.

     <u>Basic Network Control.</u> The network control or "dispatch" system in the central contr(l station has two fundamental functions. It must present important telemetered and calculated operating parameters to the dispatcher in concise and useful form, and it must provide a secure form of telecontrolling the network. With respect to these objectives, we can summarize the required *basic process monitoring and control functions* which are based on periodic, or time-of-day, or event-driven data processing:
- data acquisition including marking or elimination of obviously incorrect values,
- data storage handling or database management,
- data display,
- alarm condition monitoring (typically two or three levels),
- remote communication and control of network equipment,
- hardware fault (CPU, SCADA) handling.

Low-level application programs such as needed for the *basic management* of the network are:
- contract monitoring,
- inventory control (compressors, underground storages, etc.),
- generation of weather profiles,
- prediction of total gas consumption,
- daily dispatch guidelines.

Another class of basic application programs are needed for editing purposes, such as the generation of display screens (mask editor) or the alteration of alarm condition limits. Today, all these basic functions are available as *off-the-shelf software*.

5. ADVANCED DISPATCHING SUPPORT

Basic process monitoring and control functions as provided by a "simple" SCADA system are not sufficient to solve the complex dispatching problems. Two major reasons are: First, decisions are based on incomplete information concerning the actual state of the network, because pressure and flow are only measured at a few discrete locations, and, second, there is no possibility of predicting future pressure and flow exactly.

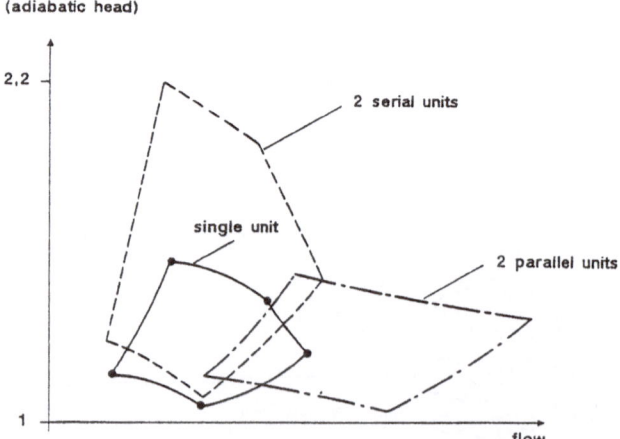

Figure 3.5. Typical envelopes of a compressor plant with two units (single -, serial -, and parallel mode operation).

Therefore, advanced process monitoring and control tools are required. A mathematical model of the transient gas flow process in a HPGTN is the common basis of various advanced monitoring and control aids. Digital simulation, which provides the real state of pressure and flow depending on variable supplies, offtakes and network control inputs, is a key-function in the solution of network planning, control and monitoring tasks.

State-of-the-art network monitoring and control systems are based on modelling and simulation. The general structure and the functional relationships of such a system are shown in Fig. 5.1. Essential off-line and on-line subsystems are:
- simulation based planning and design (CAD/CAE; not shown in Fig. 5.1),
- network state estimation,
- enhanced supervision and fault detection,
- detailed load prediction,
- look-ahead simulation (predictive simulation and supervision),
- what-if simulation (predictive control strategy development, contingency analysis),
- predictive optimization,
- man-machine-interface.
The monitoring and control system includes a CAD/CAE subsystem which is based on transient simulation. This is due to the fact that controllability of the network can become rather restricted through the results of design activities (e.g. network configuration, gas supply contracts). Therefore, such long term activities should take into account the real transient process dynamics and the short-term requirements of on-line control.

The functional relationship between the subsystems shown in Fig. 5.1 is obvious. Periodically with the data sampling rate (e.g. 5 min), the state estimator updates its calculation for the total state of network pressure and flow. This actual estimated state is used for supervision purposes and as an initial state for predictive simulations over a future period of time (e.g. 4 ... 48 hrs) taking into account the offtake forecasts provided by the load prediction module. Using actual network control data, we can start a predictive simulation run and check the future state under the assumption that no operational changes will be made. If the results of this first

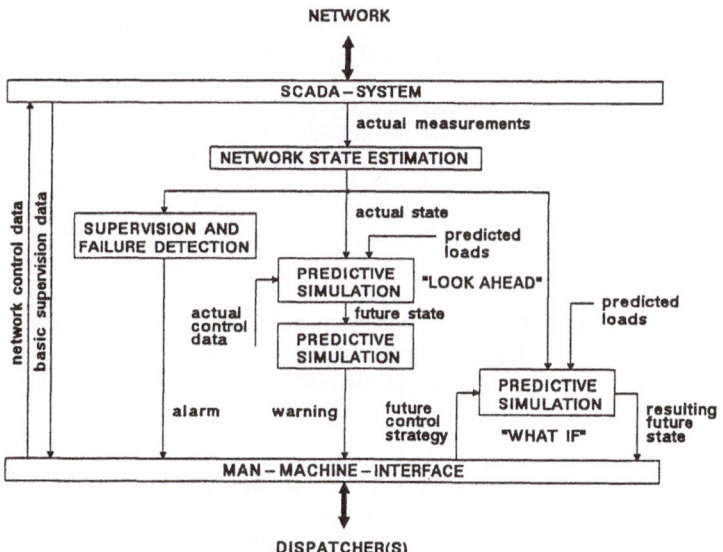

Figure 5.1. Process monitoring and control scheme including advanced dispatching support.

predictive simulation and supervision indicate that we must change network control data, we can perform a further predictive simulation using an alternate network control strategy. The predicted state is checked again in order to ensure a fail safe and economical network operation. Thus, by fast, repetitive, predictive "what-if" simulation runs with varying control data, an appropriate control strategy for the following control period (e.g. 1 hr) can be planned by the dispatcher.

We note that this process monitoring and control system already functions similarly to a simple decision support system (see section 12), though predictive on-line optimization is *not* yet available, today. Major benefits of such a network monitoring and control system are:

off-line simulation:
- improved design of pipeline network extensions,
- design of favorable gas purchase and sales contracts,
- improved design of normal gas dispatching,
- design of time and space coordinated control actions for maintenance outages,
- improved contingency analysis of possible upsets and design of appropriate counter strategies,
- dispatch training without any risk.

on-line state estimation:
- provides the instantaneous state of pressure and flow of the total network,
- indirect supervision of the SCADA and telemetry system permitting fault detection and
    equipment maintenance only when required,
- detection and location of leaks (with some additional evaluation procedures).

predictive simulation and control strategy development:
- predictive supervision of future pressure and flow, assuming the present control inputs to be
    constant (look-ahead tool),
- pre-testing of operational changes and interactive development of reasonable control strategies
    (what-if tool).

Industrial applications of such state-of-the-art process monitoring and control schemes for large gas transmission networks are discussed, for instance, by Homann (1981), Ady and Rauch (1984), Kersting and co-workers (1984), Formaggio and Luciano (1984), Hass (1984), Weimann and co-workers (1985), Van der Hoeven and Baker (1985) and Lappus (1987).

The following sections discuss the individual tools in more detail. We begin with modelling and simulation because these functions are prerequisites for the design of advanced monitoring and control systems.

6. SIMULATION

We will begin with a short review of the modelling and simulation of gas transmission networks. Today, various efficient simulation programs are available. We will refer to some well-known programs and then, as an example, discuss in more detail a program developed in our laboratory.

6.1. Review

Before 1975, it was common practice to design (and operate) gas transmission systems on the basis of steady-state models. This was not satisfactory because, generally, gas demand is not steady but can vary considerably from hour to hour and day to day. Today, it is widely accepted that dynamic models of transmission network behaviour are required as an aid to the design and operation of most transmission networks. Dynamic effects, resulting from varying supplies and offtakes or operational changes, usually occur in minutes or hours. Rapid transients (e.g. shock waves) rarely occur under normal operational conditions. To emphasize this point of interest, some authors have used the term "slow transient model" to differentiate between slow and fast-acting dynamic effects. Fast transients occur, for example, in the case of sudden pipeline ruptures.

Many network analysis programs have been developed over the years. In general, all models result from the same equations, though some authors choose to neglect terms which others include. The basic equations of isothermal pipeline flow are two simultaneous, partial differential equations. One equation relates pressure drop in the pipe to frictional losses and the rate of change in gas momentum. The other equation expresses the law of conservation for the gas mass. These are the two most commonly used basic equations. However, some authors do not assume isothermal gas flow, but include a third state equation (balance of energy) into their model. There are, however, considerable differences in the techniques used to solve the model equations. The numerical techniques used can be divided into three main classes:
- method of characteristics,
- finite difference approximations for space with explicit or implicit finite
    difference schemes for time derivatives,
- finite element approximations for functions of the space variable and implicit
    difference schemes for time derivatives.

Various modelling assumptions and numerical solution schemes are discussed in detail by Fincham (1971), Weimann (1978), Fincham and Goldwater (1979), Tiley and Thorley (1985), Osiadacz (1987). These authors also give reference to some of the already classical papers on this subject which are not cited here.

Some more recently developed comprehensive models and simulation programs are GANESI (see below), STAG (British Gas), FALCON (British Gas), DIGINET (SSI-Intercomp), GASUS (Stoner Ass.), REGVAR (Gaz de France, Michon and Sorine, 1978), SIMONE (Czech Gas, Kralik and co-workers, 1984), XTRAN (New Zealand Gas, Hester, 1984).

6.2. Simulation Program GANESI

Now let us discuss in more detail the network model and simulation program presented by Weimann and Schmidt (1977). This model was developed with major emphasis on on-line, real-time applications. Today, this model, together with the simulation program GANESI, are approved by many gas companies in Europe. According to a survey by Fischer-Uhrig (1986) GANESI has become a de-facto standard in the German gas industry.

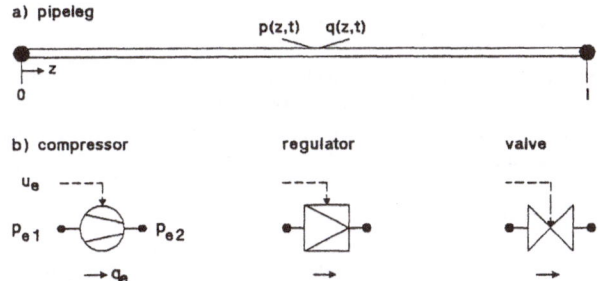

Figure 6.1. Distributed (a) and lumped (b) parameter subsystems of a HPGTN.

For modelling purposes, the gas transmission network under consideration is partitioned into n_d pipelegs, and n_l regulator, compressor, and valve stations, see Fig. 6.1. Weimann assumes one-dimensional, isothermal flow. It is modelled by a nonlinear, hyperbolic-type set of partial differential equations for the space and time-dependent variables pressure $p_i(z_i,t)$ and mass flow $q_i(z_i,t)$.

$$\begin{bmatrix} \partial p_i(z_i,t)/\partial t \\ \partial q_i(z_i,t)/\partial t \end{bmatrix} = \begin{bmatrix} -a_{1i}(p_i) \cdot \partial q_i/\partial z_i \\ -a_{2i} \cdot \partial p_i/\partial z_i \quad - \quad a_{3i}(p_i) \cdot q_i|q_i|/p_i \end{bmatrix} \qquad (6.1)$$

$$a_{1i}(p_i) = c^2(p_i) \cdot K(p_i)/A_i \; ; \qquad\qquad a_{2i} = A_i \; ;$$

$$a_{3i}(p_i) = b_i \cdot c^2(p_i)/(2 \cdot d_i \cdot A_i) \; ; \qquad c^2(p_i) = R \cdot T \cdot K(p_i) \; ;$$

$$K(p_i) = 1+k \cdot p_i \; ; \qquad\qquad i = 1,..,n_d \; .$$

In these equations z_i represents the spatial coordinate along the i-th pipeleg, and t is time. A_i, d_i, and b_i are the pipeleg cross section area, diameter, and drag coefficient, respectively. Gas properties are represented by the individual gas constant R, the temperature T and the compressibility factor K. We are considering only horizontal pipelegs in order to simplify notations. For non-horizontal pipelegs we merely have to add a further term on the right-hand side of the second equation (6.1). GANESI actually includes this term. Weimann (1978) shows that the assumption of isothermal gas flow (T = constant) is adequate for underground pipelines under normal operational conditions. Varying temperatures along a pipeleg (e.g. after a compressor station) can easily be included in this model by spatial temperature profiles (Lappus, Scheibe, Weimann, 1980).

In addition, we have to specify boundary and initial conditions. Boundary conditions at $z_i=0$ and $z_i=l_i$ are provided by pressure or flow-defining functions of time and/or coupling conditions between adjacent network elements. Initial conditions at time t_o are formally given by specification of space-dependent functions for the pressure and flow

$$p_i(z_i,t_o) = p_{io}(z_i) \; ; \quad q_i(z_i,t_o) = q_{io}(z_i) \; ; \qquad (6.2)$$

$$i = 1,..,n_d \; .$$

Spatial extensions and the dynamic effects of compressor, regulator, and valve stations are usually negligible when compared to a long pipeleg. These elements are included in the network model using linear or nonlinear algebraic equations of the form

$$h^j(p^j_{e1}, \; p^j_{e2}, \; q^j_e, \; u^j_e) = 0 \; ; \qquad j = 1,..,n_1 \; . \qquad (6.3)$$

These equations describe the steady-state relationship between inlet and outlet pressure p^j_{e1}, p^j_{e2}, the flow q^j_e, and some control input variable u^j_e, see Fig. 6.1. Let us explain this notation with a simple example shown in Fig. 6.2. In the case of a pressure regulator station, we assume the controlled outlet pressure $p_1(t)$ to always be equal to the setpoint value $u_e(t)=p_s(t)$,

$$p_1(t) - u_e(t) = 0 \; . \qquad (6.4)$$

However, we have to take into account additional constraints which are due to the technical properties of such stations, e.g. flow limitations. We can formally describe such constraints by

$$p^j_{e1,min} < p^j_{e1} < p^j_{e1,max} \; ;$$

$$p^j_{e2,min} < p^j_{e2} < p^j_{e2,max} \; ; \qquad\qquad\qquad (6.5)$$

$$q^j_{e,min} < q^j_e < q^j_{e,max} \; .$$

It is important to note that Eq. (6.3) provides boundary conditions for adjacent pipeleg equations and that these boundary conditions may change their structure if a constraint, as described above, becomes active. The exact handling of such structural changes seems to be a problem in all known simulation programs!

The mathematical model of the total gas transmission network is formed by collecting and coupling all subsystem models according to the above equations. The coupling conditions at the connection points, i.e. the network nodes, follow from mass and force balancing equations (generalized Kirchhoff's laws).

The numerical solution to all equations is performed by transforming them into a set of algebraic equations using space and time discretization according to a modified Crank-Nicholson method. The resulting, usually high-dimensional system of nonlinear algebraic equations is solved iteratively by using the Newton-Raphson method and sparse matrix techniques.

Let us add some aspects of the system of linearized algebraic equations used in this solution scheme. The dimension of the system corresponds to the number of nodes and spatial discretization points of all pipelegs. Typical spatial discretization steps, which are adjusted by the program, are between 5 and 20 km. Pipelegs shorter than twice this step are discretized into two sections. If all pipelegs are shorter than a given discretization step, then the total dimension of the system of equations is about the number of nodes plus the number of pipelegs. Typical dimensions are 250, 500 or 1500 for small, medium and large-scale networks, respectively.

The iterative solution of the nonlinear algebraic equations by the Newton-Raphson method implies a repeated set-up and solution of the linearized system of equations. The convergence of the iteration process is rather fast, due to the special type of nonlinearities. In most cases, two or three iteration cycles already produce a sufficiently accurate solution. The number of iterations is automatically controlled by the program using an accuracy criterion for all node pressures. Provided that the starting point of the iteration process is not too far away from the real solution, the Newton-Raphson method even converges with a constant Jacobian matrix. In many cases the Jacobian can therefore be held constant for all iterations during one timestep, or even for a sequence of time steps (this again is automatically controlled by the program checking divergence and accuracy).

The application of an implicit numerical method has the advantage that the choice of the time step is not limited by numerical stability requirements. The time step can be adjusted freely to the particular network, excitations and accuracy required by the simulation. Typical time steps are between 5 min for on-line applications and 1 hr for off-line simulations.

GANESI was validated by special experimental field tests (Eibl and Weimann, 1977) and by several industrial users. We may conclude that it is a sound basis for process monitoring and control applications. A version of this program also exists which runs on a personal computer

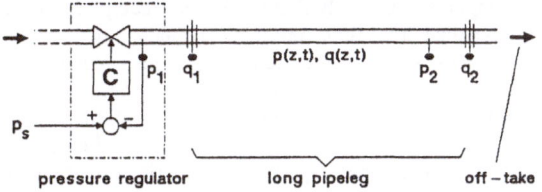

pressure regulator   long pipeleg   off − take

Figure 6.2. Example of a network subsystem.

of the IBM PC/XT/AT type, providing a low cost simulation tool. Program extensions (for details see Lappus, 1987) made by industrial companies are, for example:
- space and time varying gas qualities,
- more detailed modelling of stations,
- inclusion of special stations,
- individual man-machine interfaces.

6.3 Simulation on Multi-Microcomputer Systems

Based on the GANESI model, Schmidt, Maier and Lappus (1986) presented a completely parallel scheme for the numerical solution of the network equations. The method is based on a network-oriented partitioning of the large set of nonlinear equations. A parallel block-iteration scheme is formulated by multiple utilization of certain coupling equations. The block-iteration is based on the Newton-Raphson method combined with a linear block-iteration of the Jacobi type. All blocks of equations can be treated concurrently while the amount of data exchange between the different blocks remains rather restricted. This numerical solution scheme is well suited for the efficient implementation on a parallel computer system of the MIMD type (Maier, 1988). Osiadacz and Salimi (1987) describe another approach to parallel simulation of HPGTN. We expect that one of these rapid parallel simulation schemes will form a key-function for the on-line optimization of network operation, see section 11.

7. STATE ESTIMATION

Process monitoring, predictive simulation and predictive control strategy development require *precise* information about the total, instantaneous network state. It would be very expensive to measure pressure and flow everywhere in the network. Therefore, a tool is required which estimates the total state of pressure and flow by using only commonly available measurements from a few fixed locations in the network. This is a well-known task in the field of automatic control. In our case it can be solved by the application of
- on-line simulation using appropriately modified models,
- modified least square estimation techniques,
- filter techniques, e.g. a Kalman-type filter,
- state observation methods, e.g. a Luenberger-type observer.
From our own as well as industrial experience, we recommend using the state observer approach, which will be discussed in more detail. Let us begin with a short review on the development of state estimation techniques for gas transmission networks.

7.1 Review

Probably the first approach in tackling the state estimation problem for gas transmission networks was made by two American companies (CRC Bethany and Intercomp) in the early 70s. They promoted the application of so-called real-time models. Though not using the term "state estimation", they actually did estimate the network state by on-line simulation for purposes of process monitoring and control, as well as fault detection (Goldberg, 1978, Covington, 1979, Dupont and co-workers, 1980). SSI-Scientific Software Intercomp still offers this approach today. Lappus (1984) discusses this state estimation technique in more detail. Let us just note that this on-line simulation technique is none other than a *special case* of the Luenberger-type observer. However, this special case observer is rather *sensitive* to measurement errors.

Hasting-James (1976) proposed an extended Kalman filter to solve the state estimation problem. However, he used quite a rough lumped-parameter model, permitting only pour results.

A distributed-parameter, extended Kalman filter based on the GANESI model was developed and investigated by Girbig and Lappus (1981). When compared in tests with field data, the application of the Kalman-type filter proved not to be superior to the observer. On the contrary, the computational load involved was uneconomically high. BRITISH GAS also investigated the application of Kalman Filters in gas transmission networks (Whitley and John, 1982 a). They included models of measurement errors in the network model to estimate these errors, too. Although special factorization techniques (Whitley and John, 1982 b) were applied, the results of Girbig were finally confirmed (Lappus, 1987).

Fischer-Uhrig tried to tackle the state estimation problem by a modified least square approach (Fischer-Uhrig and Weimann, 1977). However, his results were applied only once to a small gas transmission network in Germany. An interesting modified least square approach was presented recently by van der Hoeven (1985). GASUNIE of the Netherlands is going to implement this method for on-line state estimation purposes. Van der Hoeven based his approach on physical considerations and explicitly included assumptions about measurement errors. At a closer look, the resulting state estimation scheme seems to be transformable into a Luenberger-type observer as discussed below.

A Luenberger-type observer for the reconstruction of pressure and flow in a gas transmission pipeline was presented by Weimann and Lappus (1977). This approach was then extended to the leak detection problem (Schmidt, Weimann, Lappus, 1978). A complete network observer system for the reconstruction of pressure and flow in complex gas transmission networks was then presented by Lappus (1982). Today, there are several industrial on-line applications of this network observer system in small, medium and large-scale networks with *more than 1200 state variables*.

Carmichael, Pritchard and Fincham (1981), Andersen and Farso (1983), Goldfinch (1984) and Parkinson and Wynne (1986) also investigated observers for gas transmission lines or networks.

7.2 Network observer program GANBEO

State observers were originated by Luenberger (1964). Although his basic idea can be applied to a gas transmission network model, certain modifications and adaptions have to be made.

As a first step, we will consider a single line gas transportation process. The transportation system consists of a single pipeline. The supply $y_{B1} = q(0,t)$ at the beginning and the offtake $y_{B2} = q(l,t)$ at the end of the pipeline are measured. Furthermore, we assume one additional pressure measurement $y_F = p(z_m,t)$ at an arbitrary location z_m, $0 < z_m < l$, along the pipeline. Then a distributed parameter observer of the Luenberger type is formed by the following equations

$$\begin{bmatrix} \partial \hat{p}(z,t)/\partial t \\ \partial \hat{q}(z,t)/\partial t \end{bmatrix} = \begin{bmatrix} -a_1 \cdot \partial \hat{q}/\partial z \\ -a_2 \cdot \partial \hat{p}/\partial z - a_3 \cdot \hat{q}|\hat{q}|/\hat{p} \end{bmatrix} + \begin{bmatrix} g_1(z) \\ g_2(z) \end{bmatrix} \cdot (y_F(t) - \hat{p}(z_m,t)) \qquad (7.1)$$

The boundary conditions of the observer are prescribed by the available flow measurements

$$\hat{q}(0,t) = y_{B1}(t) \ , \ \hat{q}(l,t) = y_{B2}(t) \ . \qquad (7.2)$$

The initial conditions of the observer can be chosen arbitrarily

$$\hat{p}(z,t_0) = \hat{p}_0(z) \ , \ \hat{q}(z,t_0) = \hat{q}_0(z) \ . \qquad (7.3)$$

$\hat{p}(z,t)$, $\hat{q}(z,t)$ are the calculated estimates of the real pressure and flow along the pipeleg. $g_1(z)$,

$g_2(z)$ are space-dependent gain functions.

Let us discuss the design of a *network observer* ready for industrial applications in a more detailed and systematic way. We define a state and parameter vector for each pipeleg

$$x_i(z_i,t) = [p_i(z_i,t), q_i(z_i,t)]^T ,$$

$$a_i = [a_{1i}, a_{2i}, a_{3i}]^T,$$

$$(7.4)$$

and an additional vector $v_i(t)$ of exogenous variables comprising given fixed values or functions of time, such as regulator setpoints, source flows and offtake flows. The elements of $v_i(t)$ form part of the boundary conditions of the i-th pipeleg. In addition, we define the vector valued function $f(\cdot)$ for the right-hand side of Eq. (6.1) and summarize all subvectors $x_i(z_i,t)$ in the network state vector x_{net}. With these definitions and appropriate matrices or matrix-valued operators D_i, V_i, K_i, the total network model including its initial functions $x_i^o(z_i)$ can be represented in a very compact form by

$$\partial x_i(z_i,t)/\partial t = f(a_i, x_i, \partial x_i/\partial z_i) ,$$

$$(7.5)$$

$$D_i x_i + V_i v_i + K_i x_{net} = 0 \quad \text{for} \quad z_i = 0 \text{ and } z_i = 1_i ,$$

$$x_i(z_i,t_o) = x_i^0(z_i) , \qquad i = 1,..,n_d .$$

Available Measurement Data. Discrete point measurement data from the inside or border of the network are available. From a control theoretical viewpoint, we usually differentiate between three types of measurement data (refer to Fig. 6.2 for the examples given below):
- control inputs,   e.g. $p_s(t)$,
- impressed disturbances, e.g. $q_2(t)$,
- outputs,   e.g. $p_1(t)$, $q_1(t)$, $p_2(t)$.

Each type of measurement data has a specific significance for purposes of network simulation and network state reconstruction. While control inputs and impressed disturbances are included in the *simulation model* as boundary conditions, the outputs are not required in the model description. On the other hand, for *state reconstruction purposes*, it is favorable to start with uniform treatment of all types of measurement data. Therefore, we collected them all into one (generalized) measurement vector $y_{net}(t)$. This provision simplifies the handling of "locally redundant" measurements often present in networks. For example in Fig. 6.2, such locally redundant measurements are the pairs q_1, p_1 and q_2, p_2.

In addition, the switching state of each compressor (on/off = 0/1) and valve (open/closed = 0/1) is available. This binary information is summarized in a vector describing the actual network switching state

$$y_{sw}(t) = [0/1, 0/1, ...]^T .$$

$$(7.6)$$

With respect to the observer design for a *large scale industrial system*, we must be aware of transient measurement faults (disconnections or gross errors) which are usually detected by modern data acquisition systems. We will handle this binary information (0/1 = failed/not failed) systematically by means of a vector describing the actual fault state of the measurements $y_{net}(t)$

$$y_{fail} = [0/1, 0/1, ...]^T .$$

$$(7.7)$$

Structure of the Network State Observer System. A prerequisite for the application of a state observer is the reconstructability, or at least detectability, of the entire network state from available measurements. It can be shown (Lappus, 1984) that this requirement is met (at least from a practical viewpoint) by the discrete point measurements commonly available in a network. Based on the well-known general structure of a Luenberger-type observer, we set up the following observer equations for the reconstruction of the total network state

$$\partial \hat{x}_i(z_i,t)/\partial t = f(a_i,\hat{x}_i,\partial \hat{x}_i/\partial z_i) + G_i(z_i,t) \cdot (y_{Fi} - \hat{y}_{Fi})$$

$$\hat{D}_i \hat{x}_i + \hat{V}_i y_{Bi} + \hat{K}_i \hat{x}_{net} = 0$$

$$\hat{x}_i(z_i,t_0) = \hat{x}_i^0(z_i) \ , \quad y_{Fi} = F_i y_{net} \ , \quad y_{Bi} = B_i y_{net} \ ,$$

$$\hat{x}_{net} = [\hat{x}_1(z_1,t),\dots,\hat{x}_{nd}(z_{nd},t)]^T \ , \qquad i = 1,\dots,n_d \ . \tag{7.8}$$

The structure of this observer system, which consists of n_d coupled distributed parameter subobservers, is shown in Fig. 7.1. The boundary conditions, given via $\hat{D}_i$, $\hat{V}_i$, $\hat{K}_i$, will be specified as part of the observer design procedure. y_{Bi} denotes the subset of measurements which is used to prescribe the boundary conditions of the i-th subobserver. These *boundary measurements* are not necessarily identical with those needed for modelling or simulation of the real technical process, i.e y_{Bi} could be structurally different from v_i in Eq. (7.5). The vector y_{Fi} comprises the m_i *feedback measurements* which are used within the linear feedback term $G_i(z_i,t) \cdot (y_{Fi} - \hat{y}_{Fi})$ of the i-th subobserver. In general, these measurements are not restricted to belonging only to the i-th subsystem. $\hat{y}_{Fi}$ denotes the equivalent measurements reconstructed by the observer system. The specification of the elements of the gain matrices

$$G_i(z_i,t) = \begin{bmatrix} g_{11}^i(z_i,t),\dots\dots,g_{1mi}^i(z_i,t) \\ g_{21}^i(z_i,t),\dots\dots,g_{2mi}^i(z_i,t) \end{bmatrix} \tag{7.9}$$

is part of the observer design procedure. The initial observer states are assumed to be arbitrary, but physically meaningful functions.

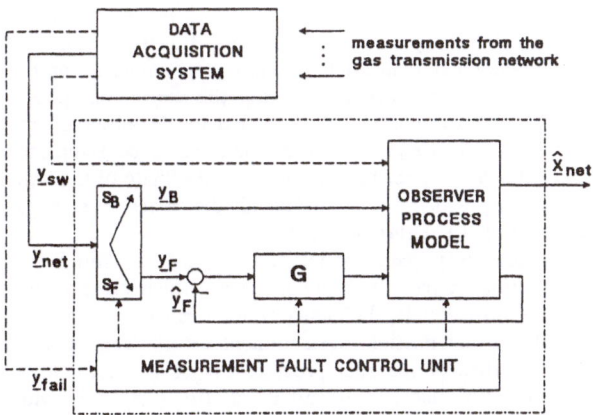

Figure 7.1. Structure of network state observer.

Design of the Observer System. The design of this observer system can be based on the following considerations and specifications:

1) The resulting observer must be *insensitive* to the presence or sudden occurrence of rather constant, but unknown model parameter errors and constant as well as time-varying measurement disturbances.

2) The reconstruction errors should fulfill the realistic objective

$$|\tilde{x}_{i,k}(z_i,t)| < eps_{i,k} \quad \text{for} \quad t > (t_0 + t_s)$$

$$i = 1,\ldots,n_d \quad , \quad k = 1,2 \ .$$

(7.10)

The reconstruction error bounds $eps_{i,k}$ and the settling time t_s may be considered as part of observer specifications.

3) The reconstructed state is supposed to be consistent, i.e. a *physically meaningful state* in $0 < z_i < l_i$. For example, the reconstructed flow should never be directed *against* the reconstructed spatial pressure drop along a pipeleg.

4) The basic function of an observer is simply to produce a real-time estimate of the actual process state, i.e. once the observer has been started at some initial state, the reconstructed state must first of all converge to and then keep track of the transient state of the process. We call the first operational phase the *converging phase* $(t_0 < t < t_0 + t_s)$ and the second one the *tracking phase* $(t > t_0 + t_s)$. By designing special observer gain matrices G_i for each of these phases and switching from G_i^{conv} to G_i^{track} at time $t = (t_0 + t_s)$, the observer system is adapted to specific objectives within these phases. Of course, t_s is not known beforehand, but we can acquire an estimate of the settling time by experience or by on-line monitoring of the available error signal $\tilde{y}_F(t) = (y_F(t) - \hat{y}_F(t))$.

5) Though often neglected, the selection of an appropriate model of the process to be observed plays an essential role in the design of a Luenberger-type observer. We call the particular model on which the observer equations (7.8) are based the *observer process model* in order to distinguish it from the common process model, Eq. (7.5), which describes the real technical process for simulation purposes.

Observer process model. A favorable simplification of the network model is to neglect the inequality constraints resulting from the lumped parameter subsystems, Eq. (6.3). This simplification reduces problems with the computational realization of these constraints as part of the observer equations. It causes an increased convergence rate during the converging phase, while during tracking the "neglected" constraints are automatically satisfied by the observer system, provided the network state is observable. In passing we note that, as a consequence of this model simplification, the reconstructed state can not be considered an estimate of the actual state during the converging phase. However, in practice, we can avoid this deficiency by starting the observer operation "well in advance" using filed measurement data.

To complete the observer process model in Eq. (7.8), we still have to specify those boundary conditions of the distributed parameter subobservers which are not directly given by the network structure. Choosing flow defining boundary conditions wherever possible has advantages with respect to observability and possible measurement errors. This choice is made independently of the technical properties of the real process. Let us illustrate this by the example shown in Fig. 6.2. Using coordination matrices B and F, cf. Eq. (7.8), we divide the available measurements $y_{net}(t)$ into boundary $y_B(t)$ and feedback measurements $y_F(t)$,

$$y_B(t) = By_{net}(t) = [q_1(t), \quad q_2(t)]^T \, ,$$

$$y_F(t) = Fy_{net}(t) = [p_1(t), \quad p_2(t)]^T \, ,$$

(7.11)

i.e. for state reconstruction purposes we prescribe the boundary conditions at $z=0$ by flow $q_1(t)$ instead of pressure $p_1(t)$ as needed for simulation purposes. In other words, in our observer process model we replace the actually existing pressure regulator by an "artificial" flow regulator.

Observer gains for the converging phase. During this phase all observation errors, Eq. (7.10), should converge *close* to zero, starting from possibly large initial errors. Since measurement and model parameter errors are obviously of minor importance in this phase, we design the corresponding observer gains based on the nominal model parameters and measurement data assumed to be free of errors.

Lappus (1984) describes an easy-to-handle procedure for deriving appropriate observer gains which are functions of space and directly related to the physical process, i.e. the assumed spatial pressure drops along the pipelegs at initial time t_o. One of the advantages of this process-oriented design technique for observer gains is the fact that it can easily be extended from simple pipelines to complex network configurations. Figure 7.2 illustrates the design procedure for a simple subsystem consisting of one pipeleg and two feedback measurements

$$y_F(t) = [p(z_{m1},t) \, , \, p(z_{m2},t)]^T \, .$$

(7.12)

An appropriate feedback term of the corresponding subobserver is given by

$$\begin{bmatrix} g_{11}(z,t) & g_{12}(z,t) \\ g_{21}(z,t) & g_{22}(z,t) \end{bmatrix} \cdot (y_F(t) - \hat{y}_F(t))$$

(7.13)

$$g_{11} = k_o k_1(z) \, , \quad g_{12} = k_o k_2(z) \, ,$$

$$g_{21} = 0 \, , \qquad g_{22} = 0 \, , \qquad k_1(z) + k_2(z) = k(z) \, .$$

Figure 7.2 b shows that the linear profile $k(z)$ is derived from the pressure profile $p^*(z)$, which is, in turn, a rough approximation of the (unknown) real pressure profile $p(z,t_o)$ at initial time. The constant k_o mainly determines the rate of convergence for the reconstruction error along the total pipeleg. Although k_o can be chosen arbitrarily, we should avoid unnecessarily rapid observer dynamics because of problems involving the *numerical solution* of the observer equations. Due to the setting of $g_{21} = g_{22} = 0$, we can expect the reconstructed state to be consistent as defined above.

Figure 7.3 illustrates the result of this design procedure in the case of a small network consisting of four coupled pipelegs with a single feedback measurement $y_F(t)$.

Observer gains for the tracking phase. During this theoretically infinitely long phase, the observer system is expected to track the actual network state as precisely as possible. For this reason, the insensitivity of the observer to modelling and measurement errors is most important. This requirement is met by using Dirac-type gain functions as depicted in Fig. 7.4 for a simple case.

Provisions against measurement failures. Having designed the observer system with two well defined sets of measurements (set S_B, being specified by the coordination matrices B_i, is required to define the boundary conditions, set S_F, being specified by matrices F_i, is needed in the feedback loops of the observer system, see Eq. (7.8) and Fig. 7.1, every disconnection or serious disturbance of a measurement obviously cause problems for the on-line state reconstruction process. Such gross measurement faults occur rather frequently in practice. However, they are detected by the data acquisition system. Consequently, the observer includes a "measurement fault control unit" to deal with such transient errors, which may last only a few minutes or several days. In most gas transmission networks, certain pressure and flow measurements are located next to each other. By using such "locally redundant" measurements or, if necessary, predefined default values (functions of time), the measurement sets S_B and S_F are reconfigured whenever a measurement fails or is restored. The gain matrices $G_i(z_i,t)$ are also adapted in order to eliminate or at least reduce the effects of such measurement faults on the reconstructed state information.

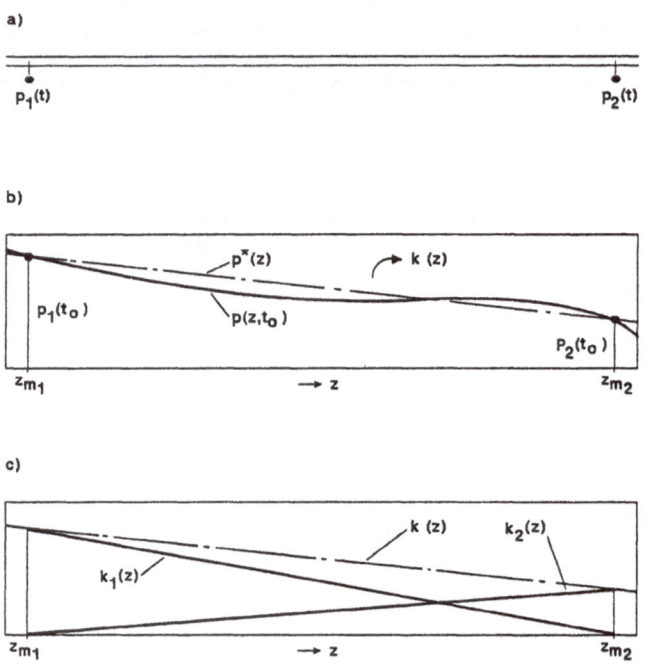

Figure 7.2. Illustration of the design procedure of the observer gain functions $k_1(z)$ and $k_2(z)$ for the converging phase in case of a pipeleg with two pressure measurements p_1 and p_2.

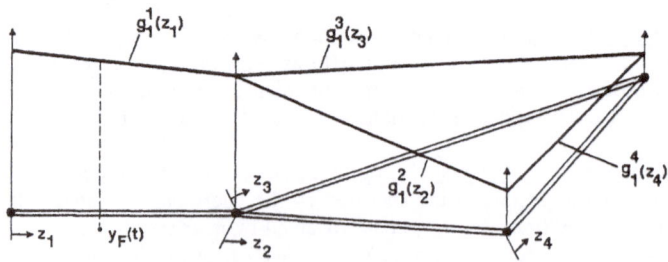

Figure 7.3. Observer gain functions for the converging phase in case of a simple network.

Implementation of the observer system. The resulting large-scale observer system as shown in Figure 7.1, was implemented as a portable FORTRAN program consisting of about 7000 statements (Lappus, 1982). This program, named GANBEO, can be applied to on-line state observation of arbitrarily meshed HPGTN. It proved to be a very reliable tool because of its comprehensive error handling and recovery provisions. Based on a graph-oriented description of the network topology (input data), the observer system equations, Eq. (7.8) including the gain matrices, are automatically generated by the program and then solved numerically by the same means as used in GANESI, see section 6. The time required for the observer computations is comparatively small, e.g. on a modern 32-bit minicomputer the reconstruction of 100 pressure and flow variables only took about 0.05 seconds per sampling interval (normally ranging from 5 to 10 minutes).

Typical Results. The observer system has been tested with field data from a section of the gas transmission system in the Netherlands. The considered network consists of about 50 pipelegs with a total length of about 400 km. The measurement data resulted from a typical transient flow process disturbed by time variable off-takes and by the control actions of three valves. Input data fed to the observer are the boundary measurements $y_B(t)$ (2 pressures, 20 measured off-take flows; 4 estimated off-take flows) and the feedback measurements $y_F(t)$ (5 pressures). Based on these input data, the observer reconstructs about 100 flow and pressure states. The convergence properties were tested by using arbitrary initial conditions for the observer states. The settling time t_s proved to be approximately 0.5 hr, which is sufficiently small for practical applications. Figure 7.5 shows a sample result of the most important tracking behaviour. Note that the measured node pressure p was not included into the reconstruction process; p was only used for checking the accuracy of the reconstructed pressure $\hat{p}$. The absolute tracking error of max. 0.4 bar lies within the accuracy of the measurement data.

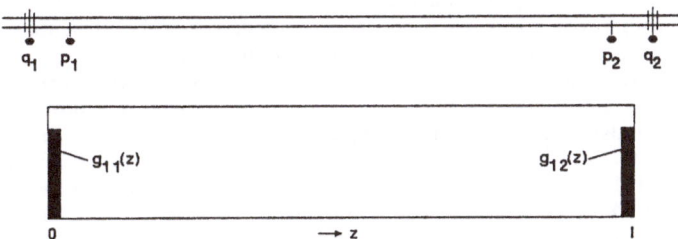

Figure 7.4. Observer gain functions for the tracking phase in case of a single pipeline.

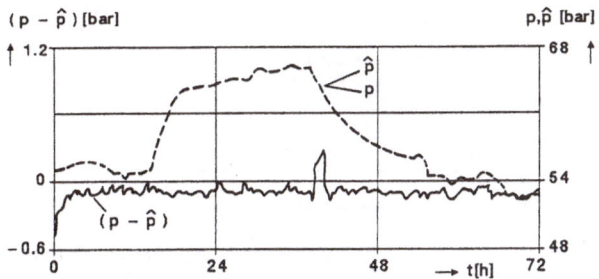

Figure 7.5. Observer test with industrial field data: Typical tracking of a node pressure.

420

<u>Applications.</u> Today, several *industrial on-line applications* of this observer system can be found ranging from medium-scale to large-scale networks with about 600 pipelegs and 600 nodes. Most of them have already been in successful operation for years. The overall accuracy of state reconstruction is usually twice as good as the accuracy of the measurements used. This is due to the good filtering properties of the observer in the tracking phase.

8. FAULT DETECTION AND DIAGNOSIS

This section discusses process monitoring and fault detection for HPGTN. The basic supervision of network equipment, such as control and measurement devices, is a task of the SCADA system as discussed in section 4. However, the application of more sophisticated schemes permits major enhancements. We discuss a fault detection scheme which is based essentially on the above network state observer system GANBEO. Using a fault-sensitive approach, fault detection itself is then performed by evaluating the estimated network state and the estimation errors. Two examples are discussed in more detail: detecting and locating leaks and sensor fault detection.

A dynamic network state observer provides a sound basis for instrument and process fault detection. This approach is well known from literature (e.g. Isermann, 1984; Tylee, 1986). The basic idea of this approach is rather simple. On one hand, we have data derived from measurements (either directly measured or calculated from measurements) and, on the other hand, we have the same type of data provided by a process model or state estimator. Then, we may conclude by comparison that some fault has occurred if there are any discrepancies greater than the prespecified limits. However, a straight forward application of this method to large-scale industrial processes imposes some problems in handling and using the vast amount of data as well as in discriminating between effects due to abnormal faults and normal, inaccurate measurement data, or normal operations, such as repair work generating abnormal measurement data. A decision support or expert system, which includes "intelligent" knowledge-based functions for network monitoring and fault detection, can help to overcome these difficulties.

When revising data, information and decisions involved in the dynamic process of monitoring a network in real time, we get a characteristic scheme as shown in Fig. 8.1. For simplicity, all static data (e.g. network topology) are omitted. *Basic data* for all further considerations are the actual measurement data, which are continuously updated. (Example: some hundred pressure and flow variables and some thousand other data are typical for a large gas transmission network.) These data provide partial information, only. Therefore, only *basic decisions* with respect to monitoring and control should be based on them. However, such basic decisions are required for rapid reaction in emergency situations. Next, we will use a network state esti-

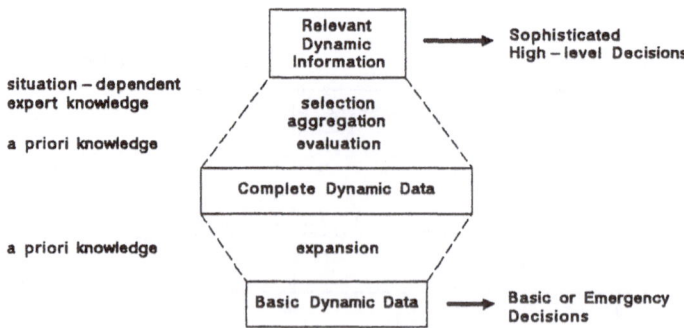

Figure 8.1. Dynamic process of network monitoring: data, information, and decisions.

mator to perform data expansion. In other words, we will merge basic data with a priori knowledge of the physical process and its dynamics in order to receive complete information about the instantaneous network state. (Example: the result may consist of 1000 to 2000 pressure and flow values.) Eventually, we will perform another data expansion by running a predictive simulation for the purpose of predictive supervision. (Example: if we choose a look-ahead period of 24 hrs and hourly intervals only, then 24,000 up to 48,000 pressure and flow values will become available.) It is difficult to base profound decisions on this vast amount of data which is, in addition, updated every few minutes. Therefore, we have to apply *data evaluation* and *data aggregation* schemes in order to extract the relevant information required for sophisticated high-level decision-making. At this end we require not only prespecified algorithmic operations, but also the situation-dependent expert knowledge and expert interaction of an experienced dispatcher.

8.1 Detection of Measurement Faults

Let us first illustrate the capability of the network observer system to detect sensor and data transmission faults by results obtained from an *industrial* on-line application of GANBEO. In this application, the observer is part of an advanced process monitoring and control system (Kersting et al, 1984). The network consists of about 50 nodes, 50 pipelegs, and 1 underground storage. Observer input data are 25 flow measurements from network boundaries (forming set S_B, see Fig. 7.1) and 12 pressure measurements (forming set S_F). Some results of on-line state estimation are shown in Fig. 8.2. The time history of the daily average of the reconstruction error at a typical measurement point is shown in Fig. 8.2 a. With one exception, this error is less than 0.5 bar. The exception was caused by a *fault in the data acquisition system*. Note that the measurement failure control unit as discussed in section 7.2 was not activated at that time.

In Fig. 8.2 b we see the average of reconstruction errors over a period of 40 days at all 12 measurement points (ordered by value, not by spatial location!). All absolute mean values, with one exception, are less than approximately 0.5 bar. This exceptionally large estimation error at point no. 12 clearly indicates a fault. Indeed, the gas company checked this measurement device and detected an *sensor fault* already existing a long time before its detection by on-line state estimation. Three other faults in the telemetry system were detected during the first months of on-line state estimation. All faults were first detected by *unusually* large deviations of the mean

b)

a)

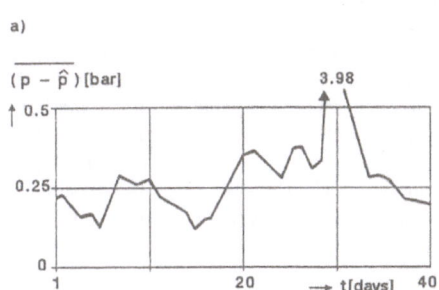
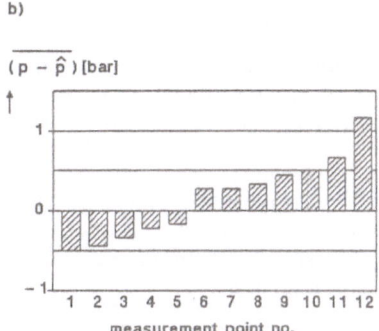

Figure 8.2. Fault detection by means of on-line state reconstruction:
a) daily average of reconstruction error at a typical measurement point,
b) 40-days-average of reconstruction errors at all 12 measurement points.

values of reconstruction errors, and then located with the help of expert knowledge of the observer function and telemetry system.

8.2 Leak Detection

The detection of suddenly occurring leaks is a special task of network monitoring. In this context, we are interested in rapidly detecting and locating leaks with a flow rate ranging from 2 to 100 % of the average gas flow (note, it is easy to detect, but difficult to quickly locate a leak of 100 % , i.e. a pipeleg rupture).

In literature we find proposals to use mathematical models of transient flow for leak detection purposes (Goldberg, 1978; Covington, 1979 a,b, Dupont and co-workers, 1980). Schmidt, Weimann and Lappus (1978) proposed a leak detection scheme for gas transport pipelines based on the observer approach and evaluation of estimation errors. Chapman and Pritchard (1982), as well as Billmann and Isermann (1984), basically used the same approach (with some modifications). All methods mentioned do not include the leak in the model, i.e. they are based on a "fault sensitivity approach". Digerness (1980) presented a "fault model approach" for leak detection in oil pipelines. He used a bank of Kalman filters, each of which models a leak at a predefined location in the pipeline. The innovations of the filters were then used by a multiple model, hypothesis probability test to decide whether there is a leak and to locate it. Benkherouf, Allidina and Singh (1984) presented a leak detection scheme for gas pipelines which is also based on a Kalman filter. However, they used a single model, but included artificial leak states at predefined locations along the pipeline. A critical assessment of various leak detection techniques can be found in Lappus and Schmidt (1987).

We will discuss a two-stage method for detecting and locating leaks in gas transmission networks using the GANBEO network observer system and a fault sensitive approach. The same method can also be used to estimate offtake flows (e.g. in case of an sensor fault or missing measurement device). Reliable results are due to the subsequent use of two basically different schemes, first, a method which we call "dynamic mass monitoring" for *detecting* a leak situation, and, second, a method which we call "estimation error processing" for *locating* the leak position. Reliability is the most crucial point of leak detection: every wrong alarm and automatic shutdown of a network segment is as bad as no alarm in an actual leak situation.

8.2.1. Dynamic Mass Monitoring. We assume that the instantaneous pressure and flow state in a gas transmission network is continuously estimated using the GANBEO network observer. In the tracking phase, the estimated pressure is known to be rather a good estimate of the real pressure. This is true not only at the discrete measurement locations, but everywhere in the network. In other words, the network observer continuously provides the time-varying spatial pressure profiles along all pipelegs. Based on these pressure profiles we can compute the actual amount of gas being stored in the total network. By comparing this value with the net flow rate as given by all time-varying supply and offtake flows, we can detect the occurrence of a leak. A similar method for fluid pipelines, however without observer, was discussed by Thompson and Skogman (1987).

Assume $dm_{in}(t_k)$ to be the total mass of gas flowing into the network during a certain time interval $dt = t_k - t_{k-1}$ (e.g. the measurement sampling interval) and $dm_{out}(t_k)$ the corresponding offtake mass. Then, the net rate of mass flowing into or out of the network during the interval dt is given by

$$dm_{ext}(t_k) = dm_{in}(t_k) - dm_{out}(t_k) \ . \tag{8.1}$$

Assume $m_{st}(t_k)$ and $m_{st}(t_{k-1})$ to be the mass of gas stored in all pipelegs of the network at time t_k and t_{k-1}, respectively. Then, the difference

$$dm_{st}(t_k) = m_{st}(t_k) - m_{st}(t_{k-1}) \tag{8.2}$$

describes the variation in gas mass stored during time interval dt. Now, leak detection at time t_k can be based on the simple, but rather effective formula

$$dm_{ext}(t_k) - dm_{st}(t_k) = \begin{cases} 0 & \text{without leak} \\ dm_L(t_k) & \text{with leak.} \end{cases} \tag{8.3}$$

$dm_L(t_k)$ is the leak rate, i.e. the mass of gas flowing through the leak during time interval dt. If a sudden leak occurs, Eq. (8.3) will, of course, not become immediately true. We must wait until dm_L indicates the actual mass leakage rate. However, this delay seems rather small when compared to other known leak detection techniques being applicable to *industrial, transient* gas transmission problems.

Fig. 8.3 shows the results of a simulation study. Data from the pipeline under consideration and time histories of supply $q_1(t)$ and offtake $q_2(t)$ are depicted in Fig. 8.3 a. A leak flow of 5 percent of the supply was assumed to occur at 8.30 hrs. Observer data input is represented by q_1, q_2 (forming set S_B) and p_1, p_2 (forming set S_F). The time history of $dm_{leak}(t)$ is shown in Fig. 8.3 b. This figure also shows $dm_L(t)$ when no leak is present. The time interval dt was chosen equal to the simulated data sampling interval (5 min). We recognize that $dm_L(t)$ clearly indicates the occurrence of the leak two or three time steps (10 ... 15 min in this example) after leak flow has begun. One hour after the occurrence of the leak, the signal $dm_L(t)$ approximately reaches the actual value of the mass leakage rate (700 kg/5 min).

Detailed analysis and *industrial field tests* proved that leak detection by dynamic mass monitoring, as described above, is rather insensitive to measurement and modelling errors. This is due to the good tracking behaviour of the network observer system and the fact that no absolute, but only relative values (variations of differences over a certain time interval) are used to detect a leak situation.

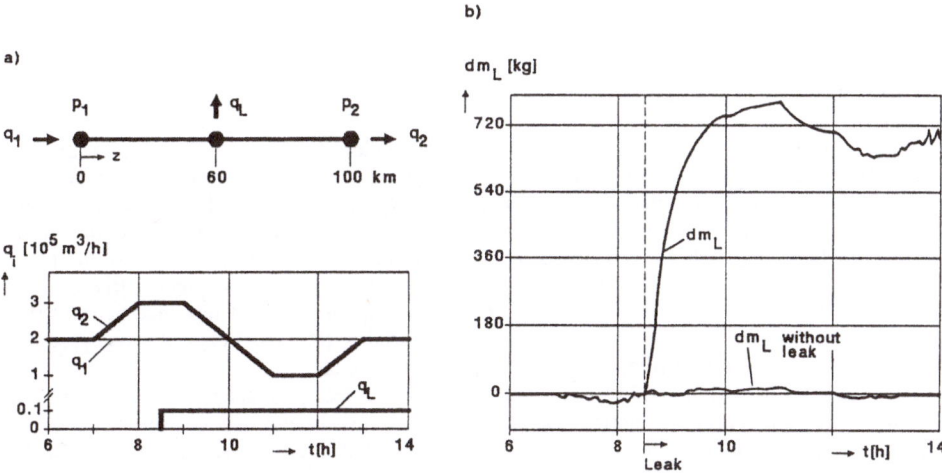

Figure 8.3. Leak detection by dynamic mass monitoring:
a) pipeline with supply q_1, offtake $q_2(t)$, and leak flow rate q_L,
b) estimated leak mass rate dm_L.

Of course, we have to specify appropriate thresholds for $dm_L(t)$ in order to indicate a leak situation. Though choosing these thresholds appears uncomplicated because of the insensitivity of this method to modelling and measurement errors, network-specific expert knowledge is required to select appropriate thresholds and situation-dependent expert knowledge is needed to adapt these thresholds and to react in the case of a leak alarm due to this algorithmic evaluation.

Finally, we note that this leak detection method can be applied to arbitrary gas transmission networks, provided all supply and offtake flows but the one(s) to be estimated are known. Preferably, we apply this technique to the total system *and* to subsystems at the same time in order to obtain additional information about the leak location. Once having detected a leaking segment in a transmission network, we propose applying an estimation error processing technique in order to locate the leak position more precisely.

8.2.2. Estimation error processing. Let us assume that the instantaneous state of pressure and flow is continuously estimated in a segment of the network under consideration. In the tracking phase $(t > t_0 + t_s)$ the errors between the estimated and measured pressures are usually very small. Now, assume a suddenly occurring leak. As a consequence, the estimation errors will increase until they reach a steady state value provided appropriate boundary measurements and observer gain functions are used. All of this information, e.g. the space and time dependency of estimation errors, their absolute values and their steady state values can be used to detect the occurrence of a leak and to estimate its location. However at the second step of this two-stage approach, we only use the latter property.

In practice, we have to apply standard signal processing algorithms (e.g. signal filtering and/or correlation techniques) in the evaluation of estimation errors to derive signals indicating leak position. Nevertheless, in case of a small leak, these signals are rather small when considering the typical accuracy of measurement data available in industry today. In particular, this holds true for the most interesting period of time of, say, 1 hr immediately following the occurrence of a leak. However, since we have already detected a leak by "dynamic mass balancing", it is now reasonable to assume that deviations in estimation errors are caused by the leak and not by other disturbances. But again, additional expert knowledge is required to efficiently implement this tool when applied in practice (c.f. section 12).

9. MODEL PARAMETER ESTIMATION

The model of transient gas flow as required for simulation (section 6) and state estimation (section 7) includes several parameters which determine the results. However, in practice, it seems not to be a serious problem to choose appropriate parameter values because most parameters represent real physical quantities. This holds true for the gas properties and the parameters of the lumped subsystems, such as regulator or compressor stations, as well as for most of the parameters of the pipelegs, e.g. length and diameter. The only exception is the pipeleg friction loss parameter (b_i in Eq. (6.1)). These n_1 parameters must be experimentally determined because they represent several physical phenomena, which are not included in the model explicitly, e.g. pressure loss due to the pipeleg wall friction, pipeleg bendings, filters, etc.

The traditional method of estimating these friction loss parameters is one that relys on the steady state equation of gas motion. This method uses pressure and flow measurements obtained from the system. The data are averaged in time and applied in an inverse way to compute the friction loss parameters. The drawback of this method is clearly the assumption of steady state gas flow which is never really true.

Stoner and Karnitz (1974) presented a *least square estimation scheme* for computing the pipeleg friction loss parameters. Their method (and all others discussed below) eliminates the necessity of attempting to hold the system in a near steady state condition during the experimental parameter determination. Fincham (1981) applied this method to real networks. The performance proved to be good if the networks were not too complex.

Sedykh and co-workers (1979), as well as Billmann (1984) presented an on-line adaptation scheme for computing the pipeleg friction loss parameters. However, their assumption of non-biased measurements with additive noise is not realistic for HPGTN.

Girbig and Lappus (1980) investigated the estimation of pipeleg friction loss parameters by means of an augmented state observer scheme. They included the friction parameters as additional state variables. By this combined state and parameter on-line estimation technique one can even compute spatially varying friction loss parameters between two pressure measurements. The results of this estimation method are excellent provided there are no biased measurements. However, this assumption is not true in reality.

In industry today, pipeleg friction loss parameters are chosen *manually* by means of on-line simulation or state estimation. The parameters are determined such that the average of the differences between measured and calculated pressures is minimized. This task has usually to be done only once, because these parameters are rather constant in time. This simple manual parameter estimation technique seems to be sufficient according to the results reported by industrial companies.

10. LOAD PREDICTION

The control of a HPGTN always requires the prediction of the expected loads. *Off-the-shelf tools* for this task are available in industry, today. These tools are typically based on combined deterministic and statistic approaches, such as Box-Jenkins time series analysis, regression methods, linear or non-linear extrapolation techniques. The method chosen for a particular offtake depends on the type of customer (municipal, industrial plant). Detailed weather forecasts are important for the prediction of loads with an essential part of heating gas. The typical accuracy of predicting *individual loads* for the next 24 hrs is in the range of 1 ... 10 % . However, in case of some industrial plants, the predicted load may be completely wrong (100 % error) during certain time intervals.

Load prediction methods are discussed, for instance, by Ady and Rauch (1984), Hass (1984), Formaggio and Luciano (1984), Weimann and co-workers (1985), Bowler (1987).

11. OPTIMAL NETWORK CONTROL

In literature we find various attempts to apply optimization techniques to the design and management of HPGTN. These approaches include off-line tasks, such as the optimal design of a pipeleg, the optimal location of a compressor station, the pre-calculation of optimal control strategies, and on-line tasks, such as optimal daily operation. As the computational load of most of these methods is rather high, powerful superminicomputers or even mainframes are still required to obtain solutions. In this volume on microprocessor applications we will therefore only outline some general aspects of optimal network control.

The primary objective of a gas transmission company is, of course, to maximize the company's revenue by buying and selling gas from a few suppliers to many customers. This goal includes the most important objective of a secure supply, i.e. the company must always meet the actual demand. Any failure in doing so results in additional short term losses (penalties) and long term cost (loss of confidence and customers). Gas supply and delivery are subject to contracts including economical as well as technical clauses. While the company's marketing department has to negotiate favorable prices, the engineers have to investigate the transmission capacity of the network and provide means for the cost-efficient daily operation permitting the transmission of the

426

contracted gas. This mixed eco-technical problem of network management proves to be a "many-headed monster" (Francis, 1982).

A primary difficulty is the existing *imbalance of supply and demand* on every time horizon. Fig. 11.1 shows the typical variations in total gas demand over a year and two typical patterns of total contracted supply. A sample pattern of daily gas demand is shown in Fig. 11.2. This well-known diurnal swing shows a variation of up to $\pm$ 50 % of the average flow in a regional distribution network, and up to $\pm$ 25 % in a transportation system. On a monthly or weekly basis we find similar patterns which are caused by commercial and industrial gas consumers (production start-up and shut-down, weekend, holidays, etc.). Sudden weather changes, equipment failures, or maintenance outages cause non-regular variations. On the other hand, for various reasons, gas wells have to be operated at rather a constant load. Therefore, gas producers prefer to deliver gas at constant flow rates, cf. Fig. 11.1. Maximal variations in gas flow of up to $\pm$ 10 % are usually forced by pricing. The total cost of supply depends not only on the total quantity of gas, but also on the maximal supply flow rate per time unit. A time unit is either a day or an hour. (Actual pricing is much more complicated, however, we can restrict our considerations to this important, basic dependency.)

Thus, the management of gas storage over various periods of time is an essential problem in network design and control. Storage facilities may be classified by their operational horizon. Aquifers or depleted gas fields are used as *seasonal storages. Middle term storage facilities* are used for peak shaving during only a few weeks each year. Examples are: liquified natur-

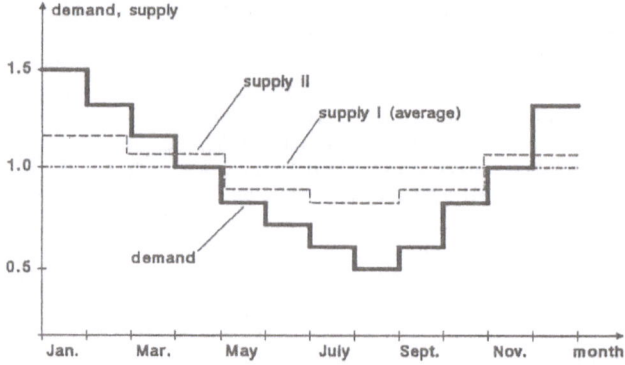

Figure 11.1. Typical (normalized) pattern of annual gas demand and supply.

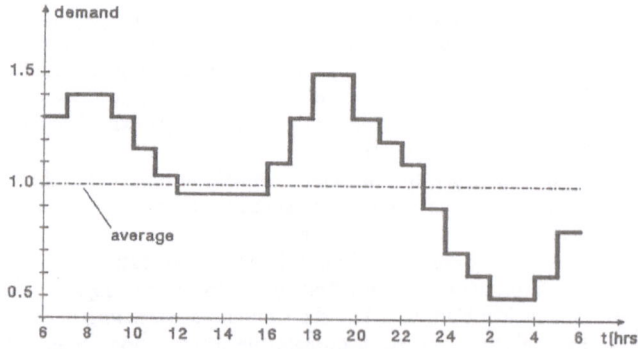

Figure 11.2. Typical (normalized) pattern of daily gas demand in a regional HPGTN.

algas (LNG) tanks, salt cavity storages. Examples of *diurnal storage facilities* are: low- and high-pressure tanks, and "line-pack". Line-pack is the volume of gas held in the total network. Whenever pressures exceed the minimum some of this gas can be used as "storage". This is an important fact and should be considered in network control. An "inverse" form of gas storages are *interruptible customers*, which are used for short term peak shaving. Bowler and Connor (1987) describe properties of different storage facilities in more detail.

The second basic problem in the management of a HPGTN is the *necessity of predictions* over various time horizons and the resulting uncertainty. A gas transmission company usually buys gas from producers on the basis of long term contracts covering 10 to 30 years. The contracted quantity of gas is based on the predicted future gas demand and the company's specific marketing strategy which may include an extension of gas supply. Network extensions enlarging its maximal capacity and additional storage facilities have also to be planned well in advance, e.g. 5 to 10 years. Short term predictions are required to forecast the diurnal swing which is mainly influenced by weather (temperature and wind). In general, the company has always to predict the total amount of gas required over a certain period of time and the maximal flow rate during this period. As prediction errors are inevitable, both, network design and operational control, have to provide adequate security margins which cause, of course, additional cost. The security needed must be carefully judged. It is usually expressed as a certain margin of pressure above the minimum essential at any offtake point of the network. Some large offtakes which have a limited line-pack upstream may need a considerable margin of pressure to accommodate the unanticipated variations of demand by giving time to adapt supply rates.

The third general aspect are *operational costs*, e. g. compressor fuel costs. Such costs are significant. Francis (1982) states a sum of 25 million pounds for British Gas in the year 1982. Therefore one obvious optimization goal is to minimize network wide fuel usage of compressors. This goal can be expanded to total running cost minimization by including factors such as marginal maintenance costs and compressor startup or shutdown costs due to special operations.

Considering gas flow from its well through the networks of two or more intermediate transmission and distribution companies to the end consumer, the cost of providing the total amount of required pressure energy and storage capacity is more or less shared by all partners involved. The "boundary conditions" between the different companies are subject to contractual agreements. These contracts prescribe or assure certain limits, such as
- minimal and/or maximal pressures,
- the total quantity of gas being delivered over a period of time,
- maximal and minimal flow rates,
- gas quality, etc.

These constraints are essential in network control; they are enforced by substantial penalties.

The combined problem of network design and operational control can be formulated as optimization problem, such as

minimize

over a certain time horizon T

the cost of (gas supply + gas storage + gas compression + security margins + ...)

subject to

the physical properties and various technical and contractual constraints.

We note that we cannot expect a unique, but only pareto-optimal solutions depending on the desired degree of security of supply. The whole problem is extremely complex when taking into account all essential requirements and constraints. We obviously have to apply *decomposition techniques* in order to obtain at least partial solutions. However, we must be very careful in minimizing isolated parts, e.g. compressor costs, because there would be no point in it if we then failed to supply the required gas.

We recognize two underlying, natural hierarchies. The first one is a *time-scale hierar-*

chy related to the time horizon T under consideration:
- strategic planning (decades),
- long term planning (years),
- short term planning (weeks, months)
- operational control (hours, days).
The *second hierarchy* is given by physical extensions:
- network level,
- station or storage level,
- single machine level.
In literature, we find various approaches to optimal network design and operational control. These attacks are usually concerned with one or sometimes two levels of the hierarchies, above. Traditionally, optimal planning as well as operational control were based on steady state flow models. However, there is a distinct trend to apply transient flow models and dynamic optimization methods on all planning horizons and levels. Let us just mention a few references: Larson and Wismer (1971), Lewandowski (1971), Lo and Brameller (1971), Piekarski (1982), Villon and Yvon (1982) investigate fuel minimization of compressor stations along a pipeline or within a network. Predictive, moving horizon on-line optimization techniques were investigated by Lappus and co-workers (1984) and Marques (1985). Rather a comprehensive transient optimization scheme was presented by Mantri and co-workers (1985). They use a three-level approach: first, initialization by steady state optimization, second, network-wide fuel gas minimization using transient network models, and third, optimal selection of compressor units at the station level.

12. OUTLOOK TO KNOWLEDGE-BASED PROCESS MONITORING AND CONTROL

State-of-the art process monitoring and control schemes are, as we have seen, essentially based on a comprehensive SCADA system (collection, pre-analysis, management and display of static and dynamic data in real time), advanced algorithmic tools (transient network simulation, dynamic network state estimation, load prediction, predictive optimization), and, last not least, the knowledge of experienced dispatchers.

For the purpose of *network monitoring and fault detection* we have discussed the algorithmic tools mentioned above for the expansion of basic measurement data to complete information about the network state, which can then be evaluated to detect malfunctions. This demonstrates essential progress when compared to fault detection on the basis of measurement data, only. However, we have also seen that additional expert knowledge is required to efficiently implement these tools.

We know from experience that the same conclusion holds true for *network control*. The task of continuously developing a safe and economic strategy for on-line, real-time network control can neither be solved by pure, algorithmic tools (e.g. predictive simulation and optimization with a moving horizon) nor manually by the dispatcher who employs heuristical rules. Again, a combination of both, algorithmic tools and expert-knowledge, is required for safe and economic network operation.

These requirements can be met by a comprehensive *decision support system for network monitoring and control* which is composed of subsystems of two basically different types:
- modules providing algorithmic operations or problem solvers,
- modules providing knowledge-based operations and problem solvers.

A possible structure of such a system is shown in Fig. 12.1. Appropriate tools for the realization of such a system are available from the AI research field, see, for instance, Talukdar and co-workers (1986). A more detailed description of this scheme is given by Schmidt and co-workers (1986) and Lappus (1986). Let us just outline the functional relationship between the subsystems with respect to fault detection and diagnosis.

The "intelligent" unit for fault detection/diagnosis is a rule-based, on-line, real-time expert system for malfunction analysis (Tao, Lappus and Schmidt, 1988). The unit becomes periodically active (e.g. in the measurement sampling interval). It sends a message to the "blackboard" (common communication handler) asking for the latest, complete network state, which is provided by the algorithmic unit for state estimation. Then, based on this actual state and transmitted data, it performs fault detection and diagnosis in a manner similar to a human expert. That is, for the purpose of leak detection, it uses both algorithmic leak detection and location schemes as described above and network and situation-dependent expert knowledge, e.g. the knowledge about ongoing construction or repair work. The unit tries first to detect the occurrence of a leak situation, second to locate the leaking segment of the network, third to locate the leaking element of the segment and last, to inform the dispatcher and recommend appropriate action. The dispatcher then asks the unit to display its "reasoning" or he performs additional tests (either on the network simulator or on the real network) in order to be sure that there really is a leak. The entire procedure is essentially based on graphic (and in the future perhaps on natural language) communication between the dispatcher and his decision support system.

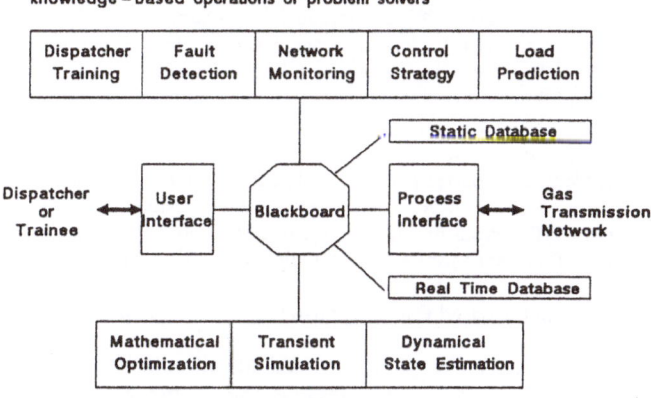

Figure 12.1. Basic structure of a decision support system for network monitoring and control.

13. REFERENCES

Ady, E. and E. Rauch (1984). 'Das System Prosit zur vorausschauenden Gasnetzsteuerung'. In: Trauboth, H. and A. Jaeschke (Ed.) *Prozessrechner 1984*, Karlsruhe, Sept. 26-28. Springer, Berlin.

Andersen, T.R. and H.F. Farsoe (1983). 'Estimation of the pressure in a natural gas pipeline'. Paper no. 10, ACI 83, Copenhagen.

Baur, P.S. (1987). 'Radio telemetry challanges wire in remote monitoring and control'. *InTech*, **34**, p. 9-14.

Benkherouf, A., Allidina, A.Y. and M.G. Singh (1984). 'Detection and location of leaks in gas pipelines'. UMIST Control Systems Centre Report No. 603, Manchester.

Billmann, L. and R. Isermann (1984). 'Leak Detection Methods for Pipelines'. 9th IFAC World Congress, Budapest, Vol.3, 213-218.

Binder, U.W. (1970). 'Problems of automation and remote control of complex plants in gas transmission networks'. 11th World Gas Conference, Moscow, IGU-C-19.

Binder, U.W. and W. Brandt (1974). 'Automatisierte Verdichterstationen'. *Erdöl-Erdgas-Zeitschrift*, **90**, p. 42-50.

Bowler, A.L. (1987). 'Gas Supply Systems: Load Forecasting'. In: Singh, M.G. (Ed.) *Systems & Control Encyclopedia*, Vol. 3, p. 1979-1982, Pergamon Press, Oxford.

Bowler, A.L. and N.E. Connor (1987). 'Gas Storage Systems'. In: Singh, M.G. (Ed.) *Systems & Control Encyclopedia*, Vol. 3, p. 1969-1975, Pergamon Press, Oxford.

Bray, K. and N.D.O. Scadding (1984). 'Interactive computer simulation models for diurnal storage planning'. Computing 84, April 9-13, Gatwick.

Carmichael, N., A.J. Pritchard and A.E. Fincham (1981). 'Observers and Application to Gas Distribution'. IEE Int. Conf. on Control and its Appl., March 23-25, Warwick, U.K. p. 332-334.

Chapman, M.J., Jones, R.P. and A.J. Pritchard (1982). 'State observers for monitoring gas pipelines'. 3rd IFAC/AFCET Symposium on Control of Distributed Parameter Systems, Toulouse.

Covington, M.T. (1979). 'Transient Models Permit Quick Leak Identification'. *Pipeline Industry*, August, 71-73.

Desaultes, F. (1981). 'Computer Based Energy Dispatching System'. Paper 81-t-20, AGA Conference 1981.

Digerness, T. (1980). 'Real Time Failure Detection and Identification Applied to Supervision of Oil and Transport in Pipelines'. *Modelling, Identification and Control*, 6, 39-49.

Dupont, T., Rachford, H.H. Jr., McDonald, R.E., Gould, T.L. and H.J. Heinze (1980). 'A transient remote integrity monitor for pipelines using standard scada measurements'. Interpipe Conference, 6. February 1980, Houston, Texas.

Eibl, K. and A. Weimann (1977). 'Experimentelle Validierung des Programmsystems GANESI'. Report KFK-PDV 99, Kernforschungszentrum, Karlsruhe.

Eichner, M. (1985). 'Prozeßrechner in Netzleitsystemen - eine systemtechnische Betrachtung'. *Elektrotechnische Zeitschrift*, 105, no.7.

Fincham, A. E. (1971). 'A Review of Computer Programs For Network Analysis'. Gas Council Research Communication GC189, London.

Fincham, A.E. and M.H. Goldwater (1979). 'Simulation Models For Gas Transmission Networks'. *Transactions Measurement and Control*, 1, 3-13.

Fincham, A.E. (1981). Private Communication.

Fischer-Uhrig, F. and A. Weimann (1977). 'On-line Anwendung von Prozessmodellen bei der Führung von Gasverteilnetzen'. *Interkama Kongress 1977*, Springer Verlag, Berlin, 206-220.

Fischer-Uhrig, F. (1986). 'Netzführungsaufgaben in der Gasversorgung und ihre Lösung mit Prozeßrechnern'. BGW/DVGW Seminar III - Prozeßautomation in der Gas- und Wasserversorgung, Nov. 12 - 14, Sasbachwald, F.R.G.

Formaggio, M. and Z. Luciano (1984). 'Grid Control in SNAM: Past, Present, Future'. Computing 84, April 9-13, Gatwick.

Francis, R.F. (1982). 'The efficient management of the bulk transmission of gas'. *Gas Engineering & Management*, p. 123-133.

Gagne, R.E. et al (1975). 'A fuel minimization procedure for stations of a gas pipeline'. National Research Council Canada, report LTR-AN-20.

Girbig, P. and G. Lappus (1980). 'Combined state reconstruction and parameter estimation applied to a gas transport line' (in german). Internal LSR report, Technical University of Munich.

Girbig, P. and G. Lappus (1981). 'Distributed-parameter Kalman filter for a gas transportation system' (in german). Internal LSR report, Technical University of Munich.

Goldberg, D.E. (1978). 'On-Line Real Time Beneath the Gulf of Mexico'. Annual PSIG Meeting 1978.

Goldfinch, M.C. (1984). 'A study of observer methods for gas network state estimation'. Report LSR T678, British Gas, London.

Hass, P. (1984). 'Network Guidance and Control Programs in Long Distance Gas Supplies'. Computing 84, April 9-13, Gatwick.

Hasting-James, R. (1976). 'A recursive filter for gas pipeline networks'. *IEEE Trans. on Industrial Electronics and Control Instr.*, 32, 455-461.

Hester, D.B. (1984). 'Gas engineering computing in New Zealand'. Computing 84, April 9-13, Gatwick.

Hoffmann, M.K. and W. Ferenz (1981). 'Man-machine interaction in the control room of a regional distribution network using computers'. 3rd Data Processing Conference, Zürich, Suisse.

Homann, K. (1981). 'Rechnerführung eines ausgedehnten Gashochdrucknetzes'. *gwf/gas-erdgas*, 122, pp. 295.

Isermann, R. (1984). 'Process Fault Detection Based on Modelling and Estimation Methods'. *Automatica*, 20, 387-404.

Jones, G.A. and J.G. Wilson (1979). 'Optimal scheduling of jobs on a transmission network'. *Management Science*, 25, 98-104.

Kersting, R. et al. (1984). 'On-Line-GANESI zu Unterstützung der Gasnetzführung und Gasnetzüberwachung'. *gwf-gas/erdgas*, 125, 283-288.

Kaier, T. (1987). 'Netzleitsystem sichert Gasversorgung für Hamburg'. *Energie & Automation*, 9, 40-42.

Kralik, J., Stiegler, P., Vostry, Z. and Zavorka J. (1984). 'Modelling the dynamics of flow in gas pipelines'. *IEEE Trans. on Systems, Man, And Cybernetics*, 14, p. 586-596.

Laduzinsky, A. J. (1986). 'Would SCADA by any other name still be the same?' *Control Engineering*, 33, 72-75.

Lappus G., Scheibe, D. and A. Weimann (1980). 'Instationäre Gasrohrnetzberechnung unter Berücksichtigung der geodätischen Höhe und der Gastemperatur'. *gwf-gas/erdgas*, 121, 230-237.

Lappus, G. (1982). 'Programmsystem GANBEO'. Reports KFK-PDV-E 153,154,155, Kernforschungszentrum Karlsruhe, Karlsruhe.

Lappus, G. (1984). 'Analysis and Synthesis of a State Observer System for Large Scale Gas Transmission Networks'. Doctoral Dissertation, Technical University of Munich.

Lappus, G., Maier, R., Paijo, B. and J. Rehn (1984). 'Prädiktive Steuerungs- und Regelungsverfahren für technische Großsysteme'. Internal Report no. LLSR-86, Lehrstuhl für Steuerungs- und Regelungstechnik, Technische Universität München.

Lappus, G. (1985). 'Gas Dispatching by Means of Applied Automatic Control Theory'. PSIG Annual Meeting, Albuquerque, New Mexico.

Lappus, G. (1986). 'Knowledge-based process monitoring and control of gas transmission networks'. In: Tzafestas, S., M. Singh and G. Schmidt (Eds.): *System Fault Diagnostics, Reliability and Related Knowledge-Based Approaches*, Vol. 2, 167-182, D. Reidel Publishing Company, New York.

Lappus, G. and G. Schmidt (1987). 'Critical assessment of various leak detection techniques for gas pipeline networks'. 2st European Workshop on Fault Diagnostics, Reliability, and Related

432

Knowledge-Based Approaches. April, Manchester, U.K.

Lappus, G. (1987).(Ed.) *GANESI-GANBEO Meeting 1987*, March 26 - 27, LSR, Technical University Munich, F.R.G.

Larson, R.E. and D.A. Wismer (1971). 'Hierarchical Control of transient flow in natural gas pipeline networks'. IFAC Symp. on the Control of Distributed Parameter Systems, Banff, Canada.

Lewandowski, A. (1971). 'The optimization of gas transmission systems'. IFAC Symp. on Multivariable Technical Control Systems, Manchester.

Liaugaudas, A.P. (1979). 'New Master Station System for Control of Automatic Compressor Stations'. Paper 79-t-52, AGA Conference 1979.

Lo, K.L. and A. Brameller (1971). 'Optimal operation of gas supply systems'. *Proceedings IEE*, 118, no. 8.

Luenberger, D.G. (1964). 'Observing the state of linear systems'. *IEEE Trans. on Military Electronics*, 8, 74-80.

Maier, R. (1988). 'Ein parallelrechnerorientiertes Verfahren zur Simulation dynamischer Strömungsvorgänge in großen Gasverteilungsnetzen'. Doctoral Dissertation, Technical University Munich.

Mantri, V.B., Preston, L.B. and C.S. Pringle (1985). 'Transient optimization of a natural gas pipeline system'. PSIG Annual Meeting, Oct. 24-25, Albuquerque, New Mexico, U.S.A.

Marques, D. (1985). 'On-line optimization of large dynamic systems'. Ph.D. Thesis, University of Wisconsin.

Matsumura, J., Meguro, K., Ogawa, M. and Y. Uchiyama (1979). 'Control and dynamic simulation of a booster station'. *Hitachi Review*, 28, 21-26.

Michon, R. and M. Sorine (1978). 'Methode de Simulation et de controle d ` un reseau de transport de gaz'. Journée AFCET-IRIA sur la maitrise des systèmes complexes, 14-15 Dec., Rocquencourt, France.

Moellenkamp, G.E. (1976). 'Mini computer in compressor station operation'. AGA Transmission Conference, Las Vegas.

Osiadacz, A.J. and D.J. Bell (1986). 'A simplified algorithm for optimization of large-scale gas networks'. *Optimal Control Applications & Methods*, 7, p. 95-104.

Osiadacz, A.J. (1987). *Simulation and analysis of gas networks*. E. and F.N. Spon, London.

Osiadacz, A.J. and M.A. Salimi (1987). 'Comparison between sequential and hierarchical simulation of gas networks'. Control Systems Centre Report no. 676 and 677, UMIST, Manchester.

Parkinson, J.S. and R.J. Wynne (1986). 'Reduced order modelling and state estimation applied to gas distribution systems'. *IEEE Trans. on Automatic Control*, 31, 701-709.

Piekarski, M. (1982). 'On the optimal control of gas pipeline network of complex and time varying structure'. 15th World Gas Conference, Lausanne, Suisse.

Schmidt, G.; Weimann,A. and G. Lappus (1978). 'Application of Simulation Techniques to Planning, Supervision and Control of Natural Gas Distribution Networks'. In: Carver, M.B. and M.H. Hamza (Ed.) *Simulation, Modelling and Decision in Energy Systems*, p. 404-409, Acta Press Anaheim.

Schmidt, G.; Meier, R. and G. Lappus (1986). 'Parallel Simulation Techniques as Kernel of a Decision Support System for Control of Gas Transmission Networks'. In: Tzafestas, S., M. Singh and G. Schmidt (Eds.): *System Fault Diagnostics, Reliability and Related Knowledge-Based Approaches*, Vol. 2, 199-213, D. Reidel Publishing Company, New York.

Sedykh, A., Mirzandjanzade, A. and B.L. Kuchin (1979). 'New methods of estimating and diagnosing in gas transportation systems'. World Gas Conference.

Stoner, M.A. and M.A. Karnitz (1974). 'Estimating natural gas pipeline friction parameters'. *Transportation Engineering Journal of ASCE*, 100, p. 757-768.

Talukdar, S.N.; Cardozo, E.and L.V. Leao (1986). 'TOAST: The Power System Operator's Assistant'. *Computer*, 19, p. 53-60.

Tiley, C. and A.R.D. Thorley (1985). 'Unsteady Flow in High Pressure Dense Phase Gas Pipelines with Friction and Thermal Effects'. Report of the Thermo Fluids Engineering Research Centre, The City University, London.

Tao, L., G. Lappus, and G. Schmidt (1988). 'Combined Knowledge- and Model-Based Fault Detection and Diagnosis for Gas Transmission Networks'. 12th IMACS World Congress '88 on Scientific Computation, July 18-22, Paris.

Turner, E.B. (1974). 'Computer Based Supervisory Control Systems'. *IEEE Transactions on Industry Applications*, 10, 305-315.

Tylee, J.L. (1986). 'Model-Based Approaches to Instrument Failure Detection'. *Intech*, 33, 59-62.

Van der Hoeven, T. (1985). 'Gas network state estimation with equal error fraction method'. IMACS World Congress, Oslo, August 1985.

Van der Hoeven, T. and T.A.G. Baker (1985). 'A Gas Transmission System Simulator'. Annual PSIG Meeting 1985, Albuquerque, New Mexico.

Villon, P.F. and J.P. Yvon (1982). 'Methodes Numeriques pour la Resolution d'un Probleme de Commande Optimale dans un Reseau de Transport de Gaz'. 3rd IFAC/AFCET Symposium on Control of Distributed Parameter Systems, Toulouse.

Weimann, A. (1978). 'Modelling and Simulation of Gas Distribution Networks'. Doctoral Thesis, Technical University of Munich.

Weimann, A. and G. Lappus (1977). 'Instationäre Gasrohrnetzberechnung, Möglichkeiten und Chancen'. DVGW-Schriftenreihe Gas Nr. 20, p. 92-105, ZFGW Verlag, Frankfurt.

Weimann, A. and G. Schmidt (1977). 'Transient Simulation of Natural Gas Distribution Networks by Means of a Medium-Sized Process Computer'. In: Van Nauta Lemke, H.R. and H.B. Verbruggen (Ed.) *Digital Computer Applications to Process Control.* p. 315-320, North Holland Publ. Co., Amsterdam.

Weimann, A., R. Kersting, P. Sandmann, D. Scheibe and P. Schröder (1985). 'Advanced approach of supervision and control of a gas transmission system'. World Gas Conference, Munich, F.R.G.

Whitley, A.W. and R.I. John (1982a). 'The Use of Kalman Filtering to estimate grid conditions from available telemetry'. Internal Report LRS T537, British Gas.

Whitley, A.W. and R.I. John (1982b). 'Improvements to the Kalman Filtering Algorithm for Gas Network State Estimation'. Internal Report LRS T536, British Gas.

Chapter 13

REAL TIME COMPUTER CONTROL
OF CEMENT INDUSTRY

S. KAWAI
FUJI FACOM CORPORATION
1. Fujimachi Hino-shi
Tokyo, Japan

Y. KOIKE
FUJI ELECTRIC CO.LTD.
30-3 Yoyogi 4 Chome
Shibuya-ku Tokyo, Japan

ABSTRACT. Real time computer control systems are used with the cement manufacturing equipments. Among them, simple or compact methods realized by μ-computers are presented.
Actual example of Kiln control, raw material blending control, and electric power demand control are described in detail. New control methods such as auto-regressive model, physical model with long integration time constants, predictive model by the least mean square error, and auto-tuning of PID control parameters are the subjects of interesting advanced control applications. CRT panels of man-machine communication system have the important role in actual plant operation.

1. INTRODUCTION

In the cement industry, as well as in the other industries, the improvement of control systems for high reliability, energy saving or automation of manual operation has been expected.

Under these circumstances, real time digital control systems have been developed and the new control methods have been researched in order to keep the stable automatic operation for a long time.

This paper outlines the cement factory and the new control methods through μ-computers. The new methods are auto-regressive kiln control as a probabilistic method, raw material blending control as a deterministic mehtod, electric power demand control as a predictive method, and auto-tuning as a control parameter optimization method.

Auto-regressive model realizes automatic kiln control which makes the deviation from the reference smaller than manual control for the slow but large disturbances. Raw material blending control decreases the variation of powder composition remarkably. By power demand control, total plant can be operated in the contract limit and the limit can be decreased. Auto-tuning gives the optimal PID control parameters without the knowledge about the tuning technology.

Man-machine communication by using CRT has become the fundamental element and is always implemented, because it is very important for the actual operation and many kinds of panels have been applied. Because of the convenience

S. G. Tzafestas and J. K. Pal (eds.), Real Time Microcomputer Control of Industrial Processes, 435–480.
© 1990 *Kluwer Academic Publishers.*

for operation and the savings of space, graphic panels have been replaced to CRT systems of digital control.

2. CEMENT MANUFACTURING EQUIPMENT

2.1. Outline of cement manufacturing equipment

As for cement (usually portland cement) manufacturing equipment NSP (New Suspension Preheater) kiln which is superior in the enegy efficency and productivity or SP (Suspension Preheater) kiln is mainly used in Japan. And, in many cases, SP kilns have been modified to the more efficient NSP kilns. In 1963, the first SP kiln was operated in Japan was used practically. Almost all the wet type kilns used up to 1963 were renewed to SP or NSP kilns by the early 1970. However, in the case of the special use cement plant like white cement or blast furnace sediment cement, wet type kiln is still effective, because these do not require big productivity.

   At present, the main status of cement manufacturing equipment in Japan is as shown below. (36 NSP and SP factories were surveyed. The indicated values are mean values in 1979.)
1) Area of factory premise and number of employees
 215,000 sq.m/Factory, 261employees/Factory
2) Number of NSP and SP kilns and normally used ability
   (1) NSP kilns: 51;Mean clinker
       manufacturing ability per kiln:About 4000t/d
   (2) SP kilns: 24;Mean clinker
       manufacturing ability per kiln:About 2600t/d
3) Lime stone, clinker and cement storing ability (per normally used ability
   of total kilns)
 Lime stone: 7.7 days, Clinker: 43 days, Cement:5.3 days
4) Kiln driving motor output (Mean value):About 445kW
5) Main induction draft fan (IDF) driving motor power (Mean value):About
 2000kW

   Fig.1 is the principal process flow diagram of cement manufacturing plant. Although Fig.1 corresponds to the plant with the use of coal as the fuel, the coal equipment changes to heavy oil tank, etc when the plant uses heavy oil for the fuel. There is a burning equipment of the mixed fuel of coal and oil, also. At the end of the burning equipment, there is a cooler which plays an important role in the cement manufacturing factory. In the burning equipment, it is necessary to heat the material at 1300 to 1500℃ for a predetermined time. Clinker of high glass component can be obtained by rapidly cooling the material with the cooler after the heating. If the heated material is cooled gradually, components with various crystals ($3CaO \cdot Al_2O_3$, $4CaO \cdot Al_2O_3$, Fe_2O_3, Fe_2O_2, etc.)are separated, and hydration(combination of cement with water) is worsened. The clinker crushing equipment is called as finishing mill, also. The finishing mill mixes gypsum ($CaSO_4 \cdot 2H_2O$ or $CaSO_4 \cdot \frac{1}{2}H_2O$) used as a setting agent and crushes clinker in about 1 cm mean diameter and, and thus, produces cement as a product. The purpose of crushing is to increase area of reactions. As finer the cement powder is crushed, the total surface area of each particle becomes larger.

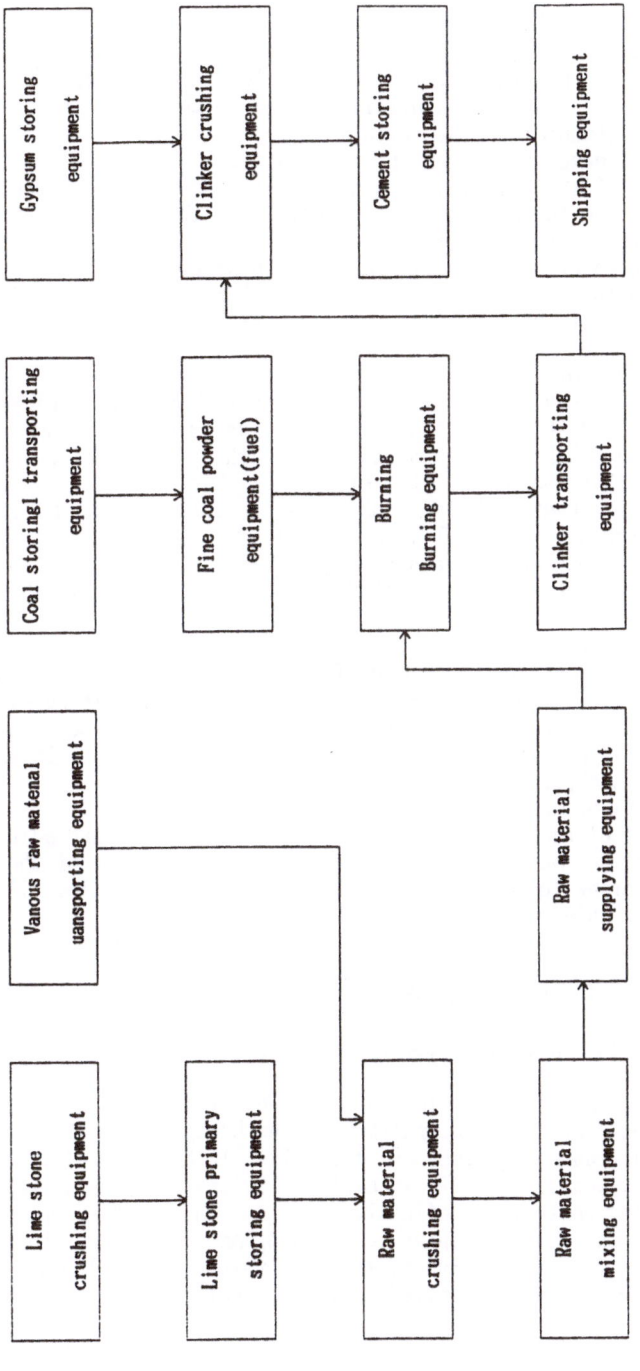

Fig.1　Principle process flow diagram of cement manufacturing plant

2.2. Wet type kiln (Semi dry process)

Untill now, semi dry process wet type kiln called as Lepaul kiln is used. In this kiln process, mixing raw materials are fed to the granulator. Water is also supplied to granulator. The ratio of water in the raw materials is 10-13%. The pellet of raw material is made by granulator.
    The chamber of the kiln inlet is called as Lepaul furnace. The raw material is preheated in the chamber, then supplied to the kiln. Therefore the length of kiln is not so long. The chamber is separated into two rooms. One is called as drying room, the other is called as heating room. The pellet of raw material is supplid to Lepaul grate in the chamber. The pellet of raw material on the Lepaul grate is heated by the kiln exhaust gas.

  Fig.2 is the control flow diagram of Lepaul kiln.
(1) Differential pressure control between chamber I and II : PbIC-1
    The chamber I and II is connected, so that gas is easy to flow from chamber I to chamber II. To prevent this flow, the pressure of chamber II is kept heigher than chamber I . The actuator of this control is damper. The pressure of chamber II is 2 mmH_2O heigher than chamber I . The static pressure of chamber I and II is about (-) 100 mmH_2O.
(2) Lepaul grate speed control and kiln revolution control
    The Lepaul grate speed is controlled by the ratio control in the relation with the kiln revolution.
(3) Hood draft control
    Gas of the kiln flows by the cooler fan and exhaust fan. The pressure balance in kiln is the most important to the flame of kiln burning zone. To prevent the unstable burning, hood draft is controlled. The actuator of this control is exhaust fan damper. The range of the sensor is usualy used (-) 5 to (+) 5mmH_2O.
(4) Ratio control of fuel and exhaust gas.
    The feed rate of fuel coal is measured by the weigher, and adjusted by the operator. If the coal feed rate is changed, the condition of the kiln is changed. In order to keep the constant coal feed rate to the clinker, ratio control is executed.

3. CONTROL SYSTEM IN CEMENT FACTORY

3.1. Outline of the control system in cement factory

When introducting a control system into a cement factory, various selections are possible. Generally, the operations can be briefly classified into a graghic operation and panel-less CRT operation. In both operations, computer and digital equipments are actively employed in the recent systems to accomplish the optimal control and effective supervision. In this paper, a graphic operation equipment is taken up as an example, and introduced. For operaions in a cement factory, the instrumentation supervisory control panel is made by using acryl or mosaic and installed in the central control room and plays the main role.
    Lamp display, controller, indicator, recorder, etc are arranged on the panel to perform supervision and control.

439

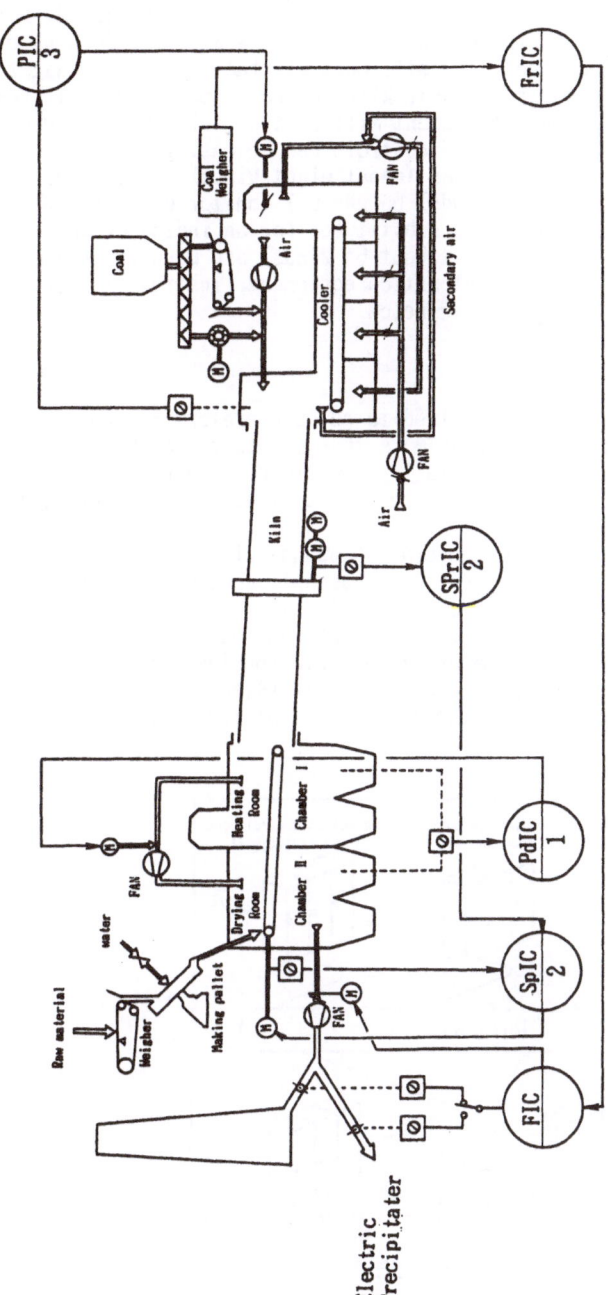

Fig.2 Control Flow diagram of Lepaul Kiln

There are many kinds of control, which are grouped and named as raw material crushing control, raw material process control, raw material feeding control, raw material blending control, raw mill control, fuel control, exhaust gas. control, kiln cooler process control, kiln control, cooler control, cement transportation process control, cement mill control, electric power demand supervision, etc. The computerized controls can be easily operated by using control status supervisory pictures (about eight kinds) and control constant setting pictures (about eight kinds) on ghe CRT monitor displays. When using the CRT monitor display, pictures substituted for an indicator (about 32 kinds) and process value trend pictures (about 64 kinds) are mounted commonly.

The main features of the automation equipment, instrumentation equipment and computer system are introduced below.

3.2. Automation

The automation equipments are applied to guide operators so that the crushing machines and transporting machines more than 200 units can be started and stopped automatically and sequentially, and stopped urgently in case of an emergency. Of course, the proper corrective actions can be taken with the help of the automatic guidance.

Up to about 10 years ago, several thousands of relays are used, which in these several years, in most cases, programmable controllers are used. In many cases, programmable controllers are installed in several separated places corresponding to the individual locations of machines and equipments in the cement factory. Wiring distance are taken into consideration for the design of a total system. An automatic sequence control system is shown in Fig.3. For the example, specifications of the Fuji's typical programmable controllers are presented in Table 1.

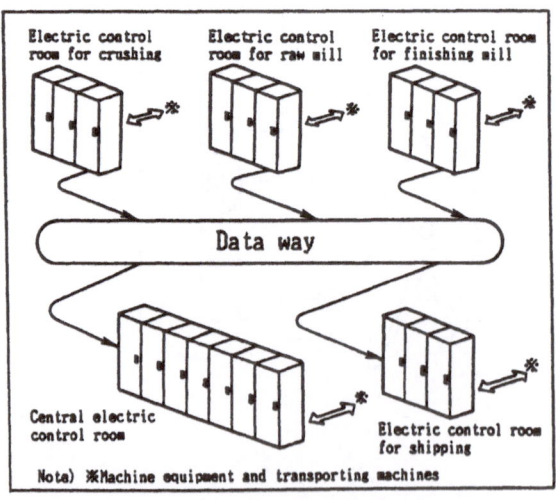

Fig.3 Automatic sequence control system

Table 1 Specifications of programmable controller (HDC-100)

Item	Specifications
Control system	Micro-program system
Instructions	
Control function	Constant syclic control Step sequence control Fixed cycle sampling control Interruption control
Storage capacity	48k words (max.) IC memory
Timer and counter	Total 240 points
Digital input Digital output Analogue input Analogue output High speed counter input	Required inputs/outputs are mounted. Remote I/O system
Power supply	AC100/110V +10~-15%, 1φ 47~63Hz or DC110V (DC90~140V)
Insulating voltage	AV1500V, 1min.
Ambient temperature, humidity	0~40°C, 20~90%RH
Allowable instantaneous power interruption time	20ms

3.3. Instrumentation

To operate cement factories properly and maintain the normal condition, it is essential to install instrumentation equipments such as sensors used to detect temperature, pressure, etc, controllers and actuators used to keep the desirable conditions, etc. Since one overall cement factory has more than 300 loops of monitoring and controlling, all of these equipment must be used easily and trouble-free. It is important that the equipment can be repaired or replaced easily if an equipment falls into a trouble.

Further, it is desirable to unify models of the used instrumentaion equipments for the preparation of spare parts. Table 2 shows major instrumention equipments used in a newest NSP cement manufacturing system and the mean quantity of each equipment. Among the many instruments with high functions, followings are the typical.

Table 2 Instrumentation for cement manufacturing indusry

Classifi-cation	Type of measurement	Name of equipment	Number
Detecting device	Temperature	Thermocouple Resistance bulb Rediation thermometer 2 color pyrometer	32 18 1 1
	Pressure	Differential pressure transmitter Pressure transmitter	53 4
	Frow	Differential pressure transmitter Area type flowmeter	7 2
	Level	Sounding level transmitter Paddle type level switch	15 13
	Analyzer, others	CO gas analyzer O_2 gas analyzer ITV	2 2 1
Instruments mounted on the control panel	Converter, calculator	—	101
	Controller, manual loader	—	76
	Supervisory meter	—	208

1) Floating cell transmitter
   Floating cell transmitters (such as FC series of Fuji Electric) are used to measure pressure, differential pressure, flow, etc, of each part, for example, in the NSP tower.
   This signal transmitter has a floating cell construction and operates under a capacitance type measuring principle. This instrument features the superior characteristics and high reliability.
2) Single loop digital controller
   The single loop digital controller by a microprocessor is highly functional. This controller is effective in forming a control loop which requires high level calculations particularly when two or more arithmetic calculations are required.
   For the practical example of the single loop digital controller, compact controller F of Fuji Electric is introduced. As long as customers are concerned; the compact controller F is similar to the conventional type control-

lers with manual-auto-remote modes, DC 1 to 5V input signal, DC 4 to 20mA output signal, etc. In addition, the controller has a number of superior features such as the employment of plasma display, wafer system program connection, built-in keyboard operator console with LED displays, built-in soft and hard manuals and non-volatile memory.

Further, with the combination of the regulating operation and logical judgement, the pulse width and pulse row output funccion is effectively controllable even for the actuators with motors. This controller is especially effective in controlling raw material feeding, mill operation.

3) CO gas analyzer (infrared type)

This is a non-dispersing type gas analyzer which uses an infrared absorbing method. By employing an IC manufacturing technique of micro flow sensor, the analyzer is outstanding in the shock resistance, stability and S/N ratio, and the operation and maintenance are easy. This analyzer is used to measure CO concentrations is the SP tower, kiln exhaust gas and coal mill.

4) O gas analyzer (paramagnetic oxygen analyzer type)

(1) Using magnetic nature of oxygen, this analyzer converts concentration of oxygen to pressure, and measures ratio of oxygen contained in measured gas.

(2) This analyzer is of a pressure measuring type, and therefore, the responese is fast.

(3) Since almost all existing gases are not affected by magnetic field, this analyzer is not affected by any coexisting gases.

With the above features, this analyzer is used to measure concentration of oxygen contained in SP tower, kiln exhaust gas, etc

3.4. Future trend of instrumentation

These years optical fiber technique is remarkably progressed. Further, the technique of the optical fiber cable transmission has been applied in the field of communication and information industry.

As this technology is very interesting and progressive, an example is presented. It is the Fuji field instrumentation(FFI) which use microcomputer and transmits the signal by optical fiber cable. The merit of optical fiber instrument is as follows.

1. Nothing of electric induction.
2. Nothing of corrosion at the joining points.
3. Cable outdiameter is small and light.
4. Nothing of spark at short trouble.
5. No need of exclusive duct or luck.
6. Easy detection of trouble points by the inteligent function.
7. Easy spare preparation because the sorts of cable are few.
8. No need of the explosion proof works.

Near future, many optical fiber field instruments will be used.

3.5. Distributed DDC system

With backgrounds of the digitalizations of the instrumentation machines and equipments developed during the recent years and improvements of the data transmission technologies, control equipment of cement factories are varying to a distributed DDC system. The distributed DDC system connects the automatic sequence controller, digital instrumentation system and computer with the data way which transmits a large volume data so that the data can be mutually used among the individual machines and equipments. The factory operations can be controlled from the operator console located in the central control room.

The major components of the distributed system are;
(1) Distributed type DDC process station (PCS-100)
(2) High speed programmable controller (HDC-100)
(3) Centralized type operator station (OCS-200/1100)
(4) High speed data way (DPCS-E)

These are composed of micro processors.

Names in ( ) corresponds to the products of Fuji Electric.

Fig.4 shows the overall system configuration. Each system configuration may vary slightly depending on the scale of a plant, idea of the plant operation, redundancy of the system or emergency operability.

The calculable controllers are installed separately in the electric rooms in the plant of raw material crushing, burning and finishing which are dispersely located with the distance of several hundred meters. The instrumentasion controller and overall controller are joined with the controller with the man-machine interface located in the central control room through one data way. Thus, composing the system by the distribution architecture, hazard caused by each process can be dispersed, maintenance ability is improved, the cable between the site and central control room can be reduced and the construction cost can be greatly reduced. Further, with the single man-machine interface, the instrumentation and overall motor operations can be controlled.

Table 3 shows the components of the system and functions of each equipment required for the plant operations. The distributed DDC systems have been delivered from Fuji Electric with the name of MICREX to cement factories. Fig.5 shows that the total cement plant with the installation of distributed digital system has the exellent functions.

3.6. Computer

Computers for control are used to compensate the portions which cannot be executed by automatic sequence controller and instrumentation equipment or portions which cannot controlled thoroughly with the automatic sequence controller or instrumentation equipments. Sometimes, a computer is used only for the optimal control. At present, almost all newly built cement factories adopts computers. The convenient softwares are prepared for various controls, CRT displays, printing process data, data transfer processing with the other computers. With processings by the operators taken into considerations, CPU, various peripheral equipment, input/output interface equipment, operator console, simulation equipment, etc are prepared so that the functions of the prepared softwares can be thoroughly displayed.

The controlcomputer software is capable of covering all the processes of the raw material crushing, the burning and finishing. The data required for the control are transmitted and received via the instrumentation equipment. The major software functions are analyzer processing, raw material blending control, mill control, kiln control, electric power demand monitoring, CRT display, printing and operator console processing. It is also possible to correspond to the product input/output managements.

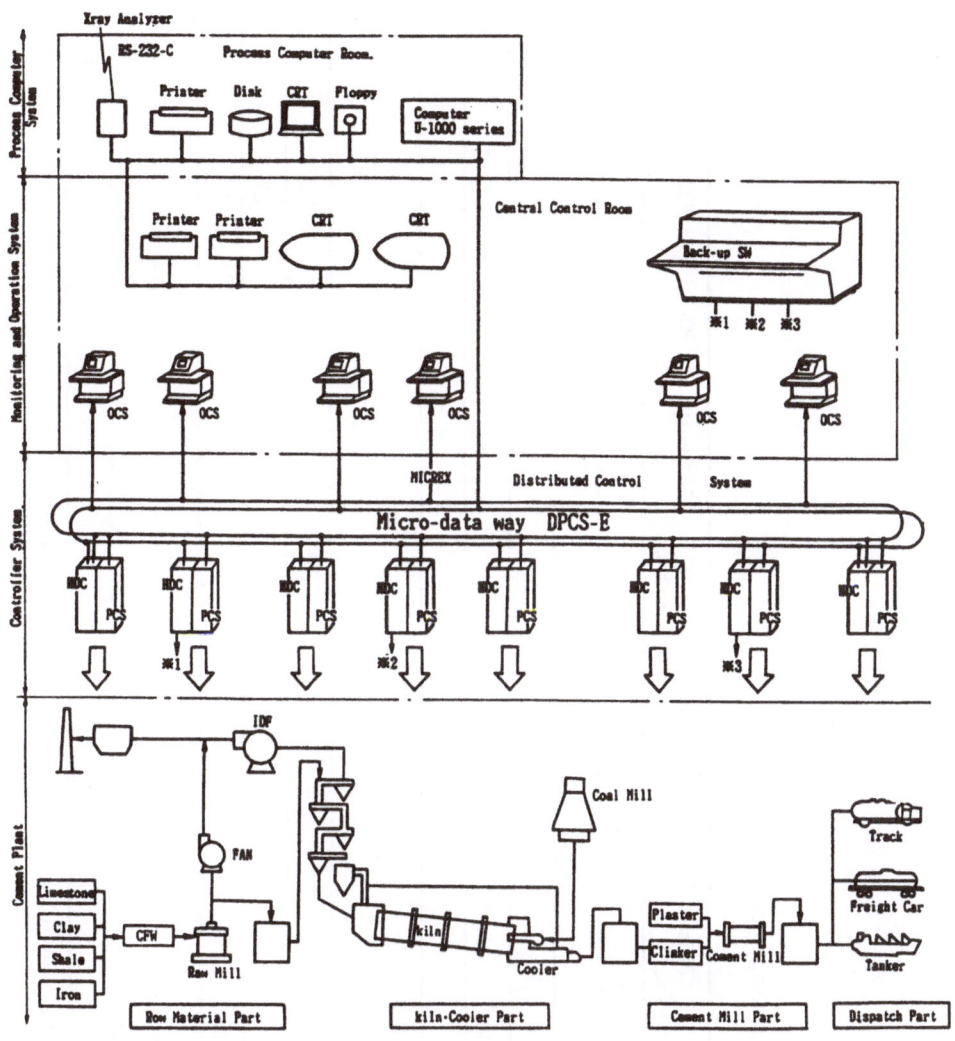

HDC is the name of Programable Logic Controller for Motor sequence control.
PCS is the name of Programable Logic Controller for Process control.
DCS is the name of CRT Operator Console.

Fig.4   Overall system configuration of Distributed control system

Table 3　System function distribution

Device \\ Function	CRT display	Mimic lamp	CRT keyboard	Desk switch	PCS	HDC	TM
Display for monitoring analogue loop	O						
Analogue loop setting/control	O		O				
Analogue loop control					O		
Analogue loop signal input/output					O		
Analogue loop alarm display	O						
Analogue loop alarm logging	O						O
Motor controlling for supervision and display	O*1	Δ*1				Δ*1	
Motor controlling for sequential starting	O		O*2	Δ*2		Δ*2	
Motor control						O	
Motor control signal input/output						O	
Motor control alarm display	O*3	Δ*3				Δ*3	
Motor control alarm logging	O						O
Data logging	O						O

(Note)　*1, *2 and *3 indicate that either O or Δ can be selected by an operation method.

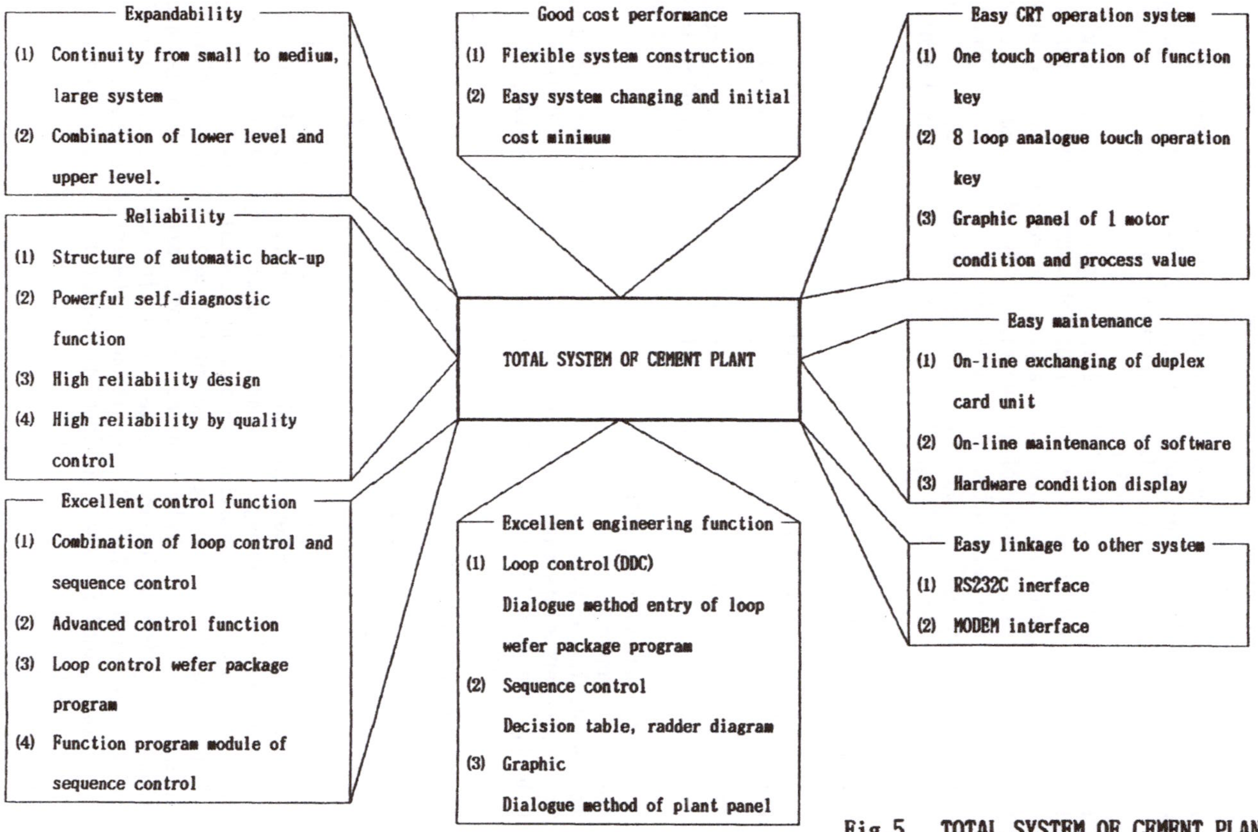

Fig.5 TOTAL SYSTEM OF CEMENT PLANT

TOTAL SYSTEM OF CEMENT PLANT

— Easy CRT operation system
(1) One touch operation of function key
(2) 8 loop analogue touch operation key
(3) Graphic panel of 1 motor condition and process value

— Easy maintenance
(1) On-line exchanging of duplex card unit
(2) On-line maintenance of software
(3) Hardware condition display

— Easy linkage to other system
(1) RS232C inerface
(2) MODEM interface

— Good cost performance
(1) Flexible system construction
(2) Easy system changing and initial cost minimum

— Excellent engineering function
(1) Loop control (DDC)
Dialogue method entry of loop wefer package program
(2) Sequence control
Decision table, radder diagram
(3) Graphic
Dialogue method of plant panel

— Expandability
(1) Continuity from small to medium, large system
(2) Combination of lower level and upper level.

— Reliability
(1) Structure of automatic back-up
(2) Powerful self-diagnostic function
(3) High reliability design
(4) High reliability by quality control

— Excellent control function
(1) Combination of loop control and sequence control
(2) Advanced control function
(3) Loop control wefer package program
(4) Function program module of sequence control

4. OPTIMAL TEMPERATURE CONTROL OF KILN BY AR MODEL

Kiln is the most important component in the cement plant. However, the characteristics of a Kiln is difficult to represent by the physical model based on the physical or chemical laws. Therefore, an auto-regressive model (abbreviated as AR model) is applied.

In order to obtain a practical AR model, suitable period of the kiln is selected from the actual operation, transient behaviors of the process variables and the manipulated variables are sampled and recorded, the model of the process, i.e. variables and order of the model, are decided, and the coefficients of the model are calculated by using a off-line large scale computer. Finally, the model is implemented in on-line digital control system, together with the interface elements between the model and the conventional control units.

4.1. Outline of Kiln Process

The purpose of kiln control is to keep the burning state in a kiln stable for the various disturbances such as the change of raw material component, fuel calorie, and inner state of a kiln.

Fig.6 is the diagram of kiln control. As this figure shows, the control system contains many variables and the variables have inter-relation to each other complicatedly. Therefore, automatic control of kiln is very difficult and it is said that only skillful operator can control it based on the empirical operation standard.

For the design of kiln control system, physical models had been studied. However, their parameters cannot be estimated during the design stage and cannot be fixed even after the plant operation starts. Accordingly, we applied the regressive model to this process.

4.2. Alogorithm of kiln control

Kiln control is consisted of the following two kinds of control mode.
   ⌜ Stationary control
   ⌞ Unstationary control
In the large part of operation, kiln is controlled stably with in the protected range and the process variables vary a little according to the small disturbance of feed rate change of clinker product or quality change of raw materials. This range of operation corresponds to stationary control. For the "stationary control" mode, we apply the optimal control based on AR-model.

In the unusual state, kiln sometime receives large disturbances and the operational condition varies to a large extent. In such a case, "unstationary control" is applied to return the condition to the range of stationary control by the special method for the individual processor the set point change in the simple situation.

449

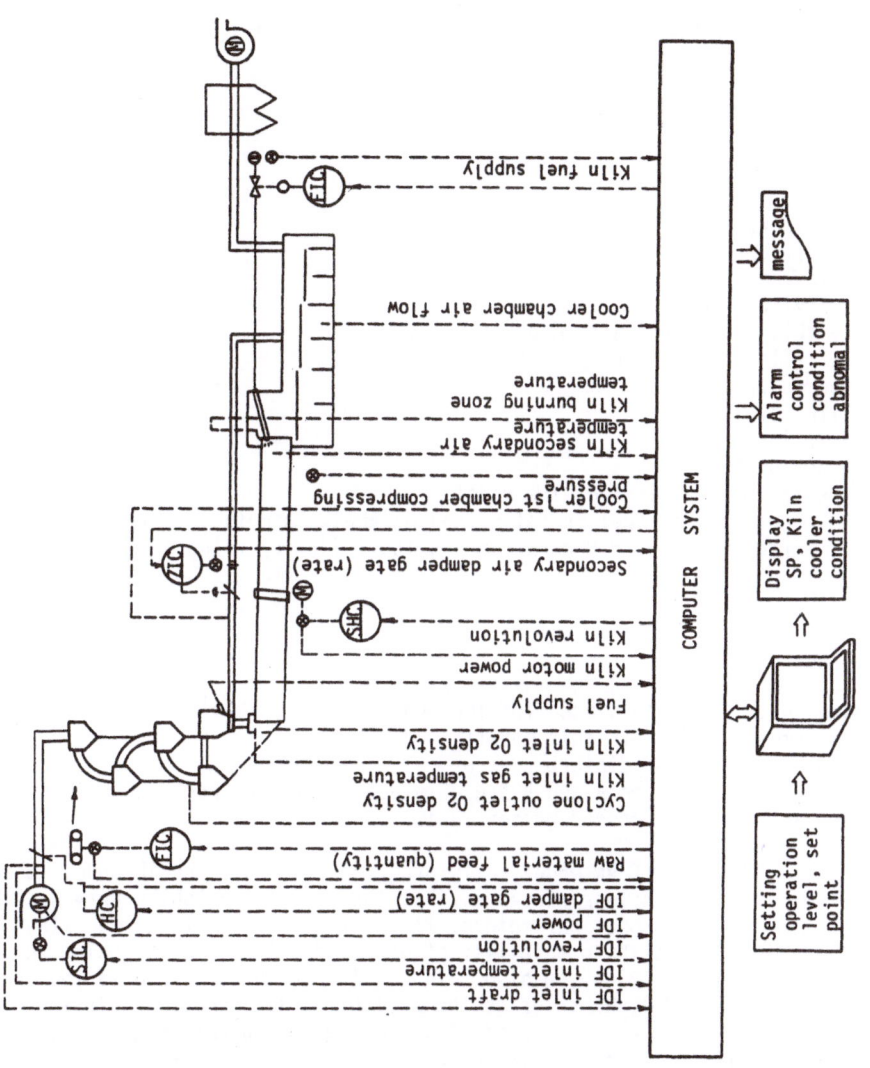

Fig.6 Kiln multi-variable control.

4.2.1. AR Model of Kiln Control.   The discrete AR model for output variables of kiln is expressed as

$$x(k) = \sum_{i=1}^{M} \{A(m)x(k-m)+B(m)u(k-m) \} + w(k) \qquad (1)$$

where

  $x = \text{col} \ (x_1, \ x_2, \ x_3)$
  $u = \text{col} \ (u_1, \ u_2, \ u_3)$
  $w = \text{col} \ (w_1, \ w_2, \ w_3)$
  $x_1, \ w_1$;kiln inlet temperature
  $x_2, \ w_2$;kiln secondary air temperature
  $x_3, \ w_3$;kiln moter power
  u_1 ;raw material feed
  u_2 ;kiln revolution
  u_3 ;kiin fuel supply
  M   ;order of AR model
  A,B ;coefficient matrices of the model

The order of the model is determined by Akaike's method, which defines the most appropriate order M as minimizes the following value of MFPE.

(Akaike, 1974)

$$MFPE(M) = (\frac{N+M\cdot K+1}{N-M\cdot K-1})^{K} \qquad det(D) \qquad (2)$$

where

N ;number of descrete data used for the model
K ;dimension of x (K=3 in this model)
D ;covariance matrix of disturbances w

Coefficient matrices A and B are obtained by Yule-Walker's equation constructed with the recorded process data.

4.2.3. Identification of AR model.   In order to construct the AR-model, the kiln should be operated long enough so that the process variables vary suffi- ciently to obtain the model parameters. It is often said that the excellent feature of AR model is the possibility of model identification by using the data under normal operation. We also intended to use this feature in the development of AR model. We selected the suitable data among the actual opera- tion considering amplitude and speed of the change of operatiog conditions and executed data processing to obtain the model.

However, no successful result was obtained from the data of the practical operation. From the viewpoint of calculation, frequency distribution of the change of variables is not wide enough for AR model, because the actual opera- tion depended on the operator's experiences and intuitions, and remained in a small range due to their own skill.

Based on these early results, presently we disturb the kiln operation from the very stationary state intentionally in order to get the good data for autoregressive calculation. As the intentional disturbance, we adopt the random signal of Maximam Period Sequence.

The Maximam Period Sequence should be examined before, it is applied to the real processes. The frequency distribution must be uniform and distur- bances to the different process variables must be independent of each other.

The intentional disturbances for modelling must be allowable in the practical operation. For the purpose of estimating the influence of disturbances and assuring the allowability, the actual behavior is simulated by using other models, which may be temporary AR models obtained by the data of practical operation of the same kiln or models of the similar kilns.

The example of the specific detail of the signals are shown in Table 4

Samplling period		1 min
Period of data collectio		540 min
Time constant of filter		1 min
amptitude of disturbance (% of full scale)	raw matrial feed	1.5 %
	kiln revolution	4 %
	kiln fuel supply	2 %

Table 4   Example of disturbance signals

After recording the process responses to these signals, the following process variables are selected among those in Fig.6.
(1) The variables which have clear and important meaning for the process.
(2) The variables which are necessary to represent the process state.
(3) The variables whose recorded data contain small noises compared with the process responses.
(4) The variables which vary differently from others.

The coefficient matrices of AR-model corresponding to the selected variables are calculated. The independence of each component of w and the power contribution in the frequency domain are examined.

The justice of the selection must be reviewed repeatedly. The dependent components of w are eliminated exclusive of one variable. The variables of which values have small power contribution to the others are also eliminated. In the opposite situation, when the power of self-contribution of the important variable are large, we increase the number of variables. The AR model obtained are evaluated by computing the step responses to input variables and the time series of residual vector between the process responses and the output of AR-model.

4.2.4. Desigh of optimal control.   In order to apply the discrete optimal regulator theory to the kiln control, AR model (1) should be transformed into the state equation as follows.

$$z(k+1) = \Phi z(k) + \Gamma u(k) \qquad (3)$$

$$x(k) = \left[I, \ 0 \cdots 0 \right] z(k) \qquad (4)$$

where
$$z=(k) = \begin{bmatrix} x(k) \\ \cdot \\ \cdot \\ x(k-M+1) \\ u(k-1) \\ \cdot \\ \cdot \\ u(k-M+1) \end{bmatrix}$$

Φ, Γ ; coefficient matrices constructed by A, B

We consider the performance index of a quardratic form

$$JI = \sum_{k=1}^{I} \{ \ x'(k) \ Qx(k) + u'(k-1) Ru(k-1) \ \} \qquad (5)$$

where

Q, R ; weighting matrices

I ; interval for evaluation

Then the controller is calculated so as to minimize the performance index as follows.

$$u(k) = - \ [G(1) \cdots G(2M-1) \ ] \begin{bmatrix} x(k) \\ \cdot \\ \cdot \\ x(k-M+1) \\ u(k-1) \\ \cdot \\ \cdot \\ u(k-M+1) \end{bmatrix} \qquad (6)$$

Next we must determine the weighting matrices Q, R to obtain the optimal feedback gain G. At first G is calculated by Q and R as follows.

$$Q = \begin{bmatrix} 1/d_{11} & 0 & 0 \\ 0 & 1/d_{22} & 0 \\ 0 & 0 & 1/d_{33} \end{bmatrix}$$

$$R = \begin{bmatrix} \sigma_1^{-2} & 0 & 0 \\ 0 & \sigma_2^{-2} & 0 \\ 0 & 0 & \sigma_3^{-2} \end{bmatrix}$$

where

$\{ d_{ii}, \ i=1 \sim 3 \}$: the diagonal component of D

$\{ \sigma_i, \ i=1 \sim 3 \}$: the standard deviation of the manipulated values

Substituting the resultant A, B and G to (1) and (4) the closed loop response of each variables can be obtained by the computer simulation, Q and R and adjusted until the controller outputs change not so often or not so much and are not so sensitive to the difference of AR-models obtained by using the data for the other period. The typical example of A, B. Q, R and G are shown Table 5, 6 and 7.

4.2.5. Practical Improvement of Control Algorithm.

In the period of data recording, many kinds of undesirable noise have the influence on the behavior of kiln. Some process variables may not recover the former staitionary values in the long time even after all of the manipulated variables return to the former values. Moreover, unreasonable changes may appear before the intentional disturbances are applied. These undesirable disturbances cause the slow chages of kiln condition.

In order to remove the large trend corresponding to the slow changes of kiln condition, the actual process variables of the model is replaced to the perturbation variables by using conversion formula;

$$\tilde{x}(k) = x(k) - \bar{x}$$

where

$\tilde{x}(k)$; perturbation variable of the equation (1)

$\underline{x}(k)$; sampled actual process variable

$\bar{x}$; standard variable

Although the values of x are commonly regarded as constant, it is recommended to regard x as variable.

That is :

$$\bar{x} = \bar{x}(k)$$

$\bar{x}(k)$ should be varied slowly both in the recorded data for making model and in the real time data for actual control.

Though the kiln process is non-linear, the process characteristic can be regarded as almost linear at the neighbourhood of $\bar{x}(t)$. Thus, for the small change of $\bar{x}(t)$ in actual control, "stationary control" can work using the constant control parameters. However, optimal control has a limit accoding to nonlinearity of kiln because AR-model is linear.

More complicated operation is needed in "unstationary control". "Unstationary control" is divided into two stage which are the "first stage" and the "second stage".

In the first stage, automatic logical control as shown in Fig.7 are executed. Very simple operation in the first stage is the change of set-point to the value which has been remaining constant for a long time. However, in more complicated situation, the plural variables, for example fuel flow and kiln revolution, must be changed simultaneously according to logical instructions. These operations on the first stage of unstationary control make kiln return to stationary control state.

In the second stage, operators must decide the operation from observasion and exeperiances.

The interval of these control actions are compared with the others in Table 8.

$$A(1)=\begin{bmatrix} 0.97634E+00 & 0.11413E+00 & -0.10196E-01 \\ 0.70408E-01 & 0.12250E+01 & 0.52549E-01 \\ -0.79881E-01 & 0.28240E-01 & 0.95670E+00 \end{bmatrix} \qquad B(1)=\begin{bmatrix} -0.10729E+00 & -0.65437E-02 & 0.10142E-01 \\ -0.47440E-01 & -0.98272E-03 & -0.54847E-03 \\ 0.14745E+00 & 0.38933E-01 & -0.36103E-02 \end{bmatrix}$$

$$A(2)=\begin{bmatrix} -0.30235E+00 & -0.51513E-01 & 0.17240E+00 \\ -0.35926E-01 & -0.31578E+00 & -0.43573E-01 \\ 0.10138E+01 & -0.19958E-01 & -0.36455E+00 \end{bmatrix} \qquad B(2)=\begin{bmatrix} 0.22189E+00 & 0.27296E-01 & -0.62665E-03 \\ -0.25762E-01 & -0.15365E-00 & 0.27887E-02 \\ 0.16649E+00 & 0.97951E-01 & 0.39060E-02 \end{bmatrix}$$

$$A(3)=\begin{bmatrix} 0.38611E-01 & 0.12179E+00 & -0.18237E+00 \\ 0.28945E-01 & 0.12706E-01 & 0.95647E-02 \\ 0.57005E-01 & 0.63771E-01 & 0.23060E+00 \end{bmatrix} \qquad B(3)=\begin{bmatrix} 0.24085E-01 & -0.24402E-01 & 0.25568E-02 \\ 0.19660E-01 & 0.94674E-02 & -0.20935E-02 \\ 0.14046E+00 & -0.14341E-01 & 0.14690E-02 \end{bmatrix}$$

$$A(4)=\begin{bmatrix} -0.46634E-01 & -0.34008E+00 & -0.65684E-02 \\ -0.38406E-01 & -0.10487E+00 & -0.94189E-02 \\ -0.86990E-01 & 0.16173E-00 & -0.14644E+00 \end{bmatrix} \qquad B(4)=\begin{bmatrix} -0.92879E-01 & 0.11794E-01 & -0.51816E-02 \\ -0.68882E-01 & -0.15372E-01 & 0.10953E-03 \\ -0.22187E+00 & -0.64030E-01 & 0.47071E-02 \end{bmatrix}$$

$$A(5)=\begin{bmatrix} 0.10111E+00 & 0.55097E-00 & 0.63451E-01 \\ 0.69849E-01 & 0.17037E-01 & 0.38904E-01 \\ -0.25049E-01 & -0.18043E+00 & 0.51693E-02 \end{bmatrix} \qquad B(5)=\begin{bmatrix} -0.35729E+00 & -0.24803E-01 & 0.77556E-02 \\ 0.36778E-02 & 0.13864E-01 & -0.10260E-02 \\ 0.93453E-02 & -0.73974E-02 & -0.53702E-02 \end{bmatrix}$$

Table 5   The result of matrix A and B

$$Q = \begin{bmatrix} 10 & 0 & 0 \\ 0 & 4 & 0 \\ 0 & 0 & 2 \end{bmatrix} \qquad R = \begin{bmatrix} 50 & 0 & 0 \\ 0 & 1060 & 0 \\ 0 & 0 & 180000 \end{bmatrix}$$

Table 6   Using values of marices Q and R

$$G x (1) = \begin{bmatrix} 0.23266E+00 & -0.64673E+00 & -0.13227E+02 \\ 0.10559E+01 & -0.37170E+01 & -0.30856E+01 \\ -0.54192E+00 & -0.33982E+01 & 0.48646E+01 \end{bmatrix}$$

$$G x (2) = \begin{bmatrix} -0.36466E-01 & -0.83806E+00 & 0.50597E+01 \\ -0.65118E+00 & 0.17884E+01 & 0.70772E+00 \\ 0.23418E+00 & 0.81380E+00 & -0.29596E+01 \end{bmatrix}$$

$$G x (3) = \begin{bmatrix} 0.10406E+00 & -0.25528E+00 & 0.32143E+00 \\ -0.13576E+00 & -0.62650E+00 & 0.16739E+01 \\ -0.51488E-01 & -0.67569E+00 & 0.43799E+01 \end{bmatrix}$$

$$G x (4) = \begin{bmatrix} 0.11146E+00 & 0.32872E+00 & -0.64915E+00 \\ -0.19494E+00 & 0.45270E+00 & 0.62644E+01 \\ 0.13827E+00 & 0.52497E+00 & -0.12884E+01 \end{bmatrix}$$

$$G x (5) = \begin{bmatrix} 0.12405E+00 & -0.25752E+00 & -0.17421E+01 \\ 0.14894E+00 & 0.79399E+00 & -0.17285E+01 \\ 0.57009E-01 & -0.24223E+00 & -0.98997E+00 \end{bmatrix}$$

$$G u (2) = \begin{bmatrix} -0.37106E+00 & -0.99586E+00 & -0.14822E+01 \\ -0.18646E-01 & -0.58109E+00 & 0.57353E+00 \\ -0.21334E-02 & -0.31867E-01 & -0.59694E-02 \end{bmatrix}$$

$$G u (3) = \begin{bmatrix} -0.13990E+00 & 0.68531E+00 & 0.40636E+01 \\ 0.37325E-01 & 0.30933E+00 & 0.65682E+00 \\ -0.36898E-02 & -0.20052E-01 & -0.25494E-01 \end{bmatrix}$$

$$G u (4) = \begin{bmatrix} 0.66000E-01 & 0.16644E+01 & 0.40917E+01 \\ 0.47803E-01 & 0.36593E+00 & -0.19922E+00 \\ -0.34905E-02 & -0.30786E-03 & 0.44458E-01 \end{bmatrix}$$

$$G u (5) = \begin{bmatrix} -0.32922E-01 & 0.10016E+00 & 0.55896E+01 \\ 0.22231E-01 & -0.18332E-01 & 0.33940E+00 \\ 0.28198E-02 & 0.28191E-01 & -0.14525E+00 \end{bmatrix}$$

Table 7   Resuls of optimal gain G

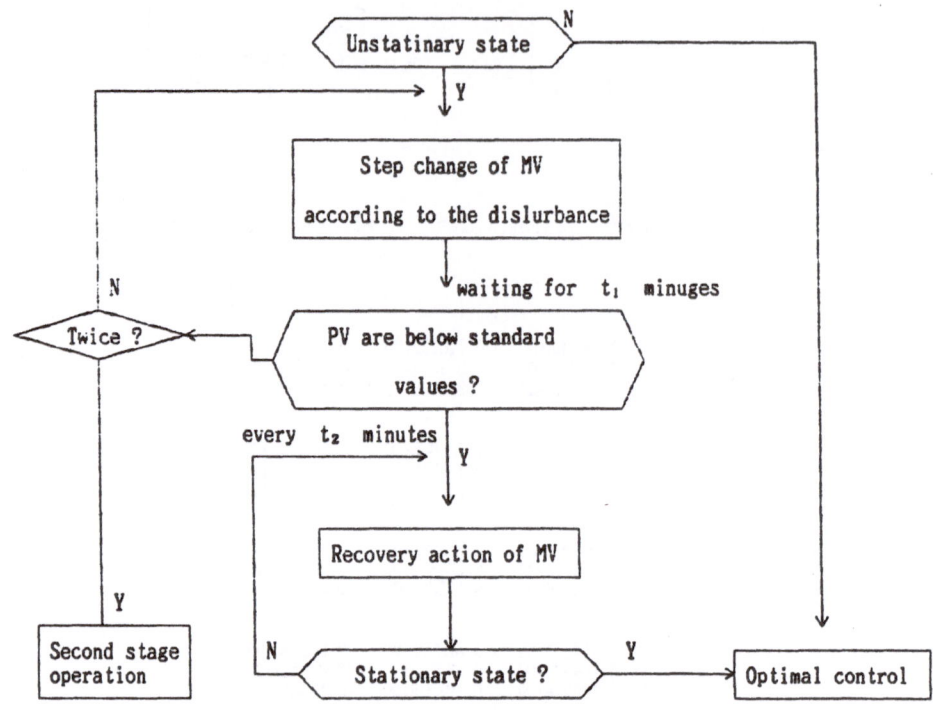

Fig.7   Flow diagram of the logical control at the first stage

Control action	Interval
AR-model control in stationary control	Once for 1 minutes
Logical control of the first stage in unstationary control	Once for 5~15 minutes
Skillful operation, of the second stage in unstationary control	Several times for 1 month

Table 8   Intervals of kiln control operation

While the computer control is executed, the estimated values of all of the process variables are calculated. The integrated values of the error between the estimated values and the actual values are used for discovering the accident of the process and the change of the process parameters. In the latter case, the coefficient of A, B and G should be recomputed by the same procedures as described here.

4.3. System configuration of kiln control.

Fig.8 shows the System configuration of kiln control. Using the large scale computer, the coefficient matrices which is necesary for stationary control are calculated. The process data for modelling are gathered by use of digital control system.

Kiln control, both "stationary control" and "unstationary control", are executed by digital control sistem using calculated optimal gain by the large scale computer.

4.4. Effect of kiln control.

In Fig.9 and Fig.10 the transients and the power spectra of three process variables controlled by AR-model are compared with those controlled manually. In low frequency domain, the effect of AR model appears. In high frequency domain, AR-model varies more largely than the manual control which manipulate very little and let the kiln free.

The control by AR model realizes auto-matic control, decreases the amplitude of process variable deviation from the expected value, and keeps the extreme operation away. As the result, the operational margin can be made small, two or three percent of fuel is saved, and kiln can be operated stably not only by the special skillful men who have much experience and know-how.

4.5. Future trend of kiln control

In the other viewpoint of kiln control, fuzzy control is applied. It is reasonable to use the optimal control for stational control and fuzzy control for unstational control.

In the application of the optimal control, it is important to develop the CAD system by which the procedures of identification and the deseigh of opti mal control by AR model can be executed conveniently.

458

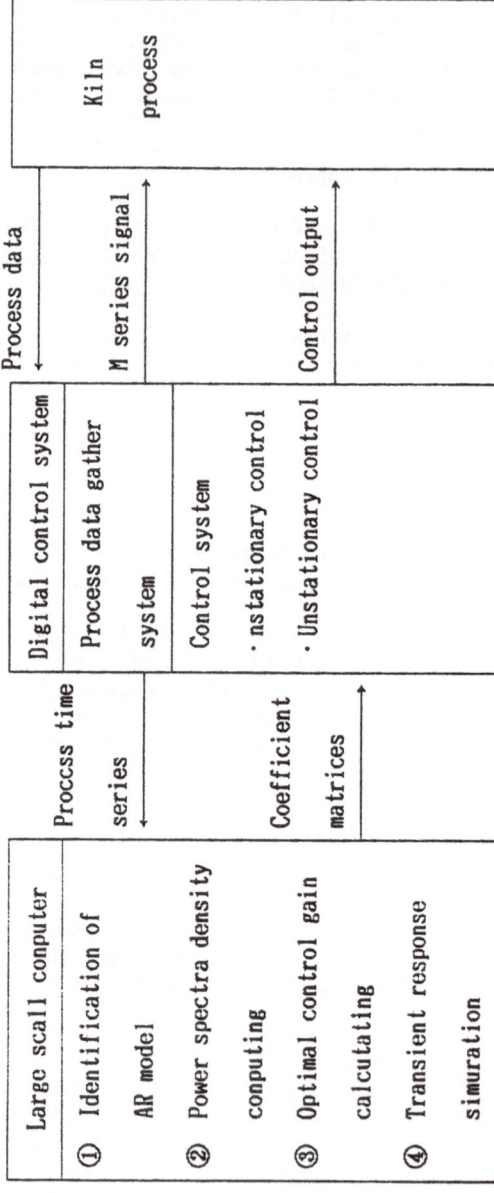

Fig.8 System configuration of kiln control

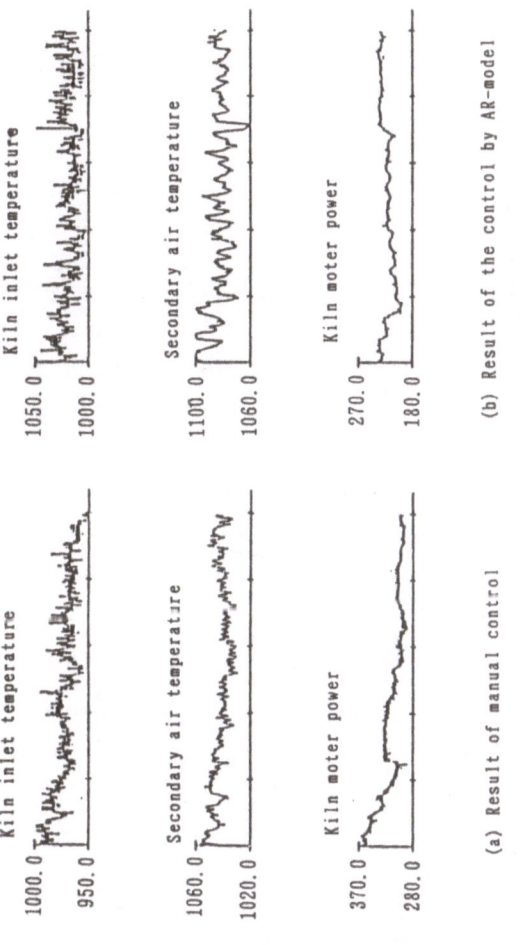

Fig.9 Transients of controlled results.

460

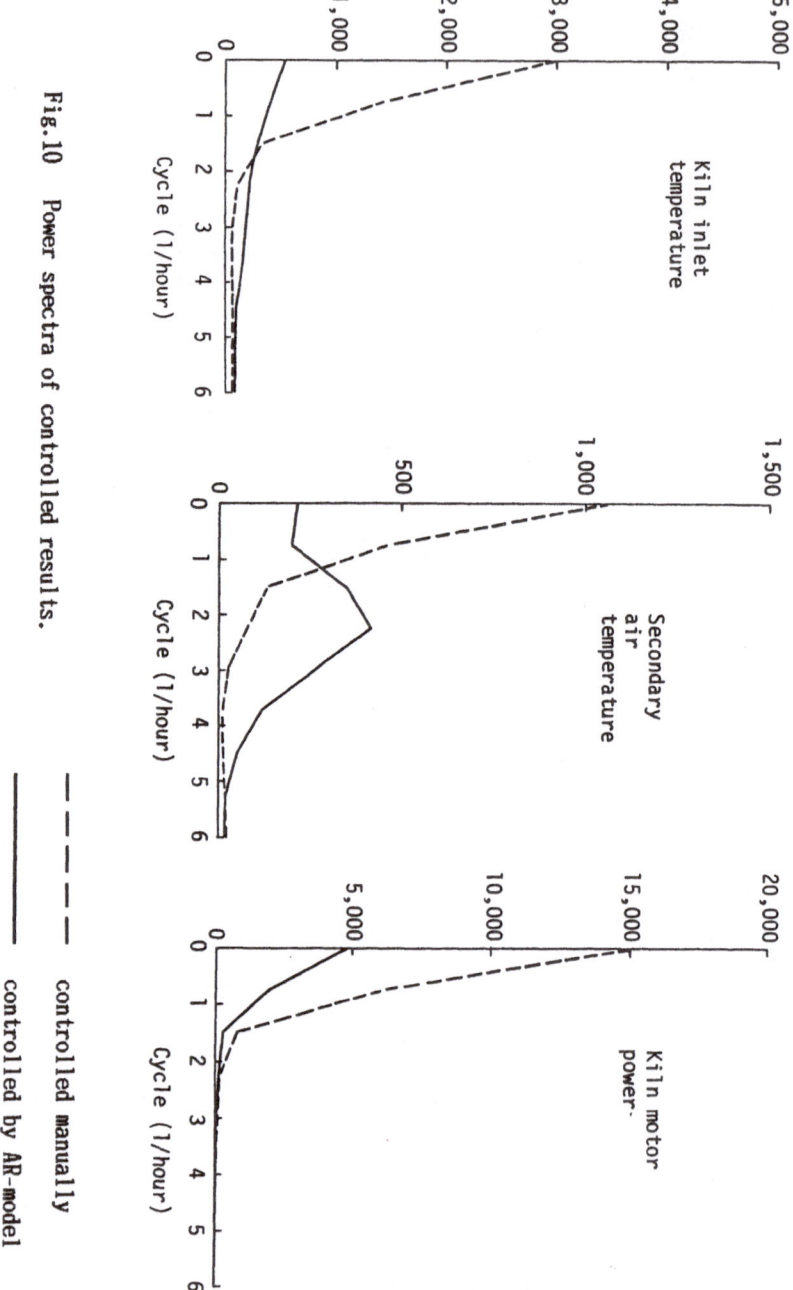

Fig. 10   Power spectra of controlled results.

− − − −   controlled manually

─────   controlled by AR-model

5. RAW MATERIAL BLENDING CONTROL WITH STATISTICAL MODIFICATION

Raw material blending control is the most effective of the computer control in cement industry. The quality of blended raw material affects kiln process and product quality. Nonlinear calculation for modulus or composition and integration for long time range are implemented easily in digital computer but very difficult in conventional analog control system.

5.1. Outline of Raw Material Blending Process

Fig.11 is the diagram of raw material blending control. Raw materials are fed according to the reference values by the individual local feed controller and the total amount is controlled so as to agree with the quantity given from the raw mill control system by digital computer. Quality of the powder from raw mill is analysed by X-ray analyzer and controlled by changing the rates of raw materials to the total feed. These rates are calculated through digital computer.

Raw materials with different composition come from the place of production They are mixed in raw mill for the first stage and blended in homogenizing silo again.

The important indexes of material composition are hydraulic modulus (HM), silica modulus (SM) and iron modulus (IM) which are defined as

$$HM = c_1 / (c_2 + c_3 + c_4) \quad (7)$$
$$SM = c_2 / (c_2 + c_4) \quad (8)$$
$$IM = c_3 / c_4 \quad (9)$$

where c_1, c_2, c_3, and c_4 are weight rates of components CaO, SiO_2 Al_2O_3, and Fe_2O_3, respectively. As these notation shows, CaO is thought as No.1 component, SiO_3 as No.2, Al_2O_3 No.3, and Fe_2O_3 as No.4, in this section. The purpose of the blending control is to make the output of homogenizing silo agree with the reference quality.

5.2. Algorithm of Raw Material Blending Control

5.2.1 Basic control based on weight balance.   The weight rates and flow rates of raw materials satisfy the following equations.

$$a_{11}w_1 + a_{12}w_2 + a_{13}w_3 + a_{14}w_4 = c_1w \quad (10)$$
$$a_{21}w_1 + a_{22}w_2 + a_{23}w_3 + a_{24}w_4 = c_2w \quad (11)$$
$$a_{31}w_1 + a_{32}w_2 + a_{33}w_3 + a_{34}w_4 = c_3w \quad (12)$$
$$a_{41}w_1 + a_{42}w_2 + a_{43}w_3 + a_{44}w_4 = c_4w \quad (13)$$
$$w_1 + w_2 + w_3 + w_4 = w \quad (14)$$

where

w_1, w_2, w_3, w_4 ; flow rates of Limestone, Clay, Shale, and Iron
  w ; total flow rate of raw material
  $a_{i1}(i=1\sim4)$; weight rates of component No.i of Limeston
  $a_{i2}(i=1\sim4)$; weight rates of component No.i of Clay
  $a_{i3}(i=1\sim4)$; weight rates of component No.i of Shale
  $a_{i4}(i=1\sim4)$; weight rates of component No.i of Iron

If the composition of all the natural raw materials are known exactly, they can be mixed by solving equations (7)~(14) in the way that the composition coincides with the reference moduli. However, as the composition of natural raw materials varies, the actual values of these moduli are different from the

462

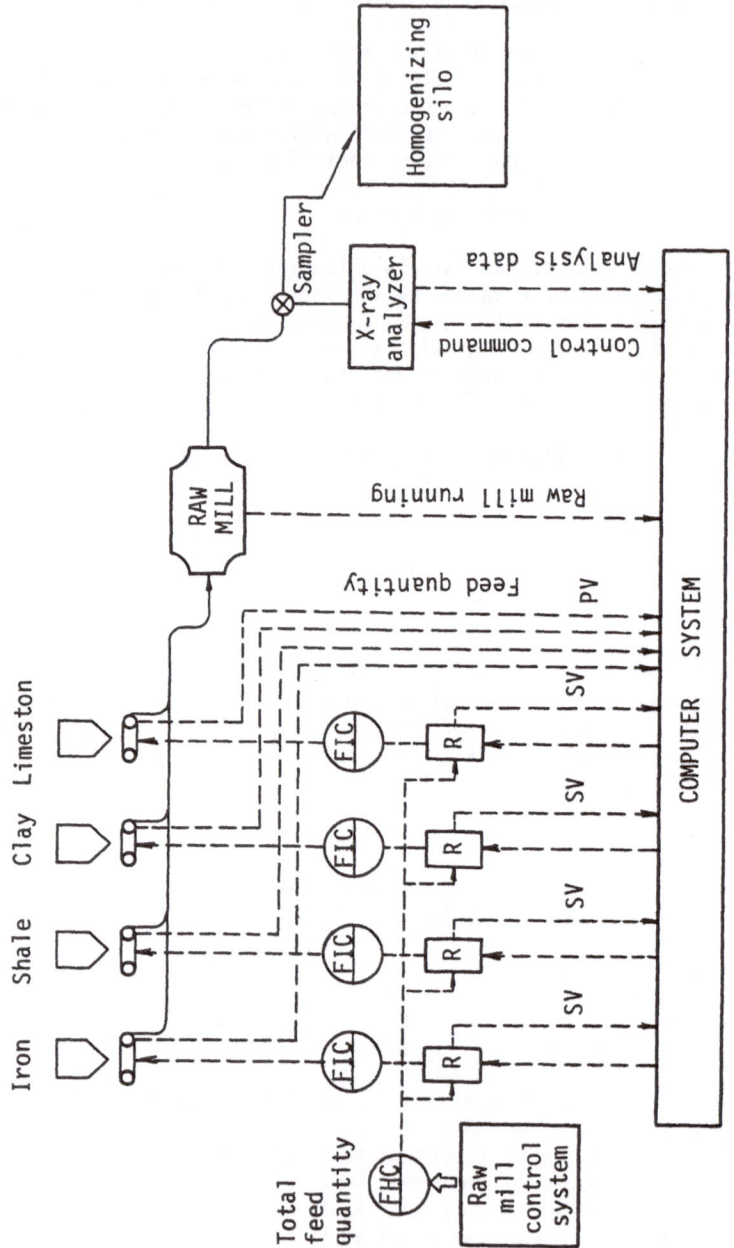

Fig. 11   Raw material blending control.

reference and their deviations are accumulated during homogenizing time. As the result, the deviations of these moduli at the outlet of thesilo remains. Therefore references of feed flowcontrol of raw materials should be changedin the computer under the consideration of mixing effect and dead time of mill and silo.

For this purpose, the reference of HM, SM and IM at the outlet of the raw mill are calculated using the past actual values of $c_1 \sim c_4$ and w during homogenizing time. For the example of hydraulic modulus, HM* (the reference of hydraulic modulus at the outlet the raw mill is inversely obtained by (7) from HM of (15) and HM* (homogenized value of hydraulic modulus)

$$HM = \frac{\overline{HM}(k)w(k) + HM^* Wz}{w(k) + Wz} \tag{15}$$

$$HM^* = \frac{\displaystyle\sum_{\ell=1}^{L} c_4(K-\ell)w(K-\ell)}{\displaystyle\sum_{\ell=1}^{L} \{c_2(K-\ell)+c_3(K-\ell)+C_4(k-\ell)\}\, w(K-\ell)} \tag{16}$$

Where
L : Sampling number during homogenizing time.
Wz : Weight of homogenized material

$$Wz = \sum_{\ell=1}^{L} w(K-\ell)$$

SM and IM are also obtained by the same way. HM, SM and IM of the equations (7)$\sim$(9) are written to replace HM, SM and IM. Then unknowm variable $c_1 \sim c_4$, $w_1 \sim w_4$ can be obtained by solving equation (7)$\sim$(14) $c_1 \sim c_4$ are replaced by the measured values afterwards.

5.2.2 Modification of raw material components. In conventional control, feed rates of natural raw materials are given so as to decrease the error of mixed composition. In the improved digital blending control, feed flow rates a: calculated exactly from the material balance equations. The constants ai; for this calculation are given from the original data and modified both with the measured compositions of mixed material and with the feed rates antecedent by the process dead time. These feed rates are memorized in the computer.

Theoretically, all the values of aij can be modified by using different sets of feed flow rates and compositions, but actually only diagonal elements {aii, i=1$\sim$4 } are modified, by the arithmetic calculation based on the equation (10)$\sim$(14) because it is too sensitive to calculate all the values of aij from measured values of four compositions and four flow rates.

Theoretically the non diagonal elements of aij should be changed, because the corresponding weight rates varies together with the variation of aij. However in the practical sense, diagonal elements have a larger value and the more important role in (10)$\sim$(11) than non-diagonal elements. That is, the simple modification of aii is effective unless aij does not varies in big amount. To these special condition, a_{12}, a_{13}, and a_{14} must be modified as well as all by calculating the generalized inverse matrix. As the same control algorithm can

be realized easily in offline large scale computer, it is possible to simulate
the transient behavior of mixed composition. It is possible to design therange
ability of feed rate change of feeder, that is, it is possible to estimate how
fast and how much feed rates must be changed for the predicted change in
natural raw material compositions. It isvery important and actually very
effective to exercise this simulation before the process operation begins or
when the composition of natural raw material varieswidely from the design
value.

5.3. System Configuration of Raw Material Blending Control

Block diagram of raw material blending control is shown in Fig.12. In this
control, arithmetic calculation based on the equation   (7)~(14) and memori-
zation by digital computer bring about big effect. Ttus, the composition of
mixed raw material after silo becomes nearer to the reference values and the
operation on the following stages is stabilized.
   The blending control uses real time signal of mixed raw material composition bymeans of the single X-ray·analyzer and uses for statical constants of
natural raw material compositions, which are individually different corresponding to their place of origin and varies depending on the depth or location even in the same production area.

5.4. Effects of Digital Raw Material Blending Control

The results of blending control by digital computer is compared with those of
no control regarding to the three moduli in Fig.13. In this example, as three
feed flow rates of four raw materials are changable and the number of unknown
variables is seven, one of the equations (1)~(3) must be removed. Therefore,
only two moduli, that is HM and SM are controllable and IM is left in no
control. However, the deviation of IM does not go out of its limit, because
larger fluctuation of IM is allowable than HM or SM. As Fig.13 shows, the
deviation of raw material blending control is smaller than half of no control.
Digital blending control enables the adjustment offered rates corresponding to
these variation, and the same control algorithm can be applied in the off-line
computer for the simulation on the stage of design and evaluation.

5.5. Future trend of raw material blending control

The advanced control based on the control theory has not been applied to the
raw material mixing blending control. The neck of it is be considered as the
time delay of measurement of mixed raw material composition by X-ray. Thus
this control method is effective in the practical sence.

6. CRT DISPLAY APPLICATIONS

6.1. Outline of CRT display applications

Man-machine communication is very important.
   These years CRT display has become popular tool for plant operation. 20
inch CRT display is usually used. Common CRT has 8 display colours (red, green,

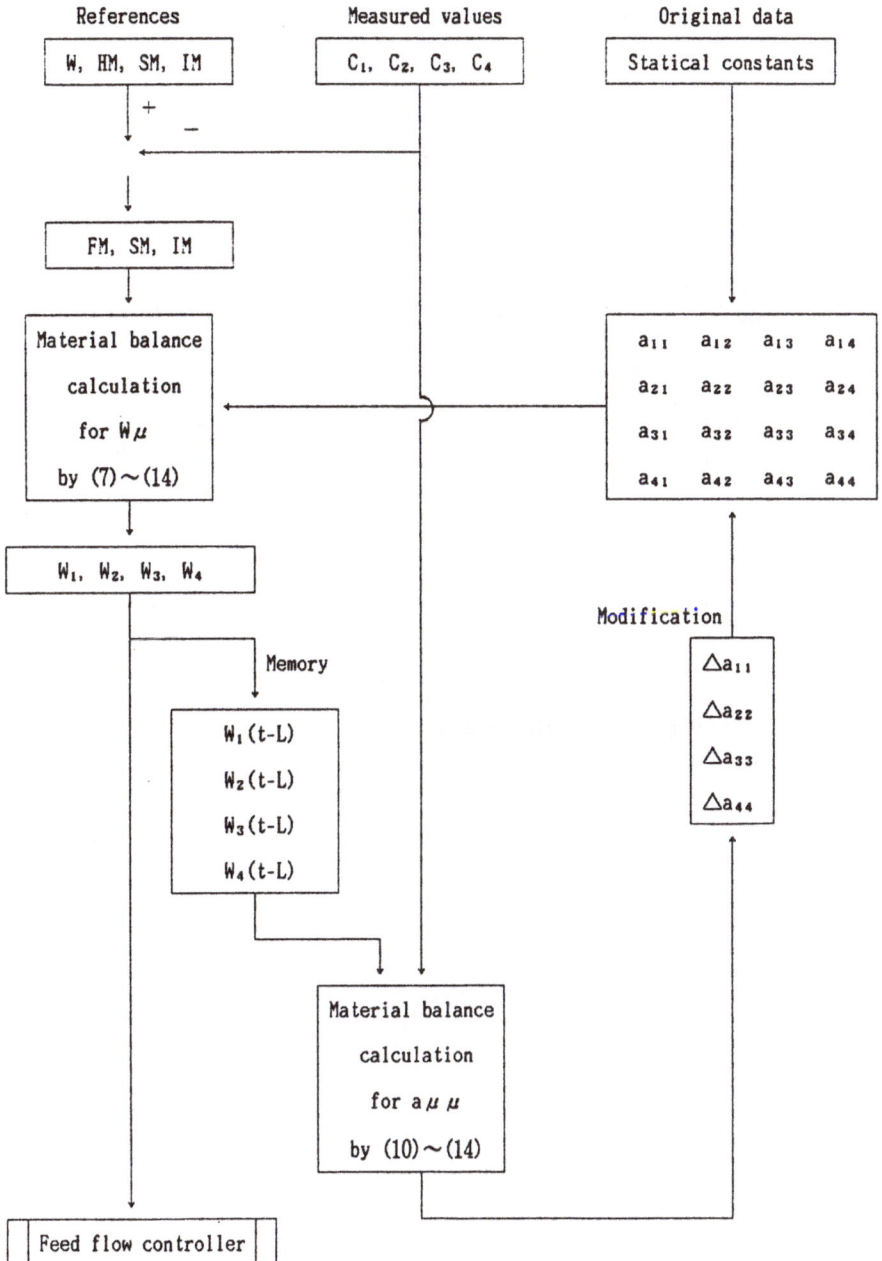

Fig.12 Diagram of digital blending control.

466

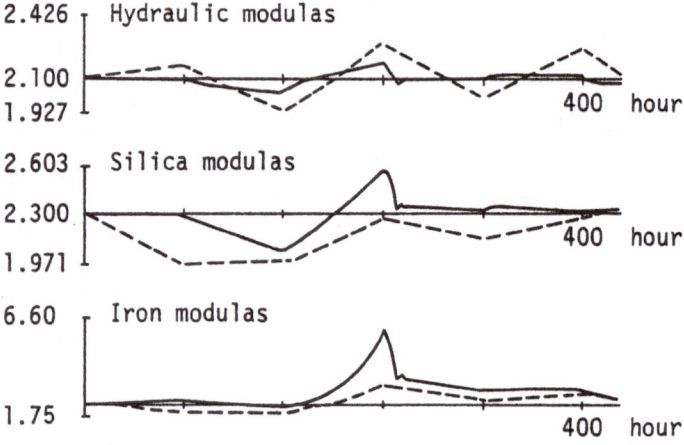

Fig.13   Simulated modulas transients

blue, yellow, cyan, magenta, white and black). And color hard copy unit is adaptable. The number of CRT displays is depend on the scale of the plant.

In an example of total cement plant control system, minimum 5 CRT displays are used. The first of this example is used for the raw mill. The second is used for the kiln and cooler. The third is used for the cement mill. Hotback up CRT for the first to the third is the fourth. The fifth is used for the supervisory computer control.

If the CRT display for a engineer and a factory manager is required, more CRT displays are necesary.

Especialy in large scale control system, operation is sometimes stabilized by human adjustment which is guided by the concentrated information through computer processing of plant data.

It is known widely that CRT display is very effective.

6.2. CRT display of supervisory computer control

In one example of large computer systems, following kinds of display are applied with the indicated numbers of the different patterns.

Display of machine operation
condition ···················· 3 pages
Display of transportation line
condition ···················· 12 pages
Supervising display of controlled
state ······················· 8 pages
Display for setting control
parameters ················ 8 pages
Supervising display of process
variable ················· 32 pages
Trend display of variable
······ 48 pages
Display of conditions after alarm
···· 2 pages

6.3. CRT display of distributed control

In addition 6.2, in case of CRT operation of distributed controk system for process control and motor control, following kinds of display are added to each CRT display.

Chronological display of the process new alarm (16 points/page) ···· 64 pages
Display of total control trend overview (64 points/page) ········ 384 pages
Operation group display for processing (8 points/page) ·········· 384 pages
Loop display for adjustment of parameter (1 points/page) ········ 384 pages
Operation guidance display at the occurrence of alarm ·············· 16 pages
Multi points indicating and setting display (16 points/page) ···· 2048 pages
Chronological trend display of the data (4 points/page) ············· 64 pages
Multi pattern setting display (1 kind/page and64 points/page) ······· 66 pages
Concentrated indicating of the related data and setting display ····256 pages
(64 points/page)
Display of plant mimic diagram ································· 32 pages
Chronological display of the system condition (16 points/page) ···· 64 pages

468

Display of the hardware system condition ···· 1 page of overall display plus
1 page/controller display
Chronological historical message display of manipulation ········ 128 pages
(16 points/page)
    In general case, minimum 4 CRT displays are applied.
    The CRT display of the first to the third are used for raw mill, kiln
cooler and cement mill. The fourth isused to back-up of hotstand by for each
CRT display.

6.4. The detail of the CRT display of distributed control.

The detail of the display panels in the practical application is described.
(1) Chronological display of the process new alarm
The newest data of the process alarm are shown on the uppermost line of the
1st pages. The TAG No. of the module of an alarm is displayed in different
colors according to the alarm level (heavy. medium or light).
− Heavy alarm   ─────────── Red
− Medium alarm   ─────────── Magenta
− Light alarm   ─────────── Yellow
− Recovery to normal condition ── Green
    Table 9 shows the contents of the alarm to be displayed. The level of
these alarms are previously assigned by self support function.
(2) Overview display of total control trend
This panel is used to check the process condition of the total plant. The
present operation status can be is confirmed by a sweeping view.
(3) Operation group display for processing
    This panel is used for process operation. Data of 8 modules per page are
displayed on the panel. It is possible to operate under the same image as the
conventional analog controller. The display is performed in order to improve
the monitoring ability. In addition to the conventional operation like analog
controller, detailed numerical input is also possible, thus the plant is
exactly operated. The panel has the following 3 display modes.
    (a) Absolute value display mode
    (b) Deviation display mode
    (c) Rader-chart display mode
(4) Loop display for adjustment of parameter
This panel is used for tuning parameters such as PID control parameters, upper
limit and lower limit in a module. On the panel the instrument face of one
module is displayed with the parameters and other instrument face of relative
module (reference module). Furthermore real time trends of PV, SV and MV are
displayd on the panel ; thus making parameter tuning easily. Realtime trend
can display from 4 minutes (1 second intervals) to 3 weeks after this display
is called out.
(5) Operation guidance display at the occurrence of alarm
    This panel is used for the reference in operation corresponding to proc-
ess alarm. In a complexed process it is possible that the operator does wrong
operation because of a sudden alarm. Misoperation for major alarm may sometimes lower the safety of the plant tremendously. On this panel the operation
guide to major alarm is shown so that the operatorcan operate smoothly and
adequately for the alarm. Thus the safty of the plant in emergency is increased much more. Description area is 70 degits × 16 lines per page and

Table 9 List of alarm display codes

Alarm display code	Contents
SH	Upper limit alarm of SV
SL	Lower limit alarm of SV
PH	Upper limit alarm of PV
PL	Lower limit alarm of PV
DPH	Change ratio upper limit alarm of PV
DPL	Change ratio lower limit alarm of PV
DHH	Upper limit alarm of deviation
DHL	Lower limit alarm of deviation
MH	Upper limit alarm of MV
ML	Lower limit alarm of MV
HH	Upper-upper alarm of PV
LL	Lower-lower alarm of PV
DMH	Upper limit of MV output change ratio
DIH	Status change module (ON)

Remarks : PV   Process Variable

SV   Set point Variable

MV   Manipulated Variable (or the opening ratio of the valve)

available up to 16 pages.
(6) Display of multi points indication and setting
This panel displays data in the controller in a given format. Setting is also possible. The panel is mainly used for setting data which the operator changes very often such as the constant in the controller.
(7) Chronological trend display of the data
This panel is used for the chronologically registeration and displays longterm process data. Up to 4 points per page are displayed, this panel is used to analyze the dynamic characteristics or mutual relationship. Any of PV, SV or MV displayed on the loop panelcan bedcalt with on the trend panel.

(8) Multi pattern setting display
This panel is used for en bloc setting of constants and parameters in plural controllers.
(9) Display of concentrated indication of the related data and setting
This panel is used for the en bloc display of various kinds of process variables or for the display or/and setting or related data in several controllers.
(10) Display of plant mimic diagram
This panel displays the process data to be controlled using a graphical process flow chart showing the process data in the forms of numerical values, bar graphs, patterns, etc. It displays the process status together with a process flow chart so that the actual process conditions can be checked for smooth supervision and operation.
(11) Chronological display of the system condition
This panel Chronologically displays data of hardware alarm in the controllers connected to the CRT display.
(12) Display of the hardware system condition
This panel is used for monitoring the condition of the control unit of the controller connected to the CRT display. When a hard alarm occurs, it is possible to respond to the alarm quickly and to take an adequate procedure by calling out this panel.
(13) Chronological message display of manipulation
This panel is used to display the operation message and to record them. It can be used as the opration diary.

7. ELECTRIC POWER DEMAND CONTROL

7.1. Outline of electric power demand control

The electric power should be consumed in the contract limit and the overcharge is required if the actual consume exceeds the contract limit. Therefore the power demand is controlled so as to be maintained in this limit.
To detect the electric power, integrated power meter is used.

7.2. The scheme of electric power demand control

The limit is contracted not for the instantaneous power but for the accumulate power between a certain interval. Accordingly, the demand control by digital computer is carried out by the following scheme corresponding to Fig.14.
The contract interval (one hour or 30 minutes for example) is devided into severalperiods (ten minutes) for calculation. In this period, the instantaneous power is sampled (at each one minute), the accumulated value is integrated, and the final value of the contract interval is predicted by least mean square error method. The actual power and the predicted power are compared with the reference curve or the contract limit, espectively.
In the case when the power demand prediction exceeds over the contract limit the messageis printed on typewriter, the buzzer sounds and the lamp is turned on, I order to inform operators. Operators communicate with computer by using CRT display which shows the curves like Fig.14.
Then the conditions of the accumulated power are classified by the importance

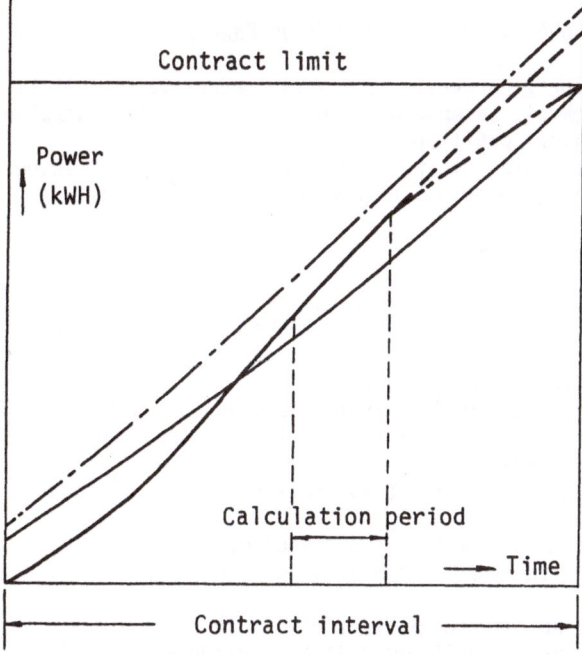

———— Actual power demand

– – – – – – Predicted demand

— · —— · · Adjusted demand

———— Reference curve

— · —— · – Upper limit

Fig.14   Power demand control scheme

grade. Each machines or electric feeders are monitored and operator can decide whether some machines should be cut off or not.

7.3. Effect of electric power demand control

The electric power demand control prevents the over consumption beyond the contract limit. In some application examples, the contract power could be decreased to 90 or 95 percent of the power without demand control.

8. AUTO-TUNING AND ADAPTATION OF PID CONTROL PARAMETERS

Nowadays, auto-tuning is prevailing. However, adaptive control is expected for the further improvement, because auto-tuning adjusts PID control parameters only for one operation condition.

The word "auto-tuning" is used when tuning results are inspected by operator. "Self-tuning" does not need operator's judgement.

8.1. Outline of auto-tuning

Auto-tuning method of PID control parameters is developed and implemented in several kinds of digital control systems such as single loop digital controller, multiple loop DDC unit, process cumputer system, and supervisory component of hierarchical DDC system.

There are many kinds of auto-tuning methods. Among them, a man-machine conversational auto-tuning system is described in this paper. (Nishikawa, Sannomiya, Ohta, and others 1981)

The auto-tuning is used for a single input and single output control system, as shown in Fig.15.

The procedure of the system is cosisted of
(1) generation of pulsewise test signal for process parameter estimation in open loop mode
(2) data sampling of the process response to the test signal
(3) determination of the self/non-self-regulation quality of the process
(4) calculation of characteristic areas defined newly by using the process step response
(5) calculation of optimal PID control parameters by using simple polynomials of the approximated relation between estimated process paramcters and optimal settings
(6) display of the results of the estimation and the optimal settings
(7) computer simulation of controlled results with estimated process characteristics

Auto-tuning is started by the operator's command, because process operation is disturbed a little. The obtained optimal settings are actually used after the operator's confirmation, because the process response may be affected by the large noise and the estimation error may be not small.

8.2. Algorithm of auto-tuning

In order to get optimal settings, a new form of performance index, i.e. the weighted ISE is adopted. That is

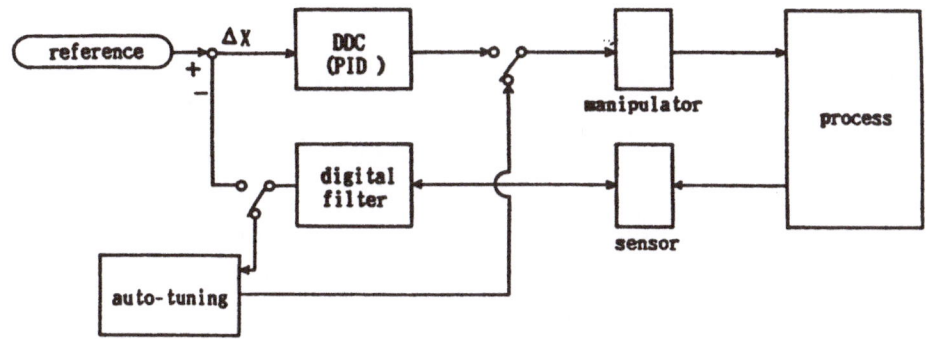

Fig.15   Black diagram of auto-tuning

$$J(B) = \int_0^w \{\Delta x(t) e^{\beta t}\}^2 dt \qquad \beta > 0 \qquad (17)$$

β is given from the period of ultimate oscillation of the loop. This weighted ISE is easily calculated and this optimization algorithm can be applied to wide variety of the transfer functions, even to those with long dead time.

Weighting function of equation (17) increases exponentially, as shown is Fig.16 (a). Accordingly, this performance index determines that the stable controlled result with bigger beginning deviation and larger damping factor like curve- I of Fig.16(b) is better than the oscillatory one with smaller deviation and less damping like curve $-$ II. The adjustment of curve $-$ I should be recommended as the practical settings of process control, because curve $-$ I is less sensible than curve $-$ II for the change of process characteristics and keeps enough margine for the estimation error.

8.3. Configuration of auto-tuning system

Very simple auto-tuning without operator's confirmation can be implemented in a single loop digital controller. However, the present auto-tuning system is implemented in the supervisory component in a hierarchical DDC system, as shown in Fig.17.

8.4. Effects of auto-tuning

The effects of auto-tuning is summarized as follows.
Before the usable settings are obtained
   (a) optimal parameters for one operation condition can be obtained easily
   (b) the difference of optimal settings among the various operations becomes clear

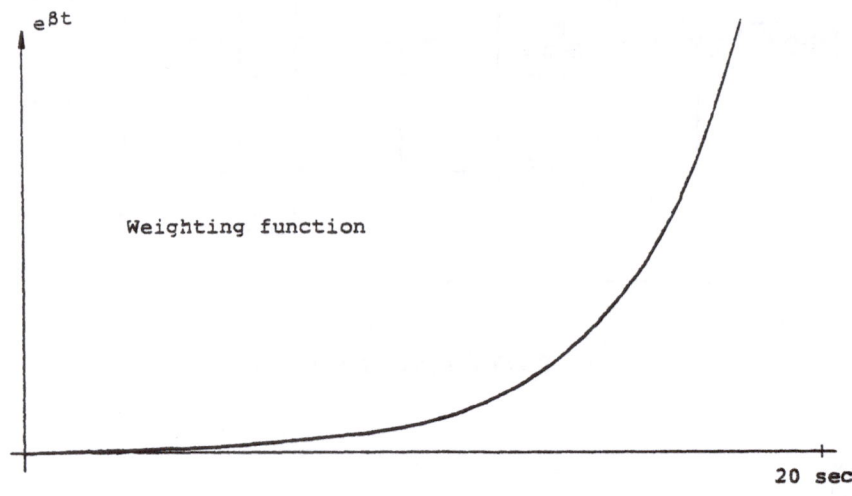

Weighting function

$e^{\beta t}$

20 sec

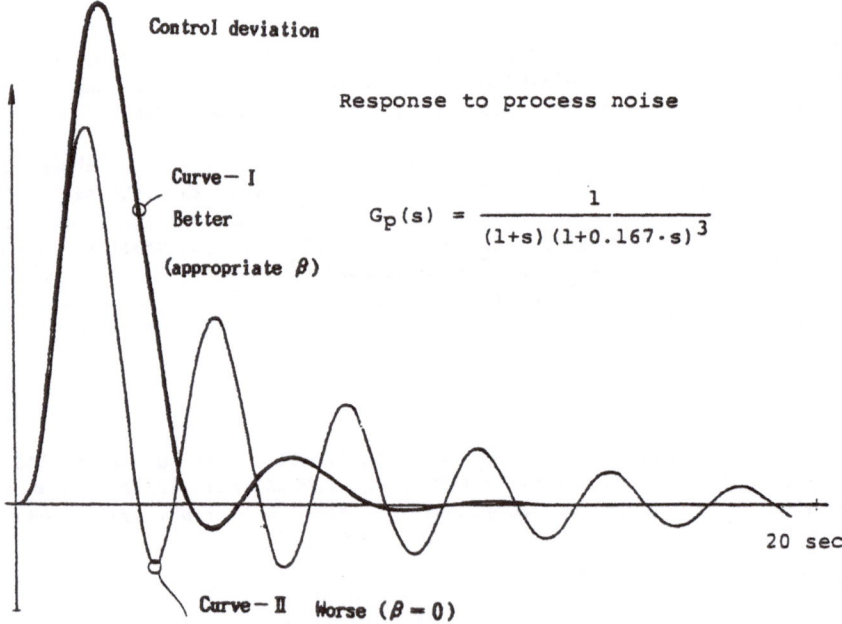

Control deviation

Response to process noise

Curve— I
Better
(appropriate β)

$$G_p(s) \;=\; \frac{1}{(1+s)(1+0.167\cdot s)^3}$$

20 sec

Curve— II   Worse ($\beta = 0$)

Fig.16   Weighting function and result of performance index.

CRT operator's station

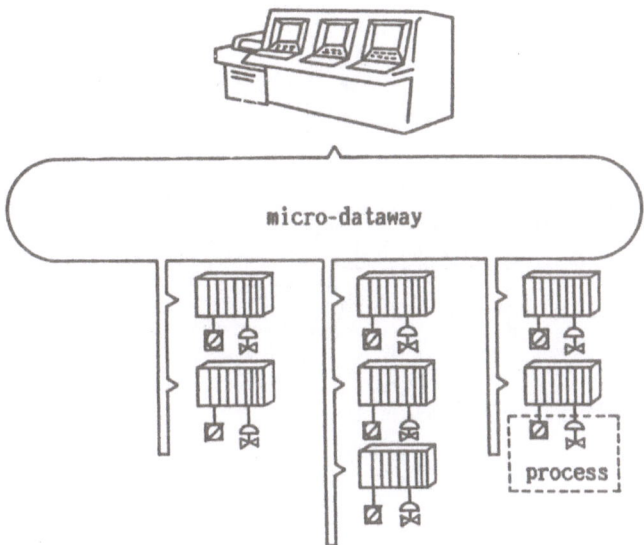

Fig.17   System configuration

After some usable settings are found
   (c) the better adjustment is possible, the control deviation becomes smaller
   and the control action becomes stabler.

8.5. Examples of auto-tuning results

The auto-tuning system has worked satisfactorily in actual processes. Typical
results are as follows.
      In cement industry, there are many kinds of material transportations,
which contain dominant dead time. Therefore, auto-tuning is very helpful,
because it is difficult to adjust the control parameters of processes with
long dead time.
      A clinker feed process as shown in Fig.18 (a) is one of self-regulating
processes. To this process, auto-tuning was applied in the initial tuning
period of the control system. The test signal and the process response recorded are shown in Fig.18 (b). From the two auto-tuning tests of this figure,
almost same PID paramenters are obtained without no a priori information about
the process dynamic characteristics. One of them is set and the process was
controlled well as shown in Fig.18 (c).

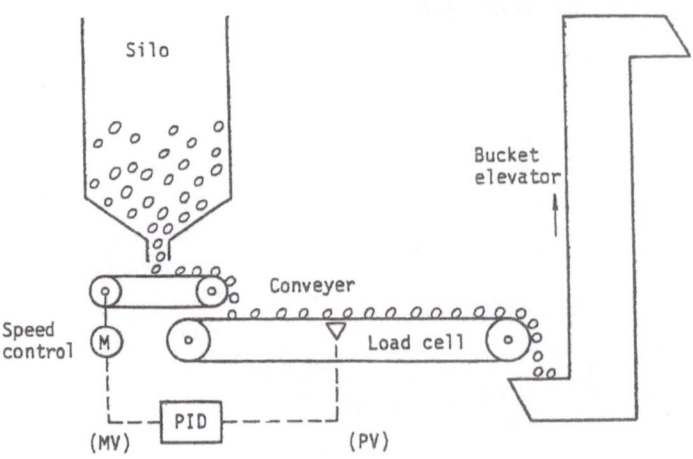

(a) Clinker-feed control loop

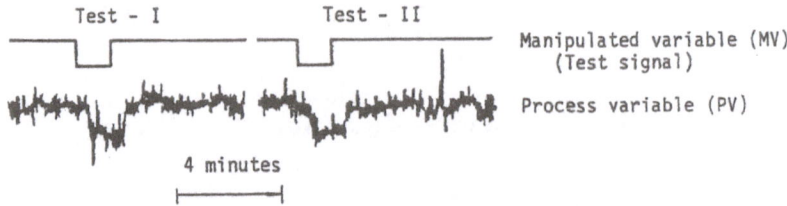

(b) Test signal and open-loop process response

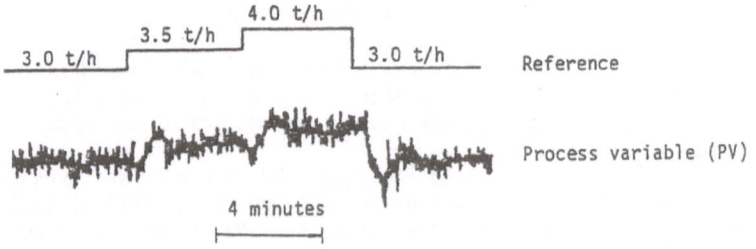

(c) Result of the DDC by the settings of the auto-tuning

Fig.18   Auto-tuning applied to a feed control loop.

8.6. Future trend to the adaptive control

Optimal tuning makes the control action nearer to the stability boundary. As almost all of the processes have some nonlinearity more or less, it is very possible that the control system with the well-adjusted parameters for one condition goes into unstable states in the other conditions. In order to solve the problem, adaptive action in expected.

The purposes of the adaptation of control parameters are considered as the optimal adjustments of control parameters for ① the unknown (but constant) process characteristics and ② varying optimal control parameters.

The variation of optimal control parameters is caused by (1) the substantial changes of process characteristics, such as by exhaustion or corrosion and (2) operation level changes of the nolinear process, such as the generated power of a steam plant or the arm angles of a robot, which effect the process characteristics.

For the industrial use of adaptive control, the adaptation to (2) is most frequently applied and is thought most effective. For this purpose, gain scheduling adaptive control is very suitable. However, it cannot adapt to (1) and the initial adjustment needs a lot of a priori information about process characteristics.

From author's experiences, it is considered that the estimation of the process characteristics is necessary because substantial changes (1) occur and they are not predictable, and that the estimation for substantial changes is possible because the changes proceed very slowly.

It is also considered that the gain scheduling adaptive control is necessary and useful because the changes of operation level (2) are quick but predictable. For these reasons, a new gain scheduling adaptive control scheme is developed. It is implemented on a micro-computer system (FASMIC G500). This new system is called as GISTLAX, shortly. (Gain Scheduling of Two direction parameter Learning for Adaptive Control System ; CS →X)

The adaptation scheme of GISTLAX is shown in Fig.19.

GISTLAX adapts control parameters to the operation level changes by gain scheduling and also adapts to the process characteristics changes by fitting the gain schedule curve based on the estimation. The estimated results are accumulated in a computer system and effective results are used for the gain schedule curve after the inspection by operator. In this procedure, conversational auto-tuning is very suitable.

The proposed adaptive scheme is very intuitive. It is based on the human knowledge about the process. However, usual adaptive control action is performed automatically. Operators need not always pay attention to the control action, but they simply judge whether the results of process estimation are reliable or not by looking at the display panels of mam-machine communication. An estimated result is thought as unreliable if the process is disturbed in the period of estimation of it or if is very different from others on the plot of accumulated data.

For the automatic learning in closedloop operation, the system will executes the direct optimization which adjusts control parameters in small steps and accumulates better parameters judged on the basis of the performance index composed of the error functions of actual process variables.

478

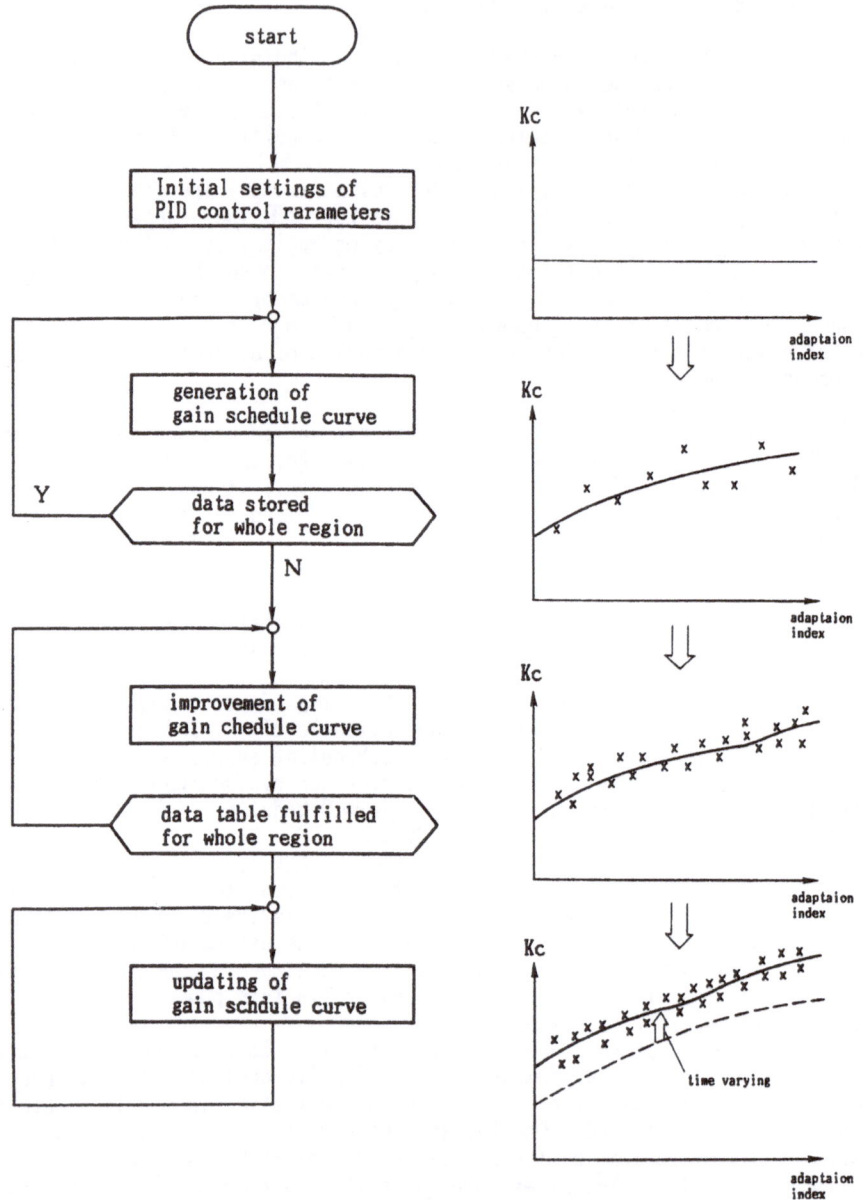

Fig.19   Data accumulation scheme of the adaptive control
system in two direction

8.7. Comparison with self-tuning

Self-tuning regulator(STR) is well known but it has many problems in its practical applications.

Perfect automatic adaptative action is called "self-" adaptive and is expected for the easy application procedure. However, the question how an adaptive system can distinguish the effective responses for estimation among many kinds of unexpected disturbances. Unless some method can discriminate effective signals, unnecessry change of control parameters takes place and the process is disturbed by adaptive action itself.

STR cannot work so quickly as the operation level change (2), because the estimation of the process characteristics cannot converge in a short time as the operation level change. From the theoretical consideration, the estimation needs some disturbances, but it is desired to decrease the disturbance as small as possible in practical processes. Therefore, the information(or the result of convergence) of the process characteristics should be memorized and the disturbance for the estimation of the same process conditions as before should be avoided.

9. CONCLUSION

New functions of real time digital control systems are described. Data processing for modeling and prediction is particularly important compared with conventional analog systems. For new control algorithms, digital system executes complicated, nonlinear, accurate, and modern theoretical calculation. Man-machine communication is very convenient. As estimation of process parameters is possible, automatic tuning of control parameters is also applicable.

For the application of a new method, hardware and software of the system must be rich enough. At the same time, the characteristics of the process should be understood.

Process model and computer simulation are useful not only for the analysis or synthesis of control systems, but also for the test or evaluation of new digital control schemes. Data processing for the development of model and prediction of application result is particularly important compared with conventional analog systems. On account of new control algorithms, digital system executes complicated, nonlinear, accurate, and modern theoretical calculation.

In cement industry, digital control extends the range of automatic control, so that manual operation becomes standard and easy and that an average operator can control plants stably. Digital control improves dynamic behavior of controlled results with the putpose of making the quality of products fine, changing operational level in short time, and operating nearer to upper or lower limit. Digital control combines informations between process and management in order to save energy, to optimaze operational conditions corresponding to production plan, tounderstand the state of process, and to respond to emergency quickly.

The described control methods can be applied by μ-computers, today. The more complicated control methods, such as Fuzzy control and total optimal management of the plant operation, are executed not by μ-computers but by mini-computers or super mini-computers. It is important that the various kinds

of control methods are applied and are connected with each other through data
way and that the more highly advanced control is accomplished.

Acknowledgement. The authors are very grateful to IHI(Ishikawajima harima
Heavy Industry) Co.,Ltd. for the cowork of practical applications.
    The authors also would like to thank Mr. Yoshio Takeyama of Fuji Facom
Corporation, and Mr. Kozo Ishida and Mr. Ken Takata of Fuji Electric Co.Ltd.,
for working together in developing new digital control methods.

REFERENCES

Akaike, H. (1974) A New-Look at the statistical Model Indentification.
IEEE Trans, AC-19-16, 716-723.

Nisikawa, Y., N. Sannomiya, T. Ohta, H. Tanaka, (1984) A Method for Auto-
tuning of PID-control Parameters. Automatica, Vol.20, No.3, 321-332.

Ohta, T., K. Ishida, Y. Koike, (1983) Real time digital control system for
the cement industry. REAL TIME DIGITAL CONTROL APPLICATIONS (Proceedings of
the IFAC/IFIP Symposium, Guadalajara, Mexico, 1983) Pergamon Press. Oxford,
197-202

Chapter 14

CURRENT STATUS OF MICROCOMPUTER APPLICATIONS ON RAILWAY TRANSPORTATION SYSTEMS

Hirokazu Ihara* and Makoto Nohmi **

* Space Systems Division, Hitachi, Ltd.
  216 Totsuka-machi Totsuka-ku, Yokohama 244, Japan
** Systems Development Laboratory, Hitachi, Ltd.
  1099 Ohzenji Asao-ku, Kawasaki 215, Japan

ABSTRACT. Microcomputers have become powerful and reliable enough to be applied to the transportation systems. There are many microcomputer applications on the railway transportation systems, such as Shinkansens, subways, monorails, private railways and medium size guideway rapid transits in Japan. Typical examples of them are introduced here.

1. Introduction

The transportation system is one of the most familiar systems for human life. Therefor it means that the transportation system is a typical social system and has to have high qualities of serviceability, reliability and safety. Nowadays on the transportation system, especially on the railway system, modernization of the system has been promoted to provide it with good service and high efficiency by introducing the computer control and management system. Then the computer system becomes to perform important roles in the system and the system becomes not to be realized without the computer system.

On the other hand, the microcomputer which was developed more than ten years ago has been applied in many fields, of course in the railway system. In the early stage of the microcomputer application, the microcomputer was applied to the system as an auxiliary component which did not need high reliability and performance because its reliability had not been sure yet and its capability of processing was not sufficient for the system requirement. At present, progress of LSI technologies makes the microcomputer powerful and reliable, and then it becomes difficult to find out any system without microcomputer in the modern railway transportation systems such as Shinkansens, subways, monorails, private railways and medium size guideway rapid transits.

In this paper, typical examples of microcomputer applications for the railway transportation systems in Japan, which include train traffic control systems, automatic train operation systems, propulsion systems, power substation control systems, passenger addressing systems, and maintenance support systems are collected from reference papers.

481

S. G. Tzafestas and J. K. Pal (eds.), Real Time Microcomputer Control of Industrial Processes, 481–508.
© 1990 *Kluwer Academic Publishers.*

2.   Total System

2.1.   Outline of the system (Ref.1)

The railway business is divided into the following three areas.
(1) Management and control of various transportation equipment; for example, rolling stocks
(2) Service for passengers; for example, ticket vending and checking
(3) Management; for example, statistical data processing

Operation is performed through cooperation among these three areas. It becomes necessary to introduce a total control system especially in the subway, making use of modern computer, communication and microcomputer technologies, in order to achieve wide scale improvements effectively and rapidly, with few operators maintaining adequate security.

In designing the total system, for example in Fukuoka City Rapid Transit System in Japan, the following seven points are considered:
(1) Realization of rational and rapid control operation by concentrating operations such as train traffic control, electric power supply control, and fire alarm and extinguishing control
(2) Reducing the number of station employees by systemizing various station operations
(3) Effective utilization of data transmission lines by unifying different data such as ticket data, fire monitoring data and so on, into one line
(4) Rationalization of various management data, such as business income and budgeting
(5) High reliability by adopting a dual configuration of equipment and a functional distribution concept
(6) Labor saving by on-line linkage among subsystems
(7) Rationalization of maintenance operations through adoption of non-mechanical equipment, automatic fault detection, and automatic inspection

2.2   System configuration

The total system owing to the philosophy mentioned above, consists of those subsystems, which are the train traffic control system, automatic train operation system, electric power supply control system, information transmission system, automatic car inspection system, and supporting business management system. Its configuration is indicated in **Fig. 2.1.**   Each subsystem has the following functions and roles.
(1) **Train traffic control system:** Observing the status of train traffic, quality improvement and labor saving in control operations, and improvement of service for passengers are performed by route control, bulletin display, automatic public address, and information transmission to the train according to the train schedule.
(2) **Automatic train operation system:** Automation and labor saving in train operation, plus maintenance of safety, are performed by automatic operation control of such things as acceleration, constant speed, stopping at predetermined points at stations, and in car public address.
(3) **Electric power supply control system:** Quality improvement and labor saving are performed by observing the status of equipment in substations and electric equipment rooms, by schedule control and power monitoring.

(4) Information transmission system: Unification of transmission lines, linking substations, and automation of issued ticket data processing are performed by transmitting the various data produced at stations to the control center and distributing them to related subsystems.

(5) Automatic car inspection system: Automation and labor car maintenance operations are performed by inspection of car mounted equipment without breaking the train formation and managing the historical data of the cars.

(6) Supporting business management system: Automation and labor saving in business management are performed by a wide variety of data processing functions, such as making various reports of issued ticket data, accounting data processing and budget management.

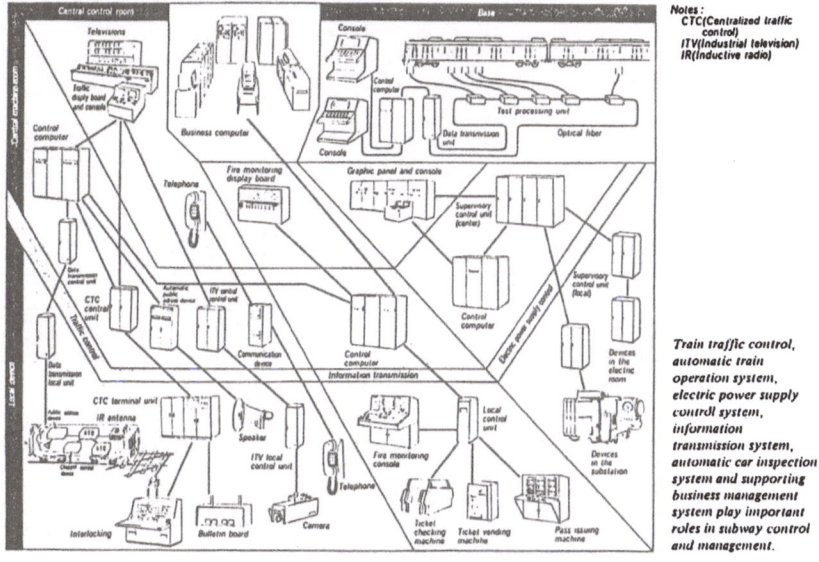

Fig. 2.1 Total System Configuration

These subsystems carry out their respective functions and exchange various data with each other. This constitutes the total system. In this total system, microcomputers are used as components of the subsystems and perform very important roles. Their capability of data processing and cost effectiveness make the system move useful and economical. Moreover, by using microcomputers, each subsystem becomes more intelligent and compact, so the total system is able to be constructed easily in the form of distribution as decentralization which makes the system more reliable and flexible.

3. Ground System

3.1. Train Traffic Control System (Ref.2)

The train traffic control system for Kobe City Subway was installed in June 1983. The main feature of this system is its configuration of decentralization (Ref. 4,5) and the optical ADL– Net (Autonomous Decentralized

484

Loop:Ref.4) which is employed as the data transmission subsystem. The concept of the system and ADL-Net is based on autonomous decentralization, which is derived from a biological model. In ADL- Net, every element of the network is uniform and equal. Each element has the same control information as others and cooperates with them in order to make the network highly fault tolerant, yet flexible and easy to enhance. Since every element is uniform, ADL-Net has neither a master station nor a physical address on each element as conventional network systems.

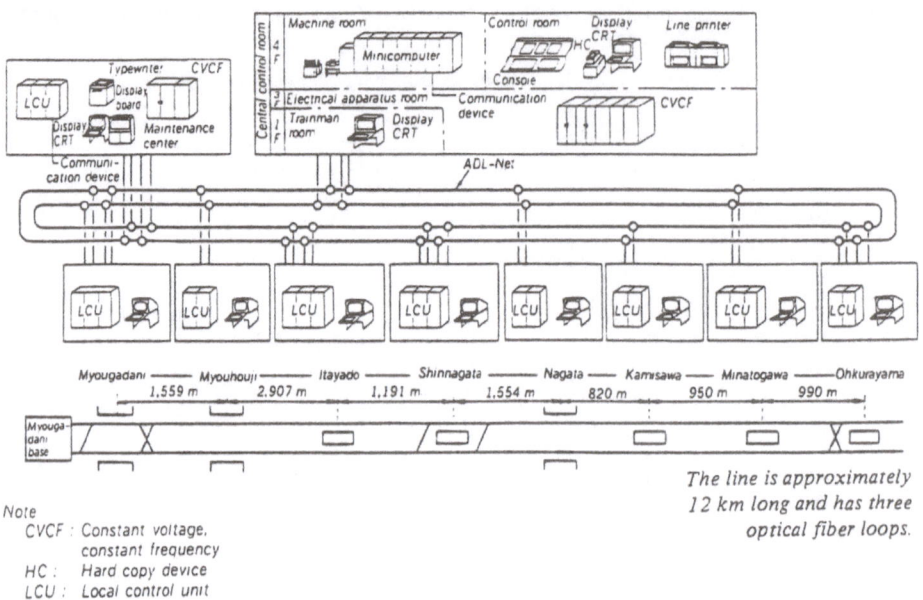

Note
CVCF : Constant voltage,
           constant frequency
HC : Hard copy device
LCU : Local control unit

The line is approximately 12 km long and has three optical fiber loops.

Fig. 3.1   Overall System Configuration for the Kobe City Subway

The overall system for the Kobe City Subway is shown in **Fig. 3.1.** The line is approximately 12 km long and has three optical fiber loops; two are for ADL-Net and one is for a spare (back-up). These optical fibers link the local control unit (LCU) in each station with a minicomputer system located at the center via network control processors (NCPs). The central computer system consists of four process control computers, one medium size minicomputer and three small size minicomputer in communication with each other. Functions such as train scheduling and man-machine communication are performed by this system, which handles communication between LCUs and itself. As shown in **Fig. 3.2,** the LCU consists of two local control processors: LCPs, one external interface processor: EXP and four network control processors: NCPs each of which has an 8bit or 16bit microprocessor and which configurates a multimicroprocessor. Each microprocessor has its own function.

For instance, the NCP performs data transmission, the LCP performs process control, the EXP performs input/output device control, and the LCP performs public address. The LCU interfaces with interlockings and bulletin boards. A CRT display and IC public address voice recorder are built in.

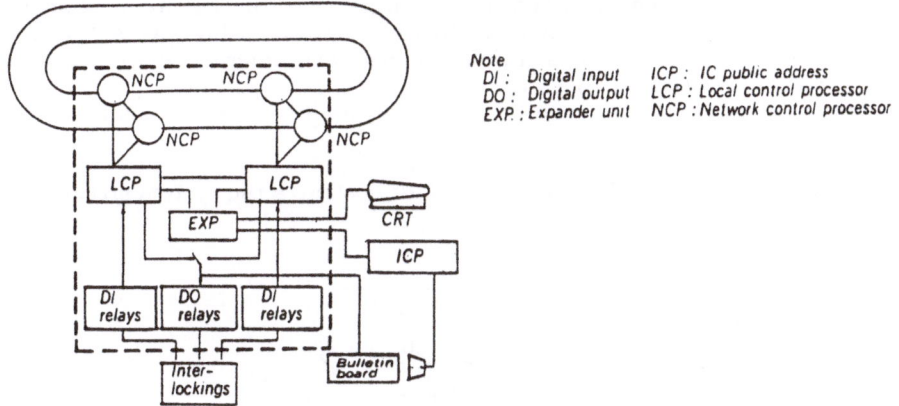

Note
DI : Digital input   ICP : IC public address
DO : Digital output   LCP : Local control processor
EXP. : Expander unit   NCP : Network control processor

The LCU consists of seven microprocessors, each of which has its own function. The two LCPs perform in dual operation.

Fig. 3.2 Configuration of the Local Control Unit (LCU)

The train traffic control system performs train schedule management, route control, public address and bulletin board control in stations, statistical processing, corrective strategies, and man-machine communication.

These functions are performed by LCUs and the central mini-computer in accordance with the basic design concept of "distributed control and centralized supervision". This is realized by scattering the functions among the central minicomputer and the LCUs at each station, as shown in **Fig. 3.3**.

By so doing, higher control responsiveness and higher fault tolerance for the whole system compared with conventional centralized control systems are made possible.

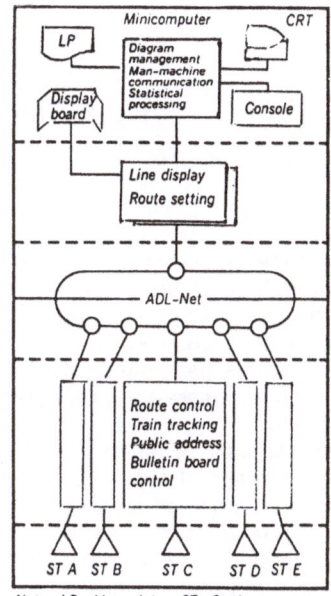

Note LP : Line printer ST : Station

Fig. 3.3 The Function of the Train Traffic Control System

3.2. Net work system (Ref.3,4,5)

The autonomous decentralized loop network (ADL-net) system, is applied and developed as a loop transmission network system using the autonomous decentralized control concept mentioned above and microcomputers for the network control processors (NCPs).

The ADL-net system architecture and maintenance techniques are described below.

3.2.1. ADL-net system architecture   Fig. 3.4 illustrates the system structure.

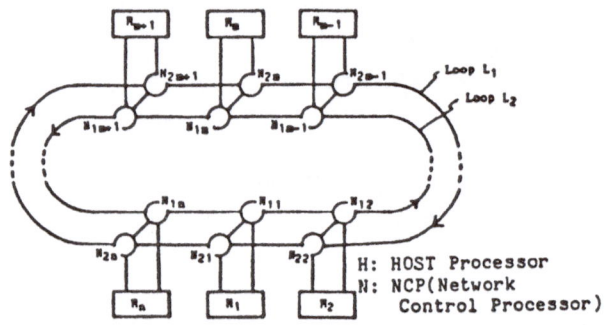

Fig. 3.4 Structure of the Network

The ADL-net system features are as follows:

(1) The network is a double loop consisting of two unidirectional loops. They are assigned to transmit messages in opposite directions.

(2) Each NCP is connected to an adjacent NCP on the same loop and with a partner NCP on the other loop by two unidirectional bypassing links.

(3) Each pair of NCPs is connected to one Host processor.

(4) Content coded communications method: NCPs can broadcast messages with a content code signal that corresponds to the contents of transmission data. Messages have no destination address. Receive controllers can receive content code messages predetermined on the basis of functions which they should play.

(5) Simultaneous broadcast method: There are no transmission NCPs which occupy all loops. Therefore, whenever each transmission NCP wants to send messages, it can send messages to the loop.

(6) Fault detection and recovery: It follows after the autonomous decentralization concept that every NCP is designed to have the same fault detection and recovery mechanism(**Fig. 3.5**).

3.2.2. Fault tolerant techniques

(1) **Fault detection (Fig. 3.5 (a)–(b)):** The NCP first detects a fault in the loop when a message originating from it does not return even after several retransmissions . Next, the NCP performs a check operation, which is called a

minor loop check, to determine whether it can transmit the message along the minor loop consisting of the adjacent NCPs and itself. Furthermore, each NCP receiving the minor loop check signal originating from the adjacent NCP initiates its own minor loop check signal in succession. When the NCP does not receive the minor loop check signal originating by itself, it detects the existence of a fault somewhere in the minor loop concerned. This detection permits the NCP to restructure the network by taking an alternate route. Thereafter, the NCP sends messages via the alternate route. Here it is not necessary for the fault detecting NCP to inform any other NCP of the detected fault or of the reconfiguration of the network.

(2) Fault recovery (Fig. 3.5(c)-(d)): The existence of an omission by the minor loop check or the fact that failed NCPs or links have been repaired is automatically detected. The NCP, which has detected the faults and has restructured the transmission route, periodically originates minor and major loop check signals, but need not inform the result of the check to the other NCPs.

During this process of fault detection and recovery, all of the other NCPs as well as the fault detecting and recovering NCP continue normal operation.

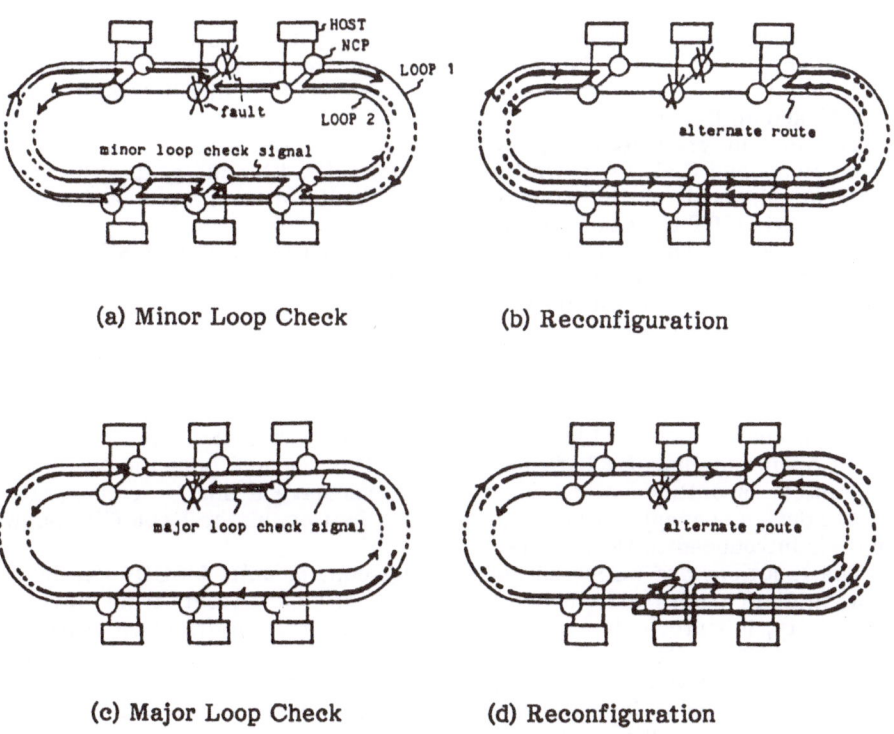

(a) Minor Loop Check   (b) Reconfiguration

(c) Major Loop Check   (d) Reconfiguration

Fig. 3.5 Fault detection and Recovery

As shown in **Table 3.1,** the performance and fault-tolerance of the ADL-net is enough to apply to the control system of subway and other guideway rapid transit.

Table 3.1 Specification of the ADL-Network

Item	Specification
Network Topology	Double Loop
No. of NCP	Max.256
Distance between Stations	Max. 4 km
Transmission Rate	64 kbps, 1Mbps
Protocol	HDLC, Broadcasting
Synchronization	Autonomous Synchronization
Fault Detection	Autonomous Detection
Fault Recovery	Retransmission and Loop Back

3.2.3. Maintenance techniques in the ADL   The system tester for online fault diagnosis and maintenance is designed also on the basis of the autonomous decentralization concept. This online tester consists of a Built-In Tester (BIT) in each NCP and an External Tester (EXT) in some Host.

The BIT in each NCP has the function of fault diagnosis concerning adjacent NCPs and their links. On the other hand, the EXT integrates the results of the fault diagnosis of the NCPs to specify faults in the system and provides instructions via CRT for a repairman.

These testers do not interrupt or stop the operation of any NCPs, so the network and the system are able to continue their operation, and are not interfered by maintenance activity.

3.3. Passenger Addressing System (Ref.2)

New passenger guidance information systems were introduced for the Marunouchi and Yurakucho Lines of the Teito Rapid Transit Authority in Tokyo Japan in April and June 1983, respectively. In this section, only the Marunouchi Line's system is outlined since the two systems have a similar configuration. The same design concept, autonomous decentralization, as the Kobe City Subway system's is introduced to these systems.

The functions are dispersed among local control units (LCUs) which are same as those of train traffic control system. Each LCU detects train movement in its vicinity, initiates public address and controls train information destination indicator.   Thus, the LCU provides accurate and precise information to passengers on the platforms.   As shown in **Fig. 3.6,** the data transmission line, ADL, which is a double inverted loop network and linked to each LCU, utilizes existing metal cables. This system has no central controller (master station), so it can localize failures. In addition, the ADL has alternate routing functions to maintain high availability even in the case of transmission failures.

Two LCUs which are located at each terminal are called parent stations and provide line supervision and train schedule maintenance functions. Each parent station can take over these functions alternately. A nonmechanical PARCOR IC voice recorder is used in the public address unit to increase reliability.

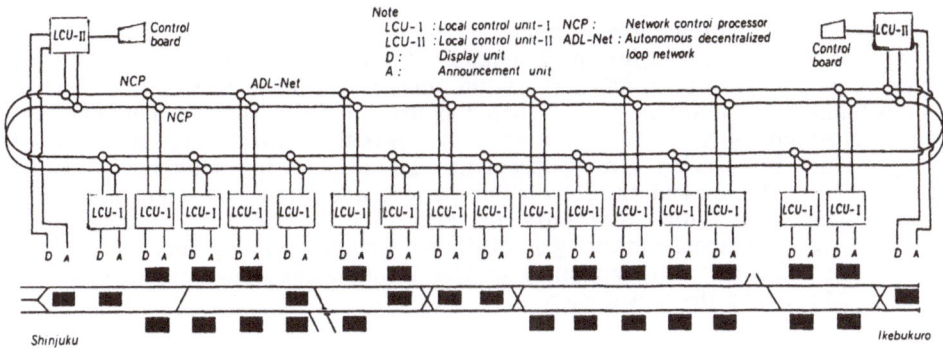

Note
LCU-1 : Local control unit-1   NCP :   Network control processor
LCU-II : Local control unit-II ADL-Net : Autonomous decentralized
D :   Display unit   loop network
A :   Announcement unit

Shinjuku   Ikebukuro

The system has gone into service at six stations so far. The system began operation at the remaining twelve stations in 1984.

Fig. 3.6 Configuration of Passenger Guidance Information System

A liquid crystal display unit, shown in **Fig. 3.7**, is used in the train information destination indicators in lieu of the conventional flap-type unit. These components are all solid state to provide improved reliability and maintainability.

Each character is formed from a 24 x 24 array of circular dots and is approximately 10 cm x 10 cm in size.

Fig. 3.7 Liquid Crystal Destination Indicator

The system has gone into service at six of the eighteen stations on the 16.6 km line between Ikebukuro and Shinjuku. The system began operation at the remaining twelve stations in 1984 without stopping the system under operation.

3.4.   Power Substation Control System (Ref.2)

The new electric power supply control system named DECS has been developed and introduced for the Tohoku-Joetsu Shinkansen in Japan. The design concept of the system is different from that of its predecessor, the Tokaido-Sanyo Shinkansen electric power supply control system. In the Tokaido-Sanyo system, which is a centralized control system, dispatching tends to be complicated, interfering with appropriate global decision making concerning train traffic control and facility maintenance, and it becomes an obstacle to prompt recovery from accidents or calamities. For the Tohoku-Joetsu Shinkansen, operational authority is dispersed among five regional offices. This fact, in addition to the incapability of the Tokaido-Sanyo system to cope with abnormality, led to the new design.

The basic idea behind the new system is hierarchy distribution. The lower level, which is the local subsystem, performs almost all automatic functions, including safety preservation, to conserve the upper level's load and to ensure expandability. This concept, the three-layer hierarchy, is depicted in **Fig. 3.8.**

(1) Electric power sequence control system: In lieu of conventional switchboards, a "neighboring interlocking system" has been developed. This neighboring interlocking system performs safety preservation functions and line configuration functions automatically at the substation level. Self-diagnostic functions and maintenance data preservation functions are performed by microprocessors.

(2) Remote control system: A new double-loop, highly reliable data trans-mission system has been developed. This subsystem can convey communication data between substations or between multiple control centers. For the second loop, the transmission direction is reversed. By adopting this configuration, switching over from a failed loop to the stand-by loop or taking an alternative route can secure stable data transmission.

(3) Electric power data processing system: At the central operation office and at each regional operation office, computer systems are installed to aid power supply dispatching. The primary feature is the display of graphic diagrams depicting the current status of the power line configuration, train traffic, and their interrelations on CRTs, which are the main interface between operators and computer systems (see **Fig. 3.8**).

Each computer system is composed of two central processing units, constituting a stand-by duplex system.

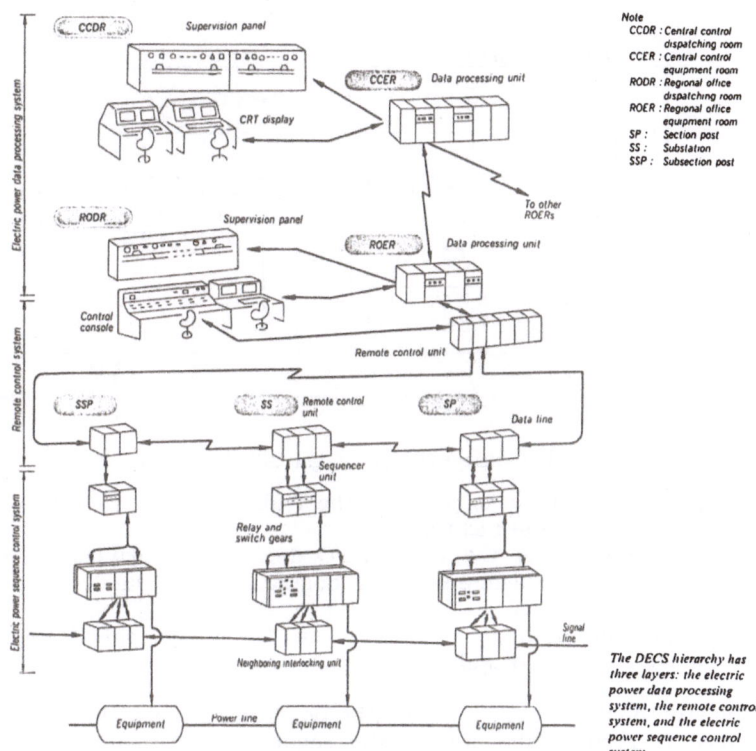

Note
CCDR : Central control dispatching room
CCER : Central control equipment room
RODR : Regional office dispatching room
ROER : Regional office equipment room
SP : Section post
SS : Substation
SSP : Subsection post

The DECS hierarchy has three layers: the electric power data processing system, the remote control system, and the electric power sequence control system.

Fig. 3.8 Overall DECS Configuration

4. On-Board System

4.1. Microcomputer based Automatic Train Operation System (Ref.6)

In recent years, electronics has been applied to many on-board systems with the improvement in performance and reliability. Automatic train operation (ATO) is one suitable application for the microcomputer. Furthermore the microcomputer's flexibility is expected greatly to increase the performance of the systems.

4.1.1. Outline of ATO system   The on-board system of the automatic train operation ATO is a subsystem of the automatic train control system (ATC). ATC has two other subsystems. These are the automatic train supervision system (ATS) and the automatic train protection system (ATP) as shown in **Fig. 4.1.**

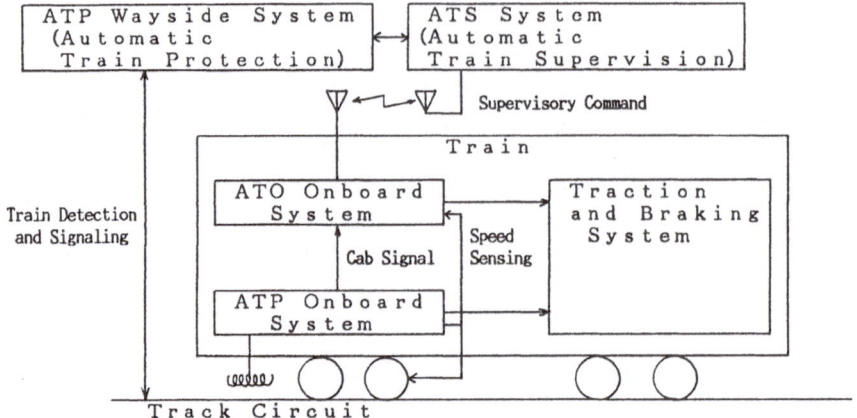

Fig. 4.1 Configuration of the ATC System

The ATO operates the train in place of the train driver as shown in **Fig. 4.2**. Major functions of the ATO are as follows:

(a) Performance control of the train: ATO decides the driving performance level of the train in accordance with the command from the ATS through the train communication system.

(b) Speed maintaining control: ATO generates the target speed according to the performance level and regulates the train speed so as to follow the target speed.

(c) Station stop control: ATO detects the position of the train, generates the target speed to stop at the station and regulates the train speed so as to follow the target speed.

(d) Dwell and departure control: ATO operates the doors of the train for passengers to get on and off the train. After the doors are closed ATO starts the train.

These functions of the ATO are specified according to the requirements of the system.

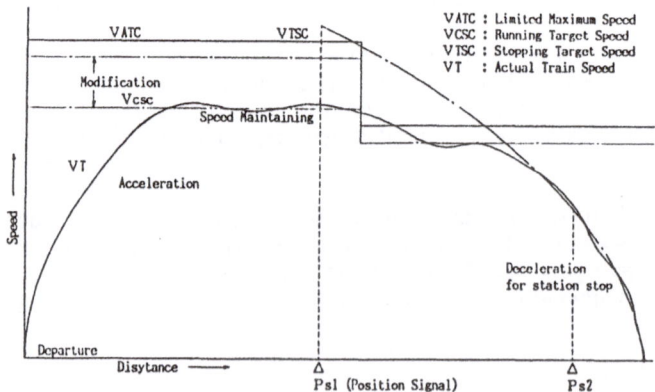

Fig. 4.2 Outline of the Automatic Train Operation

4.1.2. Requirements for ATO   Structural requirements for the ATO are as follows:

(1) Reliability and safety: Failure of the ATO system may disrupt the traffic and cause injury to the passengers. Therefore the system must be free of defects, have a long life and a fault tolerance.

(2) Portability: The on-board equipment must be small and light, and must be easy to carry and maintain.

(3) Productivity and maintainability: To reduce initial and operating cost, productivity and maintainability are important factors. The system must be easy to design, assemble, test and repair.

(4) Flexibility and extendability: The system must be flexible and extensible to cope with the various specifications and technological progress.

4.1.3. Configuration of ATO on-board system   ATO is divided into two units as shown in **Fig. 4.3.**   One is the navigation control unit (NVC) and the other is the driving control unit (DVC).   They have M-6800 microprocessors, and are connected with each other by serial communication lines.

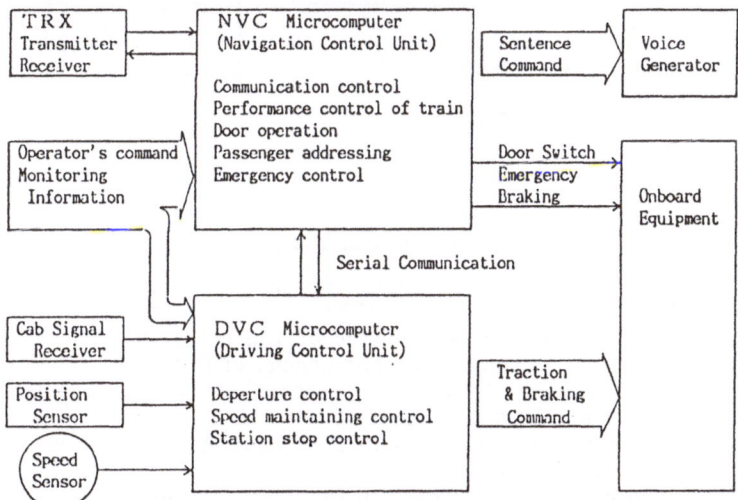

Fig. 4.3 Configuration of the On-board ATO

(1) Functions of the NVC: The functions of the NVC unit are related to supervisory control, and are as follows:

(a) Communication control to/from ATS
(b) Performance control of the train
(c) Passenger addressing
(d) Emergency control
(e) Door control

(2) Functions of the DVC: The functions of the DVC unit are related to dynamic control of the train, and are as follows:

(a) Departure control
(b) Speed maintaining control
(c) Station stop control

494

4.1.4. Structure of the software To increase the productivity the programs in the units are structured as follows:

(1) Standardization of the structure: In this system the software structure is standardized in the form of the 4-level cluster by using a nesting technique as shown in **Fig. 4.4.**

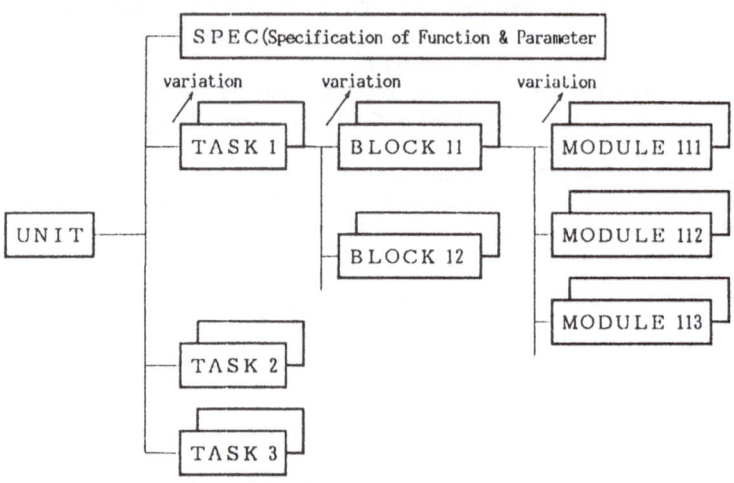

Fig. 4.4 Structure of the Software

(a) UNIT: A UNIT is the top cluster of the programs which includes all the functions of the unit, and consists of TASKs.
(b) TASK: A TASK is the second cluster of the programs. These tasks can operate individually and concurrently, and consist of BLOCKs.
(c) BLOCK: A BLOCK is the third level cluster of the programs which realizes the subfunction of the unit, and consists of MODULEs.
(d) MODULE: A MODULE is the fourth level cluster of the programs which realizes elementary function.
 (2) Interfaces and composition: The interfaces of each cluster are defined by the higher level cluster above it. Each cluster has functional variations. Therefore, when necessary program packages are prepared in a library according to the variations in each cluster. Program specifications are made by SPEC. SPEC declares the selection of the source programs in each cluster and control parameters according to the required specifications.

The source programs are written in structured programming language PL/H (Programming language of Hitachi for microcomputer). This system is applied to automatic train operation of subways, monorail and medium size guideway rapid transit.

4.2. Automatic train operation system by predictive fuzzy control (Ref.7)

As shown in the previous examples, automatic controllers using a microcomputer instead of a human operator have been developed for transportation systems and so on. In many cases, a computer control gives quick response and accurate control, but inferior quality of control to a skilled human operator. A fuzzy logic control method which can make up an algorithm from control knowhow of a skilled human operator by fuzzy sets, is proposed by Mamdani (1973), and applied to a plant, a traffic junction, a cement kiln, a water treatment and so on.

4.2.1. Conventional Control and Fuzzy Control   The conventional controller is able to give accurate control for systems which are matched to the linearized system models and/or the desired state are constant. However, this method cannot provide adequate control for systems which have time-varying parameters, unknown structures and multiobjects for the control. So, the controller is designed using a complex structured algorithm with the designer's knowledge. But reconstructions of the controller, according to the alternation of a subsystem or change of the system structure, devote a designer's effort every time. **Fig. 4.5** illustrates graphically the conventional control sequence.

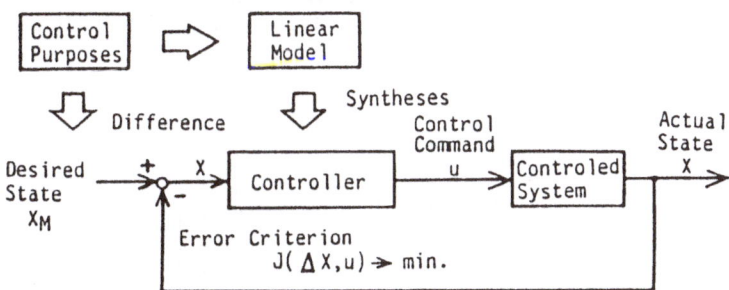

Fig. 4.5 Conventional Control Sequence

4.2.2. Predictive (objects evaluation) fuzzy control   A skilled human operator has extensive experience through many experiments with the system's operation. And he can perform high-quality control satisfying the system objectives. The predictive fuzzy controller which decides a control action u from the objects evaluation of the control results by the control actions has been proposed thanks to the high performance microcomputer. In the proposed method,
   (1) the control rules R(R1,R2,...Rn) is described as "Ri: If (u is Ci - x is Bi) then u is Ci",
   (2) the control rule Cj is selected from the predictive results(x,y) that show the highest likelihood.
"When the control notch is not changed, if the train stops in the predetermined allowance zone, then the control notch is not changed." is a typical example of train operation rules. **Fig. 4.6** illustrates graphically the control sequence.

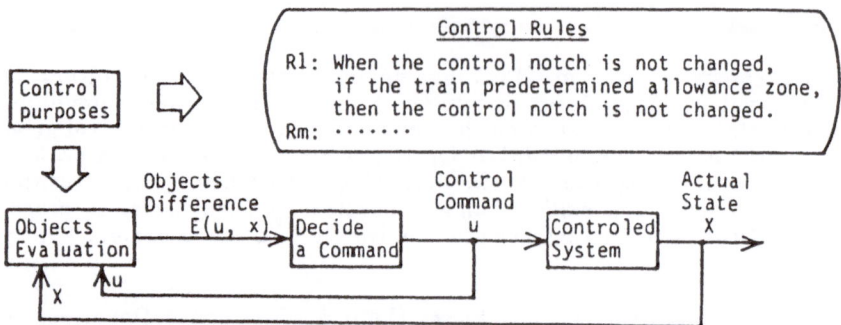

Fig. 4.6 Predictive (Objects Evaluation) Fuzzy Control Sequence

4.2.3. Simulations   By the simulation the predictive fuzzy control ATO (Fuzzy ATO) is compared with the conventional PID control ATO (PID ATO). simulation results are as follows.

(1) **Riding comfort and stop gap accuracy:** Fig. 4.7 shows a summary of simulations for riding comfort and stop gap accuracy. As the results of the simulation, the number of notch changes in the fuzzy ATO is about a half, and the stop gap is about a third compared with those of the PID ATO. Therefore, the developed fuzzy ATO is able to control a train with good riding comfort and stop gap accuracy.

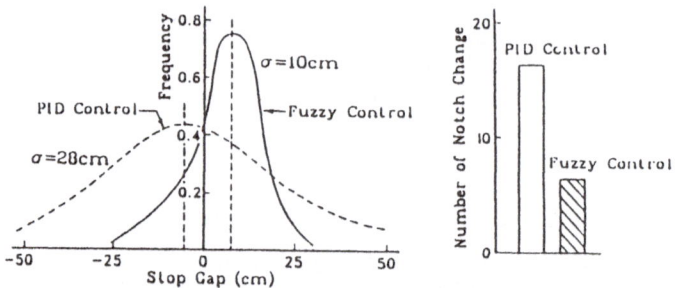

· The deceleration to maximum brake notch is 70%, 100% and 130%
· The gradient is -5°/₀₀, 0°/₀₀ (level) and +5°/₀₀

Fig. 4.7 Summary of Simulation for Riding Confort and Stop Gap Accuracy

497

(2) Energy consumption: The **Fig. 4.8** shows a summary of simulation results on running time and energy consumption. As the results of the simulation, the fuzzy ATO is able to control trains over 10% of energy saving and/or shorten a running time compared with the PID ATO.

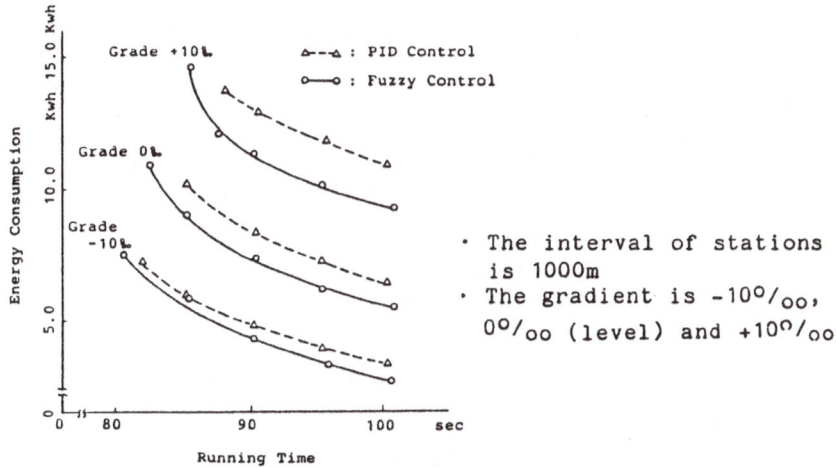

· The interval of stations is 1000m
· The gradient is -10⁰/oo, 0⁰/oo (level) and +10⁰/oo

Fig. 4.8 Summary of Simulation of Energy Consumption

4.2.4. Field test for a Subway System A field test was performed at a subway system. **Fig. 4.9** shows a result of the field test. The field test shows that the developed Fuzzy ATO can operate trains as skillfully as human experts.

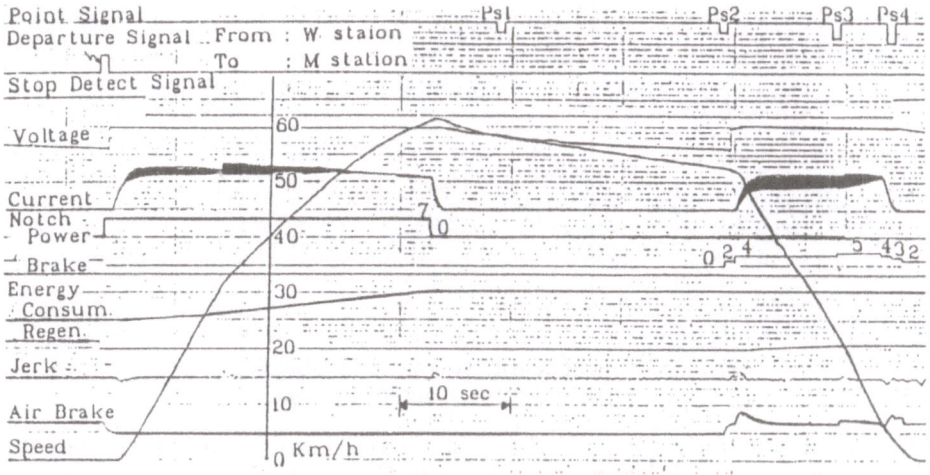

Fig. 4.9 A result of the Field Test

4.3. Propulsion system

A propulsion system for a rolling stock is also one of suitable applications of the microcomputer. Several years ago, the control system of the propulsion system was utilized using analogue devices for the reasons of lack of processing capability of the microcomputer. In recent years, the performance of the microcomputer has been increased rapidly, and became enough to apply to the control of the propulsion system. The other advance of technology is a development of a high power gate turn off (GTO) thyristor. The microcomputer and the GTO thyristor make the control electronics and the main driving circuit of the propulsion system compact and reliable.

The propulsion system for the rolling stock has two types of driving method. One of them is a chopper driven D.C. motor type and the other is a inverter driven induction motor type. Both types of propulsion system are used for now in the rail rapid transit system in Japan. The microcomputer is used as a thyristor gate controller of the chopper and the inverter to regulate tractive and braking force of the motors. Examples of microcomputer application in the chopper and the inverter are described here.

4.3.1. Chopper system (Ref.8)   The GTO thyristor chopper for shunt motor driving with field and armature control is one of the newest chopper created by integrating the current technology of a micro-electronics and power electronics, and control technologies based on experience with both of armature chopper control and field chopper control.

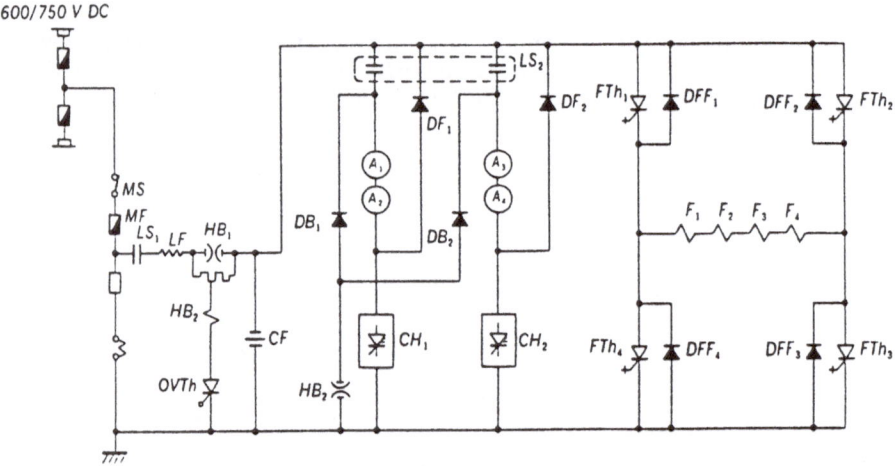

**Fig. 4.10   Schmatic Power Circuit Diagram of
Thyristor Chopper Control System**

(1) Main circuit of chopper: Fig. 4.10 shows the power circuit diagram of the field controlled shunt motor chopper.   The GTO thyristor which is used for chopper switch has a self turn off feature, so the GTO thyristor chopper circuit does not need a commutating capacitor or auxiliary thyristor, thus giving it the following advantages:

(a) The chopper frequency is increased due to the decrease of the minimum conduction ratio. Therefor, change is minimum during pulse mode transition in the chopper and the inductance of the main smoothing reactor is reduced.

(b) Electromagnetic interference to the signaling circuit is low because commutating current does not exist in the chopper. Since there is no commutation loss, high power efficiency is obtained, even if the loss in the gate circuit is increased slightly.

(2) Control mode of field controlled chopper: This chopper has several operation modes in powering and braking state. **Fig. 4.11** shows the characteristics of speed vs. traction motor current for each operation mode described below.

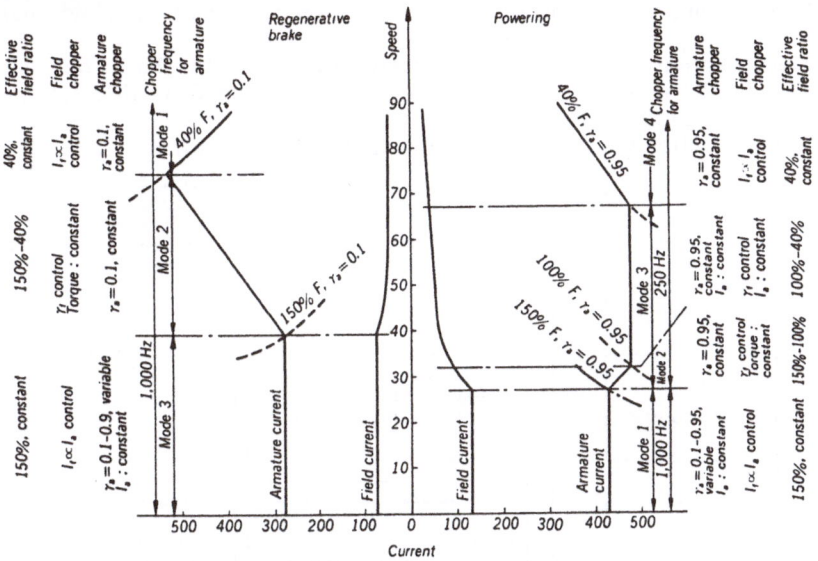

Fig. 4.11 Characteristics of Speed vs. Motor Current at each Operation Mode

(a)Powering mode 1: The conducting ratio of the armature chopper is controlled from 0.1 to 0.95 between starting and medium speed (approx. 30 km/h) in order to keep the armature current constant. The the field current is proportioned to the armature current with a 150% effective field ratio.

(b)Powering mode 2: After the conduction ratio of the armature chopper reaches the maximum of 0.95, the field current is decreased by the field chopper control so that the tractive effort is held constant with the armature current increased. Thus the effective field ratio varies from 150% to 100%.

(c)Powering mode 3: The conduction ratio of the armature chopper is fixed at the maximum of 0.95 and the field current decreases to keep the armature current constant until the effective field ratio is reduced to 40%.

(d)Powering mode 4: This mode is the free running mode with a 40% effective field ratio.

(e)Braking mode 1: In the high speed zone, the conduction ratio of the armature chopper is fixed at 0.1, and the field current is controlled so as to keep the effective field ratio at 40%.

(f)Braking mode 2: In the medium speed zone, the regenerative brake fully covers the required braking effort of the train.

(g)Braking mode 3: In the low speed zone, the armature current is kept constant by increasing the conduction ratio to the maximum of 0.9, and then the field current is controlled , maintaining the effective field ratio of 150%.

As a result , a stable regenerative brake is obtained completely by means of a varying effective field ratio. During regenerative braking, the frequency of the armature chopper is 1000 Hz and the frequency of the field chopper is 250 Hz.

(3) Configuration of the chopper controller: Fig. 4.12 show the schematic diagram of the chopper controller which controls the GTO thyristor gates of the armature and field chopper.

The controller consists of microcomputers and input/output (I/O) devices

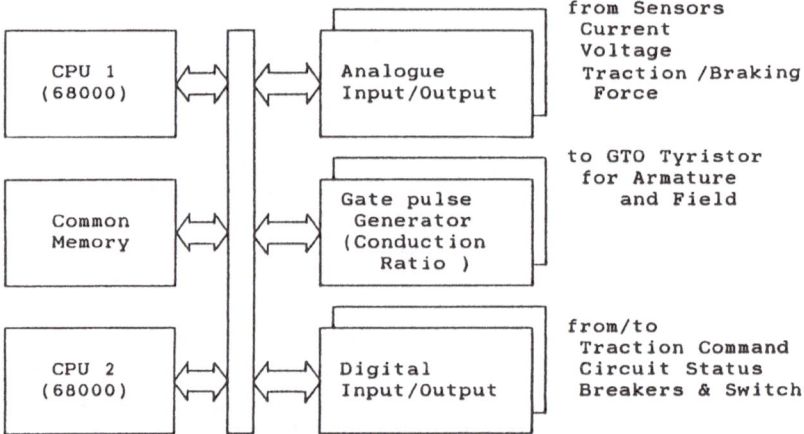

Fig. 4.12 Configuration of the Chopper Gate Controller

as follows:

(a) Microcomputer: A multi-microprocessor architecture is applied in this system to cope with a requirement of processing capability for the chopper gate control. Two processor boards, each of which consists of 16 bit microprocessor, memory, timer and some peripheral I/O devices, are installed in the controller, and they are connected into common bus to communicate with each other and to access I/O devices commonly.

(b) I/O Devices: Several types of I/O board are used in this system as follows.

(i) Analogue input board: Analogue input board is used for getting continuous data from sensor such as current and voltage of the power circuit and other circuitry and consists of AD converters with common bus interface.

(ii) Digital I/O board : Digital I/O board is for getting and putting digital data from/to the power circuit, command bus line from cab and so forth, and consists of noise filters, registers and common bus interface circuit.

(iii) Memory board : Memory board is used for common storage for the processor to communicate with each other and storing various data which are used for monitoring of the chopper equipment.

(c) Gate pulse generator: Gate pulse generator gives the gate control pulse to the GTO thyristors in accordance with the command of conduction ratio from the processors via the common bus.

(4) **Processing in each processor:** Two processors share the control processing of the chopper and cooperate with each other. The task of each processor is as follows.

(a) Task of processor 1: The Main task of the first processor is a high speed gate control of the armature chopper, which executes feed back control processing of armature current regulating conduction ratio of the GTO thyristors according to given current command value. This task is activated on each 1 ms period.

(b) Task of processor 2: One of tasks of the second processor is provided for calculating the command values of the armature current and the field current value in accordance with the command from cab and the mode described above and for regulating of field current same way as the armature control but less frequent than that for the reason that the time constant of the field circuits Is larger than that of the armature circuit. The other task is provided for monitoring processing of the chopper for the maintenance and the trouble shooting of the equipment.

The cars which are equipped this field controlled shunt motor chopper has been put into service on Ginza Line of Teito Rapid Transit Authority in Tokyo Japan, and good performance of powering and regenerative braking are obtained.

4.3.2. Inverter system (Ref.9)   The inverter is another type of propulsion system of the rail rapid transit which drives induction motor by means of converting D.C. voltage to A.C. voltage. The inverter driven induction motor has such advantages that induction motor has no commutator and brush, therefor, reliability and maintainability of the traction motor are increased, and cost and weight of the motors are decreased, even if the inverter circuit becomes expensive. **Fig. 4.13** show the schematic diagram of the inverter system which drives four induction motor.

(1) **Characteristics of the inverter system:** This inverter converts D.C. voltage given via the catenary to three phase A.C. voltage to drive three phase cage type induction motors. The values of frequency and voltage of the output from the inverter are regulated in proportion to the motor speed so that the current and torque of the motor are kept required values. Therefor this type of inverter is called variable voltage variable frequency (VVVF) inverter. The characteristics of the inverter is as follows:

(a) Torque constant mode: In the torque constant mode, the current and slip frequency are kept constant, so the volts per hertz are kept constant, and the same constant torque characteristics as in the full field control mode of a D.C. motor are obtained.

(b) Power constant mode: When the output voltage of the inverter reaches

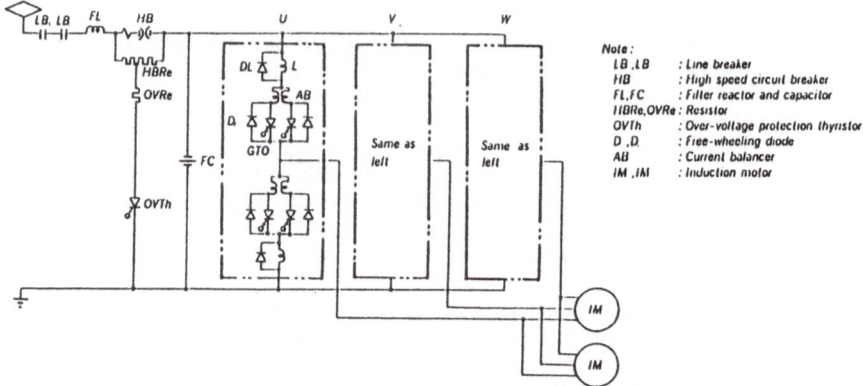

**Fig. 4.13 Schematic Power Circuit Diagram
of the Inverter Control AC Drive System**

the maximum value, the slip frequency is increased, keeping the current constant. This mode is the power constant mode, which is similar to the field weakening mode of a D.C. motor.

(c) Constant slip frequency mode: After the slip frequency reaches the maximum value limited the maximum stalling torque, only the exciting frequency is increased, keeping a constant slip frequency. In this mode, both the flux and the current are decreased, and the same characteristics as those of a series excited D.C. motor are obtained.

By controlling the slip frequency negatively, the phase of the rotor current is reversed, and negative torque, i.e. braking torque, is caused; regenerative braking effort is obtained.

(2) Voltage and frequency control: As mentioned above, it is necessary to control the voltage and the frequency in a wide range for induction motor application to the rolling stock. In the case of a D.C. line voltage system, a pulse width modulation (PWM) inverter is suitable because of the following reasons:

(a) A PWM inverter is able to control both voltage and frequency with a one stage inverter. The resulting construction of the inverter is very simple. Also, it is possible to change the rotational direction and the mode of powering or braking of the traction motor without mechanical switches.

(b) High efficiency is expected because a PWM inverter does not need a current smoothing reactor.

(c) Low torque ripple can be achieved by high modulation frequency of the PWM inverter.

The principal waveforms of the PWM inverter are shown in **Fig. 4.14.** The inverter can be simulated by three pairs of switches.

(3) Configuration of the gate controller: The configuration of the gate controller is similar as those of the chopper system because the same devices of power electronics, i.e. GTO, are used and PWM modulation of the output voltage is the same as in the chopper except the differences that the output of the inverter is the three phase A.C. voltage and the output of the chopper is

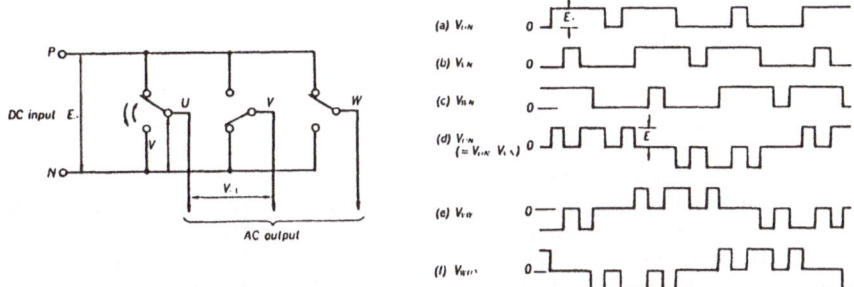

Fig. 4.14 Waveformes of the PWM Inverter

the D.C. voltage. Therefor almost same configuration of the control circuitry as those of chopper are used for the inverter system as shown in **Fig.4.15.**

(a) Microcomputer: The multi-microprocessor architecture is applied for the processing of the gate control same as the chopper system mentioned above. Two processors share the processing of the gate control in the same way as in the chopper.

(b) Gate control pulse generator: The gate pulse generator of the inverter provides the three phase PWM gate pulse to three pairs of GTO thyristors to generate the three phase A.C. voltage in accordance with the command from the processor.

(c) I/O devices: I/O devices for the inverter are almost the same as those of chopper but the circuit of the inverter is more complex than that of the chopper and the current and the voltage of the inverter are the alternative values. Therefor the I/O devices are more complex.

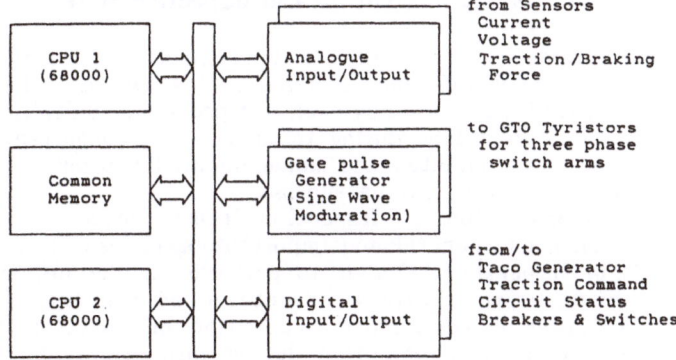

Fig. 4.15 Configuration of the Inverter Gate Controller

The inverter system has been applied to several railways in Japan, and being tested to evaluate its effectiveness and feasibility on the practical use.

5. Maintenance Support System

5.1. Automatic car inspection system (Ref.1)

As is indicated in **Fig. 5.1** , this system is a distributed system which consists of a single process control computer in the operation room and test processing units with microcomputers distributed along the car inspection line. An optical fiber cable is used between the central computer and the test processing units to avoid electrical noises and to enable fast mass transmission.

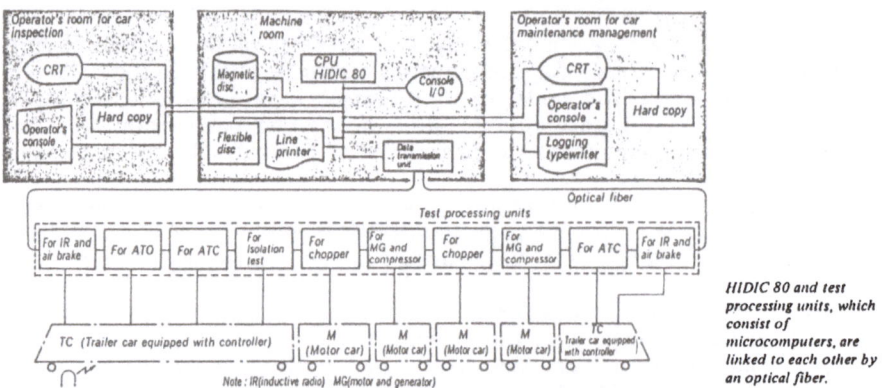

HIDIC 80 and test processing units, which consist of microcomputers, are linked to each other by an optical fiber.

Fig. 5.1 Configuration of the Car Inspection System

To perform automatic car inspection, the control center manages global test processing, data and manmachine communication. On the other hand, all actual operation control is performed by each test processing unit. Therefore the central computer has a light load during testing. It can concurrently perform other functions, such as car maintenance management. Functions of this system can be divided into the following two categories.

(1) Automatic car inspection   Inspection items consist of two levels: bimonthly inspection, and general and important parts inspection. There are also two kinds of inspections. One is to perform all the inspection items automatically in accordance with a predetermined sequence (continuous inspection). The other is to perform inspection on specific items (unit inspection). When there is a rejection during inspections, detection and adjustment of defective parts can be indicated by displaying reject data and input data on a CRT.

(2) Car maintenance management Informations of the train parameters, maintenance history, and the defects of car mounted equipment are input from the operator's console for this purpose, so that the required maintenance information can be retrieved anytime. There are other functions, such as train traffic simulation and on-line program production, to be used for confirmation of train performance and planning when the line is extended.

5.2.   Maintenance system by conversational voice I/O system (Ref.10)

5.2.1. Outline   In the conversational voice I/O system, data or operational instructions are given by voice and the operator performs the work while listening to talk back from the computer. The main components of this system are a voice recognizer, a speech synthesizer, and their controllers.

The operator has a headphone and microphone on his head. A wireless microphone is provided in some cases for greater mobility.

Equipment and installations in the transportation field have to be inspected periodically to maintain safety and correct operation. Inspection data are entered into the maintenance management data base, which is used to plan the supply of spare parts, the maintenance schedule etc. In the case of railroad coaches, inspections are carried out while the operator creeps under the body and climbs up on the roof, and in the case of elevators or escalators, inspections are carried out on each floor.

In these conventional inspection systems, in which the inspector moves around, inspects, and writes data on the inspection sheets, the total efficiency of the inspection is not very good. For example, in railroad coach maintenance shops, it is difficult for the inspector to write the data for the measured wheel shape on the inspection sheets when his hands are wet with oil. In this case, two persons are necessary, one for measuring and one for recording. After the work in the shop, the data are input to the computer through a keyboard; reading the recorded data from the sheets is necessary.

Application of a conversational voice I/O system can save manpower, because the data input can be made immediately at the inspection site where the data are obtained. A missing inspection item or abnormal test data can be picked up to improve the reliability of the test. Correct tests can be made by using the conversational voice I/O system, which gives work guidance and warnings of incorrect data to the worker.

The range of maintenance management systems using computers will be extended by this system, and the application of this conversational voice I/O system will be extended as well.

5.2.2. Considerations for voice recognition system   There are various systems for both single-person and multi-person use. System construction differs according to the use of wireless microphones, type of voice synthesizer, type of memory, etc.

(1) **Voice recognition type:** There are two speech recognition types, the speaker dependent type, which uses previous examples of the speakers, voices in memory, and the speaker independent type. The speaker dependent type has excellent recognition performance. For application in the transportation field, the speaker dependent type is currently adopted to obtain correct data and rapid reaction.

(2) **Noise filter:** In cases where the background noise normally exceeds 80 dB, it is necessary to take this noise into consideration. By performing a frequency analysis of noise, it was found that in many cases the main frequency bands of the human voice are different from the frequency bands of background noise, and noise energy is frequently concentrated in a narrow frequency band. Thus the noise filter is effective in reducing the negative influence of noise.

506

5.2.3. Application for diesel electric locomotive testing system This is an improved system, which makes it possible to make input by voice in background noise over 100 dB from a diesel engine. In this system, the conversational voice I/O system is applied for adjusting the engine control system and for the entry of measured data.

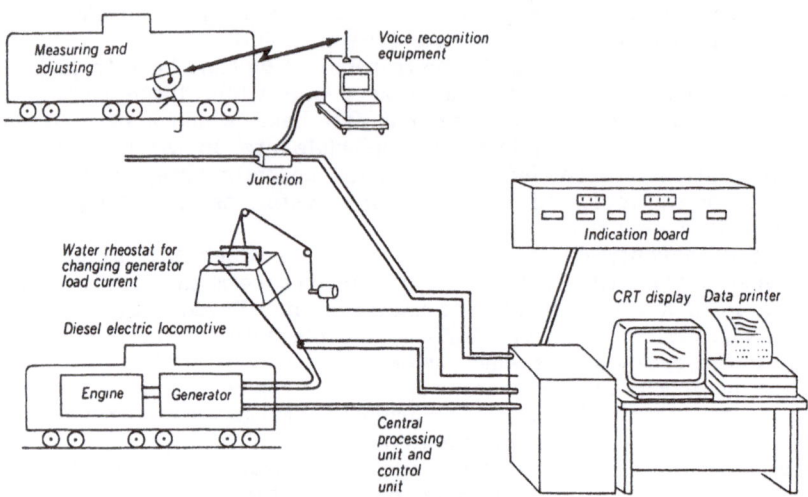

Fig. 5.2 Configuration of the Diesel Electric Locomotive Testing System

Fig. 5.2 shows the system construction of the conversational voice inspection system for electric coaches.

This system is used for data entry by voice of measured data for wheel shapes, body sizes, etc., for electric coaches during periodical inspection. In this system, a maximum of eight persons can input data by voice simultaneously to the control unit.

This system is stationary, and the speech synthesizer for the test guidance system is of the ADPCM type. **Table 5.1** outlines the system. **Fig. 5.3** shows its block diagram.

Table 5.1 Specification of the Conversational Voice Inspection System

Item	Specification
System construction	Voice Controller: Microcomputer Speech recognizer Speech synthesizer Console display, Printer, Floppy disk drive, Wireless microphone
Interface with host computer; 　Transmission Rate 　Electrical specification	 200 - 9,600 BPS RS 232-C
Wireless communication system; 　Wave type 　Frequency range 　Talk system 　Service area	 F3 100 - 140 MHz Duplex simultaneous talk Approc. 100m
Voice recognizer; 　Speeker condition 　Form of speech 　Number of recognized word 　Word length 　Recognition time	 Speeker dependent Isolated word recognition 40 - 100 words / unit 0.1 - 1.3 sec Approc. 0.3 sec
Speech synthesizer; 　Word length Number of words	 ADPCM 1.5 sec max. 128

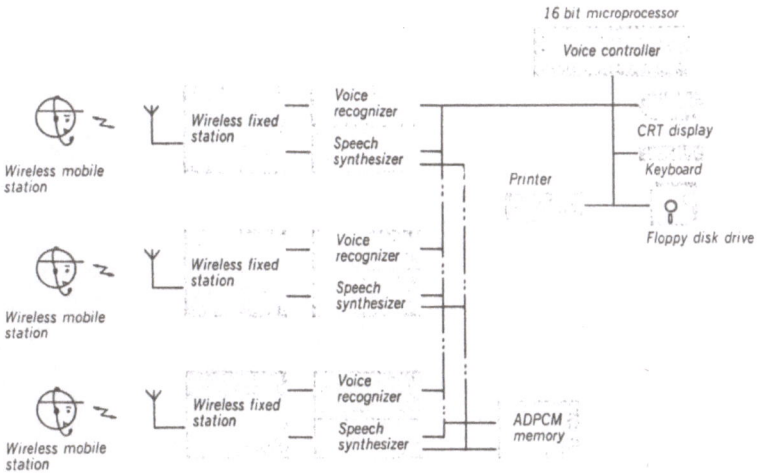

**Fig. 5.3 Block Diagram of the Conversational Voice Inspection System
for Electric Corches**

508

REFERENCES

(1) Sumi,Y., et al., 'Total subway system computer', *Hitachi Review*, **vol.31, no.1,** 28-33, 1982

(2) Matsumaru, H., et al., 'Recent Computer Application Systems for Railways', *Hitachi Review*, **vol.33, no.1,** 1-6, 1984

(3) Yabushita, M., et al., 'Autonomous Decentralization Concept and its Application to Railway Control Systems', *Proceedings of 34th IEEE Vehicular Technology Conference*, 255-260, 1984

(4) Ihara, H., et al., 'Highly Reliable Loop Computer Network System based on Autonomous Decentralization Concept', *Proceedings of 12th Annual International Symp. on Fault-Tolerant Computing*, 187-194, 1982

(5) Mori, K., et al., 'Autonomous Controlability of Decentralized System Aming at Fault-Tolorance', *Proc. of IFAC 8th World Congress*, 129-134, 1981

(6) Nohmi, M., et al., 'Microcomputer based On-line Real-time Control System for Automated Guideway Rapid Transit', *Proc. of IFAC Real Time Programing*, 43-48, 1981

(7) Yasunobu, S., et al., 'Automatic Train Operation System by Predictive Fuzzy Control', *International Application of Fuzzy Control*, 1-18, 1985

(8) Yamada, Y., et al., 'Introduction of Newest Hitachi Choppers', *Hitachi Review*, **vol.33, no.1,** 11-16, 1984

(9) Tsuboi, T., et al., 'GTO Inverter Controlled Traction Drives', *Hitachi Review*, **vol.31, no.1,** 23-27, 1982

(10) Higuchi, T., et al., 'Conversational Voice Input/Output System Applied the Transportation Feild', *Hitachi Review*, **vol.33, no.1,** 21-24, 1984

AUTHOR INDEX

SUBJECT INDEX